HIGHER SPIN
GAUGE THEORIES

HIGHER SPIN GAUGE THEORIES

Editors

Lars Brink
Chalmers University of Technology, Sweden

Marc Henneaux
Université Libre de Bruxelles, Belgium &
International Solvay Institutes, Belgium

Mikhail A Vasiliev
Lebedev Physical Institute, Russia

World Scientific

NEW JERSEY • LONDON • SINGAPORE • BEIJING • SHANGHAI • HONG KONG • TAIPEI • CHENNAI • TOKYO

Published by

World Scientific Publishing Co. Pte. Ltd.
5 Toh Tuck Link, Singapore 596224
USA office: 27 Warren Street, Suite 401-402, Hackensack, NJ 07601
UK office: 57 Shelton Street, Covent Garden, London WC2H 9HE

Library of Congress Cataloging-in-Publication Data
Names: Brink, Lars, 1945– editor. | Henneaux, Marc, editor. | Vasiliev, Mikhail A., editor.
Title: Higher spin gauge theories / edited by Lars Brink (Chalmers University of Technology, Sweden), Marc Henneaux (Université Libre de Bruxelles, Belgium & International Solvay Institutes, Belgium), Mikhail A. Vasiliev (Lebedev Physical Institute, Russia).
Description: Hackensack, NJ : World Scientific, [2016] | Includes bibliographical references.
Identifiers: LCCN 2016044198| ISBN 9789813144095 (hardcover) | ISBN 9813144092 (hardcover)
Subjects: LCSH: Gauge fields (Physics) | Field theory (Physics) | Nuclear spin.
Classification: LCC QC793.3.G38 H54 2016 | DDC 530.14/35--dc23
LC record available at https://lccn.loc.gov/2016044198

British Library Cataloguing-in-Publication Data
A catalogue record for this book is available from the British Library.

Typeset by Stallion Press
Email: enquiries@stallionpress.com

Printed in Singapore

Preface

In 1922 Otto Stern and Walther Gerlach let silver atoms travel through a magnetic field. Instead of some random spreading of the atoms they found that they split into two beams. This was before the advent of quantum mechanics and the result was difficult to understand. With the introduction of the Pauli Principle by Wolfgang Pauli in early 1925 it was clear that the electron had very special properties. With this and with the knowledge of the existing spectroscopic data two young Dutch physicists Samuel Goudsmit and George Uhlenbeck suggested that the electron should have an intrinsic spin of $1/2\hbar$. (In the future we will leave out $\hbar$.) It was not taken seriously by the establishment and it was only thanks to their professor Paul Ehrenfest that the paper was published. The idea had problems and it was not really clear until Paul Dirac's groundbreaking paper on his equation that the electron indeed has spin-1/2.

In the beginning of the 1930's only a handful of particles were known, the electron, the proton and the neutron (discovered in 1932) with spin-1/2 and the photon with spin-1. With a stretch of imagination there was also a possible particle mediating gravity, the graviton with spin-2. Nothing else seemed to be necessary. That situation was changed when Hideki Yukawa in 1935 suggested that the strong interaction was mediated by spin-0 particles to be called pions. Even so there were no signs that particles of other spins should exist and no theory demanded them.

One mystical consequence of the Dirac equation was the negative-energy solutions. In order to avoid those Ettore Majorana suggested in 1932 a variant of the equation. The solutions arranged themselves into unitary representations of the Lorentz group, with an infinite number of particles with spin-j and masses $M_j = \frac{2M}{2j+1}$. With the discovery of the positron, Majorana's work was forgotten.

In 1935 Eugene Wigner tried to publish a mathematically rigorous paper on the irreducible representations of the Poincaré group. The paper was finally published in 1939 in a mathematical journal. (This paper was eventually the main motivation for his 1963 Nobel Prize in physics.) From his analysis it was clear that there do exist representations with higher spins than the ones seen in Nature, and the year after Dirac published a seminal paper on wave equations for higher spin particles. Also Markus Fierz and Pauli published papers on higher spin wave equations. They discussed spin-3/2 but it is the work by William Rarita and Julian Schwinger from

1941 that has survived up to today. The Igor Tamm school in Moscow was also interested in such equations and his junior colleagues Aleksandr Davydov and Vitaly Ginzburg published a series of papers on spin-3/2 starting from 1942. Efim Fradkin who later joined that group once told John Schwarz the he was thinking about higher spin theories in the trenches of Stalingrad. His first paper where he discussed the spin-5/2 wave equation was published in 1950.

The group theoretical analysis was completed in 1942 with the work of Pauli and Jozef Lubanski who found the second Casimir operator for the Poincaré group then showing that a relativistic particle is characterised by its mass and its spin.

In the 1950's the large accelerators started to produce new particles, resonances, and some of them had higher spins. Eventually they were organised into linear Regge-trajectories in Chew–Frautschi plots, where the spin is plotted versus the mass^2. Little was done though to find equations of motions for such particles. Instead the efforts went into finding the S(cattering)-matrix among these states. This trend continued into the 1960's when the quarks were introduced. Gauge field theory was not a very active field and few thought about massless particles. A massless particle leads to long-range forces and no such force had been found except for the electromagnetic and the gravitational ones. However, in 1964, Steven Weinberg showed, using S-matrix methods, that higher spin massless particles cannot generate long-range forces, further closing the window to higher-spin gauge field theories.

A further interest in higher spins arose after the discovery of supergravity in 1976 by Daniel Freedman, Sergio Ferrara and Peter van Nieuvenhuzen and by Stanley Deser and Bruno Zumino, which made it clear that massless fields of spins greater than 1 are probably more interesting than the massive ones, since they are responsible for gauge symmetries. After that the theory of massless higher spin fields was taken up in earnest by Christian Fronsdal and collaborators (Fang, Flato) in late 1970's. Fronsdal clarified the free field theories and derived both the field equations and Lagrangians for massless higher spin fields and clearly established their nature of being gauge fields. Could there also be supermultiplets with spins higher than 2? These questions were asked by Bernard de Wit and Daniel Freedman and also by Thom Curtright.

The equations constructed were free equations of motion and the question now was if one could introduce interaction terms also, Fronsdal and Fang were the first to raise the question of self-interactions for higher spin gauge fields generalising the gauge field theories for spin 1 and 2. They phrased the problem as one of deforming the free Lagrangian and the free field gauge transformations. Fronsdal is also the first to realise that an infinite tower of higher spin fields is needed. The first explicit construction of interaction terms was so done in 1983 by Anders Bengtsson, Ingemar Bengtsson and Lars Brink using a light-front formulation in which the higher-spin fields are represented only by their degrees of freedom (2) and hence can be written as a complex field. In this way they could get a uniform formulation for all spins. They managed to construct self-interactions for these fields to the first order in the coupling constant and also found similar interactions for higher-spin supermultiplets. This was later generalised to all possible cubic interactions between fields of different spin on the light-front. Soon after this F.A. Berends, G.H.J. Burgers and H. van Dam obtained a covariant gauge invariant cubic self-interacting vertex for spin-3 fields. One of the important conclusions of these papers was that the order of higher derivatives increases with spins of the fields φ in the vertex. Appearance of higher derivatives in interactions demanded a dimensionful coupling constant ρ in the action, whose origin in a theory involving only massless fields was obscure.

The resolution of this puzzle was proposed in 1987 by Efim Fradkin and Mikhail Vasiliev by going to the maximally symmetrical curved (anti-)de Sitter background space followed by identification of the parameter ρ with its radius. In this setup, the higher derivative vertices allow no meaningful flat limit in agreement with numerous no-go statements due to Weinberg (1966), Coleman and Mandula (1971), Aragone and Deser (1979) and others, which had ruled out consistent interactions of massless higher spin fields in Minkowski space. In 1986 Fradkin and Vasiliev identified the non-Abelian higher spin symmetry algebras underlying four-dimensional higher spin gauge theories in very much the same way as usual supersymmetry algebra underlies supergravity. The higher spin (super)algebras contain the usual four-dimensional supersymmetries as subalgebras. On the other hand, being infinite-dimensional, they imply that starting from four space-time dimensions higher spin gauge theories describe infinite towers of massless fields of infinitely increasing spins. Complete nonlinear field equations for higher spin theories in four space-time dimensions were constructed by Vasiliev in 1992. In 1998 these results were extended to the higher spin theories in three dimensions by Prokushkin and Vasiliev and in 2003 to any dimension by Vasiliev.

The peculiar feature that consistent higher spin interactions demand a non-zero cosmological constant acquired a deeper interpretation after the discovery of the AdS/CFT correspondence in 1997 by Juan Maldacena, Steven Gubser, Igor Klebanov and Aleksandr Polyakov, and Edward Witten. The fact that higher spin theories are most naturally formulated in the AdS background was conjectured to play a role in the context of the higher spin holography in 2000 by Bo Sundborg and independently by Witten. The precise conjecture on the AdS_4/CFT_3 correspondence was put forward by Klebanov and Polyakov in 2002 while its first explicit check was performed by Simone Giombi and Xi Yin in 2009. This research triggered sharp increase of interest in higher spin theories and higher spin holographic duality.

In particular, the study of CFT_2 duals of $3d$ higher spin symmetries was initiated independently in 2010 by Marc Henneaux and Soo-Jong Rey and Andrea Campoleoni, Stefan Fredenhagen, Stefan Pfenninger and Stefan Theisen for the pure Chern-Simons sector of the three-dimensional higher spin theory. The conjecture on AdS_3/CFT_2 higher spin holography proposed later in 2010 by Matthias Gaberdiel and Rajesh Gopakumar on the duality of the minimal two-dimensional conformal models to the three-dimensional higher spin theory of Prokushkin and Vasiliev, that includes the nontrivial matter sector, also formed an active research direction in the recent years.

One of the driving forces of the study of interacting higher spin theories is the hope that higher spin holography may help to uncover the origin of AdS/CFT. However, the subtlety is that despite significant progress in the construction of actions during the last thirty years after the work of Bengtsson, Bengtsson and Brink, including the complete classification of vertices by Ruslan Metsaev in 2005, a full nonlinear generalisation of the Fronsdal action is still unavailable. This complicates the standard construction of the generating functional of boundary correlators. A new construction for invariant on-shell functionals proposed by Vasiliev in 2015 is anticipated to work both for the generating functional for boundary correlators and for black hole charges. Along with the application of the so-called unfolded formalism underlying the formulation of nonlinear higher spin equations it has a potential to make the holographic duality manifest at least in higher spin theory.

Higher Spin Theory continues today to be an extremely interesting and fertile research area, with a lot of impressive results and challenging open problems whose solution may affect our understanding of the foundations of Nature. A very important achievement of the last years resulted from the analysis of loop effects in higher spin theory. In particular in 2014 and 2015 it was shown for various higher spin models by Giombi, Klebanov and Arkady Tseytlin that the one-loop corrections for the infinite towers of fields associated with higher spin theories turn out to be finite hence indicating that higher spin gauge theories can be well defined (free of divergencies) as quantum theories.

One of the most important problems to explore is to investigate exact solutions of nonlinear higher spin equations. Only a few of them are available now in the four-dimensional theory including solutions of Sezgin and Sundell (2008) and black-hole-like solutions of Vyatcheslav Didenko and Vasiliev (2009) and Carlo Iazeolla, Sezgin and Sundell (2011). Understanding exact solutions might shed crucial light on the geometry of higher spins that must supersede Riemannian geometry given that the spacetime metric is not invariant under higher spin gauge transformations.

A problem of fundamental importance is to understand the relationship between higher spin gauge theories and the zero tension limit of String Theory. In particular, one central question is to find an extension of the higher spin gauge theory to a string-like theory rich enough to describe all Regge trajectories of String Theory as well as to make it possible to spontaneously break higher spin symmetries to allow an independent mass scale parameter analogous to the parameter α' of String Theory. Promising conjectures in this direction were put forward by Vasiliev in 2012 and Gaberdiel and Gopakumar in 2013.

An international workshop on Higher-Spin Gauge Theories was held at the Institute of Advanced Studies at Nanyang Technological University in Singapore from 4 to 6 November 2015. Most of the leading scientists in the field were gathered for this meeting and gave talks both with some historic reviews as well as presenting the latest results in the subject. This book represents the proceedings of the workshop but is also aimed as both an introduction as well as a present overview of the subject.

The Editors

NANYANG
TECHNOLOGICAL
UNIVERSITY

Institute of Advanced Studies

Contents

Preface v

1. Symmetries and Invariants in Higher-Spin Theory 1
M. A. Vasiliev

2. The Higher Spin Square 17
Matthias R. Gaberdiel and Rajesh Gopakumar

3. Higher Spin Representations of $K(E_{10})$ 25
Axel Kleinschmidt and Hermann Nicolai

4. Scattering Via Conformal Higher Spin Exchanges 39
A. A. Tseytlin

5. Note on Higher-Spin Equations in Four Dimensions 51
V. E. Didenko, N. G. Misuna and M. A. Vasiliev

6. Pseudo-local Theories: A Functional Class Proposal 59
Massimo Taronna

7. Finite Temperature Maps in Vector/Higher Spin Duality 85
Antal Jevicki, Kenta Suzuki and Junggi Yoon

8. On (Un)Broken Higher-Spin Symmetry in Vector Models 103
E. D. Skvortsov

9. Extended Anti-de Sitter Hypergravity in $2 + 1$ Dimensions and Hypersymmetry Bounds 139
Marc Henneaux, Alfredo Pérez, David Tempo and Ricardo Troncoso

10. Quasiconformal Group Approach to Higher Spin Algebras, Their Deformations and Supersymmetric Extensions 159
Murat Günaydin

11. BMS Modules in Three Dimensions 187
A. Campoleoni, H.A. González, B. Oblak and M. Riegler

12. Action Principles for Higher and Fractional Spin Gravities 213
Cesar Arias, Roberto Bonezzi, Nicolas Boulanger, Ergin Sezgin, Per Sundell, Alexander Torres-Gomez and Mauricio Valenzuela

13. Deriving Field Theories for Particles of Arbitrary Spin With and Without Supersymmetry 255
Sudarshan Ananth

14. Reducible Higher-Spin Systems in Metric-like and Frame-like Formulation 273
Alessandro Agugliaro, Francesco Azzurli and Dmitri Sorokin

15. Bulk Quartic Vertices from Boundary Four-Point Correlators 291
X. Bekaert, J. Erdmenger, D. Ponomarev and C. Sleight

16. The Sky is the Limit: Free Boundary Conditions in AdS_3 Chern-Simons Theory 305
Luis Apolo and Bo Sundborg

17. Higgs Mechanism and Holography of Partially Massless Higher Spin Fields 317
Seungho Gwak, Jaewon Kim and Soo-Jong Rey

18. Quartic Amplitudes for Minkowski Higher Spin 353
Anders K. H. Bengtsson

19. Solutions in Bosonic String Field Theory and AdS Higher Spin Algebras 371
Dimitri Polyakov

20. Comments on Higher-Spin Fields in Nontrivial Backgrounds 381
Rakibur Rahman and Massimo Taronna

Symmetries and Invariants in Higher-Spin Theory

M. A. Vasiliev

I.E. Tamm Department of Theoretical Physics,
Lebedev Physical Institute,
Leninsky prospect 53, 119991, Moscow, Russia

General aspects of higher-spin gauge theory and unfolded formulation are briefly recalled with some emphasize on the recent results on the breaking of $sp(8)$ symmetry by current interactions and on the construction of invariant functionals relevant to the higher-spin holography.

1. Introduction

Higher spin (HS) gauge theory is based on higher symmetries associated with HS massless fields. HS gauge symmetries are expected to play a role at ultra high energies possibly beyond the Planck energy. Since such energies are unreachable by modern accelerator devices the conjecture that a fundamental theory exhibits the HS symmetries at ultra high energies provides a unique chance to explore properties of this regime. HS symmetries severely restrict the structure of HS theory.

The study of HS fields has long history starting from seminal papers of Dirac [1], Fierz and Pauli [2], and others including the Tamm group [3–5]. The role of HS gauge symmetries for massless fields in four dimensions was originally appreciated at the linearized level for spin 3/2 by Rarita and Schwinger [6] and for any spin by Fronsdal [7].

Extension to the interacting level was not simple encountering difficulties of combining nonAbelian symmetries of different types. First positive results were obtained in eighties of the last century in the papers by A. Bengtsson, I. Bengtsson, Brink [8, 9] and Berends, Burgers, van Dam [10, 11] who found that the action consistent with HS gauge symmetries in the cubic order contains higher derivatives in interactions

$$S = S^2 + S^3 + \cdots , \qquad S^3 = \sum_{p,q,r} (D^p \varphi)(D^q \varphi)(D^r \varphi)\rho^{p+q+r+\frac{1}{2}d-3}.$$

Since full HS theory necessarily involves infinite towers of HS fields, such a theory is somewhat nonlocal (note however that no higher derivatives appear at the quadratic level within the expansion around AdS background). Of course some kind of nonlocality beyond Planck scale should be expected of the theory anticipated to capture the quantum gravity regime.

Appearance of higher derivatives in interactions requires a dimensionful coupling constant ρ in the action, whose origin in a theory involving only massless fields was obscure. Resolution of this paradox was proposed in [12, 13] via identification the parameter ρ with the radius of the background (anti-)de Sitter space. In this setup, the higher derivative vertices do not allow a meaningful flat limit. This is consistent with numerous no-go statements ruling out consistent interactions of massless HS fields in Minkowski space [14, 15].

In eighties, the fact that consistent HS interactions require non-zero cosmological constant looked like a peculiarity. It acquired a much deeper interpretation after discovery of the AdS/CFT correspondence [16–18]. The fact that HS theories are most naturally formulated in the AdS background was conjectured to play a role in the context of the HS holography in [19–21]. The precise conjecture on the AdS_4/CFT_3 correspondence was put forward by Klebanov and Polyakov [22] while its first explicit check was performed by Giombi and Yin in [23]. This research triggered a great increase of interest in HS theories and HS holographic duality [24]–[37]. CFT_2 duals of $3d$ HS symmetries were studied in [38, 39]. The conjecture on AdS_3/CFT_2 HS holography proposed by Gaberdiel and Gopakumar [40] also formed an active research direction in the recent years.

One of the driving forces of the study of interacting HS theories is the hope that HS holography may help to uncover the origin of AdS/CFT. However, the subtlety is that despite significant progress in the construction of actions during last thirty years [8]–[13],[41]–[48] the full nonlinear generalization of the Fronsdal action is still unavailable. This complicates the standard construction of the generating functional. One of the aims of this talk is to discuss a recently proposed construction for invariant on-shell functionals [49] conjectured to give rise to both the generating functional for boundary correlators and black hole charges.

2. Unfolded Dynamics

2.1. *General setup*

Modern formulation of nonlinear HS theory [50] is based on the so-called unfolded approach [51] which reformulates dynamics in question in terms of differential forms. The unfolded form of dynamical equations provides a covariant generalization of the first-order form of differential equations

$$\dot{q}^i(t) = \varphi^i(q(t)),$$

which is convenient in many respects. In particular, since initial data can be given in terms of the values of variables $q^i(t_0)$ at any given point t_0, the number of degrees of freedom in the first-order formulation equals to the number of dynamical variables.

Unfolded formulation is a multidimensional generalization achieved via the replacement of the time derivative by the de Rham derivative

$$\frac{\partial}{\partial t} \to \mathrm{d} = dx^\nu \partial_\nu$$

and the dynamical variables q^i by a set of differential forms

$$q^i(t) \to W^\Omega(dx, x) = dx^{\nu_1} \dots dx^{\nu_p} W^\Omega_{\nu_1 \dots \nu_p}(x)$$

to reformulate a system of partial differential equations in the first-order covariant form

$$\mathrm{d}W^\Omega(dx, x) = G^\Omega(W(dx, x)). \tag{2.1}$$

Here dx^ν are anticommuting differentials ($dx^\nu dx^\mu = -dx^\mu dx^\nu$; the wedge product symbol is implicit) and $G^\Omega(W)$ are some functions of the "supercoordinates" W^Ω

$$G^\Omega(W) = \sum_n f^\Omega{}_{\Lambda_1 \dots \Lambda_n} W^{\Lambda_1} \dots W^{\Lambda_n}.$$

Since $\mathrm{d}^2 = 0$ at $d > 1$ the functions $G^\Lambda(W)$ have to obey the compatibility conditions

$$G^\Lambda(W) \frac{\partial G^\Omega(W)}{\partial W^\Lambda} \equiv 0. \tag{2.2}$$

(Not that all products of the differential forms $W(dx, x)$ are the wedge products due to anticommutativity of dx^ν.) Let us stress that these are conditions on the functions $G^\Lambda(W)$ rather than on W.

As a consequence of (2.2) system (2.1) is invariant under the gauge transformation

$$\delta W^\Omega = d\varepsilon^\Omega + \varepsilon^\Lambda \frac{\partial G^\Omega(W)}{\partial W^\Lambda}, \tag{2.3}$$

where the gauge parameter $\varepsilon^\Omega(x)$ is a $(p_\Omega - 1)$-form for a p_Ω-form W^Ω. Strictly speaking, this is true for the class of *universal* unfolded systems in which the compatibility conditions (2.2) hold independently of the dimension d of space-time, *i.e.*, (2.2) should be true disregarding the fact that any $(d+1)$-form is zero. Let us stress that all unfolded systems appearing in HS theories are universal.

The unfolded formulation can be applied to description of invariant functionals. As shown in [52], the variety of gauge invariant functionals associated with the unfolded equations (2.1) is described by the cohomology of the operator

$$Q = G^\Omega(W) \frac{\partial}{\partial W^\Omega}, \tag{2.4}$$

which obeys

$$Q^2 = 0$$

as a consequence of (2.2). By virtue of (2.1), Q-closed p-form functions $L_p(W)$ are d-closed, giving rise to the gauge invariant functionals

$$S = \int_{\Sigma^p} L_p.$$

In the off-shell case they can be used to construct invariant actions while in the on-shell case they describe conserved charges. (For more detail and examples see [52].)

2.2. *Properties*

The unfolded formulation of partial differential equations has a number of remarkable properties.

- First of all, it has general applicability: every system of partial differential equations can be reformulated in the unfolded form.
- Due to using the exterior algebra formalism, the system is invariant under diffeomorphisms, being coordinate independent.
- Interactions can be understood as nonlinear deformations of $G^\Omega(W)$.
- Unfolded formulation gives clear group-theoretical interpretation of fields and equations in terms of modules and Chevalley-Eilenberg cohomology of a symmetry algebra h. In particular, background fields are described by a flat connection of h. Dynamical fields are described in terms of h-modules while equations of motion have a form of covariant constancy conditions.
- Local degrees of freedom are carried by 0-forms $C^i(x_0)$ at any $x = x_0$ (as $q(t_0)$), valued in the infinite-dimensional module dual to the space of single-particle states: $C^i(x_0)$ are moduli of solutions replacing initial data in the conventional Cauchy problem. It is worth to mention that this property of unfolded dynamics gives a tool to control unitarity in presence of higher derivatives via the requirement on the space of zero-forms like $C^i(x_0)$ to admit a positive-definite norm preserved by the unfolded equations in question.
- The most striking feature of the unfolded formulation is that it makes it possible to describe one and the same dynamical system in space-times of different dimensions. Unfolded dynamics exhibits independence of the "world-volume" space-time with coordinates x. Instead, geometry is encoded by the functions $G^\Omega(W)$ in the "target space" of fields W^Ω. Indeed, the universal unfolded equations make sense in any space-time independently of a particular realization of the de Rham derivative d. For instance one can extend space time by adding additional coordinates z

$$\mathrm{d}W^\Omega(x) = G^\Omega(W(x)), \quad x \to X = (x, z),$$

$$\mathrm{d}_x \to \mathrm{d}_X = \mathrm{d}_x + \mathrm{d}_z, \quad \mathrm{d}_z = dz^u \frac{\partial}{\partial z^u}.$$

The unfolded equations reconstruct the X-dependence in terms of values of the fields $W^\Omega(X_0) = W^\Omega(x_0, z_0)$ at any X_0. Clearly, to take $W^\Omega(x_0, z_0)$ in space

M_X with coordinates X_0 is the same as to take $W^\Omega(x_0)$ in the space $M_x \subset M_X$ with coordinates x.

Such uplifting becomes most interesting provided that there is a nontrivial vacuum connection along the additional coordinates z. This is in particular the case for AdS/CFT correspondence where the conformal flat connection at the boundary is extended to the flat AdS connection in the bulk with z being a radial Poincaré coordinate.

3. Field-Current-Field Correspondence

The AdS_4/CFT_3 HS holography [22] relates the HS gauge theory in AdS_4 to the quantum theory of conformal currents in three dimensions. To see how it works from the unfolded dynamics perspective, let us first discuss the unfolded equations for free massless fields and currents on the $3d$ boundary.

The unfolded equations of $3d$ conformal massless fields is formulated in terms of 0-forms $C(y|x)$ [53, 54] which depend on the coordinates $\mathbf{x}^{\alpha\beta} = \mathbf{x}^{\beta\alpha}$ of $3d$ space-time and auxiliary commuting spinorial variables y^α ($\alpha, \beta = 1, 2$ are $3d$ spinorial indices). Unfolded equations for conformal massless fields are

$$\left(\frac{\partial}{\partial \mathbf{x}^{\alpha\beta}} \pm i \frac{\partial^2}{\partial y^\alpha \partial y^\beta}\right) C_j^\pm(y|\mathbf{x}) = 0, \quad \alpha, \beta = 1, 2, \tag{3.1}$$

where $j = 1, \ldots, \mathcal{N}$ is a color index.

The meaning of these equations is twofold. Firstly, they express all terms of degree two or higher in the y variables via x-derivatives of the fields $C^\pm(\mathbf{x}) := C^\pm(0|\mathbf{x})$ and $C^\pm_\alpha(\mathbf{x}) := \frac{\partial}{\partial y^\alpha} C^\pm(y|\mathbf{x})|_{y=0}$. The latter are the usual scalar and spinor fields which obey, respectively, Klein-Gordon and Dirac equations by virtue of (3.1). More precisely, the $\pm$ components should be identified with positive- and negative-frequency parts of the solutions of free field equations. Note that the fields $C^\pm(\mathbf{x})$ and $C^\pm_\alpha(\mathbf{x})$ are primaries of the conformal modules underlying equations (3.1).

The unfolded equations for $3d$ conformal conserved currents have the rank-two form [55]

$$\left\{\frac{\partial}{\partial \mathbf{x}^{\alpha\beta}} - \frac{\partial^2}{\partial y^{(\alpha} \partial u^{\beta)}}\right\} J(u,\, y|\mathbf{x}) = 0. \tag{3.2}$$

$J(u,\, y|\mathbf{x})$ contains all $3d$ HS currents along with their derivatives.

Elementary $3d$ conformal currents, which are conformal primaries, contain conserved currents of all spins

$$J(u, 0|\mathbf{x}) = \sum_{2s=0}^{\infty} u^{\alpha_1} \ldots u^{\alpha_{2s}} J_{\alpha_1 \ldots \alpha_{2s}}(\mathbf{x}), \quad \tilde{J}(0, y|\mathbf{x}) = \sum_{2s=0}^{\infty} y^{\alpha_1} \ldots y^{\alpha_{2s}} \tilde{J}_{\alpha_1 \ldots \alpha_{2s}}(\mathbf{x})$$

along with the additional scalar current

$$J^{asym}(u, y|\mathbf{x}) = u_\alpha y^\alpha J^{asym}(\mathbf{x}).$$

Their conformal dimensions are

$$\Delta J_{\alpha_1\ldots\alpha_{2s}}(\mathbf{x}) = \Delta\tilde{J}_{\alpha_1\ldots\alpha_{2s}}(\mathbf{x}) = s+1 \quad \Delta J^{asym}(\mathbf{x}) = 2.$$

The unfolded equations express all other components of $J(u,y|\mathbf{x})$ in terms of derivatives of the primaries, also imposing the differential equations on the latter, which are just the conservation conditions

$$\frac{\partial}{\partial \mathbf{x}^{\alpha\beta}}\frac{\partial^2}{\partial u_\alpha \partial u_\beta}J(u,0|\mathbf{x}) = 0, \qquad \frac{\partial}{\partial \mathbf{x}^{\alpha\beta}}\frac{\partial^2}{\partial y_\alpha \partial y_\beta}\tilde{J}(0,y|\mathbf{x}) = 0$$

for all currents except for the scalar ones that obey no differential equations.

The rank-two equation is obeyed by

$$J(u,\, y\,|\mathbf{x}) = \sum_{i=1}^{\mathcal{N}} C_i^-(u+y|\mathbf{x})\, C_i^+(y-u|\mathbf{x}).$$

This simple formula gives the explicit realization of the HS conformal conserved currents in terms of bilinear combinations of derivatives of free massless fields in three dimensions.

Generally, the rank-two fields and hence conserved currents can be interpreted as bi-local fields in the twistor space. In this respect they are somewhat analogous to space-time bi-local fields also used for the description of currents (see e.g [29] and references therein).

To relate $3d$ currents to $4d$ massless fields we extend the $3d$ current equation to the $4d$ massless equations. This is easy to achieve in the unfolded dynamics via the extension of the $3d$ coordinates $\mathbf{x}^{\alpha\beta}$ to the $4d$ coordinates $x^{\alpha\dot{\beta}}$, extending $3d$ equations (3.2) to

$$\left(\frac{\partial}{\partial x^{\alpha\dot{\alpha}}} + \frac{\partial^2}{\partial y^\alpha \partial \bar{y}^{\dot{\beta}}}\right) C(y,\bar{y}|x) = 0. \tag{3.3}$$

These are just the free unfolded equations [51] for $4d$ massless fields of all spins in Minkowski space, *i.e.*, at $\Lambda = 0$.

The analysis in AdS_4, which is also simple, is performed analogously. In this case, $\mathbf{x}^{\alpha\beta} = \frac{1}{2}(x^{\alpha\beta} + x^{\beta\alpha})$ are boundary coordinates, while $z^{-1} = x^{\alpha\beta}\epsilon_{\alpha\beta}$ is the radial coordinate. (For more detail see [25].) At the non-linear level, the full HS theory in AdS_4 turns out to be equivalent to the theory of $3d$ currents of all spins interacting through conformal HS gauge fields [25].

A rank-two field (current) in AdS_3 is equivalent to a rank-one field in a larger space with ten coordinates $X^{AB} = X^{BA}$

$$\left(\frac{\partial}{\partial X^{AB}} + \frac{\partial^2}{\partial y^A \partial y^B}\right) J^3(y|x) = 0, \quad A,B = 1,\ldots,4, \quad X^{AB} = X^{BA},$$

$$X^{AB} = (x^{\alpha\dot{\alpha}}, x^{\alpha\beta}, \bar{x}^{\dot{\alpha}\dot{\beta}}), \quad x^{\alpha\dot{\alpha}} = (\mathbf{x}^{\alpha\dot{\alpha}}, \varepsilon^{\alpha\dot{\alpha}}\mathbf{z}).$$

Reduction to Minkowski coordinates $x^{\alpha\dot{\alpha}}$ gives $4d$ massless equations for all spins with $J^3 \to C^4$. Mathematically this is the manifestation of the Flato-Fronsdal theorem [56] stating that the tensor product of unitary modules associated with $3d$ massless fields gives the unitary module associated with all $4d$ massless fields:

$$(3d, m=0) \otimes (3d, m=0) = \sum_{s=0}^{\infty} (4d, m=0). \tag{3.4}$$

The full system of $4d$ massless fields of all spins exhibits $sp(8)$ symmetry [54,57–59]. A rank-two field in four dimensions describes $4d$ conserved currents equivalent to a rank-one field in six dimensions [59,60]

$$C^4 C^4 \sim J^4 \sim C^6.$$

Dualities of this type can be called field-current-field correspondence.

4. From Free Massless Equations to Current Interactions and Holography

4.1. *Central on-shell theorem*

The infinite set of $4d$ massless fields of all spins $s = 0, 1, 2, \ldots$ is conveniently described by a 1-form $\omega(y, \bar{y} \mid x)$, and 0-form $C(y, \bar{y} \mid x)$

$$A(y, \bar{y} \mid x) = i \sum_{n,m-0}^{\infty} \frac{1}{n!m!} y_{\alpha_1} \ldots y_{\alpha_n} \bar{y}_{\dot{\beta}_1} \ldots \bar{y}_{\dot{\beta}_m} A^{\alpha_1 \ldots \alpha_n}{}_{,}{}^{\dot{\beta}_1 \ldots \dot{\beta}_m}(x).$$

The central fact of the analysis of free massless fields in four dimensions known as Central on-shell theorem is that unfolded system for free massless fields has the form [51]

$$R_1(y, \bar{y} \mid x) = \overline{H}^{\dot{\alpha}\dot{\beta}} \frac{\partial^2}{\partial \bar{y}^{\dot{\alpha}} \partial \bar{y}^{\dot{\beta}}} \bar{C}(0, \bar{y} \mid x) + H^{\alpha\beta} \frac{\partial^2}{\partial y^\alpha \partial y^\beta} C(y, 0 \mid x), \tag{4.1}$$

$$\tilde{D}_0 C(y, \bar{y} \mid x) = 0, \tag{4.2}$$

where

$$R_1(y, \bar{y}|x) = D^{ad} \omega(y, \bar{y}|x) = D^L \omega(y, \bar{y}|x) - \lambda e^{\alpha\dot{\beta}} \Big(y_\alpha \frac{\partial}{\partial \bar{y}^{\dot{\beta}}} + \frac{\partial}{\partial y^\alpha} \bar{y}_{\dot{\beta}} \Big) \omega(y, \bar{y}|x), \tag{4.3}$$

$$\tilde{D} C(y, \bar{y}|x) = D^L C(y, \bar{y}|x) + \frac{i}{2} \lambda e^{\alpha\dot{\beta}} \Big(y_\alpha \bar{y}_{\dot{\beta}} - \frac{\partial^2}{\partial y^\alpha \partial \bar{y}^{\dot{\beta}}} \Big) C(y, \bar{y}|x), \tag{4.4}$$

$$D^L A(y, \bar{y}|x) = dA(y, \bar{y}|x) - \Big(\omega^{\alpha\beta} y_\alpha \frac{\partial}{\partial y^\beta} + \overline{\omega}^{\dot{\alpha}\dot{\beta}} \bar{y}_{\dot{\alpha}} \frac{\partial}{\partial \bar{y}^{\dot{\beta}}} \Big) A(y, \bar{y}|x). \tag{4.5}$$

Here the background AdS_4 Lorentz connection $\omega_{\alpha\beta}$, $\overline{\omega}_{\dot{\alpha}\dot{\beta}}$ and vierbein $e_{\alpha\dot{\beta}}$ obey the AdS_4 equations

$$R_{\alpha\beta} = 0, \quad \overline{R}_{\dot{\alpha}\dot{\beta}} = 0, \quad R_{\alpha\dot{\alpha}} = 0, \tag{4.6}$$

where λ^{-1} is the AdS_4 radius and

$$R_{\alpha\beta} = d\omega_{\alpha\beta} + \omega_\alpha{}^\gamma \wedge \omega_{\beta\gamma} + \lambda^2 e_\alpha{}^{\dot\delta} \wedge e_{\beta\dot\delta}, \tag{4.7}$$

$$\overline{R}_{\dot\alpha\dot\beta} = d\overline{\omega}_{\dot\alpha\dot\beta} + \overline{\omega}_{\dot\alpha}{}^{\dot\gamma} \wedge \overline{\omega}_{\dot\beta\dot\gamma} + \lambda^2 e^\gamma{}_{\dot\alpha} \wedge e_{\gamma\dot\beta},$$

$$R_{\alpha\dot\beta} = de_{\alpha\dot\beta} + \omega_\alpha{}^\gamma \wedge e_{\gamma\dot\beta} + \overline{\omega}_{\dot\beta}{}^{\dot\delta} \wedge e_{\alpha\dot\delta}. \tag{4.8}$$

(Two-component indices are raised and lowered by $\epsilon_{\alpha\beta}$ or $\epsilon_{\dot\alpha\dot\beta}$.) $H^{\alpha\beta} = H^{\beta\alpha}$ and $\overline{H}^{\dot\alpha\dot\beta} = \overline{H}^{\dot\beta\dot\alpha}$ are the basis 2-forms

$$H^{\alpha\beta} := e^\alpha{}_{\dot\alpha} e^{\beta\dot\alpha}, \quad \overline{H}^{\dot\alpha\dot\beta} := e_\alpha{}^{\dot\alpha} e^{\alpha\dot\beta}. \tag{4.9}$$

The 0-forms $C(Y|x)$ form a Weyl module equivalent to the boundary current module. 1-form HS connections $\omega(y,\bar y \mid x)$ contain HS gauge fields. For spins $s \geq 1$, equation (4.1) expresses the Weyl 0-forms $C(Y|x)$ via gauge invariant combinations of derivatives of the HS gauge connections. From this perspective the Weyl 0-forms $C(Y|x)$ generalize the spin-two Weyl tensor along with all its derivatives to any spin.

4.2. *Current deformation*

Schematically, for the flat connection $D = \mathrm{d} + w$ the current deformation of the free equations (4.1), (4.2) has the form

$$\begin{cases} D\omega^4 + L(C^4, w) = 0 \\ \tilde D C^4 = 0 \\ D_2 J^4 = 0 \end{cases} \qquad \Rightarrow \qquad \begin{cases} D\omega^4 + L(C^4, w) + G(w, J^4) = 0 \\ \tilde D C^4 + F(w, J^4) = 0 \\ D_2 J^4 = 0 \end{cases}$$

The sector of 0-forms of this system was analyzed in detail in [61, 62]. Here J^4 can be interpreted either as a $4d$ current or as a $6d$ massless field. As a result, $4d$ current interactions can be interpreted as a mixed linear system of $d4$ and $d6$ fields [61]. Algebraically this is the semidirect sum of a rank-one and rank-two systems.

An interesting question is what symmetry is preserved by the deformed system? When unmixed, both rank-one and rank-two system are $sp(8)$-invariant. The question whether $sp(8)$ is preserved by the deformation is equivalent to that whether formal consistency of the deformation takes place with any connection $w \in sp(8)$. The analysis of this question [62] shows that current interactions break $sp(8)$ down to the conformal algebra $su(2,2)$.

4.3. *Kinematics of AdS_4/CFT_3 HS holography*

To make conformal invariance manifest it is convenient to use the following basis

$$y^+_\alpha = \frac{1}{2}(y_\alpha - i\bar y_\alpha), \quad y^-_\alpha = \frac{1}{2}(\bar y_\alpha - i y_\alpha), \quad [y^-_\alpha, y^{+\beta}]_\star = \delta^\beta_\alpha.$$

AdS_4 can be foliated as $x^n = (\mathbf{x}^a, \mathbf{z})$, where $\mathbf{x}^a$ are coordinates of leaves ($a = 0, 1, 2,$) and the Poincaré coordinate $\mathbf{z}$ is the foliation parameter. AdS_4 infinity is at $\mathbf{z} = 0$. In these coordinates the background connection at $\lambda = 1$ is

$$W = \frac{i}{\mathbf{z}} d\mathbf{x}^{\alpha\beta} y^-_\alpha y^-_\beta - \frac{d\mathbf{z}}{2\mathbf{z}} y^-_\alpha, y^{+\alpha}$$

$$e^{\alpha\dot{\alpha}} = \frac{1}{2\mathbf{z}} dx^{\alpha\dot{\alpha}}, \quad \omega^{\alpha\beta} = -\frac{i}{4\mathbf{z}} d\mathbf{x}^{\alpha\beta}, \quad \bar{\omega}^{\dot{\alpha}\dot{\beta}} = \frac{i}{4\mathbf{z}} d\mathbf{x}^{\dot{\alpha}\dot{\beta}}.$$

Using insensitiveness of unfolded equations to the extension to a larger space, the vacuum connection can be analitically extended to the complex plane of $\mathbf{z}$ with all components containing $d\bar{\mathbf{z}}$ being zero. In these terms the generating functional for the boundary correlators takes the form

$$S = \frac{1}{2\pi i} \oint_{\mathbf{z}=0} \mathcal{L}(\omega(C), C)$$

if $\mathcal{L}(\omega(C), C)$ is an on-shell closed $(d+1)$-form for a d-dimensional boundary

$$\mathrm{d}\mathcal{L}(\omega(C), C) = 0, \quad \mathcal{L} \neq \mathrm{d}M.$$

The resulting functional is the residue at $\mathbf{z} = 0$ giving the boundary functional of the structure analogous to $\phi_{n_1 \ldots n_s} J^{n_1 \ldots n_s}$

$$S_{M^3}(\omega) = \int_{M^3} \mathcal{L}, \quad \mathcal{L} = \frac{1}{2} \omega_{\mathbf{x}}^{\alpha_1 \ldots \alpha_{2(s-1)}} e_{\mathbf{x}}^{\alpha_{2s-1}}{}_{\beta} e_{\mathbf{x}}^{\alpha_{2s}\beta} (aC_{\alpha_1 \ldots \alpha_{2s}}(\omega) + \bar{a}\bar{C}_{\alpha_1 \ldots \alpha_{2s}}(\omega)).$$

Here $C_{\alpha_1 \ldots \alpha_{2s}}(\omega)$, which have conformal properties of currents J, are expressed via the HS connections ω by Eq. (4.1). On the other hand $\omega^{\alpha_1 \ldots \alpha_{2(s-1)}}$ have conformal dimensions of the shadow sources ϕ to the currents. Being related to C via unfolded equations it does not describe new degrees of freedom however.

The C-dependent terms can be represented in the form

$$aC_{\alpha_1 \ldots \alpha_{2s}}(\omega) + \bar{a}\bar{C}_{\alpha_1 \ldots \alpha_{2s}}(\omega) = a_- \mathcal{T}_{-\alpha_1 \ldots \alpha_{2s}}(\omega) + a_+ \mathcal{T}_{+\alpha_1 \ldots \alpha_{2s}}(\omega),$$

where $\mathcal{T}_-$ describes local boundary terms while $\mathcal{T}_+$ describes nontrivial correlators via the variation of S_{M_3} over the HS gauge fields $\omega_{\mathbf{x}}^{\alpha_1 \ldots \alpha_{2(s-1)}}$

$$\langle J(\mathbf{x}_1) J(\mathbf{x}_2) \ldots \rangle = \frac{\delta^n \exp\left[-S_{M^3}(\omega, C(\omega))\right]}{\delta\omega(\mathbf{x}_1)\delta\omega(\mathbf{x}_2) \ldots}\Big|_{\omega=0}.$$

The main problem is to find an appropriate nonlinear invariant functional $\mathcal{L}$.

5. Nonlinear HS Equations in AdS_4

To explain the construction of invariant functionals we first recall the form of nonlinear massless field equations of [50]. The key element is the doubling of auxiliary Majorana spinor variables Y_A in the HS 1-forms and 0-forms

$$\omega(Y; \mathcal{K}|x) \to W(Z; Y; \mathcal{K}|x), \quad C(Y; \mathcal{K}|x) \to B(Z; Y; \mathcal{K}|x) \tag{5.1}$$

supplemented with equations which determine dependence on the additional variables Z_A in terms of "initial data"

$$\omega(Y;\mathcal{K}|x) = W(0;Y;\mathcal{K}|x), \quad C(Y;\mathcal{K}|x) = B(0;Y;\mathcal{K}|x). \tag{5.2}$$

An additional spinor field $S_A(Z;Y;\mathcal{K}|x)$, that carries only pure gauge degrees of freedom, plays a role of connection in Z^A directions. It is convenient to introduce anticommuting $Z-$differentials $dZ^A dZ^B = -dZ^B dZ^A$ to interpret $S_A(Z;Y;\mathcal{K}|x)$ as a Z–1-form,

$$S = dZ^A S_A(Z;Y;\mathcal{K}|x). \tag{5.3}$$

The variables $\mathcal{K} = (k, \bar{k})$ are Klein operators that satisfy

$$\begin{aligned} k w^\alpha &= -w^\alpha k, \quad k\bar{w}^{\dot\alpha} = \bar{w}^{\dot\alpha} k, \quad \bar{k} w^\alpha = w^\alpha \bar{k}, \\ \bar{k}\bar{w}^{\dot\alpha} &= -\bar{w}^{\dot\alpha}\bar{k}, \quad k^2 = \bar{k}^2 = 1, \quad k\bar{k} = \bar{k}k \end{aligned} \tag{5.4}$$

with $w^\alpha = (y^\alpha, z^\alpha, dz^\alpha)$, $\bar{w}^{\dot\alpha} = (\bar{y}^{\dot\alpha}, \bar{z}^{\dot\alpha}, d\bar{z}^{\dot\alpha})$.

The nonlinear HS equations are [50]

$$dW + W * W = 0, \tag{5.5}$$

$$dB + W * B - B * W = 0, \tag{5.6}$$

$$dS + W * S - S * W = 0, \tag{5.7}$$

$$S * B = B * S, \tag{5.8}$$

$$S * S = -i(dZ^A dZ_A + dz^\alpha dz_\alpha F_*(B) k v + d\bar{z}^{\dot\alpha} d\bar{z}_{\dot\alpha} \bar{F}_*(B) \bar{k}\bar{\kappa}), \tag{5.9}$$

where $F_*(B)$ is some star-product function of the field B.

Setting $\mathcal{W} = \mathrm{d} + W + S$ brings equations (5.5)–(5.9) to the concise form

$$\mathcal{W} * \mathcal{W} = -i[dZ_A dZ^A + \eta\delta^2(dz)\mathcal{B} * k * \kappa + \bar{\eta}\delta^2(d\bar{z})\mathcal{B} * \bar{k} * \bar{\kappa}], \tag{5.10}$$

$$\mathcal{W} * B = B * \mathcal{W}. \tag{5.11}$$

The simplest choice of linear functions

$$F_*(B) = \eta B, \quad \bar{F}_*(B) = \bar{\eta} B, \tag{5.12}$$

where η is some phase factor (its absolute value can be absorbed into redefinition of B) leads to a class of pairwise nonequivalent nonlinear HS theories. The particular cases of $\eta = 1$ and $\eta = \exp\frac{i\pi}{2}$ are especially interesting, corresponding to so called A and B HS models. These two cases are distinguished by the property that they respect parity [63].

The associative star product $*$ acts on functions of two spinor variables

$$\begin{aligned}(f * g)(Z;Y) = \frac{1}{(2\pi)^4}\int d^4U\, d^4V \exp[iU^A V^B C_{AB}] \\ \times f(Z+U;Y+U) g(Z-V;Y+V), \end{aligned} \tag{5.13}$$

where $C_{AB} = (\epsilon_{\alpha\beta}, \bar{\epsilon}_{\dot{\alpha}\dot{\beta}})$ is the $4d$ charge conjugation matrix and U^A, V^B are real integration variables. It is normalized so that 1 is a unit element of the star-product algebra, *i.e.*, $f * 1 = 1 * f = f$. Star product (5.13) provides a particular realization of the Weyl algebra

$$[Y_A, Y_B]_* = -[Z_A, Z_B]_* = 2iC_{AB}, \quad [Y_A, Z_B]_* = 0 \tag{5.14}$$

($[a, b]_* = a * b - b * a$).

The left and right inner Klein operators

$$\kappa = \exp iz_\alpha y^\alpha, \quad \bar{\kappa} = \exp i\bar{z}_{\dot{\alpha}}\bar{y}^{\dot{\alpha}}, \tag{5.15}$$

which enter Eq. (5.9), change a sign of undotted and dotted spinors, respectively

$$(\kappa * f)(z,\bar{z}; y,\bar{y}) = \exp iz_\alpha y^\alpha f(y,\bar{z}; z,\bar{y}), \quad (\bar{\kappa} * f)(z,\bar{z}; y,\bar{y}) = \exp i\bar{z}_{\dot{\alpha}}\bar{y}^{\dot{\alpha}} f(z,\bar{y}; y,\bar{z}), \tag{5.16}$$

$$\kappa * f(z, \bar{z}; y, \bar{y}) = f(-z, \bar{z}; -y, \bar{y}) * \kappa, \quad \bar{\kappa} * f(z, \bar{z}; y, \bar{y}) = f(z, -\bar{z}; y, -\bar{y}) * \bar{\kappa}, \tag{5.17}$$

$$\kappa * \kappa = \bar{\kappa} * \bar{\kappa} = 1, \quad \kappa * \bar{\kappa} = \bar{\kappa} * \kappa. \tag{5.18}$$

To analyze Eqs. (5.5)–(5.9) perturbatively, one has to linearize them around some vacuum solution. The simplest choice is

$$W_0(Z; Y|x) = W_0(Y|x), \quad S_0(Z; Y|x) = dZ^A Z_A, \quad B_0 = 0, \tag{5.19}$$

where $W_0(Y|x)$ is some solution of the flatness condition

$$dW_0(Y|x) + W_0(Y|x) * W_0(Y|x) = 0. \tag{5.20}$$

$W_0(Y|x)$ bilinear in Y^A describes AdS_4. Linearization of system (5.5)–(5.9) around this vacuum just reproduces free field equations (4.1), (4.2) (for more detail see [50, 64, 65]).

6. Invariants of the AdS_4 HS Theory

To explain the idea of our construction let us first consider an example of a contractible unfolded system of the form

$$\mathrm{d}w = \mathcal{L}, \quad \mathrm{d}\mathcal{L} = 0. \tag{6.1}$$

It is obviously consistent and hence is invariant under the gauge transformations (2.3)

$$\delta w(x) = \epsilon(x), \quad \delta\mathcal{L}(x) = \mathrm{d}\epsilon(x). \tag{6.2}$$

As such it is dynamically empty since the gauge transformation allows one to gauge fix $w = 0$. By virtue of (6.1) it follows then that $\mathcal{L} = 0$.

A more interesting system is

$$dw + L(W) = \mathcal{L}, \quad d\mathcal{L} = 0, \tag{6.3}$$

where $L(W)$ is some closed function of other fields W that obey some unfolded equations (2.1. In the *canonical gauge* $w = 0$ it takes the form

$$\mathcal{L} = L(W), \quad dL(W) = 0.$$

The singlet field L becomes a Lagrangian giving rise to an invariant action

$$S = \int_{\Sigma} L(W). \tag{6.4}$$

So defined functional is independent of local variations of the integration cycle and gauge invariant. Indeed, being formally consistent, the system is invariant under gauge transformations (2.3) with respect to the gauge parameter ϵ associated with w and the gauge parameters ε^{Ω} associated with W^{Ω}. In the gauge $w = 0$, the parameter ϵ is expressed by the condition $\delta w = 0$ via the gauge parameters ε^{Ω} and the gauge fields W^{Ω}

$$\epsilon = \epsilon(\varepsilon, W). \tag{6.5}$$

Though $\mathcal{L}$ is not gauge invariant under the gauge transformations of the system, it transforms by a total derivative of a function of fields W^{Ω} and gauge parameters ε^{Ω}. As a result, the action S is gauge invariant.

Note that though the system $dw + L(W) = 0$ is formally consistent it is not guaranteed that it admits a solution with regular w. In fact, the Lagrangial $\mathcal{L}$ defined by (6.3) is nontrivial for non-exact $L(W)$.

The proposal of [49] is to consider invariants resulting from the following extension of the HS unfolded equations

$$\mathcal{W} * \mathcal{W} = F(\mathcal{B}) + \mathcal{L}\, Id, \quad \mathcal{W} * \mathcal{B} = \mathcal{B} * \mathcal{W}, \quad d\mathcal{L} = 0,$$

where $\mathcal{W} = d + W$ and $\mathcal{B}$ are differential forms of all odd and even degrees, respectively (both in dx and dZ). An appropriate choice is

$$iF(\mathcal{B}) = dZ_A dZ^A + \eta\delta^2(dz)\mathcal{B} * k * \kappa + \bar{\eta}\delta^2(d\bar{z})\mathcal{B} * \bar{k} * \bar{\kappa} + G(\mathcal{B})\delta^4(dZ)k * \bar{k} * \kappa * \bar{\kappa}.$$

$G = g + O(\mathcal{B})$, where g is the coupling constant. $\mathcal{L}(x)$ are x-dependent space-time differential forms of positive even degrees since the left-hand side of (6.6) contains a product of forms of even degrees. That it enters as a coefficient in front of the unit element Id of the star-product algebra means that $\mathcal{L}(x)$ is independent of Y^A and Z^A. As a result, application of the covariant derivative to the right-hand side of (6.6) gives $d\mathcal{L}(x) = 0$.

It should be stressed that the modification of the system by the "Lagrangians" $\mathcal{L}$ does not affect the form of all equations except for the single Z, Y-independent equation proportional to Id, which just acquires the form (6.3). The form of the Lagrangian $L(W)$ (6.3) now results from the perturbative solution of the other equations, *i.e.*, nonlinear HS equations.

The density relevant to the generating functional of correlators in AdS_4/CFT_3 HS holography is a 4-form $\mathcal{L}^4$. The density relevant to BH entropy is a 2-form $\mathcal{L}^2$ (for recent progress in this direction see [66].)

7. Conclusions

A very general property illustrated by the analysis of HS theory is that the unfolding machinery makes holographic duality manifest at the level of the unfolded formulation of HS equations. Following [49], the duality extends to the level of generating functionals. The latter can be identified with integrals of differential forms of positive even degrees valued in the center of the star-product algebra. So defined functionals are gauge invariant, coordinate independent and can be evaluated for any boundaries and bulk solutions.

In $4d$ HS theory the 4-form $\mathcal{L}^4$ is conjectured to give rise to the generating functional for boundary correlators while the 2-form $\mathcal{L}^2$ gives black-hole charges opening new perspectives for the understanding of black-hole physics including the informational paradox [66].

As shown in [49] a similar construction applies to the HS theory in AdS_3. In this case the only Lagrangian density is a 2-form $\mathcal{L}^2$. An exciting peculiarity of this construction is that the boundary functional results from the integration over a one-dimensional cycle at the boundary (times a cycle over the complexified Poincaré coordinate $\mathbf{z}$). It is tempting to speculate that this property expresses holomorphicity of the $2d$ boundary conformal theory.

By virtue of unfolded dynamics usual field-current correspondence can be extended via interpretation of further nonlinear combinations of fields with linear fields in higher dimensions. An interesting subtlety here is that the mixing of fields in different dimensions representing nonlinear interactions in the original system can decrease the symmetries of unmixed fields. This is illustrated by current interactions of massless fields of all spins in $d = 4$ which break the $sp(8)$ symmetry of free fields down to the conformal symmetry $su(2,2) \subset sp(8)$ [62].

Acknowledgments

I am grateful to the Institute for Advanced Study of Nanyang Technical University for its kind hospitality during the workshop "Higher Spin Gauge Theories". This research was supported in part by the RFBR Grant No 14-02-01172.

References

[1] P. A. M. Dirac, *Proc. Roy. Soc. Lond.* **155A** (1936) 447.
[2] M. Fierz and W. Pauli, *Proc. Roy. Soc. Lond.* **A173** (1939) 211–232.
[3] V. L. Ginzburg, *JETPh* **12** (1942) 425.
[4] E. S. Fradkin, *JETPh* **20** (1950) 27; 211.

[5] V. L. Ginzburg and I. E. Tamm, *J. Phys.* **11** (1947).
[6] W. Rarita and J. Schwinger, *Phys. Rev.* **60** (1941) 61.
[7] C. Fronsdal, *Phys. Rev.* **D18** (1978) 3624.
[8] A. K. H. Bengtsson, I. Bengtsson, and L. Brink, *Nucl. Phys.* **B227** (1983) 31.
[9] A. K. H. Bengtsson, I. Bengtsson, and L. Brink, *Nucl. Phys.* **B227** (1983) 41.
[10] F. A. Berends, G. J. H. Burgers, and H. Van Dam, *Z. Phys.* **C24** (1984) 247–254.
[11] F. A. Berends, G. J. H. Burgers, and H. van Dam, *Nucl. Phys.* **B260** (1985) 295.
[12] E. S. Fradkin and M. A. Vasiliev, *Phys. Lett.* **B189** (1987) 89–95.
[13] E. S. Fradkin and M. A. Vasiliev, *Nucl. Phys.* **B291** (1987) 141.
[14] S. R. Coleman and J. Mandula, *Phys. Rev.* **159** (1967) 1251–1256.
[15] C. Aragone and S. Deser, *Phys. Lett.* **B86** (1979) 161.
[16] J. M. Maldacena, *Adv. Theor. Math. Phys.* **2** (1998) 231 [*Int. J. Theor. Phys.* **38** (1999) 1113] [arXiv:hep-th/9711200].
[17] S. S. Gubser, I. R. Klebanov and A. M. Polyakov, *Phys. Lett. B* **428**, 105 (1998) [arXiv:hep-th/9802109].
[18] E. Witten, *Adv. Theor. Math. Phys.* **2**, 253 (1998) [arXiv:hep-th/9802150].
[19] B. Sundborg, *Nucl. Phys. Proc. Suppl.* **102** (2001) 113 [arXiv:hep-th/0103247].
[20] E. Witten, talk at the John Schwarz 60-th birthday symposium, http://theory.caltech.edu/jhs60/witten/1.html
[21] E. Sezgin and P. Sundell, *Nucl. Phys. B* **644** (2002) 303 [Erratum-*ibid.* B **660** (2003) 403] [arXiv:hep-th/0205131].
[22] I. R. Klebanov and A. M. Polyakov, *Phys. Lett. B* **550** (2002) 213 [arXiv:hep-th/0210114].
[23] S. Giombi and X. Yin, *JHEP* **1009** (2010) 115 [arXiv:0912.3462 [hep-th]].
[24] J. Maldacena and A. Zhiboedov, *Class. Quant. Grav.* **30** (2013) 104003 [arXiv:1204.3882 [hep-th]].
[25] M. A. Vasiliev, *J. Phys. A* **46** (2013) 214013 [arXiv:1203.5554 [hep-th]].
[26] S. Giombi and X. Yin, *J. Phys. A* **46** (2013) 214003 [arXiv:1208.4036 [hep-th]].
[27] N. Colombo and P. Sundell, arXiv:1208.3880 [hep-th].
[28] V. E. Didenko and E. D. Skvortsov, arXiv:1210.7963 [hep-th].
[29] A. Jevicki, K. Jin and Q. Ye, *J. Phys. A* **46** (2013) 214005 [arXiv:1212.5215 [hep-th]].
[30] S. Giombi and I. R. Klebanov, *JHEP* **1312** (2013) 068 [arXiv:1308.2337 [hep-th]].
[31] S. Giombi, I. R. Klebanov and A. A. Tseytlin, *Phys. Rev. D* **90** (2014) 024048 [arXiv:1402.5396 [hep-th]].
[32] M. Beccaria, X. Bekaert and A. A. Tseytlin, *JHEP* **1408** (2014) 113 [arXiv:1406.3542 [hep-th]].
[33] R. d. M. Koch, A. Jevicki, J. P. Rodrigues and J. Yoon, arXiv:1408.4800 [hep-th].
[34] S. Giombi and I. R. Klebanov, arXiv:1409.1937 [hep-th].
[35] M. Beccaria and A. A. Tseytlin, arXiv:1410.3273 [hep-th].
[36] A. O. Barvinsky, *J. Exp. Theor. Phys.* **120** (2015) 3, 449 [arXiv:1410.6316 [hep-th]].
[37] A. Hegde, P. Kraus and E. Perlmutter, *JHEP* **1601** (2016) 176 [arXiv:1511.05555 [hep-th]].
[38] M. Henneaux and S. J. Rey, *JHEP* **1012** (2010) 007 [arXiv:1008.4579 [hep-th]].
[39] A. Campoleoni, S. Fredenhagen, S. Pfenninger and S. Theisen, *JHEP* **1011** (2010) 007 [arXiv:1008.4744 [hep-th]].
[40] M. R. Gaberdiel and R. Gopakumar, *Phys. Rev. D* **83** (2011) 066007 [arXiv:1011.2986 [hep-th]].
[41] R. R. Metsaev, *Nucl. Phys. B* **759** (2006) 147 [arXiv:hep-th/0512342].
[42] A. Sagnotti and M. Taronna, *Nucl. Phys. B* **842** (2011) 299 [arXiv:1006.5242 [hep-th]].
[43] A. Fotopoulos and M. Tsulaia, *JHEP* **1011** (2010) 086 [arXiv:1009.0727 [hep-th]].

[44] R. Manvelyan, K. Mkrtchyan and W. Ruehl, *Phys. Lett. B* **696** (2011) 410 [arXiv:1009.1054 [hep-th]].
[45] N. Boulanger and P. Sundell, *J. Phys. A* **44** (2011) 495402 [arXiv:1102.2219 [hep-th]].
[46] E. Sezgin and P. Sundell, arXiv:1103.2360 [hep-th].
[47] M. A. Vasiliev, *Nucl. Phys. B* **862** (2012) 341 [arXiv:1108.5921 [hep-th]].
[48] E. Joung, L. Lopez and M. Taronna, *JHEP* **1301** (2013) 168 [arXiv:1211.5912 [hep-th]]
[49] M. A. Vasiliev, arXiv:1504.07289 [hep-th].
[50] M. A. Vasiliev, *Phys. Lett.* B **285** (1992) 225.
[51] M. A. Vasiliev, *Ann. Phys. (NY)* **190** (1989) 59.
[52] M. A. Vasiliev, *Int. J. Geom. Meth. Mod. Phys.* **3** (2006) 37 [hep-th/0504090].
[53] O. V. Shaynkman and M. A. Vasiliev, *Theor. Math. Phys.* **128** (2001) 1155 [*Teor. Mat. Fiz.* **128** (2001) 378] [hep-th/0103208].
[54] M. A. Vasiliev, *Phys. Rev. D* **66** (2002) 066006 [hep-th/0106149].
[55] O. A. Gelfond and M. A. Vasiliev, *Theor. Math. Phys.* **145** (2005) 1400 [*Teor. Mat. Fiz.* **145** (2005) 35] [hep-th/0304020].
[56] M. Flato and C. Fronsdal, *Lett. Math. Phys.* **2** (1978) 421.
[57] C. Fronsdal, "Massless Particles, Ortosymplectic Symmetry and Another Type of Kaluza–Klein Theory", Preprint UCLA/85/TEP/10, in Essays on Supersymmetry, Reidel, 1986 (Mathematical Physics Studies, v.8).
[58] I. Bandos and J. Lukierski, *Mod. Phys. Lett* **A14** (1999) 1257, hep-th/9811022.
[59] I. Bandos, J. Lukierski and D. Sorokin, *Phys. Rev.* **D61** (2000) 045002, hep-th/9904109.
[60] M. A. Vasiliev, "Relativity, Causality, Locality, Quantization and Duality in the $Sp(2M)$ Invariant Generalized Space Time", hep-th/0111119.
[61] O. A. Gelfond and M. A. Vasiliev, *J. Exp. Theor. Phys.* **120** (2015) 3, 484 [arXiv: 1012.3143 [hep-th]].
[62] O. A. Gelfond and M. A. Vasiliev, arXiv:1510.03488 [hep-th].
[63] E. Sezgin and P. Sundell, *JHEP* **0507** (2005) 044 [arXiv:hep-th/0305040].
[64] M. A. Vasiliev, in *The Many Faces of the Superworld*, ed. M. A. Shifman, 533–610 [hep-th/9910096].
[65] V. E. Didenko and E. D. Skvortsov, arXiv:1401.2975 [hep-th].
[66] V. E. Didenko, N. G. Misuna and M. A. Vasiliev, arXiv:1512.07626 [hep-th].

The Higher Spin Square

Matthias R. Gaberdiel* and Rajesh Gopakumar†

*Institut für Theoretische Physik, ETH Zurich,
CH-8093 Zürich, Switzerland
gaberdiel@itp.phys.ethz.ch

†International Centre for Theoretical Sciences-TIFR,
Survey No. 151, Shivakote, Hesaraghatta Hobli,
Bengaluru North, India 560 089
rajesh.gopakumar@icts.res.in

We describe how the unbroken stringy symmetries of the orbifold CFT dual to strings on $AdS_3 \times S^3 \times T^4$ can be usefully viewed through the lens of higher spin symmetry. We find a novel structure – the higher spin square – which is generated by two independent higher spin symmetries. This structure appears to be a useful way to also characterise the matter content of the theory.

1. Introduction

One of the primary motivations for this work is to make headway towards understanding why the structure of string theory is so unique. Usually such rigidity follows from a large underlying invariance which is however, mostly unmanifest in string theory in flat space. Here we see only Yang-Mills gauge symmetry, diffeomorphism invariance and their SUSY partners. Nevertheless, it had been noticed many years ago that in the limit where $\alpha' E^2 \to \infty$ there seem to be additional relations between amplitudes at different levels.[1] This suggests that the tensionless limit of string theory might be a good place to look for unbroken symmetry.[1–4] However, in flat space, this limit cannot be taken uniformly on the whole theory because of the absence of any other dimensionful parameter other than the string tension. Hence, the above kinematic regime is the best we can do here.

But AdS/CFT has taught us in recent years that we can learn about string theory on such backgrounds using the dual conformal field theory. A very important observation is that we can now take the tensionless limit in a uniform way by taking $\frac{R_{AdS}^2}{\alpha'} \propto \lambda^{\beta} \to 0$. Here R_{AdS} is the radius of curvature of AdS and λ represents a ('t Hooft) coupling in the dual CFT (while $\beta > 0$ is a number that depends on the particular background) which is dual to the radius deformation parameter. Thus the tensionless limit translates to a free CFT. And in fact, free Yang-Mills theory has a single tower of higher spin conserved currents (which are bilinears in the free fields) that are dual to the higher spin gauge fields in AdS.[5,6] It is remarkable that this is precisely in correspondence with the theory of interacting higher spin gauge fields in AdS constructed by Vasiliev.[7] These latter theories are an example of how constraining the presence of higher spin gauge invariance can be.

The higher spin gauge fields of Vasiliev can be viewed as the leading Regge trajectory of strings in AdS. But this is far from a stringy spectrum of unbroken symmetries. Can we have backgrounds where we have more unbroken symmetries? It is easy to argue that in AdS_{d+1} (with $d > 2$) we cannot have more than the Vasiliev symmetry. The dual conserved currents saturate a unitarity bound on the conformal dimension $\Delta = s + (d-2)$. For $d > 2$, where s is the spin. This can be satisfied in a free theory only by the bilinears. But for $d = 2$, the bound $\Delta = s$ can be saturated by (anti-)holomorphic currents which are not just bilinears in free fields. This gives an exponentially larger unbroken symmetry algebra as we will see in the specific case of the free orbifold CFT dual to the $AdS_3 \times S^3 \times T^4$ string background.

We will describe the structure of this much larger unbroken symmetry in this specific case. We will see that we can organise the symmetry algebra into a Higher Spin Square (HSS). The name indicates that the symmetry is effectively generated by two $\mathcal{W}_\infty$ symmetries (in that all commutators can be expressed in terms of these two higher spin commutators). We will also see that a subsector of the matter fields of the string theory are nicely captured by the HSS. The entire untwisted sector of the orbifold CFT is constructed from the Fock space built from a *single* representation of the HSS (which correspond to an infinite number of representations from the point of view of the underlying Vasiliev algebra). The material presented here is based on work that has already appeared.[8,9]

2. The Symmetric Product CFT

String theory on $AdS_3 \times S^3 \times T^4$ is believed to lie on the moduli space of the orbifold CFT $(T^4)^N/S_N$ (at large N). This is a free orbifold theory whose partition function is explicitly known.[10] If

$$\tilde{Z}(X) = \sum_{h,\bar{h}} c(h,\bar{h}) q^h \bar{q}^{\bar{h}}, \tag{1}$$

is the partition function with the insertion of $(-1)^{F+\tilde{F}}$, then the generating function for the full partition function (not the index) of the orbifold CFT is

$$\sum_{N\geq 0} p^N Z(X^N/S_N) = \prod_{n>0} \prod_{h,\bar{h}} (1-(-1)^{2h+2\bar{h}} p^n q^h \bar{q}^{\bar{h}})^{-c(h,\bar{h})}. \tag{2}$$

This is essentially a formula of taking multiparticles. In particular, this formula applies to the chiral sector ($\bar{h} = 0$) of conserved currents. Thus single particle conserved current generators of the orbifold CFT X^N/S_N are in 1-1 association with the chiral algebra of X. The correspondence is

$$J^{(s)} \leftrightarrow \frac{1}{\sqrt{N}} \sum_{i=1}^{N} J_i^{(s)}. \tag{3}$$

Therefore, for the case where $X = T^4$, i.e. for the orbifold CFT $(T^4)^N/S_N$, there is a 1-1 correspondence with the chiral algebra of the (supersymmetric) T^4 theory, i.e. four free bosons and four free fermions. Note that at large N the generators are all independent. Thus $\sum_{i=1}^{N}(\partial\phi_i)^2$ is an independent generator compared to $\sum_{i=1}^{N}(\partial\phi_i)^4$. The generating function of the *single particle* chiral algebra of the symmetric product is therefore

$$\prod_{n=1} \frac{(1+yq^{n-\frac{1}{2}})^2(1+y^{-1}q^{n-\frac{1}{2}})^2}{(1-q^n)^4}. \tag{4}$$

We see from this expression that the number of currents at any given spin (or dimension) and thus massless gauge fields in the dual AdS_3 grows exponentially - it has a Cardy growth with s. This is to be contrasted with the constant (independent of s) growth of higher spin fields in a Vasiliev theory. Our goal, nevertheless, is to try and usefully view these stringy symmetries through the lens of higher spin symmetry.

3. Simpler Case: Single Boson

To strip off all the decorations of $\mathcal{N} = 4$ SUSY and see the bare bones of the symmetry algebra, we restrict to the case of the symmetric product of a single real boson. In other words, the single particle generators are in correspondence with the chiral algebra of a single boson with generating function

$$\prod_{n=1}^{\infty} \frac{1}{(1-q^n)}. \tag{5}$$

This generating function essentially counts the monomials $\prod_j (\partial^j\phi)^{k_j}$, where ϕ is the real boson. The bilinears $\sum_{j=1}^{s-1} c_j(\partial^j\phi)(\partial^{s-j}\phi)$ correspond to the (even spin) $\mathcal{W}_\infty[\lambda = 1]$ symmetry generators. Here c_j are combinatorial coefficients chosen to make these currents quasi-primary. The other monomials fall into representations of this $\mathcal{W}_\infty[\lambda = 1]$. Recalling the labelling of coset representations of $\mathcal{W}_\infty[\lambda]$ by the two sets of Dynkin labels $(\Lambda_+;\Lambda_-)$, the terms with n ϕ's transform in the representation $(\Lambda_+ = [0^{n-1}, 1, 0\ldots 0], \Lambda_- = 0)$. This decomposition is reflected in a character decomposition of the generating function

$$\prod_{k=1}^{\infty} \frac{1}{(1-q^k)} = 1 + \sum_{n=1}^{\infty} \frac{q^n}{\prod_{j=1}^{n}(1-q^j)}. \tag{6}$$

Here the individual terms on the RHS are the characters of the $\mathcal{W}_\infty[\lambda = 1]$ representation $(\Lambda_+ = [0^{n-1}, 1, 0\ldots 0], \Lambda_- = 0)$. We depict this decomposition of the algebra in Fig.1 where we have arranged the representations in columns labelled by the number of ϕ's. The $n = 2$ column consists of the generators of $\mathcal{W}_\infty[\lambda = 1]$ themselves. We should view the topmost entry in each column as the highest weight vector of the representation with $\mathcal{W}_\infty[\lambda = 1]$ descendants down the column. We will refer to this $\mathcal{W}_\infty$ algebra as the vertical higher spin algebra.

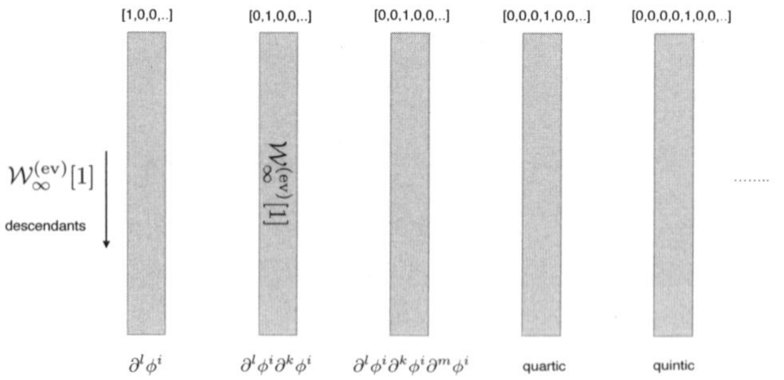

Fig. 1. Decomposition in terms of the vertical $\mathcal{W}_\infty[\lambda = 1]$.

The novel observation is that we can alternatively organise the generators in a horizontal way starting with the top row. By fermionisation, the top row are bilinears of fermions which generate a different higher spin symmetry, namely $\mathcal{W}_{1+\infty}[\lambda = 0]$. The other rows are different representations of this new horizontal $\mathcal{W}_{1+\infty}[0]$ algebra. In terms of the coset labelling these are $(\Lambda_+ = 0, \Lambda_- = [m, 0 \ldots 0, m])$. They are labelled by the number of fermion bilinears. We should view the leftmost entries in each row as the highest weight vector with $\mathcal{W}_{1+\infty}[\lambda = 0]$ descendants as you move rightwards. We again have a character decomposition

$$\prod_{k=1}^{\infty} \frac{1}{(1 - q^k)} = 1 + \sum_{m=1}^{\infty} \frac{q^{\frac{m^2}{2}}}{\prod_{j=1}^{m} (1 - q^j)^2} . \tag{7}$$

Once again the individual terms on the RHS are the characters of the $\mathcal{W}_{1+\infty}[\lambda = 0]$ representation $(\Lambda_+ = 0, \Lambda_- = [m, 0 \ldots 0, m])$. We refer to this $\mathcal{W}_\infty$ algebra as the horizontal higher spin algebra.

4. The Higher Spin Square

We now see that the vertical and horizontal algebras together generate the structure of a higher spin square. All commutators of this stringy algebra are determined in terms of higher spin commutators of the two underlying higher spin algebras. We should hasten to add that the stringy algebra is, however, not a tensor product in any sense of the underlying higher spin algebras. It is exponentially larger in size than either the horizontal or vertical $\mathcal{W}_\infty$ algebras.

This can be illustrated in a very similar toy example - what one may call the 'Clifford square'. Consider the $SO(2d)$ Clifford algebra

$$\{\gamma^i, \gamma^j\} = 2\delta^{ij} . \tag{8}$$

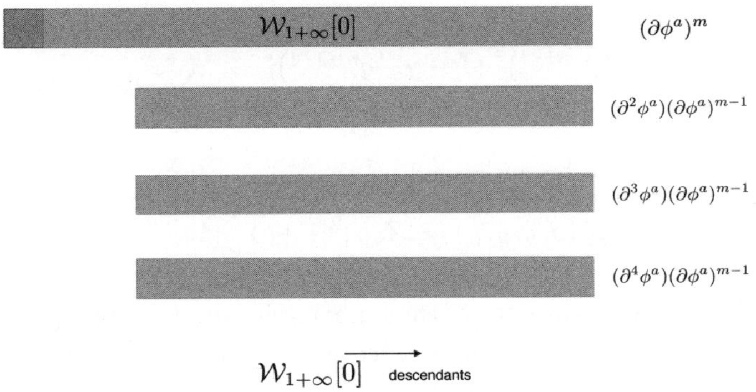

Fig. 2. Decomposition in terms of the horizontal $\mathcal{W}_{1+\infty}[\lambda = 0]$.

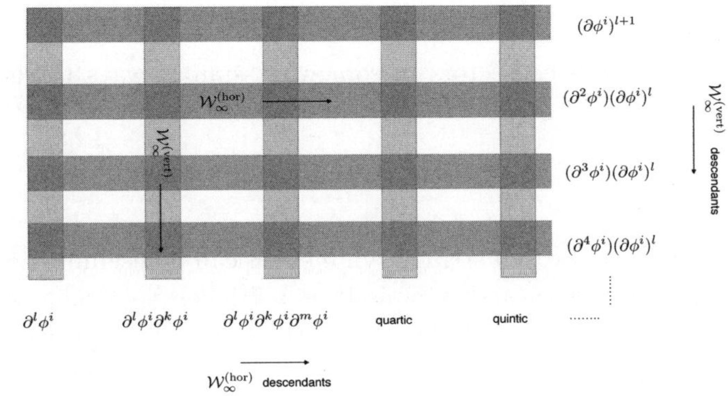

Fig. 3. Combining the vertical and horizontal slicings into a higher spin square.

The bilinears built from them $\gamma^{[i}\gamma^{j]}$ are generators of $SO(2d)$. The other multilinears with $n \leq 2d$ γ's, i.e $\gamma^{[i_i} \dots \gamma^{i_n]}$, transform in the n'th antisymmetric representation of the $SO(2d)$. Each forms a column, like in Fig.1. But because of the Clifford multiplication, there is an algebra across the columns as well. In fact, the set of all these gamma matrices and their products generates, as is well known, $SU(2^d)$. Note that this is exponentially larger in rank than $SO(2d)$.

5. Matter and the Higher Spin Square

We would like to exploit the huge symmetry of the higher spin square. Thus we can try to view the (non-chiral) spectrum of the orbifold CFT through the lens of a higher spin theory. For instance, the full untwisted sector of the orbifold CFT is given by

$$q^{\frac{N}{4}}\,\bar{q}^{\frac{N}{4}}\;Z_{\mathrm{U}}(q,y,\bar{q},\bar{y}) = \prod_{r,\bar{r}=0}^{\prime}\;\prod_{l,\bar{l}\in\mathbb{Z}}\left(1-(-1)^{2r+2\bar{r}}\,q^r y^l \bar{q}^{\bar{r}} \bar{y}^{\bar{l}}\right)^{-d(r,l)d(\bar{r},\bar{l})} \tag{9}$$

where

$$\prod_{n=1}^{\infty} \frac{\left(1 - yq^{n-1/2}\right)^2 \left(1 - y^{-1}q^{n-1/2}\right)^2}{(1-q^n)^4} = \sum_{r,l} d(r,l)\, q^r y^l. \tag{10}$$

It turns out that this can be expressed as

$$Z_{\mathrm{U}}(q,y,\bar{q},\bar{y}) = |Z_{\mathrm{vac}}|^2 [1 + \sum_R |\Phi_R|^2]. \tag{11}$$

Here Z_{vac} is the vacuum character of the higher spin square, and Φ_R is the multiparticled character of the so-called minimal irrep of the higher spin square. The sum is over all nontrivial Young diagrams capturing different ways of multiparticling. The minimal representation of the HSS has the character

$$\Phi_1^{(\mathrm{wedge})}(q,y) = \prod_{n=1}^{\infty} \frac{\left(1 + yq^{n-1/2}\right)^2 \left(1 + y^{-1}q^{n-1/2}\right)^2}{(1-q^n)^4} - 1. \tag{12}$$

One can analyse the twisted sectors in somewhat similar ways though the corresponding representations are not very well understood so far.

6. Outlook

We have seen that the unbroken stringy symmetries can be usefully viewed through the lens of a Vasiliev theory and organised as a Higher Spin Square. It is quite remarkable that the entire untwisted sector of the string theory is captured by one representation - the minimal representation of the higher spin square. A similar description is also expected to exist for the twisted sector, but we need to understand the relevant representations better. In particular, one may hope to use this viewpoint to understand better the structure of the symmetry breaking as one moves away from the orbifold point. (Recall that this symmetry is 'higgsed' upons switching on a vev to a charged field (the two cycle twisted sector). The bulk Vasiliev theory corresponding to the HSS is another thing one might hope to explicitly construct and compare with the CFT in terms of correlation functions.

Acknowledgments

One of us (R.G.) would like to thank Lars Brink for organising this meeting and the invitation to speak as well as the participants for stimulating conversations. He would also like to thank the IAS at Nanyang Technological University for their generous hospitality.

References

1. D. J. Gross, "High-energy symmetries of string theory," Phys. Rev. Lett. **60** (1988) 1229.

2. E. Witten, "Space-time and topological orbifolds," Phys. Rev. Lett. **61** (1988) 670.
3. G.W. Moore, "Symmetries and symmetry breaking in string theory," in proceedings of the SUSY '93 conference, 'Supersymmetry and unification of fundamental interactions,' (1993) 540 `[arXiv:hep-th/9308052]`.
4. A. Sagnotti, "Notes on strings and higher spins," J. Phys. A **46** (2013) 214006 `[arXiv:1112.4285 [hep-th]]`.
5. B. Sundborg, "Stringy gravity, interacting tensionless strings and massless higher spins," Nucl. Phys. Proc. Suppl. **102** (2001) 113 `[arXiv:hep-th/0103247]`.
6. E. Witten, talk at the John Schwarz 60-th birthday symposium (Nov. 2001), `http://theory.caltech.edu/jhs60/witten/1.html`.
7. M.A. Vasiliev, "Nonlinear equations for symmetric massless higher spin fields in (A)dS(d)," Phys. Lett. B **567** (2003) 139 `[arXiv:hep-th/0304049]`.
8. M. R. Gaberdiel and R. Gopakumar, "Stringy Symmetries and the Higher Spin Square," J. Phys. A **48** (2015) 185402 `[arXiv:1501.07236 [hep-th]]`.
9. M. R. Gaberdiel and R. Gopakumar, "String Theory as a Higher Spin Theory," `arXiv:1512.07237 [hep-th]`.
10. R. Dijkgraaf, G.W. Moore, E.P. Verlinde, and H.L. Verlinde, "Elliptic genera of symmetric products and second quantized strings," Commun. Math. Phys. **185** (1997) 197 `[arXiv:hep-th/9608096]`.

Higher Spin Representations of $K(E_{10})$

Axel Kleinschmidt and Hermann Nicolai
Max Planck Institute for Gravitational Physics, Am Mühlenberg 1, 14476 Potsdam, Germany

We review the recently constructed non-trivial fermionic representations of the infinite-dimensional subalgebra $K(\mathfrak{e}_{10})$ of the hyperbolic Kac–Moody algebra $\mathfrak{e}_{10}$. These representations are all unfaithful (and more specifically, of finite dimension). In addition we present their decompositions under the various finite-dimensional subgroups associated with some maximal supergravities in dimensions $D \leq 11$, and the projectors for 'spin-$\frac{7}{2}$' which have not been given before. Those representations that have not been derived from supergravity still have to find a role and a proper physical interpretation in the conjectured correspondence between E_{10} and M-theory. Nevertheless, they provide novel mathematical structures that could shed some light on fundamental questions in supergravity and on the possible role of $K(E_{10})$ as an 'R-symmetry' of M-theory, and perhaps also on the algebra $\mathfrak{e}_{10}$ itself.

1. Introduction

The hyperbolic Kac–Moody algebra $\mathfrak{e}_{10}$ has been conjectured to generate an underlying symmetry of M-theory[1,2] and its (maximal compact) subalgebra $K(\mathfrak{e}_{10})$ (fixed by the Chevalley involution) plays the role of the generalised R-symmetry transformations.[3–7] In the $\mathfrak{e}_{10}$ conjecture the constrained null motion of a spinning particle on the symmetric space $E_{10}/K(E_{10})$ is equivalent to the dynamics of supergravity or even M-theory. This conjecture thus far has only been verified for a finite set of generators of the infinite-dimensional algebra $\mathfrak{e}_{10}$ both in the bosonic and fermionic sector.[2,4,5,8–10] However, it has thus far proved impossible to construct a spinning particle action on $E_{10}/K(E_{10})$ that has one-dimensional local supersymmetry, as was explained at length in Ref. 11.

One major obstacle when constructing a supersymmetric E_{10}-model is the disparity between the bosonic and fermionic degrees of freedom that are used: The bosons are associated with the infinitely many directions of the symmetric space $E_{10}/K(E_{10})$ whereas the fermions used in Refs. 4 and 5 were constructed out of a finite-dimensional (hence unfaithful) representation of dimension 320 of the R-symmetry group $K(E_{10})$.[12] It therefore appears necessary to construct larger, preferably infinite-dimensional, fermionic representations of $K(E_{10})$ and this is the topic we will pursue in the present contribution that is partially based on our paper Ref. 13.

We develop a new formalism for constructing representations of $K(\mathfrak{e}_{10})$ and exhibit new irreducible examples of dimensions 1728 and 7040, respectively. We refer to them as 'higher spin representations' although their spin is not necessarily higher from a space-time point of view but rather when viewed from the (truncated) Wheeler–DeWitt superspace of metrics. This point will be explained in more detail below. We will see that only the **7040** contains also genuine higher spin fields from the space-time perspective. Our formalism gives the action of an infinite number of $K(\mathfrak{e}_{10})$ generators that are labelled by the positive real roots of $\mathfrak{e}_{10}$. Since the representations are finite-dimensional and therefore necessarily unfaithful, an infinite number of these generators will be represented by the same operator on the representation space.

Let us emphasize that a proper understanding of the fermionic sector will be essential for further progress with understanding the role of E_{10} in M-theory, something that is unlikely in our opinion to be achievable if one restricts attention to the bosonic sector only. On top of the (unknown) representation theory of $K(E_{10})$ this might quite possibly require some novel type of bosonisation, as is suggested by the fact that E_{10} seems to 'know everything' about the fermions of maximal supergravity that we have learnt from supersymmetry (in particular, the structure of the bosonic and fermionic multiplets). Equally important, the actual *physics* of the quantised theory with fermions is likely to differ very much from that of the purely bosonic system, as is obvious from the example of supersymmetric quantum cosmology investigated in Ref. 14.

2. $\mathfrak{e}_{10}$ and $K(\mathfrak{e}_{10})$

The (split real) Lie algebra $\mathfrak{e}_{10}$ is a hyperbolic Kac–Moody Lie algebra.[15] Its only known definition is in terms of generators and relations. There are 30 generators (e_i, f_i, h_i) for $i = 1, \dots, 10$ and each triple generates an $\mathfrak{sl}(2,\mathbb{R})$ subalgebra of $\mathfrak{e}_{10}$. The full set of defining relations is given by

$$[h_i, h_j] = 0, \qquad [h_i, e_j] = A_{ij} e_j, \qquad [h_i, f_j] = -A_{ij} f_j,$$
$$[e_i, f_j] = \delta_{ij} h_i, \qquad (\mathrm{ad}\, e_i)^{1-A_{ij}} e_j = 0, \qquad (\mathrm{ad}\, f_i)^{1-A_{ij}} f_j = 0. \tag{1}$$

Here, A_{ij} are the elements of the symmetric Cartan matrix associated with the $\mathfrak{e}_{10}$ Dynkin diagram shown in figure 1. The Cartan matrix is of Lorentzian signature and there are roots α of the algebra with norms $\alpha^2 = 2 - 2k$ for $k \in \mathbb{N}_0$. The roots with $\alpha^2 = 2$ are called real roots and they have multiplicity one; all others are imaginary and have higher multiplicity.

The subalgebra $K(\mathfrak{e}_{10})$ is generated by the 'compact' combinations

$$x_i = e_i - f_i \tag{2}$$

which are invariant under the Cartan-Chevalley involution

$$\omega(e_i) = -f_i\,, \quad \omega(f_i) = -e_i\,, \quad \omega(h_i) = -h_i \tag{3}$$

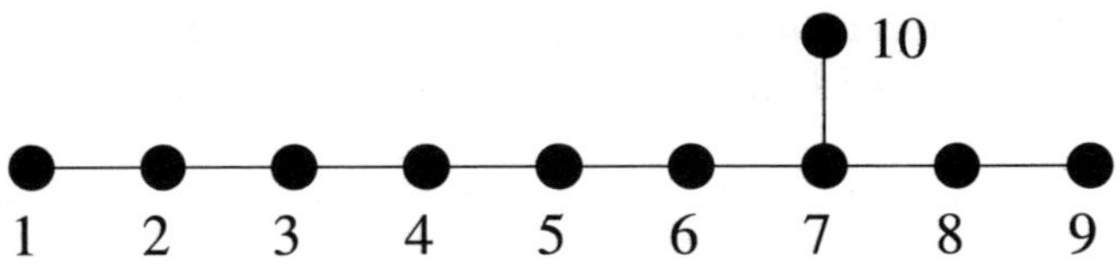

Fig. 1. The Dynkin diagram of $\mathfrak{e}_{10}$ with labelling of nodes.

The relations satisfied by these elements are in general not homogeneous (unlike the standard relations in the Chevalley–Serre presentation for the e_i and f_i above). Depending on whether two nodes i and j are connected by a line in the Dynkin diagram or not one has two cases

$$\begin{aligned} [x_i, x_j] &= 0 && \text{if } i \text{ and } j \text{ are not connected} \\ [x_i, [x_i, x_j]] + x_j &= 0 && \text{if } i \text{ and } j \text{ are connected} \end{aligned} \tag{4}$$

We will refer to these as the Berman–Serre relations; these relations were studied in a more general context in Ref. 16. The algebra $K(\mathfrak{e}_{10})$ is then defined as the free Lie algebra over the generators $\{x_i\}$ subject to the relations Eq. (4). The task of finding representations of $K(\mathfrak{e}_{10})$ is tantamount to finding matrices or operators that satisfy these relations.

Since all simple generators x_i are associated with real simple roots (of multiplicity one) one can also rephrase these relations more generally for any real roots by considering a generator $J(\alpha)$ for any (positive) real root α. Using a basis of simple roots α_i of the root lattice one then has $x_i = J(\alpha_i)$ as particular case. The relations Eq. (4) are then equivalent for real roots α and β obeying $\alpha \cdot \beta \in \{-1, 0, 1\}$

$$\begin{aligned} [J(\alpha), J(\beta)] &= \epsilon_{\alpha,\beta} J(\alpha+\beta), && \text{if } \alpha\cdot\beta = -1, \\ [J(\alpha), J(\beta)] &= -\epsilon_{\alpha,-\beta} J(\alpha-\beta), && \text{if } \alpha\cdot\beta = +1, \\ [J(\alpha), J(\beta)] &= 0, && \text{if } \alpha\cdot\beta = 0, \end{aligned} \tag{5}$$

and $\epsilon_{\alpha,\beta} \in \{-1, 1\}$ is a certain cocycle on the $\mathfrak{e}_{10}$ root lattice that satisfies

$$\epsilon_{\alpha,\beta} = -\epsilon_{\beta,\alpha} = -\epsilon_{-\alpha,-\beta}, \quad \epsilon_{\alpha+\beta,-\beta} = \epsilon_{\alpha,\beta}. \tag{6}$$

The restriction on the inner product in the commutation is to make sure that $\alpha \mp \beta$ is a real root or no root at all, such that one does not have to worry about multiplicities from imaginary roots on the right-hand side. By contrast $\epsilon_{\alpha,\beta}$ can be defined for any pair of elements (α, β) of the root lattice.

To the root lattice of $\mathfrak{e}_{10}$ one can also associate elements $\Gamma(\alpha)$ of the $\mathfrak{so}(10)$ Clifford algebra of real (32×32) matrices such that[13]

$$\Gamma(\alpha)\Gamma(\beta) = \epsilon_{\alpha,\beta}\Gamma(\alpha+\beta) = -\epsilon_{\alpha,-\beta}\Gamma(\alpha-\beta). \tag{7}$$

With these rules it is then not hard to verify that

$$J(\alpha) = \frac{1}{2}\Gamma(\alpha) \tag{8}$$

provides a representation of $K(\mathfrak{e}_{10})$ for all real roots α. This 32-dimensional representation is known as the Dirac-spinor of $K(\mathfrak{e}_{10})$. By choosing a particular basis of the root lattice, called wall basis, one could exhibit[13] that the x_i for $i = 1, \ldots, 9$ are just the usual spin representation $x_i = \frac{1}{2}\Gamma^{i\,i+1}$ of $\mathfrak{so}(10)$ but we will not use this here.

3. Tensors and Spinors on Wheeler–DeWitt Mini-superspace

The space of diagonal spatial metrics in 11 space-time dimensions is a Lorentzian ten-dimensional space in the Hamiltonian treatment of general relativity. This space is actually a finite-dimensional truncation of the full Wheeler–DeWitt 'superspace' (*alias* the 'moduli space of 10-geometries') to the finite-dimensional subspace of diagonal scale factors (the negative direction that renders this metric indefinite is associated with the scaling mode of the metric). We choose a basis e_{a} for this ten-dimensional space $(\mathsf{a}, \mathsf{b}, \ldots = 1, \ldots, 10)$ with inner products

$$e_{\mathsf{a}} \cdot e_{\mathsf{b}} = G_{\mathsf{ab}} \tag{9}$$

where G_{ab} is the Lorentzian DeWitt superspace metric restricted to the space of metric scale factors; more explicitly, it follows from the Einstein–Hilbert action that

$$G_{\mathsf{ab}} = \delta_{\mathsf{ab}} - 1 \quad \Rightarrow \quad G^{\mathsf{ab}} = \delta_{\mathsf{ab}} - \frac{1}{9} \tag{10}$$

This Lorentzian space can be identified with the Lorentzian space spanned by the roots of $\mathfrak{e}_{10}$. In the remainder we do not require the explicit form of G_{ab} of Eq. (10).

Our ansatz for fermionic representations of $K(\mathfrak{e}_{10})$ then consists in considering tensor-spinors $\phi_A^{\mathsf{a}_1\ldots\mathsf{a}_n} = \phi_A^{(\mathsf{a}_1\ldots\mathsf{a}_n)}$ that are completely symmetric in their n tensor indices and also carry a spinor index $A = 1, \ldots, 32$ of $\mathfrak{so}(10)$. The Dirac-spinor discussed in the preceding section then simply corresponds to $n = 0$. We will also consider the case when $\phi_A^{\mathsf{a}_1\ldots\mathsf{a}_n}$ is traceless in its tensor indices. Since the tensor indices are those of a Lorentzian $\mathfrak{so}(1, 9)$ space while the spinor index belongs to the Euclidean $\mathfrak{so}(10)$ subalgebra of $K(\mathfrak{e}_{10})$ our approach could be termed hybrid. Certainly one cannot take simple Γ-traces of $\phi_A^{\mathsf{a}_1\ldots\mathsf{a}_n}$ because $\mathsf{a}, \mathsf{b}, \ldots$ are not SO(10) indices, so the only option to render the tensor-spinor irreducible is to make it traceless in its indices $\mathsf{a}_1, \mathsf{a}_2, \ldots$.

The generators $J(\alpha)$ of $K(\mathfrak{e}_{10})$ are then given by combinations of an object acting on the tensor indices and gamma matrices acting on the spinor index. More precisely, we make the ansatz

$$J(\alpha)\phi_A^{\mathsf{a}_1\ldots\mathsf{a}_n} = -2X(\alpha)^{\mathsf{a}_1\ldots\mathsf{a}_n}{}_{\mathsf{b}_1\ldots\mathsf{b}_n}\Gamma(\alpha)_{AB}\phi_B^{\mathsf{b}_1\ldots\mathsf{b}_n}. \tag{11}$$

Due to the known properties Eq. (5) of the $\Gamma(\alpha)$ under commutation, checking the consistency relations Eq. (7) then can be reduced to checking the following conditions on the tensors $X(\alpha)$ for real roots[13]

$$\begin{aligned} \{X(\alpha), X(\beta)\} &= \frac{1}{2}X(\alpha \pm \beta), && \text{if } \alpha \cdot \beta = \mp 1, \\ [X(\alpha), X(\beta)] &= 0, && \text{if } \alpha \cdot \beta = 0. \end{aligned} \tag{12}$$

Note that there is no $\epsilon_{\alpha,\beta}$ in these relations as it is already taken care of by the $\Gamma(\alpha)$. The Dirac-spinor corresponds to the solution $X(\alpha) = \frac{1}{4}$ to these equations.

Another $K(\mathfrak{e}_{10})$ representation that has been known from supergravity considerations is the case $n = 1$ that corresponds to the $D = 11$ gravitino and has dimension 320.[4–6] In our language it corresponds to the solution

$$X(\alpha)^{\mathtt{a}}{}_{\mathtt{b}} = -\frac{1}{2}\alpha^{\mathtt{a}}\alpha_{\mathtt{b}} + \frac{1}{4}\delta^{\mathtt{a}}_{\mathtt{b}}, \tag{13}$$

where $\alpha^{\mathtt{a}}$ are the components of the root α with respect to the basis $e_{\mathtt{a}}$, i.e., $\alpha = \sum_{\mathtt{a}} \alpha^{\mathtt{a}} e_{\mathtt{a}}$. 'Typewriter font' indices are raised and lowered with the Lorentzian $G_{\mathtt{ab}}$.

For the gravitino (or vector-spinor) one can find a rewriting in terms of pure $\mathfrak{so}(10)$ representation by letting[17]

$$\psi^a_A = \sum_B \Gamma^a_{AB}\phi^a_B \qquad \text{(no sum on } a\text{)}. \tag{14}$$

The object on the left is then a standard vector-spinor of $\mathfrak{so}(10)$. A similar simple and explicit rewriting into $\mathfrak{so}(10)$ representations is not known for the new representations we discuss below.

We also note that due to the unfaithfulness of the representations, one obtains (infinite-dimensional) ideals in $K(\mathfrak{e}_{10})$, leading to the result that $K(\mathfrak{e}_{10})$ is not a simple Lie algebra. The quotient Lie algebras $\mathfrak{q}$ of $K(\mathfrak{e}_{10})$ by the ideals of the **32** and **320** have been analysed and are given by $\mathfrak{q}_{\mathbf{32}} \cong \mathfrak{so}(32)$ and $\mathfrak{q}_{\mathbf{320}} \cong \mathfrak{so}(288, 32)$. It may seem surprising that the 'compact' $K(\mathfrak{e}_{10})$ admits a non-compact quotient in the **320** representation but this is not a contradiction due to the infinite-dimensionality of $K(\mathfrak{e}_{10})$. For the higher spin representations below, the quotients have not been worked out.

4. Higher Spin Representations

In Ref. 13 two further solutions to Eq. (12) were found that correspond to the values $n = 2$ and $n = 3$ (corresponding to spin $s = \frac{5}{2}$ and $s = \frac{7}{2}$, respectively[18]). These representations go beyond supergravity as there appears to be no supergravity model from which they would be derivable. For spin $s = \frac{5}{2}$ ($n = 2$) the corresponding tensors are given by

$$X(\alpha)^{\mathtt{a}_1\mathtt{a}_2}{}_{\mathtt{b}_1\mathtt{b}_2} = \frac{1}{2}\alpha^{\mathtt{a}_1}\alpha^{\mathtt{a}_2}\alpha_{\mathtt{b}_1}\alpha_{\mathtt{b}_2} - \alpha^{(\mathtt{a}_1}\delta^{\mathtt{a}_2)}_{(\mathtt{b}_1}\alpha_{\mathtt{b}_2)} + \frac{1}{4}\delta^{(\mathtt{a}_1}_{\mathtt{b}_1}\delta^{\mathtt{a}_2)}_{\mathtt{b}_2} \tag{15}$$

and for $n = 3$ (spin-$\frac{7}{2}$) by

$$\begin{aligned} X(\alpha)^{\mathtt{a}_1\mathtt{a}_2\mathtt{a}_3}{}_{\mathtt{b}_1\mathtt{b}_2\mathtt{b}_3} = & -\frac{1}{3}\alpha^{\mathtt{a}_1}\alpha^{\mathtt{a}_2}\alpha^{\mathtt{a}_3}\alpha_{\mathtt{b}_1}\alpha_{\mathtt{b}_2}\alpha_{\mathtt{b}_3} + \frac{3}{2}\alpha^{(\mathtt{a}_1}\alpha^{\mathtt{a}_2}\delta^{\mathtt{a}_3)}_{(\mathtt{b}_1}\alpha_{\mathtt{b}_2}\alpha_{\mathtt{b}_3)} - \frac{3}{2}\alpha^{(\mathtt{a}_1}\delta^{\mathtt{a}_2}_{(\mathtt{b}_1}\delta^{\mathtt{a}_3)}_{\mathtt{b}_2}\alpha_{\mathtt{b}_3)} \\ & + \frac{1}{4}\delta^{(\mathtt{a}_1}_{(\mathtt{b}_1}\delta^{\mathtt{a}_2}_{\mathtt{b}_2}\delta^{\mathtt{a}_3)}_{\mathtt{b}_3)} + \frac{1}{12}(2-\sqrt{3})\alpha^{(\mathtt{a}_1}G^{\mathtt{a}_2\mathtt{a}_3)}G_{(\mathtt{b}_1\mathtt{b}_2}\alpha_{\mathtt{b}_3)} \\ & + \frac{1}{12}(-1+\sqrt{3})\left(\alpha^{(\mathtt{a}_1}\alpha^{\mathtt{a}_2}\alpha^{\mathtt{a}_3)}G_{(\mathtt{b}_1\mathtt{b}_2}\alpha_{\mathtt{b}_3)} + \alpha^{(\mathtt{a}_1}G^{\mathtt{a}_2\mathtt{a}_3)}\alpha_{(\mathtt{b}_1}\alpha_{\mathtt{b}_2}\alpha_{\mathtt{b}_3)}\right). \end{aligned} \tag{16}$$

These expressions can be found and verified analytically. We have also extended the search for solutions of this type for $n \leq 10$ with the ansatz above but have not found any additional solutions so far.

The spin-$\frac{5}{2}$ solution as given is of dimension $\frac{10\times 11}{2} \times 32 = 1760$. It turns out that this representation is reducible since the subspace spanned by the trace $G_{\mathsf{ab}}\phi_A^{\mathsf{ab}}$ is invariant. This trace transforms in the spin-$\frac{1}{2}$ representation of dimension 32, leaving an irreducible 1728-dimensional representation of $K(\mathfrak{e}_{10})$. By contrast, the spin-$\frac{7}{2}$ representation of dimension $\frac{10\times 11\times 12}{6} \times 32 = 7040$ is irreducible as given.

In the next two sections, we investigate further properties of the new higher spin representations.

5. Projectors and Weyl Group Action

The $K(\mathfrak{e}_{10})$ generators $J(\alpha)$ are defined for all positive roots α of $\mathfrak{e}_{10}$. As the roots α are space-like elements in a Lorentzian ten-dimensional space, they have a stabiliser of type $\mathfrak{so}(1,8) \subset \mathfrak{so}(1,9)$. This stability algebra can be used to decompose the 'polarisation tensor' $X(\alpha)$ into irreducible pieces for a fixed α. The irreducible $\mathfrak{so}(1,8)$ terms are given by projectors $\Pi^{(j)}(\alpha)$, such that tensor $X(\alpha)$ can be expressed in terms of these projectors.[19] This rewriting greatly facilitates the exponentiation of the corresponding matrices, and will make it easy to work out the exponentiated (Weyl) group actions.

5.1. *Projectors for spin-$\frac{5}{2}$*

For $n = 2$, the polarisation tensor $X(\alpha)^{\mathsf{a}_1\mathsf{a}_2}{}_{\mathsf{b}_1\mathsf{b}_2}$ lies in the reducible **55** of $\mathfrak{so}(1,9)$, where we work for simplicity with the reducible representation of dimension 1760 given in Eq. (15). The decomposition of $X(\alpha)$ under the regularly embedded $\mathfrak{so}(1,8)$ is

$$\mathbf{55} \to \mathbf{54} \oplus \mathbf{1} \to (\mathbf{44} \oplus \mathbf{9} \oplus \mathbf{1'}) \oplus \mathbf{1}. \tag{17}$$

The splitting of the singlets here has been done in such a way that **1** corresponds to the $\mathfrak{so}(1,9)$ singlet corresponding to the trace with G_{ab}. One can check that the following are complete orthonormal projectors on the various pieces

$$\begin{aligned}
\Pi^{(\mathbf{44})}(\alpha)^{\mathsf{a}_1\mathsf{a}_2}{}_{\mathsf{b}_1\mathsf{b}_2} &= \frac{2}{9}\alpha^{\mathsf{a}_1}\alpha^{\mathsf{a}_2}\alpha_{\mathsf{b}_1}\alpha_{\mathsf{b}_2} - \alpha^{(\mathsf{a}_1}\delta^{\mathsf{a}_2)}_{(\mathsf{b}_1}\alpha_{\mathsf{b}_2)} + \delta^{(\mathsf{a}_1}_{\mathsf{b}_1}\delta^{\mathsf{a}_2)}_{\mathsf{b}_2} \\
&\quad + \frac{1}{18}\left(\alpha^{\mathsf{a}_1}\alpha^{\mathsf{a}_2}G_{\mathsf{b}_1\mathsf{b}_2} + G^{\mathsf{a}_1\mathsf{a}_2}\alpha_{\mathsf{b}_1}\alpha_{\mathsf{b}_2}\right) - \frac{1}{9}G^{\mathsf{a}_1\mathsf{a}_2}G_{\mathsf{b}_1\mathsf{b}_2}, \\
\Pi^{(\mathbf{9})}(\alpha)^{\mathsf{a}_1\mathsf{a}_2}{}_{\mathsf{b}_1\mathsf{b}_2} &= -\frac{1}{2}\alpha^{\mathsf{a}_1}\alpha^{\mathsf{a}_2}\alpha_{\mathsf{b}_1}\alpha_{\mathsf{b}_2} + \alpha^{(\mathsf{a}_1}\delta^{\mathsf{a}_2)}_{(\mathsf{b}_1}\alpha_{\mathsf{b}_2)}, \\
\tilde{\Pi}^{(\mathbf{1})}(\alpha)^{\mathsf{a}_1\mathsf{a}_2}{}_{\mathsf{b}_1\mathsf{b}_2} &= \frac{1}{10}G^{\mathsf{a}_1\mathsf{a}_2}G_{\mathsf{b}_1\mathsf{b}_2}, \\
\tilde{\Pi}^{(\mathbf{1'})}(\alpha)^{\mathsf{a}_1\mathsf{a}_2}{}_{\mathsf{b}_1\mathsf{b}_2} &= \frac{5}{18}\alpha^{\mathsf{a}_1}\alpha^{\mathsf{a}_2}\alpha_{\mathsf{b}_1}\alpha_{\mathsf{b}_2} - \frac{1}{18}\left(\alpha^{\mathsf{a}_1}\alpha^{\mathsf{a}_2}G_{\mathsf{b}_1\mathsf{b}_2} + G^{\mathsf{a}_1\mathsf{a}_2}\alpha_{\mathsf{b}_1}\alpha_{\mathsf{b}_2}\right) \\
&\quad + \frac{1}{90}G^{\mathsf{a}_1\mathsf{a}_2}G_{\mathsf{b}_1\mathsf{b}_2}.
\end{aligned} \tag{18}$$

In terms of these, the tensor $X(\alpha)$ takes the form

$$X(\alpha)^{a_1a_2}{}_{b_1b_2} = \left(\frac{1}{4}\tilde{\Pi}^{(\mathbf{1})}(\alpha) + \frac{1}{4}\tilde{\Pi}^{(\mathbf{1}')}(\alpha) - \frac{3}{4}\Pi^{(\mathbf{9})}(\alpha) + \frac{1}{4}\Pi^{(\mathbf{44})}(\alpha)\right)^{a_1a_2}{}_{b_1b_2}. \tag{19}$$

What is important here is that the coefficients of all projectors are of the form $\frac{2k+1}{4}$ with $k \in \mathbb{Z}$. This implies that when one constructs the 'Weyl group' generator

$$w_\alpha = e^{\frac{\pi}{2}J(\alpha)} \tag{20}$$

acting in the representation is idempotent in the eighth power. Weyl group has been put into inverted commas above because this is more correctly an element of a covering of the Weyl group that has been dubbed the spin-extended Weyl group.[17,20] Acting on spinor representations, the characteristic feature is that only the eighth power $w_\alpha^8 = \mathbb{1}$ whereas one normally has the fourth power for the covering of the Weyl on bosonic representations.[15]

5.2. *Projectors for spin-$\frac{7}{2}$*

In this case, the polarisation tensor $X(\alpha)^{a_1a_2a_3}{}_{b_1b_2b_3}$ is in the (reducible) totally symmetric **220** of $\mathfrak{so}(1,9)$. This decomposes under $\mathfrak{so}(1,8)$ as

$$\mathbf{220} \to \mathbf{210} \oplus \mathbf{10} \to (\mathbf{156} \oplus \mathbf{44} \oplus \mathbf{9} \oplus \mathbf{1}) \oplus (\mathbf{9} \oplus \mathbf{1})\,. \tag{21}$$

There are two singlets and two vectors of $\mathfrak{so}(1,8)$ appearing in the decomposition and some associated freedom in constructing the orthonormal projectors. We choose a particular combination of these representations as follows[21]

$$\begin{aligned}
\Pi^{(\mathbf{156})}(\alpha)^{a_1a_2a_3}{}_{b_1b_2b_3} &= -\frac{1}{11}\alpha^{a_1}\alpha^{a_2}\alpha^{a_3}\alpha_{b_1}\alpha_{b_2}\alpha_{b_3} + \frac{15}{22}\alpha^{(a_1}\alpha^{a_2}\delta^{a_3)}_{(b_1}\alpha_{b_2}\alpha_{b_3)}\\
&\quad - \frac{3}{2}\alpha^{(a_1}\delta^{a_2}_{(b_1}\delta^{a_3)}_{b_2}\alpha_{b_3)} + \delta^{(a_1}_{(b_1}\delta^{a_2}_{b_2}\delta^{a_3)}_{b_3)}\\
&\quad - \frac{3}{44}\left(\alpha^{a_1}\alpha^{a_2}\alpha^{a_3}\alpha_{(b_1}G_{b_2b_3)} + \alpha^{(a_1}G^{a_2a_3)}\alpha_{b_1}\alpha_{b_2}\alpha_{b_3}\right)\\
&\quad + \frac{3}{22}\left(\alpha^{(a_1}\alpha^{a_2}\delta^{a_3)}_{(b_1}G_{b_2b_3)} + G^{(a_1a_2}\delta^{a_3)}_{(b_1}\alpha_{b_2}\alpha_{b_3)}\right)\\
&\quad - \frac{3}{11}G^{(a_1a_2}\delta^{a_3)}_{(b_1}G_{b_2b_3)} + \frac{3}{22}\alpha^{(a_1}G^{a_2a_3)}G_{(b_1b_2}\alpha_{b_3)}\\
\Pi^{(\mathbf{44})}(\alpha)^{a_1a_2a_3b_1b_2b_3} &= \frac{1}{3}\alpha^{a_1}\alpha^{a_2}\alpha^{a_2}\alpha_{b_1}\alpha_{b_2}\alpha_{b_3} - \frac{3}{2}\alpha^{(a_1}\alpha^{a_2}\delta^{a_3)}_{(b_1}\alpha_{b_2}\alpha_{b_3)}\\
&\quad + \frac{3}{2}\alpha^{(a_1}\delta^{a_2}_{(b_1}\delta^{a_3)}_{b_2}\alpha_{b_3)} - \frac{1}{6}\alpha^{(a_1}G^{a_2a_3)}G_{(b_1b_2}\alpha_{b_3)}\\
&\quad + \frac{1}{12}\left(\alpha^{a_1}\alpha^{a_2}\alpha^{a_3}\alpha_{(b_1}G_{b_2b_3)} + \alpha^{(a_1}G^{a_2a_3)}\alpha_{b_1}\alpha_{b_2}\alpha_{b_3}\right)
\end{aligned} \tag{22}$$

for the (unique) two biggest representations,

$$\Pi^{(\mathbf{9})}(\alpha)^{a_1a_2a_3}{}_{b_1b_2b_3} = -\frac{9}{22}\alpha^{a_1}\alpha^{a_2}\alpha^{a_3}\alpha_{b_1}\alpha_{b_2}\alpha_{b_3} + \frac{9}{11}\alpha^{(a_1}\alpha^{a_2}\delta^{a_3)}_{(b_1}\alpha_{b_2}\alpha_{b_3)}$$
$$+\frac{3}{44}\left(\alpha^{a_1}\alpha^{a_2}\alpha^{a_3}\alpha_{(b_1}G_{b_2b_3)} + \alpha^{(a_1}G^{a_2a_3)}\alpha_{b_1}\alpha_{b_2}\alpha_{b_3}\right)$$
$$-\frac{3}{22}\left(\alpha^{(a_1}\alpha^{a_2}\delta^{a_3)}_{(b_1}G_{b_2b_3)} + G^{(a_1a_2}\delta^{a_3)}_{(b_1}\alpha_{b_2}\alpha_{b_3)}\right)$$
$$+\frac{1}{44}G^{(a_1a_2}\delta^{a_3)}_{(b_1}G_{b_2b_3)} - \frac{1}{88}\alpha^{(a_1}G^{a_2a_3)}G_{(b_1b_2}\alpha_{b_3)}$$
$$\Pi^{(\mathbf{9'})}(\alpha)^{a_1a_2a_3}{}_{b_1b_2b_3} = \frac{1}{4}G^{(a_1a_2}\delta^{a_3)}_{(b_1}G_{b_2b_3)} - \frac{1}{8}\alpha^{(a_1}G^{a_2a_3)}G_{(b_1b_2}\alpha_{b_3)} \tag{23}$$

for the vectors and finally for the singlets

$$\Pi^{(\mathbf{1})}(\alpha)^{a_1a_2a_3}{}_{b_1b_2b_3} = \frac{1}{12}\alpha^{a_1}\alpha^{a_2}\alpha^{a_3}\alpha_{b_1}\alpha_{b_2}\alpha_{b_3}$$
$$+\frac{1}{24}(-1-\sqrt{3})\left(\alpha^{a_1}\alpha^{a_2}\alpha^{a_3}\alpha_{(b_1}G_{b_2b_3)} + \alpha^{(a_1}G^{a_2a_3)}\alpha_{b_1}\alpha_{b_2}\alpha_{b_3}\right)$$
$$+\frac{1}{24}(2+\sqrt{3})\alpha^{(a_1}G^{a_2a_3)}G_{(b_1b_2}\alpha_{b_3)}$$
$$\Pi^{(\mathbf{1'})}(\alpha)^{a_1a_2a_3}{}_{b_1b_2b_3} = \frac{1}{12}\alpha^{a_1}\alpha^{a_2}\alpha^{a_3}\alpha_{b_1}\alpha_{b_2}\alpha_{b_3}$$
$$+\frac{1}{24}(-1+\sqrt{3})\left(\alpha^{a_1}\alpha^{a_2}\alpha^{a_3}\alpha_{(b_1}G_{b_2b_3)} + \alpha^{(a_1}G^{a_2a_3)}\alpha_{b_1}\alpha_{b_2}\alpha_{b_3}\right)$$
$$+\frac{1}{24}(2-\sqrt{3})\alpha^{(a_1}G^{a_2a_3)}G_{(b_1b_2}\alpha_{b_3)} \tag{24}$$

The tensor $X(\alpha)$ reads as follows in this basis

$$X(\alpha)^{a_1a_2a_3}{}_{b_1b_2b_3} = \Bigg(\frac{5}{4}\Pi^{(\mathbf{1})}(\alpha) - \frac{3}{4}\Pi^{(\mathbf{1'})}(\alpha) + \frac{1}{4}\Pi^{(\mathbf{9})}(\alpha) + \frac{1}{4}\Pi^{(\mathbf{9'})}(\alpha)$$
$$-\frac{3}{4}\Pi^{(\mathbf{44})}(\alpha) + \frac{1}{4}\Pi^{(\mathbf{156})}(\alpha)\Bigg)^{a_1a_2a_3}{}_{b_1b_2b_3}. \tag{25}$$

Again, it is important that the coefficients of all the orthonormal projectors are of the form $\frac{2k+1}{4}$ such that we are dealing with a genuine fermionic representation of $K(E_{10})$.

6. Branching Under Subalgebras

The infinite-dimensional Lie algebra $K(\mathfrak{e}_{10})$ has infinitely many finite-dimensional subalgebras.[22] Of these are of particular interest to us the following, all of which can be obtained by deleting a single node from the $\mathfrak{e}_{10}$ Dynkin diagram:

(*a*) $\mathfrak{so}(10)$	deleting node 10	SUGRA in $D = 11$
(*b*) $\mathfrak{so}(2) \oplus \mathfrak{so}(16)$	deleting node 2	SUGRA in $D = 3$
(*c*) $\mathfrak{so}(9) \oplus \mathfrak{so}(2)$	deleting node 8	IIB SUGRA in $D = 10$
(*d*) $\mathfrak{so}(9) \oplus \mathfrak{so}(9)$	deleting node 9	Doubled SUGRA in $D = 10$

The last case requires some explanation. In Ref. 8 the decomposition of $\mathfrak{e}_{10}$ under its $\mathfrak{so}(9,9)$ subalgebra was studied and shown to correspond to both type IIA and type IIB theory since the Ramond–Ramond potentials occurred in a spinor representation of $\mathfrak{so}(9,9)$ that can be read either as all even or all odd forms; similarly, the fermions arrange themselves correctly for the two theories.[24] In investigations of double field theory the same structure appears[25] and we have therefore dubbed this T-duality agnostic decomposition as 'doubled SUGRA.'

There are some additional subtleties associated with the global assignment of fermionic and bosonic representations at the group level. More precisely, the $\mathfrak{so}(16)$ is the Lie algebra of $Spin(16)/\mathbb{Z}_2$. The $\mathbb{Z}_2$ is not diagonally embedded in the center $\mathbb{Z}_2 \times \mathbb{Z}_2$ but as one of the factors; this entails that the representations $\mathbf{16}_v$ and $\mathbf{128}_c$ are spinorial (that is, they transform with a factor (-1) upon rotation by 2π), whereas the $\mathbf{128}_s$ is tensorial.[26] Moreover, the $\mathbf{16}$ spinor of $Spin(9) = [Spin(9) \times Spin(9)]_{\text{diag}}$ is identified with the (spinorial) $\mathbf{16}_v$ of $Spin(16)$. The diagonal $Spin(9)$ also lies as a regular subgroup in $Spin(9)$ as it corresponds to the dimensional reduction from $D = 11$ to $D = 10$ (over a spatial direction).

The decompositions of the spin-$\frac{1}{2}$ and spin-$\frac{3}{2}$ representations were already given in Ref. 24, while the decompositions of the new higher spin representations under the various subalgebras have not been given previously. To find the relevant decompositions for spin-$\frac{5}{2}$ and spin-$\frac{7}{2}$ is actually rather involved, and can only be done on a computer. The main problem here is that the $K(E_{10})$ representations are not highest or lowest weight representations (it is not even clear whether $K(E_{10})$ admits any analog of such representations), so the customary tools of representation theory cannot be applied. However, the subrepresentations obtained after descending to any finite-dimensional subgroup *are* highest or lowest weight representations, so given any of the above subgroups, one must first identify the corresponding highest or lowest weights. For instance, for the spin-$\frac{7}{2}$ representation this requires (amongst other things) the (simultaneous) diagonalisation of various 7040×7040 matrices. It seems clear that for yet higher dimensional realisations such a procedure would become impractical very quickly unless better methods are developed.

6.1. *Branching the spin $s = \frac{1}{2}$ and $s = \frac{3}{2}$ representations*

These were already understood in previous work.[24] The fractions $\frac{1}{2}$ and $\frac{3}{2}$ in the decompositions (b) and (c) below correspond to the $\mathfrak{so}(2) \cong \mathfrak{u}(1)$ charges. In these cases all the representations form doublets of $\mathfrak{so}(2)$ that can also be thought of as complex one-dimensional representations of $\mathfrak{u}(1)$. This has to be taken into account when checking the dimension count of the decompositions.

$$\begin{aligned}\mathbf{32} &\xrightarrow{a} \mathbf{32}\\ &\xrightarrow{b} \left(\frac{1}{2}, \mathbf{16}_v\right)\end{aligned}$$

$$\begin{aligned}
&\xrightarrow{c} \left(\mathbf{16}, \frac{1}{2}\right) \\
&\xrightarrow{d} (\mathbf{16}, \mathbf{1}) \oplus (\mathbf{1}, \mathbf{16})
\end{aligned} \tag{26}$$

and

$$\begin{aligned}
\mathbf{320} &\xrightarrow{a} \mathbf{288} \oplus \mathbf{32} \\
&\xrightarrow{b} \left(\frac{1}{2}, \mathbf{128}_c\right) \oplus \left(\frac{1}{2}, \mathbf{16}_v\right) \oplus \left(\frac{3}{2}, \mathbf{16}_v\right) \\
&\xrightarrow{c} \left(\mathbf{16}, \frac{3}{2}\right) \oplus \left(\mathbf{128}, \frac{1}{2}\right) \oplus \left(\mathbf{16}, \frac{1}{2}\right) \\
&\xrightarrow{d} (\mathbf{9}, \mathbf{16}) \oplus (\mathbf{16}, \mathbf{9}) \oplus (\mathbf{1}, \mathbf{16}) \oplus (\mathbf{16}, \mathbf{1})
\end{aligned} \tag{27}$$

Since these are the 'physical' fermions of maximal supergravity, let us briefly comment on their interpretation.

The **32** representation of $K(\mathfrak{e}_{10})$ corresponds to the 32 supersymmetry generators of maximal supergravity. We see that in the decomposition (a) relevant for $D = 11$ supergravity one obtains a single generator consistent with $\mathcal{N} = 1$ supersymmetry. In the decomposition (b) one obtains an $\mathfrak{so}(2)$ doublet of sixteen generators (in the vector of $\mathfrak{so}(16)$; the $\mathfrak{so}(2)$ corresponds to the spatial part of the $\mathfrak{so}(1,2)$ Lorentz symmetry of which the doublet is the irreducible spinor and the sixteen components correspond to $\mathfrak{so}(16)$ R-symmetry of maximal $\mathcal{N} = 16$ supersymmetry in $D = 3$ dimensions. The decomposition (c) gives an $\mathfrak{so}(2)$ R-symmetry doublet of spinors of the spatial $\mathfrak{so}(9)$ Lorentz symmetry in $D = 10$ in agreement with the supersymmetry generators of chiral type IIB supergravity. More specifically, the appearance of this U(1) effectively 'complexifies' the SO(9) representation, in line with the chirality of the type IIB fermions. The last decomposition (d) is consistent with a type IIA formulation of doubled supergravity.[8,25]

The decompositions of the **320** representation of $K(\mathfrak{e}_{10})$ can be interpreted similarly.[24] For example, the decomposition (b) gives the 128 physical fermions in $D = 3$ together with components associated with the non-propagating gravitino that is needed when formulating $\mathcal{N} = 16$ supergravity in $D = 3$. We also note again that in the type IIB decomposition (c) one always obtains doublets of the R-symmetry $\mathfrak{so}(2)$, in accord with the chirality of the underlying fermionic multiplets.

6.2. *Branching of the spin-$\frac{5}{2}$ representation*

The decomposition under the various subalgebras is

$$\begin{aligned}
&\mathbf{1728} \xrightarrow{a} \mathbf{1120} \oplus 2 \times \mathbf{288} \oplus \mathbf{32} \\
&\xrightarrow{b} \left(\frac{1}{2}, \mathbf{560}_v\right) \oplus \left(\frac{1}{2}, \mathbf{128}_c\right) \oplus 2 \times \left(\frac{1}{2}, \mathbf{16}_v\right) \oplus \left(\frac{3}{2}, \mathbf{128}_{c},\right) \oplus \left(\frac{3}{2}, \mathbf{16}_v\right)
\end{aligned}$$

$$\xrightarrow{c} \left(\mathbf{432}, \frac{1}{2}\right) \oplus 2 \times \left(\mathbf{128}, \frac{1}{2}\right) \oplus 2 \times \left(\mathbf{16}, \frac{1}{2}\right) \oplus \left(\mathbf{128}, \frac{3}{2}\right) \oplus \left(\mathbf{16}, \frac{3}{2}\right)$$

$$\xrightarrow{d} (\mathbf{36}, \mathbf{16}) \oplus (\mathbf{16}, \mathbf{36}) \oplus (\mathbf{9}, \mathbf{16}) \oplus (\mathbf{16}, \mathbf{9}) \oplus$$
$$\oplus (\mathbf{128}, \mathbf{1}) \oplus (\mathbf{1}, \mathbf{128}) \oplus (\mathbf{1}, \mathbf{16}) \oplus (\mathbf{16}, \mathbf{1}) \quad (28)$$

From the $\mathfrak{so}(10)$ decomposition we see that the space-time spin of this $K(\mathfrak{e}_{10})$ representation is not really higher than 3/2 since the **1120** corresponds to an *anti*-symmetric tensor-spinor of $\mathfrak{so}(10)$ with two tensor indices. The $\mathbf{560}_v$ of $\mathfrak{so}(16)$ that arises is the anti-symmetric three-form. Similar to the $\mathbf{16}_v$ discussed above, this is actually a spinorial representation with the correct assignment when lifted to the group $Spin(16)/\mathbb{Z}_2$. The **432** of $\mathfrak{so}(9)$ that arises in case (c) is the tensor-spinor with two antisymmetric indices.

6.3. *Branching of the spin-$\frac{7}{2}$ representation*

Under the subalgebras listed above, the $K(\mathfrak{e}_{10})$ spin-$\frac{7}{2}$ representation of dimension 7040 decomposes as

$$\mathbf{7040} \xrightarrow{a} \mathbf{2400} \oplus \mathbf{1440} \oplus 2 \times \mathbf{1120} \oplus 3 \times \mathbf{288} \oplus 3 \times \mathbf{32}$$

$$\xrightarrow{b} \left(\frac{1}{2}, \mathbf{1920}_s\right) \oplus \left(\frac{3}{2}, \mathbf{560}_v\right) \oplus \left(\frac{1}{2}, \mathbf{560}_v\right) \oplus \left(\frac{3}{2}, \mathbf{128}_c\right) \oplus 2 \times \left(\frac{1}{2}, \mathbf{128}_c\right)$$
$$\oplus \left(\frac{5}{2}, \mathbf{16}_v\right) \oplus 2 \times \left(\frac{3}{2}, \mathbf{16}_v\right) \oplus 3 \times \left(\frac{1}{2}, \mathbf{16}_v\right)$$

$$\xrightarrow{c} \left(\mathbf{768}, \frac{1}{2}\right) \oplus \left(\mathbf{576}, \frac{1}{2}\right) \oplus \left(\mathbf{432}, \frac{3}{2}\right) \oplus 2 \times \left(\mathbf{432}, \frac{1}{2}\right) \oplus 2 \times \left(\mathbf{128}, \frac{3}{2}\right)$$
$$\oplus 4 \times \left(\mathbf{128}, \frac{1}{2}\right) \oplus \left(\mathbf{16}, \frac{5}{2}\right) \oplus 2 \times \left(\mathbf{16}, \frac{3}{2}\right) \oplus 4 \times \left(\mathbf{16}, \frac{1}{2}\right)$$

$$\xrightarrow{d} (\mathbf{128}, \mathbf{9}) \oplus (\mathbf{128}, \mathbf{1}) \oplus (\mathbf{16}, \mathbf{84}) \oplus (\mathbf{16}, \mathbf{36}) \oplus 2 \times (\mathbf{16}, \mathbf{9}) \oplus 2 \times (\mathbf{16}, \mathbf{1})$$
$$\oplus (\mathbf{9}, \mathbf{128}) \oplus (\mathbf{1}, \mathbf{128}) \oplus (\mathbf{84}, \mathbf{16}) \oplus (\mathbf{36}, \mathbf{16}) \oplus 2 \times (\mathbf{9}, \mathbf{16}) \oplus 2 \times (\mathbf{1}, \mathbf{16})$$
$$(29)$$

As already mentioned, it is a non-trivial task to work out these decompositions in practice. As a further test we have also checked that the further decompositions of the $\mathfrak{so}(2) \oplus \mathfrak{so}(16)$ and $\mathfrak{so}(10)$ representations under their common $\mathfrak{so}(8)$ subalgebra coincide (for $\mathfrak{so}(16)$ this subalgebra is obtained after descending first to the diagonal subalgebra $[\mathfrak{so}(8) \oplus \mathfrak{so}(8)]_{\text{diag}}$). Similarly, there is another $\mathfrak{so}(8)$ that is common to the $\mathfrak{so}(10)$ decomposition (a), to the type IIB decomposition (c) and to the $\mathfrak{so}(9) \oplus \mathfrak{so}(9)$ decomposition in (d) and that corresponds to the spatial rotations of maximal $D = 9$ supergravity. The further branching of (a), (c) and (d) to this common subgroup has been checked to be consistent. Moreover, we have verified that the common $\mathfrak{so}(9)$ of the type IIB decomposition (c) and the T-duality agnostic decomposition (d) gives the same representations.

Let us finally highlight some new features arising here, that have no analog for spin $s \leq \frac{5}{2}$.

- In the $\mathfrak{so}(10)$ decomposition (a) one sees the **2400** that corresponds to a tensor-spinor that is antisymmetric in *three* tensor indices. The **1440** is a tensor-spinor with two *symmetric* tensor indices; since the $\mathfrak{so}(10)$ is the spatial rotation group of $D = 11$ supergravity, this means that the spin-$\frac{7}{2}$ of $K(\mathfrak{e}_{10})$ contains genuinely higher spin representations also from a space-time perspective!
- Under the $\mathfrak{so}(2) \oplus \mathfrak{so}(16)$ decomposition (b) one finds the vector-spinor of $\mathfrak{so}(16)$ with 1920 components. Note that consistent with the spinorial nature of the $K(E_{10})$ representation it is the $\mathbf{1920}_s$ where the spinorial double-valued aspect of $Spin(16)/\mathbb{Z}_2$ is carried by the vector index and not by the s-type spinor index.
- The **768** appearing in the $\mathfrak{so}(9) \oplus \mathfrak{so}(2)$ decomposition (c) is the anti-symmetric three-form tensor-spinor of $\mathfrak{so}(9)$. By contrast the **576** is a tensor-spinor with two *symmetric* tensor indices and therefore this $K(\mathfrak{e}_{10})$ representation also contains fermionic higher spin fields from the type IIB perspective.
- The **84** in the $\mathfrak{so}(9) \oplus \mathfrak{so}(9)$ decomposition (d) is the anti-symmetric three-form of $\mathfrak{so}(9)$; the **36** is the anti-symmetric two-form already encountered above.

We also note that the $\mathfrak{so}(2)$ eigenvalues can become larger and larger the bigger the $K(\mathfrak{e}_{10})$ representation becomes.

7. Outlook

There are two pressing questions arising out of our work. The first concerns the possible physical role of the new $K(E_{10})$ representations. In particular, one may wonder whether they are of relevance to overcoming the difficulties in constructing a supersymmetric E_{10} model that were encountered in Ref. 11. It is conceivable that in order to make progress both the supersymmetry constraint and the propagating fermions will have to be assigned to representations of $K(E_{10})$ different from the ones used so far (and in particular incorporate spatial gradients in one form or another). Let us also note that one can easily couple the new fermion representations to the bosonic $E_{10}/K(E_{10})$ sigma model, namely by adding a Dirac-like term $\propto \Psi D_t \Psi$ to the bosonic action, where $D_t \equiv \partial_t + \sum_{\alpha,r} Q^r(\alpha) J^r(\alpha)$ is the $K(E_{10})$ covariant derivative, and $Q^r(\alpha)$ the $K(\mathfrak{e}_{10})$-connection as computed from the bosonic sigma model in the standard way. Of course, there remains the question whether one can define a new supersymmetry that makes the combined action supersymmetric at least at low levels.

Secondly, the very existence of the two new higher spin representations for $s = \frac{5}{2}$ and $s = \frac{7}{2}$ which cannot be explained from maximal supergravity, strongly suggests that these constitute only the tip of the iceberg of the unexplored representation theory of $K(\mathfrak{e}_{10})$. Although our (limited) search for new examples has not been successful so far, we expect there to exist an infinite tower of such realisations of higher and higher spin, which are less and less unfaithful with increasing spin, but which can occur only at 'sporadic' values of the spin, because the simultaneous decomposability under all the subgroups analysed in the foregoing section puts very tight constraints on such new representations.[27] We reiterate that working out these decompositions is currently a tedious task due to the lack of general methods for studying the representation theory of $K(\mathfrak{e}_{10})$. An explicit construction of further examples and, more ambitiously, a systematic understanding of their structure would afford an entirely new method to explore the root spaces associated with timelike imaginary roots, and thus one of the main obstacles towards a better understanding of $\mathfrak{e}_{10}$. One step forward might be the understanding of the decomposition of tensor products of $K(\mathfrak{e}_{10})$ representations. We thus hope that our investigations help to clarify the structure of this enigmatic object and maybe also the elusive Kac–Moody algebra $\mathfrak{e}_{10}$ itself.

Acknowledgments

AK would like to thank the organisers of the workshop on 'Higher Spin Gauge Theories' for putting together a stimulating program as well as the Institute for Advanced Study of Nanyang Technical University for its generous and kind hospitality.

References

1. B. Julia, "Group Disintegrations," in: S. W. Hawking and M. Roček (eds.), *Superspace and Supergravity*, Proceedings of the Nuffield Work-shop, Cambridge, Eng., Jun 22–Jul 12, 1980, Cambridge University Press (Cambridge, 1981) 331–350; "Kac–Moody Symmetry of Gravitation and Supergravity Theories," in: M. Flato, P. Sally and G. Zuckerman (eds.), *Applications of Group Theory in Physics and Mathematical Physics* (Lectures in Applied Mathematics **21**), Am. Math. Soc. (Providence, 1985) 355–374, LPTENS 82/22.
2. T. Damour, M. Henneaux and H. Nicolai, "E(10) and a 'small tension expansion' of M theory," Phys. Rev. Lett. **89** (2002) 221601 [hep-th/0207267].
3. S. de Buyl, M. Henneaux and L. Paulot, "Hidden symmetries and Dirac fermions," Class. Quant. Grav. **22** (2005) 3595 [hep-th/0506009].
4. T. Damour, A. Kleinschmidt and H. Nicolai, "Hidden symmetries and the fermionic sector of eleven-dimensional supergravity," Phys. Lett. B **634** (2006) 319 [hep--th/0512163].
5. S. de Buyl, M. Henneaux and L. Paulot, "Extended E(8) invariance of 11-dimensional supergravity," JHEP **0602** (2006) 056 [hep-th/0512292].

6. T. Damour, A. Kleinschmidt and H. Nicolai, "K(E(10)), Supergravity and Fermions," JHEP **0608** (2006) 046 [hep-th/0606105].
7. A. Kleinschmidt, "Unifying R-symmetry in M-theory," in: V. Sidoravičius (ed.) *New Trends in Mathematical Physics*, Proceedings of the XVth International Congress on Mathematical Physics, Springer (2009), [hep-th/0703262].
8. A. Kleinschmidt and H. Nicolai, "E(10) and SO(9,9) invariant supergravity," JHEP **0407** (2004) 041 [hep-th/0407101].
9. A. Kleinschmidt and H. Nicolai, "IIB supergravity and E(10)," Phys. Lett. B **606** (2005) 391 [hep-th/0411225].
10. T. Damour, A. Kleinschmidt and H. Nicolai, "Constraints and the E10 coset model," Class. Quant. Grav. **24** (2007) 6097 [arXiv:0709.2691 [hep-th]].
11. A. Kleinschmidt, H. Nicolai and N. K. Chidambaram, "Canonical structure of the E10 model and supersymmetry," Phys. Rev. D **91** (2015) 8, 085039 [arXiv:1411.5893 [hep-th]].
12. The precise counting of propagating degrees of freedom also hinges on the correct implementation of the constraints Ref. [10], not all of which are known.
13. A. Kleinschmidt and H. Nicolai, "On higher spin realizations of $K(E_{10})$," JHEP **1308** (2013) 041 [arXiv:1307.0413 [hep-th]].
14. T. Damour and P. Spindel, "Quantum Supersymmetric Bianchi IX Cosmology," Phys. Rev. D **90** (2014) 10, 103509 [arXiv:1406.1309 [gr-qc]].
15. V. G. Kac, "Infinite dimensional Lie algebras," Cambridge University Press (1990).
16. S. Berman, "On generators and relations for certain involutory subalgebras of Kac–Moody Lie algebras," Commun. Algebra **17** (1989) 3165–3185.
17. T. Damour and C. Hillmann, "Fermionic Kac-Moody Billiards and Supergravity," JHEP **0908** (2009) 100 [arXiv:0906.3116 [hep-th]].
18. In the remainder we will not always put quotation marks when we talk of 'spin'. We trust that readers will understand that this terminology is to be taken with a grain of salt, cf. Ref. [13].
19. The projectors obey the conditions $\Pi^{(i)}\Pi^{(j)} = \delta^{ij}\Pi^{(i)}$ and $\sum_j \Pi^{(j)} = \mathbf{1}$.
20. D. Ghatei, M. Horn, R. Köhl and S. Weiß, "Spin covers of maximal compact subgroups of Kac-Moody groups and spin-extended Weyl groups," arXiv:1502.07294 [math.GR].
21. The condition employed when fixing these projectors is that they commute with $X(\alpha)$.
22. This follows from the fact that $\mathfrak{e}_{10}$ has infinitely many (finite, affine and indefinite) subalgebras that are characterized by their simple root systems[23] (which are embedded in the $\mathfrak{e}_{10}$ root lattice). The associated subalgebras of $K(\mathfrak{e}_{10})$ are then simply obtained as the associated involutory subalgebras, as in Eq. (2).
23. A.J. Feingold and H. Nicolai, "Subalgebras of hyperbolic Kac-Moody algebras", in: Contemp. Math. 343, Amer. Math. Soc, Providence, RI (2004)
24. A. Kleinschmidt and H. Nicolai, "IIA and IIB spinors from K(E(10))," Phys. Lett. B **637** (2006) 107 [hep-th/0603205].
25. I. Jeon, K. Lee and J. H. Park, "Incorporation of fermions into double field theory," JHEP **1111** (2011) 025 [arXiv:1109.2035 [hep-th]].
26. A. Keurentjes, "The Topology of U duality (sub)groups," Class. Quant. Grav. **21** (2004) 1695 [hep-th/0309106].
27. In fact, the consistent decomposability would have to extend to *all* the (infinitely many) subgroups of $K(E_{10})$, including the ones that descend from affine or indefinite subalgebras of $\mathfrak{e}_{10}$!

Scattering Via Conformal Higher Spin Exchanges

A. A. Tseytlin*
Blackett Laboratory, Imperial College London
London SW7 2AZ, UK
tseytlin@ic.ac.uk

We consider a model of massless scalars interacting (via bilinear conserved currents) with conformal higher spin fields in flat space. We compute the tree-level four-scalar scattering amplitude using a natural prescription for summation over an infinite set of conformal higher spin exchanges and find that it vanishes. Independently, we show that the vanishing of the scalar scattering amplitude is, in fact, implied by the global conformal higher spin symmetry of this model.

1. Introduction

Higher spin theories containing infinite number of particles pose a challenge of how to define them at the quantum level in a way consistent with their large amount of symmetry. One particular issue is how to treat sums over infinite number of spins. This question was recently addressed on examples of simplest higher spin partition functions in Ref. 1 following Refs. 2–8.

Our aim will be to review the recent study of this issue in the context of S-matrix of scalars interacting via exchange of an infinite set of higher spin fields.[9] This is an analog of the Veneziano amplitude in string theory where the infinite tower of exchanged fields are massive. This set-up was originally discussed in Ref. 10 where a tree-level scalar scattering amplitude with standard massless higher spin particles exchange was considered. Since an interacting theory of massless higher spin particles ought to be not well-defined in flat space the computation of Ref. 10 is, however, hard to embed into a consistent theory.

Here instead we shall consider a model where the scalars interact through exchange of a tower of *conformal higher spin* fields. Conformal Higher Spin (CHS) theories are generalisations of $d = 4$ Maxwell ($s = 1$) and Weyl ($s = 2$) theories that describe pure spin s states off shell, i.e. have maximal gauge symmetry consistent with locality at the expense of having higher-derivative kinetic terms[11] (see also

*Also at Lebedev Institute, Moscow.

Refs. 7, 12–14). In contrast to the two-derivative massless higher spin theory, the CHS theory (that can be defined at the full non-linear level as the UV singular local part of the induced action of free scalars with higher spin background fields coupled to all conserved spin s scalar currents may be viewed as a formally consistent (but a priori non-unitary) interacting gauge theory when expanded near flat space.

To introduce a particular model which we shall study in this paper, let us first recall the basics of vectorial AdS/CFT duality (see, e.g., Refs. 6, 8, 15). Consider a free CFT$_d$ of N complex scalar fields $S = \int d^d x \ \vec{\chi}^* \cdot \partial^2 \vec{\chi}$, with primary conformal operators being on-shell-conserved traceless currents $J_{\mu_1 \ldots \mu_s}$ of dimension $\Delta = d - 2 + s$. The latter are bilinear $U(N)$ singlets (see Ref. 16) $J_s(\vec{\chi}) = \vec{\chi}^* \mathcal{J}_s \vec{\chi} \sim \vec{\chi}^* \partial^s \vec{\chi}$, $s = 0, 1, 2, \ldots$, where $\mathcal{J}_s$ is an appropriate differential operator. Introducing source fields $h_s(x)$ for all J_s and integrating out $\vec{\chi}$, one gets a generating functional for connected correlators of all currents

$$\Gamma[h] = N \log\det(-\partial^2 + \sum_s h_s \mathcal{J}_s) \, . \tag{1}$$

The d-dimensional fields h_s may be viewed as gauge fields for the symmetries of the free classical scalar theory with linearised differential and algebraic ("trace shifting") symmetries generalising the reparametrization and Weyl symmetry of the Weyl gravity. They can thus be identified with the CHS fields.[a]

The same functional $\Gamma[h]$ (1) should follow from the Vasiliev's massless higher spin theory in AdS$_{d+1}$ upon integrating over the AdS$_{d+1}$ Fronsdal fields Φ_s with Dirichlet boundary conditions. The number of scalars N then plays the role of the inverse coupling of the higher spin theory in AdS$_{d+1}$ (appearing in front of its classical action). All quantum (order $N^0, N^{-1}, \ldots$) corrections to the generating functional computed from the Vasiliev's theory should then vanish to match the boundary theory result.[b]

The quadratic term in h_s term of $\Gamma[h]$ in (1) is

$$\Gamma_2[h] = N \sum_s \int h_s K_s h_s \, , \tag{2}$$

with

$$K_s \sim N^{-1} \langle J_s(x) J_s(x') \rangle \sim \mathrm{P}_s \, |x - x'|^{4-2d-2s} \sim \mathrm{P}_s \, \partial^{2s+d-4} \, \delta^{(d)}(x - x') \, \log \Lambda + \ldots \tag{3}$$

where P_s is the transverse traceless projector and Λ is a UV cutoff (we asume d is even). Thus the UV singular part of Γ_2 is proportional to the collection of CHS kinetic terms $\int d^d x \ h_s \, \mathrm{P}_s \, \partial^{2s+d-4} \, h_s$.

Suppose now we start with $N+1$ scalar fields, $\vec{\chi}$ and ϕ, couple them to the CHS fields h_s via the currents $J_s(\vec{\chi}) + J_s(\phi)$ and integrate out only N scalars $\vec{\chi}$. The

[a]Demanding invariance under non-linear symmetries for a particular subset of fields may require introducing extra terms non-linear in h_s (like in scalar electrodynamics or in covariant coupling to a curved metric). However, being local (involving powers of h_s fields at the same point), they would not change the values of the CFT correlators of primary operators J_s at separated points.
[b]More precisely, what should vanish are corrections to derivatives of the generating functional at separated points.

resulting effective theory will contain the remaining scalar ϕ coupled to the CHS fields h_s described by the induced action, i.e.

$$S[\phi, h] = \int d^d x \Big[\phi^* \partial^2 \phi + \sum_s h_s \, J_s(\phi)\Big] + \Gamma[h] \,, \tag{4}$$

where $\Gamma[h] = N \sum_s [\int h_s K_s h_s + \mathcal{O}(h^3)]$. The UV singular local part of $\Gamma[h]$ may be identified with a non-linear CHS action.[12–14] One may then compute the S-matrix for ϕ due to the exchange of the tower of all CHS fields h_s. Assuming N (or the inverse CHS theory coupling) is large we may treat self-interactions of h_s in perturbation theory.

While a non-trivial S-matrix for ϕ is not a natural observable in the boundary CFT$_d$ (which is a free theory from the start) this set-up is in a sense a higher spin theory analog of the computation of the 4d gluon S-matrix from the AdS$_5$ point of view[17] where one first "integrates out" $SU(N)$ gauge vectors to "build" the bulk geometry, and then considers the scattering of extra gluons on a probe 3-brane.

In general, one may study the case when the CHS part $\Gamma[h]$ of the model (4) is given by either the full non-local induced action (i.e. with kinetic term $\mathrm{P}_s \, \partial^{2s+d-4} \log(\partial^2/\Lambda^2)$) or simply its local UV singular part $\mathrm{P}_s \, \partial^{2s+d-4} \log \Lambda$. The latter choice is preferable when trying to include also self-interactions of h_s: the finite part of the full induced action is a priori anomalous, breaking the classical algebraic symmetries of the CHS fields. At the same time, the local $\log \Lambda$ part of $\Gamma[h]$ is invariant under the symmetries of the CHS theory.[13,14]

In what follows we shall study the model (4) viewed as a *local* CHS theory interacting with a free conformal scalar matter, i.e. assume that only the local part of $\Gamma[h]$ defining the CHS action $S[h]$ is kept with coefficient $\kappa \sim N$ as the (inverse) coupling constant. Starting with (4) and rescaling h_s as $h_s \to \sqrt{N}\, h_s$, we get

$$S[\phi, h] = \int d^d x \Big[\phi^* \, \partial^2 \, \phi + \sum_s h_s \, \mathrm{P}_s \, \partial^{2s+d-4} \, h_s + \tfrac{1}{\sqrt{\kappa}} \Big(\sum_s h_s \, \phi^* \mathcal{J}_s \, \phi + h^3\Big) + \mathcal{O}(\tfrac{1}{\kappa} \, h^4)\Big]. \tag{5}$$

We shall explicitly compute the corresponding amplitude below.

We will start in Section 2 with a description of the model of a free scalar field coupled to a tower of CHS fields. In Section 3 we will compute the tree level amplitude using a particular regularisation prescription for the sum over all spins. The resulting amplitude will have a special scale-invariant form and will vanish due to the constraints of the massless scalar kinematics. As we shall show in Section 4 the vanishing of the four-scalar amplitude is, in fact, implied by the global CHS symmetry of the model. This will thus justify our choice of the summation over spins prescription.

2. Scalar Field Interacting with Conformal Higher Spin Fields

Let us start with a free complex massless scalar ϕ with the flat space action

$$S_{\text{free}}[\phi] = \int d^d x \, \phi^* \, \partial^2 \, \phi \,. \tag{6}$$

This free theory admits infinitely many conserved (on-shell) currents, which are traceless due to conformal invariance. A generating function for such traceless conserved currents may be defined using an auxiliary vector u_μ as (see Ref. 16)

$$J(x,u) = \sum_{s=0}^{\infty} \frac{1}{s!} J^{\mu_1\cdots\mu_s}(x)\, u_{\mu_1}\cdots u_{\mu_s}\,. \tag{7}$$

Here

$$J(x,u) = \Pi_d(u,\partial_x)\,\mathfrak{J}(x,u)\,, \tag{8}$$

where $\mathfrak{J}(x,u)$ is the generating function of traceful currents

$$\mathfrak{J}(x,u) = \phi^*(x+\tfrac{i}{2}u)\,\phi(x-\tfrac{i}{2}u)\,, \tag{9}$$

and Π_d is an operator mapping the traceful currents into traceless currents[10,14,c]

$$\Pi_d(u,\partial_x) = \sum_{n=0}^{\infty} \frac{1}{n!\,(-u\cdot\partial_u - \frac{d-5}{2})_n}\Big(\frac{u^2\,\partial_x^2 - (u\cdot\partial_x)^2}{16}\Big)^n\,. \tag{10}$$

Let us consider an infinite set of couplings of ϕ to external higher spin fields h_s through these currents:

$$S_{\rm int}[\phi,h] = \sum_{s=0}^{\infty} \frac{1}{s!}\int d^dx\, J^{\mu_1\cdots\mu_s}\, h_{\mu_1\cdots\mu_s}\,. \tag{11}$$

Introducing

$$h(x,u) = \sum_{s=0}^{\infty} \frac{1}{s!} h_{\mu_1\cdots\mu_s}(x)\, u^{\mu_1}\cdots u^{\mu_s}\,, \tag{12}$$

the coupling (11) may be written also as

$$S_{\rm int}[\phi,h] = \int d^dx\; h(x,\partial_u)\,J(x,u)\,\big|_{u=0}\,. \tag{13}$$

Due to the transversality and tracelessness of the currents on the scalar mass shell, these couplings are invariant under

$$\delta_{\rm lin}\, h_{\mu_1\cdots\mu_s} = \partial_{(\mu_1}\varepsilon_{\mu_2\cdots\mu_s)} + \eta_{(\mu_1\mu_2}\,\alpha_{\mu_3\cdots\mu_s)}\,, \tag{14}$$

provided ϕ is subject to its free equations of motion. These are linearised conformal higher spin (CHS) transformations.[11] Off the scalar mass shell, these symmetries are deformed to the nonlinear CHS ones[13,14] generalising the diffeomorphism and Weyl transformations of the Weyl gravity

$$\delta_{\rm CHS}\, h_{\mu_1\cdots\mu_s} = \delta_{\rm lin}\, h_{\mu_1\cdots\mu_s} + \mathcal{O}(h)\,, \qquad\qquad \delta_{\rm CHS}\,\phi = \mathcal{O}(\phi)\,. \tag{15}$$

For s=0 the field h_0 is a scalar coupled to $J_0 = \phi^*\phi$, for s=1 we get a coupling of a vector h_μ to $U(1)$ current, and for s=2 we get linearised metric $h_{\mu\nu}$ coupled to

[c]Here $(q)_n = \frac{\Gamma(q+n)}{\Gamma(q)}$ is the Pochhammer symbol.

energy-momentum tensor.[d] The higher spin couplings are natural generalisations of these lower spin couplings.

Next, we may supplement $S_{\rm free}[\phi] + S_{\rm int}[\phi, h]$ with the dynamical action for CHS fields h_s. The functional of h_s invariant under (15) can be identified with the local UV divergent part of the induced action found by integrating out some number of additional scalars.[13,14] The induced action (discussed already in the Introduction, see (1),(2),(3)) may be written as[14]

$$\Gamma[h] = \int d^d p\, k(p)\, (p^2)^{\frac{d-4}{2}}\, G(X,Y)\, \tilde{h}(p,u_1)\, \tilde{h}(-p,u_2)\Big|_{u_i=0} + \mathcal{O}(h^3)\,, \qquad (16)$$

where $\tilde{h}(p,u)$ is the Fourier transform of $h(x,u)$ in (12) and $k(p)$ is a spin-independent function

$$k(p) = c_1 \log \frac{p^2}{\Lambda^2} + c_2\ . \qquad (17)$$

Λ is a UV cutoff (we omit power divergences) and c_1, c_2 are simple numerical constants. The operator $G(X,Y)$ acting on u_1, u_2 is given by

$$G(X,Y) = \sum_{s=0}^{\infty} \frac{\Gamma(\frac{d-3}{2})}{2^{4s}\,\Gamma(s+\frac{d-3}{2})\,\Gamma(s+\frac{d-1}{2})}\, C_s^{(\frac{d-3}{2})}\Big(\frac{X}{\sqrt{Y}}\Big)\, Y^{\frac{s}{2}}\,, \qquad (18)$$

where $C_s^{(l)}(z)$ is the Gegenbauer polynomial and X and Y are differential operators defined by

$$X = p^2\, \partial_{u_1}\cdot\partial_{u_2} - p\cdot\partial_{u_1}\, p\cdot\partial_{u_2}\,, \qquad Y = \big[(p\cdot\partial_{u_1})^2 - p^2\,\partial_{u_1}^2\big]\,\big[(p\cdot\partial_{u_2})^2 - p^2\,\partial_{u_2}^2\big]\,. \qquad (19)$$

Keeping only the singular $\log\Lambda$ part of $k(p)$ or, equivalently, replacing it by a renormalized constant $\kappa = c_1 \log\mu^2$ (proportional to the number N of scalars that were integrated out and playing the role of the overall inverse coupling constant) we may define the local CHS action as

$$S_{\rm CHS}[h] = \kappa \int d^d p\ (p^2)^{\frac{d-4}{2}}\, G(X,Y)\, \tilde{h}(p,u_1)\, \tilde{h}(-p,u_2)\Big|_{u_i=0} + \mathcal{O}(h^3)\,. \qquad (20)$$

The quadratic part of (20) represents a collection of free conformal spin s actions[11]

$$S_{{\rm CHS},2}[h_s] \sim \kappa \int d^d x\ h_s\, {\rm P}_s\, \partial^{2s+d-4}\, h_s\,, \qquad (21)$$

where as in (3) the operator P_s is transverse traceless projector. $S_{{\rm CHS},2}[h_s]$ is invariant under (14) and in $d=4$ may be interpreted as the square of the linearised spin s analog of Weyl tensor. The important point here is that the relative normalisation of conformal spin s fields in the induced action are fixed by the coupling $S_{\rm int}[\phi, h]$ (11) (other choices of normalisation would break the CHS symmetries (15)).[e]

[d] Other standard scalar coupling terms such as $h_\mu h^\mu \phi^* \phi$ for electrodynamics and $\frac{(d-2)}{4(d-1)} R\, \phi^* \phi$ for Weyl gravity can be absorbed into a redefinition of h_0.

[e] One can also compute the h^3 term in the local CHS-invariant $\log\Lambda$ part of the induced action.[13,14] Extending the construction of the non-linear CHS action to higher orders in h_s appears to be technically non-trivial and may require a new method which is non-perturbative in number of fields.

3. Four-Scalar Tree-Level Scattering Amplitude

Given the system of CHS fields coupled to a free scalar via (11), we can study the simplest four-scalar scattering process with the exchange of all CHS fields. This provides an interesting example when the issue of definition of the sum over all spins becomes important. Reference 10 analysed a similar process where the exchanged particles were the standard massless Fronsdal higher spin ones. There, the scattering amplitude was obtained as a function of infinitely many undetermined coupling constants between the massless higher spin fields and a scalar. In the present case all the $\phi - \phi - h_s$ coupling constants are fixed up to an overall factor (the coupling constant κ^{-1} of the CHS theory) and as a result the amplitude will be given by an explicit expression in terms of a sum over spins.

3.1. *Conformal spin s exchange*

To compute the relevant four-scalar amplitude we start with the vertex (11) and consider integrating over h_s (in quadratic approximation only) while keeping ϕ as external fields:

$$\langle S_{\rm int}[\phi, h]\, S_{\rm int}[\phi, h]\,\rangle_0$$
$$= \sum_{s=0}^{\infty} \int \frac{d^d p}{(2\pi)^d} \frac{1}{(s!)^2}\, \tilde{J}^{\mu_1\cdots\mu_s}(p) \left\langle \tilde{h}_{\mu_1\cdots\mu_s}(p)\, \tilde{h}_{\nu_1\cdots\nu_s}(-p)\right\rangle_0 \tilde{J}^{\nu_1\cdots\nu_s}(-p)\,. \qquad (22)$$

Here $\tilde{J}_s$ are the Fourier transforms of the bilinear conserved currents in (7) and the free propagators of the CHS fields are (in transverse traceless gauge)

$$\left\langle \tilde{h}_{\mu_1\cdots\mu_s}(p)\, \tilde{h}^{\nu_1\cdots\nu_s}(-p)\right\rangle_0 = \frac{n_s}{2\kappa\ s!} \frac{\mathrm{P}^{\nu_1\cdots\nu_s}_{\mu_1\cdots\mu_s}(p)}{(p^2)^{s+\frac{d-4}{2}}}\,, \qquad (23)$$

where $\mathrm{P}^{\nu_1\cdots\nu_s}_{\mu_1\cdots\mu_s}(p) = \delta^{\nu_1\cdots\nu_s}_{\mu_1\cdots\mu_s} + \ldots$ is the projector to transverse traceless totally symmetric tensors and κ is the overall coefficient in (20). Since the propagators are contracted with traceless and conserved currents (the external scalar legs are assumed to be on-shell), all other terms denoted by dots in P_s will drop out.

The coefficients n_s in (23) are given by the normalisation of the quadratic part in (20). That they are completely fixed is equivalent to the fact that the $\phi - \phi - h_s$ coupling constants are all fixed. Explicitly, Eq. (20) contains different tensor structures represented by different monomials in X and Y. As we have remarked before, since the propagators are contracted with traceless conserved currents, only traceless and transverse terms are relevant. The Y operator contains at least one trace or divergence, so it is sufficient to consider only the Y-independent part of the CHS action, i.e. to expand $G(X,Y)$ in (18) as

$$G(X,Y) = \sum_{s=0}^{\infty} \frac{\Gamma(\frac{d-3}{2})}{2^{3s}\,\Gamma(s+\frac{d-1}{2})} \frac{X^s}{s!} + \mathcal{O}(Y)\,. \qquad (24)$$

As a result, one finds

$$n_s = \frac{2^{3s}\,\Gamma\big(s+\frac{d-1}{2}\big)}{\Gamma(\frac{d-3}{2})}\,. \tag{25}$$

Let us represent (22) as a sum over spins

$$\langle S_{\rm int}[\phi,h]\,S_{\rm int}[\phi,h]\,\rangle_0 = \kappa^{-1}\sum_{s=0}^{\infty} n_s\,V_s\,, \tag{26}$$

where the spin s contribution is found to be

$$V_s = \frac{1}{2\,s!}\int\frac{d^dp}{(2\pi)^d}\;\tilde{J}^{\mu_1\cdots\mu_s}(p)\,\frac{1}{(p^2)^{s+\frac{d-4}{2}}}\,\tilde{J}_{\mu_1\cdots\mu_s}(-p)$$

$$= \frac{1}{2\,s!}\int\frac{d^dp}{(2\pi)^d}\,\frac{1}{(p^2)^{s+\frac{d-4}{2}}}\,(\partial_{u_1}\cdot\partial_{u_2})^s\,\Pi_d(u_1,i\,p)\,\tilde{\mathfrak{J}}(p,u_1)\,\tilde{\mathfrak{J}}(-p,u_2)\Big|_{u_i=0}\,, \tag{27}$$

where Π_d was defined in (10). Finally, the spin-s exchange amplitude ($p = k_1 - \ell_1 = \ell_2 - k_2$)

$$A_s(k_1,k_2,\ell_1,\ell_2) = \frac{(\partial_{u_1}\cdot\partial_{u_2})^s}{2\,s!\,(p^2)^{s+\frac{d-4}{2}}}\,\Pi_d(u_1,i\,p)\,e^{\frac{1}{2}\,[u_1\cdot(k_1+\ell_1)+u_2\cdot(k_2+\ell_2)]}\Big|_{u_i=0}\,. \tag{28}$$

Using the explicit expression for Π_d in (10) the resulting t-channel amplitude due to spin s exchange is found to be

$$A^{(\mathsf{t})}_s(\mathsf{s},\mathsf{t},\mathsf{u}) = \frac{1}{2(-4)^s\,(-\mathsf{t})^{\frac{d-4}{2}}}\sum_{n=0}^{[s/2]}\frac{1}{2^{2n}\,n!\,(s-2n)!\,(-s-\frac{d-5}{2})_n}\left(\frac{\mathsf{s}-\mathsf{u}}{\mathsf{s}+\mathsf{u}}\right)^{s-2n} \tag{29}$$

$$= \frac{1}{2(-8)^s\,(\frac{d-3}{2})_s}\,\frac{1}{(-\mathsf{t})^{\frac{d-4}{2}}}\,C_s^{(\frac{d-3}{2})}\left(\frac{\mathsf{s}-\mathsf{u}}{\mathsf{s}+\mathsf{u}}\right). \tag{30}$$

Here $\mathsf{s},\mathsf{t},\mathsf{u}$ are the Mandelstam variables (with $\mathsf{s}+\mathsf{t}+\mathsf{u}=0$ in the present massless scalar case) and $C^{(\lambda)}_n(z)$ is the Gegenbauer polynomial.

Since the theory under consideration is conformal, the amplitude has a manifestly scale-covariant form. In particular, in d=4 it depends only on ratio of the Mandelstam variables (also, in d=4 the Gegenbauer polynomial reduces to the Legendre one).

The total summed over spins t-channel amplitude is thus given by

$$A^{(\mathsf{t})}(\mathsf{s},\mathsf{t},\mathsf{u}) = \kappa^{-1}\sum_{s=0}^{\infty} n_s\,A^{(\mathsf{t})}_s(\mathsf{s},\mathsf{t},\mathsf{u}) = \kappa^{-1}\,\frac{1}{2\,(-\mathsf{t})^{\frac{d-4}{2}}}\,F_d\Big(-\frac{\mathsf{s}-\mathsf{u}}{\mathsf{s}+\mathsf{u}}\Big)\,, \tag{31}$$

where the function $F_d(z)$ is given by

$$F_d(z) = \sum_{s=0}^{\infty}\frac{n_s}{2^{3s}\,(\frac{d-3}{2})_s}\,C_s^{(\frac{d-3}{2})}(z)\,. \tag{32}$$

Using the expression for n_s in (25), $F_d(z)$ simplifies to

$$F_d(z) = \sum_{s=0}^{\infty}(s+\alpha_d)\,C_s^{(\alpha_d)}(z)\,, \qquad \alpha_d \equiv \tfrac{d-3}{2}\,. \tag{33}$$

For generic values of z, the sum over spins diverges and thus needs to be defined with a certain regularisation prescription.

3.2. *Summing over spins*

In general, a particular definition of the sum over spins and thus the resulting expressions for the scattering amplitudes should be consistent with the underlying symmetries of the theory.[f] We shall return to this point below but let us first proceed formally, choosing a natural cutoff prescription to define the sum over s. Let us introduce a parameter $w = e^{-\varepsilon} < 1$ (with $\varepsilon \to 0$), compute the sum and then define (33) as a limit $w \to 1$

$$F_d(z) = \lim_{w\to 1} F_d(z,w) \ , \qquad\qquad F_d(z,w) = \sum_{s=0}^{\infty} (s+\alpha_d)\, w^s\, C_s^{(\alpha_d)}(z) \, . \tag{34}$$

We may write $F_d(z,w)$ as

$$F_d(z,w) = w^{1-\alpha_d} \frac{d}{dw}\Big(w^{\alpha_d} \sum_{s=0}^{\infty} w^s\, C_s^{(\alpha_d)}(z)\Big) \ , \tag{35}$$

and use the expression for the generating function $\sum_{s=0}^{\infty} w^s\, C_s^{(\alpha_d)}(z) = (1-2zw + w^2)^{-\alpha_d}$ for the Gegenbauer polynomials to define the regularized expression for $F_d(z,w)$ by an analytic continuation:[g]

$$F_d^{\rm reg}(z,w) = \alpha_d \frac{1-w^2}{(1-2z\,w+w^2)^{\alpha_d+1}} \, . \tag{36}$$

Notice that $F_d^{\rm reg}(z,1)$ happens to vanish for $z \neq 1$, while for $z=1$, we get

$$F_d^{\rm reg}(1,w) = \alpha_d \frac{1+w}{(1-w)^{d-2}} \, , \tag{37}$$

which diverges as $w \to 1$. Thus $F_d^{\rm reg}(z)$ is a particular distribution with support localised at $z=1$. In fact, it is just proportional to the $(d-4)$-th derivative of the delta-function, i.e.

$$F_d^{\rm reg}(z) = \frac{(-1)^{d-4}}{(d-4)!}\, \delta^{[d-4]}(z-1) \, , \qquad \text{i.e.} \qquad F_4^{\rm reg}(z) = \delta(z) \ . \tag{38}$$

The above regularisation of the sum over spins is essentially the same as the one used in Refs. 1, 5, 7 in the context of higher spin partition functions. In the case of CHS theory in d dimensions (or d-dimensional boundary theory) the sum $\sum_{s=0}^{\infty} f_d(s)$ was first replaced by the convergent sum $\sum_{s=0}^{\infty} e^{-\varepsilon(s+\alpha_d)} f_d(s)$ where $\alpha_d = \frac{d-3}{2}$ and then taking the limit $\varepsilon \to 0$ all $\frac{1}{\varepsilon^n}$ poles were dropped.

---

[f] One may draw an analogy with the Veneziano amplitude in string theory where one also sums over an infinite number of different (massive) field contributions. When computing it in string field theory context, one would also need to choose a particular summation over modes prescription. This prescription is selected automatically in the first-quantised world sheet approach in which the 2d conformal invariance and the associated space-time symmetries are built in.

[g] The radius of convergence of the series in w is not greater than 1 (it is 1 when $|z| < 1$ and e^{-x} when $|z| = \cosh x \geq 1$) so the direct evaluation of $F_d(z,1)$ gives a divergent expression.

The same result (38) is found also using another natural regularisation prescription utilizing integral representation for the Gegenbauer polynomials. For simplicity, let us focus on the $d = 4$ case where (33) reduces to

$$F_4(z) = \sum_{s=0}^{\infty} \left(s + \tfrac{1}{2}\right) P_s(z) . \tag{39}$$

Here $P_s = C_s^{(1/2)}$ is the Legendre polynomial. The idea is to use the integral representation

$$P_s(z) = \frac{1}{\pi} \int_0^{\pi} dx \left(z + \sqrt{z^2 - 1} \cos x\right)^s , \tag{40}$$

and interchange the summation over s with the integration. Performing first the sum we find the following integrand

$$\sum_{s=0}^{\infty} \left(s + \frac{1}{2}\right) \left(z + \sqrt{z^2 - 1} \cos x\right)^s = \frac{z + 1 + \sqrt{z^2 - 1} \cos x}{2\,(z - 1 + \sqrt{z^2 - 1} \cos x)^2} . \tag{41}$$

Here we have also used an analytic continuation since for any $x \in [0, \pi]$, there exists such z that the series is divergent. Performing the x-integral we get

$$F_4^{\text{reg}}(z) = \frac{1}{\pi} \int_0^{\pi} dx \, \frac{z + 1 + \sqrt{z^2 - 1} \cos x}{2\,(z - 1 + \sqrt{z^2 - 1} \cos x)^2} = \delta(z - 1) , \tag{42}$$

i.e. the same result as in (38).

3.3. *Total amplitude in* $\mathbf{d = 4}$

In the case of a complex scalar scattering $\phi\,\phi \to \phi\,\phi$ in $d = 4$ one finds the total amplitude by adding the t-channel and the u-channel contributions following from (31) and (38), (42)

$$A_{\phi\phi\to\phi\phi} = \frac{\kappa^{-1}}{4} \left[\delta\left(\frac{\mathsf{s}}{\mathsf{t}}\right) + \delta\left(\frac{\mathsf{s}}{\mathsf{u}}\right)\right] . \tag{43}$$

This unfamiliarly looking amplitude actually vanishes due to massless kinematics. Indeed, choosing the c.o.m. frame ($\vec{p}_1 + \vec{p}_2 = 0 = \vec{p}_3 + \vec{p}_4$) and introducing the scattering angle θ for which $\cos\theta = \frac{\vec{p}_1 \cdot \vec{p}_3}{|\vec{p}_1||\vec{p}_3|}$ one can show (using $E_i = |\vec{p}_i|$) that $\frac{\mathsf{s}}{\mathsf{t}} = -\frac{1}{\sin^2\frac{\theta}{2}}$, $\frac{\mathsf{s}}{\mathsf{u}} = -\frac{1}{\cos^2\frac{\theta}{2}}$.[h] Thus the arguments of the delta-functions never vanish for real θ, i.e. we get

$$A_{\phi\phi\to\phi\phi} = 0 . \tag{44}$$

For the $\phi\,\phi^* \to \phi\,\phi^*$ scattering, we find

$$A_{\phi\phi^*\to\phi\phi^*} = \frac{\kappa^{-1}}{4} \left[\delta\left(\frac{\mathsf{u}}{\mathsf{t}}\right) + \delta\left(\frac{\mathsf{u}}{\mathsf{s}}\right)\right] = \frac{\kappa^{-1}}{4} \left[\delta\left(\cot^2\tfrac{\theta}{2}\right) - \delta\left(\cos^2\tfrac{\theta}{2}\right)\right] , \tag{45}$$

[h]In general, there may be a possible subtlety in the collinear limit when $p_1^\mu = r p_2^\mu$ and one cannot go to the c.o.m. frame but this limit requires complex momenta and its significance in the present context is unclear.

where the two delta-functions correspond to the t-channel and the s-channel contributions, respectively. These two contributions cancel each other, so that again the total amplitude vanishes

$$A_{\phi\phi^*\to\phi\phi^*} = 0\,. \tag{46}$$

Thus, while the individual spin s exchange contributions are nontrivial, the total amplitude vanishes if computed with a particular prescription for summation over spins. As we shall argue below, the vanishing of the four scalar scattering amplitude is actually implied by the global CHS symmetry of the theory.

4. Constraints of Conformal Higher Spin Symmetry on Scalar Amplitudes

We have seen that the tree-level scattering amplitude vanishes when a particular regularisation is used to define the summation over all exchanged spins. The principle that should be selecting one regularization over the other should be the preservation of underlying symmetries of the theory.[i]

The system of CHS fields coupled to massless scalar has the *global* CHS symmetry which plays an analogous role to Lorentz or conformal symmetry in standard field theory. One may thus require a consistency of a prescription of summation over spins with this symmetry. For example, the introduction of the regularization factor w^s in (34) may be implemented by adding it to the CHS propagator in (23). This translates into the following modification of the quadratic part of the CHS action (20) (see (18),(19))

$$S^{\rm reg}_{\rm CHS,2}[h;w] = \int d^dp\ (p^2)^{\frac{d-4}{2}}\ G(w^{-1}X, w^{-2}Y)\,\tilde h(p,u_1)\,\tilde h(-p,u_2)\Big|_{u_i=0}\,. \tag{47}$$

One may then ask if this regularized action still preserves the global CHS symmetry.

Below we will demonstrate that the vanishing of the tree amplitude found in the previous section is actually implied by the invariance under a particular subset of global CHS symmetry transformations. This provides an evidence of a consistency of the regularization of the sum over spins used in Section 3.

Assuming that CHS symmetry is free from anomalies,[j] we would like to analyze how the global CHS symmetry of the scalar action coupled to the CHS fields constrains the correlators (and thus the scattering amplitudes) of massless scalar

[i]One possible analogy is with summation over the Kaluza-Klein modes in a 5d theory compactified on a circle. Viewed as a 4d theory it involves sum over an infinite number of KK mode contributions with manifest symmetry being only 4d Lorentz symmetry, but the requirement of preservation of the original 5d Lorentz symmetry should impose constraints on how one should perform the sum to recover the result found directly in 5d.

[j]Possible anomalies from loop graphs may cancel if one sums over all CHS fields. Indeed, it was demonstrated in[2,4] that a-coefficient of Weyl anomaly of the d=4 CHS theory vanishes assuming a particular prescription of summation over spins. The same may apply also to the c-coefficient of 4d Weyl anomaly.[1,4,5,8] As the Weyl symmetry is one of the CHS gauge symmetries, this is an indication that the same may apply to all algebraic CHS symmetries.

fields. The global CHS symmetry should constrain possible interaction terms in the effective action for the scalars (with CHS fields integrated out, i.e. appearing only on internal lines). In fact, it may prohibit any non-trivial interaction terms, i.e. may imply the vanishing of the corresponding S-matrix.

Among the infinitely many global CHS transformations (14), let us consider the "hyper-translations"

$$\delta\phi(x) = \varepsilon^{\mu_1 \dots \mu_r}\, \partial_{\mu_1} \cdots \partial_{\mu_r}\, \phi(x)\,. \tag{48}$$

Here $\varepsilon^{\mu_1 \dots \mu_r}$ is a constant parameter. For simplicity, let us restrict the discussion to the case of real scalars, so that r will take only odd values. Choosing $\varepsilon^{\mu_1 \dots \mu_r}$ proportional to a product $y^{\mu_1}...y^{\mu_r}$ where y^μ is an arbitrary vector we conclude that (48) implies also the invariance under

$$\delta\phi(x) = (e^{y\cdot\partial_x} - e^{-y\cdot\partial_x})\phi(x) = \phi(x+y) - \phi(x-y)\,. \tag{49}$$

The invariance of the scalar four-point correlation function under such symmetry translated to the momentum space becomes

$$\sin(p_{12}\cdot y)\, \sin(p_{13}\cdot y)\, \sin(p_{14}\cdot y)\, \langle \tilde\phi(p_1)\, \tilde\phi(p_2)\, \tilde\phi(p_3)\, \tilde\phi(p_4)\rangle = 0\,, \tag{50}$$

where $p_{ij} = \frac{1}{2}(p_i + p_j)$ and we have used trigonometric identities and momentum conservation, $p_1 + p_2 + p_3 + p_4 = 0$. Making special choice of the vector y^μ as

$$y^\mu = a\, p_{12}^\mu + b\, p_{13}^\mu + c\, p_{14}^\mu\,, \tag{51}$$

where a, b, c are some arbitrary parameters, and applying the condition (50) to the case of the on-shell scattering amplitude of four real scalars we get (using that $p_i^2 = 0$)

$$\sin(\tfrac{1}{4} a\, \mathsf{s})\, \sin(\tfrac{1}{4} b\, \mathsf{t})\, \sin(\tfrac{1}{4} c\, \mathsf{u})\, A_{\phi\phi\to\phi\phi}(\mathsf{s},\mathsf{t},\mathsf{u}) = 0\,. \tag{52}$$

Since a, b, c are arbitrary, Eq. (52) completely fixes the form of the amplitude up to an overall constant factor k as

$$A_{\phi\phi\to\phi\phi}(\mathsf{s},\mathsf{t},\mathsf{u}) = k\, \delta(\mathsf{s}\,\mathsf{t}\,\mathsf{u})\,. \tag{53}$$

In addition, we may use also the conformal symmetry which is also a sub-algebra of the CHS symmetry. In particular, in $d = 4$ the amplitude should be invariant under the dilatation symmetry, i.e. under the rescaling of momenta by a real constant l

$$A_{\phi\phi\to\phi\phi}(l^2\,\mathsf{s}, l^2\,\mathsf{t}, l^2\,\mathsf{u}) = A_{\phi\phi\to\phi\phi}(\mathsf{s},\mathsf{t},\mathsf{u})\,. \tag{54}$$

Combining the two constraints (53) and (54), we conclude that the only solution consistent with the global CHS symmetries is $k = 0$, i.e. the vanishing of the amplitude.

This formal argument appears to apply not only at the tree but also at the loop level if the global CHS symmetry is not anomalous. It should also apply to the complex scalar scattering case. As we have already seen in Section 3, the tree-level scalar amplitude indeed vanishes in a particular regularization of the sum under spins which should thus be consistent with the CHS symmetry.

References

1. M. Beccaria and A. Tseytlin, On higher spin partition functions, *J. Phys.* **A48**(27), 275401, (2015). doi: 10.1088/1751-8113/48/27/275401.
2. S. Giombi, I. R. Klebanov, S. S. Pufu, B. R. Safdi, and G. Tarnopolsky, AdS Description of Induced Higher-Spin Gauge Theory, *JHEP.* **1310**, 016, (2013). doi: 10.1007/JHEP10(2013)016.
3. S. Giombi and I. R. Klebanov, One Loop Tests of Higher Spin AdS/CFT, *JHEP.* **1312**, 068, (2013). doi: 10.1007/JHEP12(2013)068.
4. A. A. Tseytlin, On partition function and Weyl anomaly of conformal higher spin fields, *Nucl. Phys.* **B877**, 598–631, (2013). doi: 10.1016/j.nuclphysb.2013.10.009.
5. S. Giombi, I. R. Klebanov, and B. R. Safdi, Higher Spin $\mathrm{AdS}_{d+1}/\mathrm{CFT}_d$ at One Loop, *Phys. Rev.* **D89**, 084004, (2014). doi: 10.1103/PhysRevD.89.084004.
6. S. Giombi, I. R. Klebanov, and A. A. Tseytlin, Partition Functions and Casimir Energies in Higher Spin AdS_{d+1}/CFT_d, *Phys. Rev.* **D90**(2), 024048, (2014). doi: 10.1103/PhysRevD.90.024048.
7. M. Beccaria, X. Bekaert, and A. A. Tseytlin, Partition function of free conformal higher spin theory, *JHEP.* **1408**, 113, (2014). doi: 10.1007/JHEP08(2014)113.
8. M. Beccaria and A. A. Tseytlin, Higher spins in AdS_5 at one loop: vacuum energy, boundary conformal anomalies and AdS/CFT, *JHEP.* **1411**, 114, (2014). doi: 10.1007/JHEP11(2014)114.
9. E. Joung, S. Nakach, and A. Tseytlin, Scalar scattering via conformal higher spin exchange, *JHEP* **1602**, 125 (2016) doi: 10.1007/JHEP02(2016)125.
10. X. Bekaert, E. Joung, and J. Mourad, On higher spin interactions with matter, *JHEP.* **0905**, 126, (2009). doi: 10.1088/1126-6708/2009/05/126.
11. E. S. Fradkin and A. A. Tseytlin, Conformal supergravity, *Phys. Rept.* **119**, 233–362, (1985). doi: 10.1016/0370-1573(85)90138-3.
12. A. A. Tseytlin, On limits of superstring in $AdS_5 \times S^5$, *Theor. Math. Phys.* **133**, 1376–1389, (2002). doi: 10.1023/A:1020646014240.
13. A. Y. Segal, Conformal higher spin theory, *Nucl. Phys.* **B664**, 59–130, (2003). doi: 10.1016/S0550-3213(03)00368-7.
14. X. Bekaert, E. Joung, and J. Mourad, Effective action in a higher-spin background, *JHEP.* **1102**, 048, (2011). doi: 10.1007/JHEP02(2011)048.
15. I. Klebanov and A. Polyakov, AdS dual of the critical O(N) vector model, *Phys. Lett.* **B550**, 213–219, (2002). doi: 10.1016/S0370-2693(02)02980-5.
16. N. S. Craigie, V. K. Dobrev, and I. T. Todorov, Conformally Covariant Composite Operators in Quantum Chromodynamics, *Annals Phys.* **159**, 411–444, (1985). doi: 10.1016/0003-4916(85)90118-6.
17. L. F. Alday and J. M. Maldacena, Gluon scattering amplitudes at strong coupling, *JHEP.* **0706**, 064, (2007). doi: 10.1088/1126-6708/2007/06/064.

Note on Higher-Spin Equations in Four Dimensions

V. E. Didenko,[*,‡] N. G. Misuna[†,§] and M. A. Vasiliev[*,||]

[*] *I.E. Tamm Department of Theoretical Physics, Lebedev Physical Institute, Leninsky prospect 53, 119991, Moscow, Russia*

[†] *Moscow Institute of Physics and Technology, Institutsky lane 9, 141700, Dolgoprudny, Moscow region, Russia*

[‡] *didenko@lpi.ru*

[§] *misuna@phystech.edu*

[||] *vasiliev@lpi.ru*

A standard approach to perturbation theory of nonlinear higher-spin equations involves multiple homotopy integration that makes the whole procedure quite involved. We propose a new scheme of perturbative analysis that automatically accounts for repetitive homotopy integrals thus eliminating all intermediate steps within the standard approach. Our method gives a very efficient tool for application in perturbative higher-spin theory.

1. Introduction

The *AdS/CFT* correspondence conjecture [1–3] has become recently an active research area in higher-spin theory. Its general features are known for quite some time since [4–6] evolving into a concrete proposal by Klebanov and Polyakov [7] that states that nonlinear higher-spin (HS) gauge theory is holographically dual to vectorial conformal theories.

The conjecture was further generalized in [8, 9] and received some verification from three-point correlation functions level [10, 11] which was somewhat indirect as an appropriate full nonlinear action of HS theory is as yet unavailable (see however [12]). Interesting results have nonetheless appeared even at one-loop level [13–15].

Recently an alternative conjecture for the boundary generating functional has been put forward in [16], where an extension of the original HS system of [17, 18] was proposed. The idea behind this construction is the following. Since higher-spin equations are strongly constrained by gauge symmetry allowing no other freedom than that in the equations of motion, it may imply that the theory is fixed by symmetries at quantum level. If so, the information about the correlation functions is somehow stored inside classical equations. The 'Lagragian' extension has been obtained in [16] by introducing higher differential forms that on-shell bring a gauge invariant space-time integral over a certain top form. This integral arguably plays role of the on-shell action.

As is mentioned the generalization of nonlinear HS equations involves differential forms of higher degrees. Such an extension neither affects the dynamical field content, nor the structure of interactions as the original system appears as subsector of the lowest grade, with all higher-forms perturbatively expressed via lower forms. The only reason for a higher-form extension is that it brings invariant functionals into the HS theory. In four dimensions those are the two-form responsible for conserved surface charges and the four-form arguably playing a role of the HS on-shell action. The appearance of higher-degree forms however significantly complicates the perturbative analysis. In practice it forces one to repeatedly process similar equations, in order to resolve auxiliary functions in terms of physical space-time forms. Straightforward computation quickly gets involved. The goal of this note which is a short version of [20] is to develop a general machinery that allows one to handle those equations in a more efficient way.

2. 4d Higher-Spin Equations

Higher-spin gauge theory equations are written using the so called unfolded approach in various dimensions [19]. In this note we restrict ourselves to the case of $d = 4$ [17] underlying the AdS_4/CFT_3 HS holography. These equations have the form

$$\mathrm{d}W + W * W = -i\theta_\alpha \wedge \theta^\alpha \left(1 + \eta B * k\varkappa\right) - i\bar{\theta}_{\dot{\alpha}} \wedge \bar{\theta}^{\dot{\alpha}} \left(1 + \bar{\eta} B * \bar{k}\bar{\varkappa}\right), \tag{2.1}$$

$$\mathrm{d}B + W * B - B * W = 0. \tag{2.2}$$

Here $\mathrm{d} = dx^m \frac{\partial}{\partial x^m}$ is the space-time de Rham differential (onwards we omit wedge symbol). W and B are fields of the theory which depend both on space-time coordinates and on twistor-like variables $Y^A = \left(y^\alpha, \bar{y}^{\dot{\alpha}}\right)$ and $Z^A = \left(z^\alpha, \bar{z}^{\dot{\alpha}}\right)$ where the spinor indices α and $\dot{\alpha}$ take two values. The Y and Z variables provide a realization of HS algebra through the following noncommutative star product

$$(f * g)(Z, Y) = \int d^4U d^4V f(Z + U, Y + U)\, g(Z - V, Y + V)\, e^{iU_A V^A}, \tag{2.3}$$

where $U_A V^A = U^A V^B \epsilon_{AB} = u^\alpha v^\beta \epsilon_{\alpha\beta} + \bar{u}^{\dot{\alpha}} \bar{v}^{\dot{\beta}} \epsilon_{\dot{\alpha}\dot{\beta}}$ and $\epsilon_{AB} = \begin{pmatrix} \epsilon_{\alpha\beta} & 0 \\ 0 & \epsilon_{\dot{\alpha}\dot{\beta}} \end{pmatrix}$ is the $sp\,(4)$-invariant antisymmetric form built from the $sp\,(2)$-metrics $\epsilon_{\alpha\beta}$, $\epsilon_{\dot{\alpha}\dot{\beta}}$. It follows then

$$\left[Y^A, Y^B\right]_* = -\left[Z^A, Z^B\right]_* = 2i\epsilon^{AB}, \qquad \left[Y^A, Z^B\right]_* = 0. \tag{2.4}$$

Inner Klein operators $\varkappa$ and $\bar{\varkappa}$ are specific elements of the star-product algebra defined as

$$\varkappa := \exp\left(iz_\alpha y^\alpha\right), \qquad \bar{\varkappa} := \exp\left(i\bar{z}_{\dot{\alpha}} \bar{y}^{\dot{\alpha}}\right). \tag{2.5}$$

They have the following characteristic properties

$$\varkappa * \varkappa = 1, \qquad \varkappa * f(z^\alpha, y^\alpha) = f(-z^\alpha, -y^\alpha) * \varkappa, \tag{2.6}$$

$$f(y, z) * \varkappa = f(-z, -y)\, e^{iz_\alpha y^\alpha}, \tag{2.7}$$

analogously for $\bar{\varkappa}$.

B is a 0-form, while W is a 1-form in a space-time differential $dx^{\underline{m}}$ and in an auxiliary differential θ^A dual to Z^A. All differentials anticommute

$$\{dx^{\underline{m}}, dx^{\underline{n}}\} = \{dx^{\underline{m}}, \theta^A\} = \{\theta^A, \theta^B\} = 0. \tag{2.8}$$

In addition to the inner Klein operators of the star-product algebra there is also a pair of exterior Klein operators k and $\bar{k}$ which have similar properties

$$kk = 1, \qquad kf(z^\alpha; y^\alpha; \theta^\alpha) = f(-z^\alpha; -y^\alpha; -\theta^\alpha)\, k, \tag{2.9}$$

(analogously for $\bar{k}$), but as opposed to the inner Klein operators, k ($\bar{k}$) anticommute with (anti)holomorhic θ differentials and hence do not admit a realization in terms of the star-product algebra. The master fields B and W depend on the exterior Klein operators yielding a natural splitting of the field content into physical (propagating) and topological (nonpropagating) sectors of the theory. The sector of physical fields is represented by B linear in k or $\bar{k}$, while W instead depends on $k\bar{k}$. For topological sector this is other way around. Note that the topological sector can be truncated away, while this is not the case for the physical one. In what follows we consider propagating fields.

η in (2.1) is a free complex parameter of the theory which can be normalized to be unimodular $|\eta| = 1$ hence representing the phase factor freedom. HS theory is parity invariant in the two cases of $\eta = 1$ (A-model) and $\eta = i$ (B-model) [9].

3. Standard Perturbative Analysis

Conventional way one goes into HS perturbation is to first fix the following vacuum solution of (2.1)-(2.2).

$$\mathcal{B}_0 = 0. \tag{3.1}$$

and

$$W_0 = \omega_{AdS} + Z_A\theta^A, \tag{3.2}$$

where ω_{AdS} is the space-time 1-form of $sp(4)$-connection describing the AdS_4 background

$$\omega_{AdS} = \frac{i}{4}\left(\omega^{\alpha\beta} y_\alpha y_\beta + \bar{\omega}^{\dot{\alpha}\dot{\beta}} \bar{y}_{\dot{\alpha}} \bar{y}_{\dot{\beta}} + 2h^{\alpha\dot{\beta}} y_\alpha \bar{y}_{\dot{\beta}}\right), \tag{3.3}$$

$$\mathrm{d}\omega_{AdS} + \omega_{AdS} * \omega_{AdS} = 0. \tag{3.4}$$

(We set the cosmological constant to unity.) This vacuum solution sets the pattern to all higher orders. At free level one finds from (2.2) that

$$[\mathcal{B}_1, Z_A]_* = 0 \tag{3.5}$$

and therefore $\mathcal{B}_1$ is Z-independent

$$\mathcal{B}_1 = C(Y, |x)\,, \tag{3.6}$$

which satisfies

$$\mathrm{d}C + [W_0, C]_* = 0\,. \tag{3.7}$$

To get equation for W_1 one resolves its Z-dependence in terms of homotopy representation using (2.2)

$$W_{1\alpha} = \eta z_\alpha \int_0^1 dt C(-tz, \bar{y}) e^{itz_\beta y^\beta}\,, \tag{3.8}$$

similar for $W_{1\dot{\alpha}}$ and by substituting into (2.1) arrives at the equation that fixes the Z-dependence of space-time one-form W_1 in homotopy-like manner. Eventually after evaluation of two successive homotopy integrals we end up with what is known as the central on-mass-shell theorem

$$\begin{aligned} D_{\omega_{AdS}} W_1(y, \bar{y}|x) = &-\frac{i\eta}{4} e_\gamma{}^{\dot{\alpha}} \wedge e^{\gamma\dot{\beta}} \frac{\partial^2}{\partial \bar{y}^{\dot{\alpha}} \partial \bar{y}^{\dot{\beta}}} C(0, \bar{y}|x) \\ &- \frac{i\bar{\eta}}{4} e^\alpha{}_{\dot{\gamma}} \wedge e^{\beta\dot{\gamma}} \frac{\partial^2}{\partial y^\alpha \partial y^\beta} C(y, 0|x)\,. \end{aligned} \tag{3.9}$$

The pattern persists at higher orders. Namely one keeps searching for Z-dependence of all fields in (2.1)-(2.2) until reaches the bottom *cohomology* part of fields that does not depend on Z and describes HS dynamics at given order. The prescribed procedure gets combinatorially involved especially in the presence of higher-forms. So let us review shortly an alternative way that allows one to avoid repetitive sliding along homotopies.

4. Spectral Sequence Perturbation

It is convenient to extract the Lorentz connection and vierbein from (3.4) as follows

$$\omega_{AdS} = \frac{i}{2}\omega^{AB} Y_A Y_B = \frac{i}{2}\left(\omega_L^{AB} + h^{AB}\right) Y_A Y_B, \tag{4.1}$$

$$\omega_L^{AB} Y_A Y_B := \frac{1}{2}\omega^{\alpha\beta} y_\alpha y_\beta + \frac{1}{2}\bar{\omega}^{\dot{\alpha}\dot{\beta}} \bar{y}_{\dot{\alpha}} \bar{y}_{\dot{\beta}}, \tag{4.2}$$

$$h^{AB} Y_A Y_B := h^{\alpha\dot{\beta}} y_\alpha \bar{y}_{\dot{\beta}}. \tag{4.3}$$

Let us note that at all stages of the computation one encounters only two types of equations to be solved in the adjoint and twisted adjoint sectors

$$\Delta_{ad} f := \mathrm{d}f + [\omega_{AdS}, f]_* - 2i\mathrm{d}_Z f = J, \tag{4.4}$$

$$\Delta_{tw} \widetilde{f} := \mathrm{d}\widetilde{f} + \frac{i}{2}\left[\omega_L^{AB} Y_A Y_B, \widetilde{f}\right]_* + \frac{i}{2}\left\{h^{AB} Y_A Y_B, \widetilde{f}\right\}_* - 2i\mathrm{d}_Z \widetilde{f} = \widetilde{J}, \tag{4.5}$$

where $\mathrm{d}_Z := \theta^A \frac{\partial}{\partial Z^A}$ and f's for now are arbitrary rank differential forms. Indeed, these equations arise from the expansion of (2.1) and (2.2), respectively, around the vacuum with W_0 from (3.2) and determine f ($\widetilde{f}$) in terms of some sources J ($\widetilde{J}$) that are differential forms in $dx^{\underline{m}}$ and θ^A expressed order by order in terms of the lower order fields and/or differential forms of lower degrees. The Z-derivative d_Z emerges from star commutator with $Z_A\theta^A$ in (3.2). The anticommutator in (4.5) results from the linear dependence of $\mathcal{B}$ on k or $\bar{k}$ flipping a sign of $h^{AB}Y_AY_B$. This corresponds to the so-called twisted adjoint representation as opposed to the adjoint one in (4.4). Our aim is to find the iterative solution to equations (4.4), (4.5) in a closed form.

Both equations (4.4) and (4.5) have the following form

$$\mathrm{d}f + \mathcal{D}f = J\,, \tag{4.6}$$

where d and $\mathcal{D}$ obey the standard spectral sequence relations

$$\mathrm{d}^2 = 0, \qquad \{\mathrm{d}, \mathcal{D}\} = 0\,, \qquad \mathcal{D}^2 = 0\,, \tag{4.7}$$

f is an m-form and J is an $(m+1)$-form obeying the compatibility condition

$$\mathrm{d}J + \mathcal{D}J = 0\,. \tag{4.8}$$

System (4.6) is double graded with respect to two different differentials $dx^{\underline{m}}$ and θ^A associated with $\mathcal{D}$ and d, respectively.

Generic solution to (4.6) can be written as

$$f = \Delta^* J + \Delta\epsilon + g, \tag{4.9}$$

where Δ^* is some 'inverse' operator to $d{+}\mathcal{D}$, $\Delta\epsilon$ represents a 'gauge' part of solution, and g belongs to cohomology $H(\mathrm{d})$. Here we have not yet specified operator Δ^* and the equation for g. Formally, Δ^* is given by[1]

$$\sum_{n=0}^{m} f_n = \sum_{n=0}^{m} (-\mathrm{d}^*\mathcal{D})^n\, \mathrm{d}^* J = \Delta^* J\,, \qquad \Delta^* J := \mathrm{d}^* \frac{1}{1+\mathcal{D}\mathrm{d}^*} J\,, \tag{4.10}$$

where

$$\mathrm{d}^* f\,(Z;Y;\theta) = Z^A \frac{\partial}{\partial \theta^A} \int_0^1 dt \frac{1}{t} f\,(tZ;Y;t\theta)\,. \tag{4.11}$$

Cohomology part turns out to satisfy the following equation

$$\mathcal{D}g = H(J)\,, \tag{4.12}$$

where

$$H := \hat{h}\,(1 - \mathcal{D}\Delta^*) = \hat{h}\frac{1}{1+\mathcal{D}\mathrm{d}^*} \tag{4.13}$$

[1] Similar geometric progression like formula was obtained in [21]

and

$$\hat{h}f\,(Z;Y;\theta) = f\,(0;Y;0) \tag{4.14}$$

One of the most important properties of (4.13) is the resolution of identity

$$\{\Delta, \Delta^*\} + H = 1\,. \tag{4.15}$$

The main findings of our work is that operators (4.10) and (4.13) can be explicitly found for particular operators in (4.4) and (4.5). The final result is

$$\Delta^*_{ad}J = -\frac{1}{2i}Z^A\frac{\partial}{\partial\theta^A}\int_0^1\frac{dt}{t}J\Big(tZ;Y_B + \frac{1-t}{t}\omega_B{}^C\frac{\partial}{\partial\theta^C};t\theta\Big) \tag{4.16}$$

and

$$H_{ad}J\,(Z,Y,\theta) = \hat{h}\exp\left(\omega^{AB}\frac{\partial^2}{\partial Y^A\partial\theta^B}\right)J\,(Z;Y;\theta) = \hat{h}J\left(0;Y_A+\omega_A{}^B\frac{\partial}{\partial\theta^B};\theta\right) \tag{4.17}$$

for adjoint sector, that is for (4.4). The final result for the twisted-adjoint case (4.5) is more complicated

$$\begin{aligned}\Delta^*_{tw}\widetilde{J} \;=\; & -\frac{1}{2i}Z^C\frac{\partial}{\partial\theta^C}\int\limits_0^1 dt\frac{1}{t}\\ & \cdot\exp\left\{-\frac{i}{2}\left(\frac{1-t}{t}\right)^2\omega_L^{AB}h_A{}^C\frac{\partial^2}{\partial\theta^B\partial\theta^C} - i\frac{1-t}{t}h^{AB}Y_A\frac{\partial}{\partial\theta^B}\right\}\\ & \cdot\exp\left\{\frac{1-t}{t}\left(\omega_L^{AB}\frac{\partial^2}{\partial Y^A\partial\theta^B} - \frac{1+t}{2t}h^{AB}\frac{\partial^2}{\partial Z^A\partial\theta^B}\right)\right\}\widetilde{J}\,(tZ;Y;t\theta).\end{aligned} \tag{4.18}$$

$$\begin{aligned}H_{tw}\widetilde{J} = \hat{h}\,\exp&\left\{-\frac{i}{2}\omega_L^{AB}h_A{}^C\frac{\partial^2}{\partial\theta^B\partial\theta^C} - ih^{AB}Y_A\frac{\partial}{\partial\theta^B}\right\}\\ \cdot\exp&\left\{\omega_L^{AB}\frac{\partial^2}{\partial Y^A\partial\theta^B} - \frac{1}{2}h^{AB}\frac{\partial^2}{\partial Z^A\partial\theta^B}\right\}\widetilde{J}\,(Z;Y;\theta).\end{aligned} \tag{4.19}$$

To illustrate power of the obtained operators let us re-examine first order analysis that leads to (3.9). Recall that normally to reproduce (3.9) one scrambles along successive homotopy integrals coming from Z-dependence of all differential forms. Now that we have (4.19) the right hand side of (3.9) arises immediately as

$$DW_1 = H(-i\eta B * k\varkappa\theta_\alpha\theta^\alpha - i\bar{\eta}B * \bar{k}\bar{\varkappa}\bar{\theta}_{\dot\alpha}\bar{\theta}^{\dot\alpha})\,. \tag{4.20}$$

In this fashion any given order of HS equations can be extracted.

5. Conclusion

In this note a tool that allows one systematically reproducing perturbative contributions of HS interaction falling out from (2.1)–(2.2) was proposed. Our method is much simpler compared to the standard analysis available in the literature. Most importantly it can be directly applied to the extended HS systems that contain top-form invariant functional related to the AdS/CFT problem as opposed to the standard machinery that gets tremendously involved in practice in this case even at free level.

Our method rests on the application of the general spectral sequence formula of resolution of identity to specific HS covariant derivatives. Remarkably, in this case one can write down the explicit expressions containing a single homotopy integral instead of formal operator series in terms of homotopy integrals for the 'inverse' operator and the cohomology projection. Effectively this means that formulae presented in this paper evaluate most of multiple homotopy integrations appearing in the conventional computation.

Acknowledgments

MV and VD are grateful to the organizers of the workshop Higher Spin Gauge Theory at the Institute of Advanced Studies from Nanyang Technological University in Singapore for the opportunity to present this work. We are especially thankful to professor Lars Brink and Louis Lim for taking care of most organizational concerns. The research is supported by the Russian Science Foundation grant 14-42-00047 in association with Lebedev Physical Institute.

References

[1] J. M. Maldacena, The large N limit of superconformal field theories and supergravity, *Adv. Theor. Math. Phys.* **2** (1998) 231–252.
[2] S. S. Gubser, I. R. Klebanov and A. M. Polyakov, Gauge theory correlators from noncritical string theory, *Phys. Lett.* **B428** (1998) 105–114.
[3] E. Witten, Anti-de Sitter space and holography, *Adv. Theor. Math. Phys.* **2** (1998) 253–291.
[4] B. Sundborg, Stringy gravity, interacting tensionless strings and massless higher spins, *Nucl. Phys. Proc. Suppl.* **102** (2001) 113–119.
[5] E. Witten, talk at the John Schwarz 60-th birthday symposium.
[6] E. Sezgin and P. Sundell, Massless higher spins and holography, *Nucl. Phys.* **B644** (2002) 303–370.
[7] I. R. Klebanov and A. M. Polyakov, AdS dual of the critical $O(N)$ vector model, *Phys. Lett.* **B550** (2002) 213–219.
[8] R. Leigh and A. C. Petkou, Holography of the $N = 1$ higher spin theory on AdS(4), *JHEP* 0306 (2003) 011.
[9] E. Sezgin and P. Sundell, Holography in 4D (super) higher spin theories and a test via cubic scalar couplings, *JHEP* **0507** (2005) 044.

[10] S. Giombi and X. Yin, Higher spin gauge theory and holography: the three-point functions, *JHEP* **1009** (2010) 115.
[11] S. Giombi and X. Yin, Higher spins in AdS and twistorial holography, *JHEP* **1104** (2011) 086.
[12] N. Boulanger and P. Sundell, An action principle for Vasiliev's four-dimensional higher-spin gravity, *J. Phys.* **A44** (2011) 495402.
[13] S. Giombi and I. Klebanov, One Loop Tests of Higher Spin AdS/CFT, *JHEP* **1312** (2013) 068.
[14] M. Beccaria and A. Tseytlin, Higher spins in AdS_5 at one loop: vacuum energy, boundary conformal anomalies and AdS/CFT, *JHEP* **1411** (2014) 114.
[15] M. Beccaria and A. Tseytlin, On higher spin partition functions, *J. Phys.* **A48** (2015) 275401.
[16] M. A. Vasiliev, *Invariant Functionals in Higher-Spin Theory.*
[17] M. A. Vasiliev, More on equations of motion for interacting massless fields of all spins in $(3+1)$-dimensions, *Phys. Lett.* **B285** (1992) 225–234.
[18] M. A. Vasiliev, Higher spin gauge theories: Star-product and AdS space, in *The Many Faces of the Superworld*, ed. M. A. Shifman, pp. 533–610.
[19] M. A. Vasiliev, Nonlinear equations for symmetric massless higher spin fields in (A)dS(d), *Phys. Lett.* **B567** (2003) 139–151.
[20] V. E. Didenko, N. G. Misuna and M. A. Vasiliev, *Perturbative analysis in higher-spin theories*, e-Print: arXiv:1512.04405 [hep-th].
[21] E. Sezgin and P. Sundell, Analysis of higher spin field equations in four-dimensions, *JHEP* **0207** (2002) 055.

Pseudo-local Theories: A Functional Class Proposal

Massimo Taronna*
Université Libre de Bruxelles,
ULB-Campus Plaine C.P. 231, B-1050 Bruxelles, Belgium
massimo.taronna@ulb.ac.de

In this article, using the language of jet space, we propose a functional class space for pseudo-local functionals. We test this functional class proposal in a number of examples ranging from string-field-theory to AdS/CFT dualities. Implications of the locality proposal at the quartic order are also discussed.

1. Introduction

Locality is a fundamental concept in physics. In general terms, it is related to a classical space-time description which should emerge in the semiclassical limit of quantum gravity. Locality plays also a key role in quantum field theory and its interrelations with analyticity and unitarity at the S-matrix level have been the subject of many studies.[1] In the holographic context the challenge is to reconstruct the bulk physics from the boundary CFT observables and in particular their analyticity and singularity properties.[2–5] Nonetheless, from the perspective of quantum gravity in the bulk one expects the local semiclassical description to break down for energies high enough.

In the string theory context the scales at play are the Planck length l_p and the string length l_s. So one expects to recover a semiclassical local space-time description for length scales L such that[a] $l_p \ll l_s \ll L$. At energy scales large enough to probe the string length one can indeed excite macroscopic strings and their effect should be taken into account by the theory without breaking unitarity and causality. String scattering amplitudes for instance, predict a whole tail of corrections to the

*Postdoctoral Researcher of the Fund for Scientific Research-FNRS Belgium.
[a]Recall the relation between Planck length, string length and string coupling constant

$$g_s^2 \sim \left(\frac{l_p}{l_s}\right)^{\frac{d-2}{2}}.$$

semiclassical point-particle approximation which become relevant at length scales comparable with the string-length[6–10] $L \sim l_s$. This translates in an effective pseudo-local form of string interactions which involve derivatives of arbitrary order. This is a key ingredient for the soft behaviour of high energy scattering amplitudes in string theory. Therefore, already string cubic (off-shell) vertices are pseudo-local functionals expressed as formal series of the following type:

$$\mathcal{V} \sim \sum_{l_1,l_2,l_3=0}^{\infty} c_{l_1,l_2,l_3}(\alpha'\nabla)^{l_1}\Phi_1(\alpha'\nabla)^{l_2}\Phi_2(\alpha'\nabla)^{l_3}\Phi_3\,. \tag{1}$$

To appreciate this features more closely various high energy limits are usually studied probing different kinematic regions of string scattering. One is the high-energy and fixed momentum transfer limit: $s \to \infty$ at t fixed (fixed impact parameter) and the second is the $s \to \infty$ limit at $\frac{t}{s}$ fixed (fixed angle scattering). The main question to address in this limits is to understand which among string effects or gravitational effects dominate. Remarkably both limits show asymptotic behaviours which are quite far from those of a local field theory. The $s \to \infty$ at fixed t behaviour of the graviton amplitude is given (in $\alpha' = \frac{1}{2}$ units) by:[6,9,10]

$$\mathcal{A} \sim g_s^2\,\frac{\Gamma(-\frac{t}{8})}{\Gamma(1+\frac{t}{8})}\,s^{2+\frac{t}{4}}e^{-\frac{t}{4}}\,, \tag{2}$$

which should be compared with the pure gravity result $\mathcal{A} \sim \frac{s^2}{t}$ and differs from it by an effective form factor. The $s \to \infty$ at fixed angle scattering, which is a truly short distance limit of the theory due to $t \to \infty$, gives similarly:

$$\mathcal{A} \sim \frac{g_s^2}{(stu)^3}\,e^{-\frac{1}{4}(s\log s + t\log t + u\log u)}\,. \tag{3}$$

The exponentially suppressed behaviour above was argued to be related to a violation of polynomial boundedness of the amplitude.[7] At length scales $L \sim l_s$ or $L < l_s$ one thus expects a break-down of usual effective-field theory locality, as the whole infinite higher-derivative tail (1) present in the interactions becomes relevant.[11–17] Locality may then be reconsidered or generalised. Most importantly a key question that should be addressed in this context, if one insists in a higher-spin theory description of string theory, is which pseudo-local field redefinitions are admissible, if any. In other words, we should be able to specify a functional class of admissible pseudo-local functionals.

This questions are even more relevant in the tensionless limit of string theory: $l_s \sim \sqrt{\alpha'} \to \infty$. Higher-spin theories are conjectured to effectively describe such limits of quantum gravity[18,19] in AdS space-time and they turn out to be naturally pseudo-local.[20–22] The higher-spin algebra mixes indeed derivatives of arbitrary orders into higher-spin multiplets and requires infinitely many HS fields as well. HS vertex themselves have a number of derivatives bounded from below by the highest spin involved into the interaction.[20,23–28] On top of this one should carefully take into account the infinite sums over spins that accompany and infinite higher-derivative contributions. There should be in fact a close relation between higher-spin exchanges in the direct channel and higher-derivatives contributions in cross channels. Crossing requires the latter to reconstruct the former and vice-versa. This may also play a key role in defining a tensionless limit in string theory where all poles in Virasoro amplitude degenerate to a single massless pole[b]. Similar issues may as well arise in curved backgrounds but the status in this case is not as clear due to the luck of an handy formulation of string theory.

Here, we will try and address this questions in simple examples ranging from bosonic string cubic interactions to conserved currents in AdS space. We also analyse some of the consequences of our discussion at the quartic order and in the AdS/CFT context making use of Mellin amplitudes techniques.

This article is organised as follows. In Section 2, we introduce the main concepts to deal with locality in a field theory framework. In particular, we introduce the concept of jet space and inverse limit to define pseudo-local interactions. In Section 3 we apply the logic discussed in Section 2 to various examples ranging from string theory to conserved currents in AdS_3 and AdS_4. In Section 4 we summarise some results available on the CFT side and discuss locality in this framework. The conclusions are summarised in Section 5.

2. Pseudo-Locality and Inverse Limit

The goal of this paper is to study a framework that would allow to treat the problem of locality in the context of field theories with infinitely many derivatives. This is a difficult problem and here we will summarise and push forward some ideas recently appeared in the literature.[30] For definiteness, we shall be interested in the analysis of Noether current interactions at the equations of motion level. We shall comment on the extensions and implications of this analysis to higher orders in some examples. In this section we introduce the main ideas.

Jet space: The standard tool to address locality in field theory is jet space. Given a set of fields $\phi^I(x)$, which are usually tensors on some tensor bundle $\pi : E \to M$, the central idea is to promote them, together with all of their derivatives of arbitrary order, to new independent coordinates. Using the notation:

$$[\phi(x)]^k = \left\{C^I(x), C^I_\mu(x), \dots, C^I_{\mu(k)}(x)\right\} = \{\phi^I(x), \partial_\mu\phi^I(x), \dots, \partial_{\mu(k)}\phi^I(x)\}\ , \quad (4)$$

[b]See e.g. the some recent results on conformal higher-spin theories around flat space[29]

one first defines the space V^k of local functionals of order-k. This is the space whose coordinates are given by $[\phi]^k$:

$$V^k = \left\{ [\phi(x)]_k \mid \phi(x) : M \to E \right\} . \tag{5}$$

Then, one defines the jet space of order k as $J^k = M \times V^k$. The jet space of order zero coincides with the original fibre bundle E. Each section on E induces a section over $J^k(E)$ through the identification $C^I_{\mu(k)} \sim \partial^k \phi^I$. We thus have a sequence J^k of topological spaces. Furthermore, the projection maps $\pi_{k,l}$ $(k > l)$ defined as:

$$\pi_{k,l}([\phi]_k) = [\phi]_l \, , \tag{6}$$

are smooth fibre bundles.

A sequence of topological spaces as above endowed with maps $\pi_{k,k-1}$:

$$\cdots \longrightarrow J^{k+1} \xrightarrow{\pi_{k+1,k}} J^k \longrightarrow \cdots \longrightarrow J^1 \xrightarrow{\pi_{1,0}} E \xrightarrow{\pi} M \, , \tag{7}$$

such that:

$$\pi_{i,j} \circ \pi_{j,k} = \pi_{i,k} \, , \quad \forall i > j > k \, , \tag{8}$$

is called in mathematical terms an *inverse sequence* and it defines the infinite jet space J^∞ as the following commutative diagram:[31]

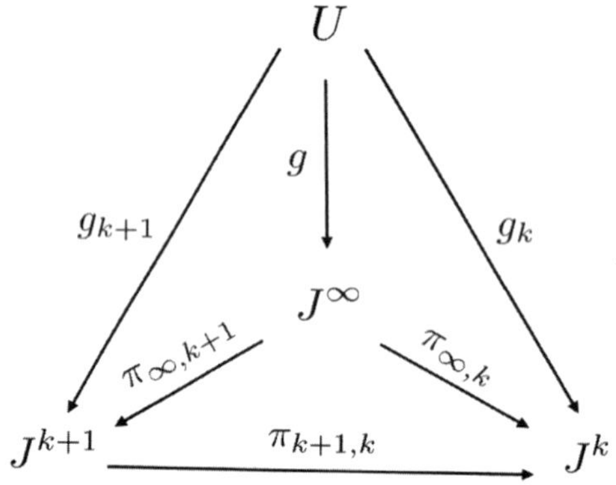

. (9)

The above is *unique* and *universal* when it exists.

For later convenience it is also useful to introduce the space of quadratic functions Q_k over J^∞ with at most k derivatives. This will be the natural space where to study Noether currents. The space Q_k is simply obtained by decomposing the tensor products of two jets and considering the corresponding linear vector space. A choice of basis for Q_k is for instance:

$$[\phi \times \phi]_k = \left\{ C^I_{\mu(m)}(x) C^J_{\nu(n)}(x) \right\}_{m+n \leq k} \, , \tag{10}$$

which allows again to define continuous $\pi_{k,l}$ maps as:

$$\pi_{k,l}([\phi \times \phi]_k) = [\phi \times \phi]_l \, . \tag{11}$$

One can then proceed to define the limiting space Q_∞ of quadratic functionals of infinite order in derivatives. The space Q_∞ is the main object we aim to study in this article. Elements $j \in Q_\infty$ can be defined starting from the following formal series:

$$\tilde{j}_{\mu(s)} = \sum_{n,m} (f_{JK})_{\mu(s)}{}^{\nu(n),\rho(m)}\, C^J_{\nu(m)}(x) C^K_{\rho(n)}(x)\,, \tag{12}$$

where the tilde $\tilde{j}$ notation is meant to stress that the corresponding object one starts with is a formal series. Here, $(f_{JK})_{\mu(s)}{}^{\nu(n),\rho(m)}$ are structure constants parametrising the decomposition of the tensor product of two jets. The maps $\pi_{\infty,k}$ are then defined as:

$$\pi_{\infty,k}(\tilde{j}_{\mu(s)}) = \sum_{m+n\leq k} (f_{JK})_{\mu(s)}{}^{\nu(n),\rho(m)}\, C^J_{\nu(m)}(x) C^K_{\rho(n)}(x)\,, \tag{13}$$

so that the actual definition of $j \in Q_\infty$, from (12), is the following limit:

$$j_{\mu(s)} = \lim_{k\to\infty} \pi_{\infty,k}(\tilde{j}_{\mu(s)})\,. \tag{14}$$

If the above limit exists the inverse-limit construction guarantees its uniqueness.

So far we have introduced the off-shell jet space space J^∞ on which we have defined quadratic functionals Q_∞. One can also define an *on-shell* jet space often called in the literature *stationary surface* $\Sigma^\infty \subset J^\infty$. Given some equations of motion $L^I([\phi]) = 0$, all of their higher-derivative consequences $[L^I([\phi])] = 0$ define an algebraic equation on J^∞. One first defines:

$$\Sigma^k = \left\{[\phi]^k \in J^k \,\middle|\, [L^I([\phi])]^k = 0\right\} \subset J^k\,, \tag{15}$$

as the subspace of the order-k jet J^k defined by the equations $[L^I([\phi])]^k = 0$. One then take the limit $k \to \infty$. In terms of the tensorial coordinates which define the off-shell jet, the stationary surface becomes an algebraic condition which amounts to some irreducibility projection for the original tensorial coordinates. For instance, in the case of a scalar on Minkowski space, the irreducibility condition simply amounts to a tracelessness condition:

$$[\phi(x)]^k \approx \left\{C(x), C_\mu(x), \ldots, C_{\mu(k)}(x)\right\}\,, \qquad C^\nu{}_{\nu\mu(k-2)} = 0\,. \tag{16}$$

In the presence of gauge symmetries it is also important to take into account that not all equations of motion $L^I([\phi])$ are independent. As a consequence of gauge symmetries the equations should satisfy some relations which are called Noether or Bianchi identities:

$$[\mathcal{B}^I([L])] = 0\,. \tag{17}$$

They parametrise all redundancies among the equations $L^I([\phi])$ themselves and all higher-derivative consequences thereof $[L^I([\phi])]$. In the following we shall use the week equality symbol $\approx$ whenever an equality is assumed to hold on the stationary surface. Furthermore, given the stationary surface defined by the *linear* equations of motion: Σ^∞, one can consider the on-shell space $\bar{Q}_\infty$ of quadratic functionals defined on such stationary surface $\bar{Q}_\infty \approx Q_\infty$.

Unfolding and Jet Space: Unfolded equations acquire a natural interpretation from the point of view of jet space.[32–35] One can indeed consider a decomposition of J^∞ into modules of the rigid symmetry algebra of the background. Geometry of the jet bundle is then encoded into appropriate nilpotent form-degree one covariant derivatives $\mathcal{D}$, $\mathcal{D}^2 = 0$. The linear dynamical equations take the form of covariant constancy conditions:

$$\mathcal{D}C^{\mathsf{A}} = (d - \omega^{ab}\hat{L}_{ab} - h^a\hat{P}_a)C^{\mathsf{A}} = 0\,, \tag{18}$$

directly in terms of the stationary surface variables (15). For ease of notation, and to be more uniform with the literature on the unfolded formalism, we have renamed the jet variables as $C^{\mathsf{A}} = [\phi]$.

In the unfolded language one identifies the stationary surface coordinates at the linear level with 0-forms, owing to their covariance under rigid transformations. On the other hand, the gauge dependent components can be introduced as connections gauging the (HS-)global symmetries. This generalise the spin-connection ω^{ab} and the vielbein h^a in the gravity case.[36–38] HS connections $\omega^{\mathcal{A}}$ transform in a different module of the isometry group and their gauge dependence is naturally encoded into the following gauge transformations:

$$\delta\omega^{\mathcal{A}} = \mathcal{D}\epsilon^{\mathcal{A}}\,. \tag{19}$$

Above $\epsilon^{\mathcal{A}}$ is a gauge-symmetry parameter. The module in which it transforms defines the adjoint module of the HS algebra.

To summarise, the unfolding formalism relies on a reformulation of the dynamics directly in terms of the stationary surface coordinates. These are decomposed in gauge-dependent and gauge-covariant components transforming as modules under the (HS) rigid symmetries of the theory. In the following we will mostly concentrate our attention on gauge-invariant current interactions which are build out of 0-forms only.

The "current" interaction problem: At the equations of motion level we consider HS-currents $j_{\mu(s)}(C,C) \in \bar{Q}_\infty$ as quadratic sources to linear HS equations:

$$\mathcal{F}_{\mu(s)}(\phi) \equiv (\nabla^\mu\nabla_\mu - m^2)\phi_{\mu(s)} + \ldots = j_{\mu(s)}(C,C)\,. \tag{20}$$

Depending on the eventual Bianchi identities[c]:

$$\mathcal{B}(\mathcal{F}_{\mu(s)}) = 0\,, \tag{22}$$

satisfied by the kinetic operator, the current should also satisfy analogous compatibility conditions:

$$\mathcal{B}(j_{\mu(s)}(C,C)) \approx 0\,. \tag{23}$$

[c]For instance the Bianchi identity satisfied by the Fronsdal operator on AdS is:

$$\nabla^\mu\left(\mathcal{F}_{\mu(s)} - \frac{1}{2}g_{\mu(2)}\mathcal{F}'_{\mu(s-2)}\right) = 0\,. \tag{21}$$

where the stationary surface is defined by the linear equations of motions for the jet C. In the standard examples of massless fields the compatibility condition is *conservation.* In addition, non-trivial current interactions are naturally defined as equivalence classes:

$$j_{\mu(s)}(C,C) \sim j_{\mu(s)}(C,C) + \mathcal{F}_{\mu(s)}(\mathcal{I}(C,C))\,. \tag{24}$$

Here $\mathcal{I}(C,C) \in \bar{Q}_\infty$ parametrises the most general (admissible) redefinition of the field $\phi_{\mu(s)}$ quadratic in the jet C:

$$\phi_{\mu(s)} \to \phi_{\mu(s)} + \mathcal{I}_{\mu(s)}(C,C)\,. \tag{25}$$

Notice that the eventual Bianchi identities of the kinetic operator ensure that $\mathcal{F}_{\mu(s)}(\mathcal{I}(\phi,\phi))$ is an off-shell compatible current. Such currents are usually referred to in the physics literature as *improvements* and parametrise the freedom in choosing different field frames of the theory.

To summarise, current interactions are defined by the following "cohomological" problem:

$$\mathcal{B}(j_{\mu(s)}(C,C)) \approx 0\,, \qquad j_{\mu(s)} \sim j_{\mu(s)} + \mathcal{F}_{\mu(s)}(\mathcal{I}(C,C))\,, \tag{26}$$

where with a small abuse of notation we include also the $s = 0$ case where the current degenerates to a scalar bilinear. In standard situations one restricts the above problem to local functionals: $j_{\mu(s)}(C,C)\,,\ \mathcal{I}_{\mu(s)}(C,C) \in \bar{Q}_{k\leq\infty}$. A standard example of non-trivial current interaction is then given by the standard electromagnetic current $j_\mu = i\phi^* \overleftrightarrow{\partial} \phi$ or the stress tensor, which encodes the backreaction of the scalar on the gravitational sector. Generalisations of these currents are known also for HS and follow from the Metsaev classification in flat space[24] and generalisations thereof to AdS space.[25–27,39]

From local to non-local: Given the setting described here-above, the main problem to address in a theory involving couplings with infinitely many derivatives like string theory or Vasiliev's equations is to characterise the most general admissible field redefinitions $\mathcal{I}_{\mu(s)}(C,C) \in \tilde{Q}_\infty$, beyond the strict locality condition.

Redefinitions of this type are indeed unavoidable if one wants to bring string-field theory interactions to their canonical Metsaev frame. Enlarging the functional space can however have undesired consequences like for instance the loss of any non-trivial solution. It has been already observed some time ago, for the stress tensor,[40] and proved more recently in general,[21,22,30] that allowing arbitrary pseudo-local redefinitions in maximally symmetric backgrounds erases the difference between non-trivial local currents and improvements. The prototypical example of such a scenario is a system of equations of the type:

$$(\Box - m^2)\Psi(x) = \Phi^2(x)\,, \tag{27a}$$

$$(\Box - \widetilde{m}^2)\Phi(x) = 0\,. \tag{27b}$$

In this case:

- Any local redefinition of Ψ quadratic in Φ will contribute with interaction terms involving at least two derivatives. Hence the interaction term Φ^2 is a non-trivial interaction when restricting the functional space to $\bar{Q}_{k<\infty}$.
- Any term of the type $C_{\mu(s)}C^{\mu(s)}$ which can be inserted on the right-hand side of (27a) admits a completion to a local improvement of the type

$$C_{\mu(s)}C^{\mu(s)} - C_s\Phi^2\,, \tag{28}$$

where the coefficient depends on the mass m of Ψ.
- Allowing $\mathcal{I} \in Q_\infty$ one can find a trivial representative also for the Φ^2 current:

$$\mathcal{I} = \frac{1}{\Box - m^2}\,\Phi^2 = -\frac{1}{m^2}\sum_{i=0}^{\infty}\left(\frac{\Box}{m^2}\right)^i \Phi^2 \in Q^\infty\,. \tag{29}$$

What we have described above in a simple example is a generic feature for current interactions whenever the theory involves some mass parameter. In particular, one can show in various relevant cases on both AdS and flat space[21,22,30] that:

$$\mathcal{B}(j_{\mu(s)}(C,C)) \approx 0\,, \quad \Longrightarrow \quad j_{\mu(s)} = \mathcal{F}_{\mu(s)}(\mathcal{I}(C,C))\,, \tag{30}$$

with

$$j_{\mu(s)}(C,C) \in \bar{Q}_\infty\,, \qquad \mathcal{I}(C,C) \in \bar{Q}_\infty\,, \tag{31}$$

The above statement also extends to massive fields on flat space[d].

A consequence of the above statement is that in a large class of theories, including string theory and Vasiliev's equations, any solution to the conservation condition:

$$\mathcal{B}(j_{\mu(s)}(C,C)) \approx 0\,, \tag{32}$$

can be written as an infinite sum of *local* conserved currents. By definition:

$$\begin{aligned}&\mathcal{I}_{\mu(s)}(C,C) \in \bar{Q}_\infty\\ &\qquad\Longrightarrow \mathcal{I}_{\mu(s)}(C,C) = \sum_{n,m}(f_{JK})_{\mu(s)}{}^{\nu(n),\rho(m)}\, C^J_{\nu(m)}(x)C^K_{\rho(n)}(x)\,.\end{aligned} \tag{33}$$

and one easily concludes that

$$\begin{aligned}j_{\mu(s)} &= \mathcal{F}_{\mu(s)}\left(\sum_{n,m}(f_{JK})_{\mu(s)}{}^{\nu(n),\rho(m)}\, C^J_{\nu(m)}(x)C^K_{\rho(n)}(x)\right)\\ &= \sum_{n,m}\underbrace{\mathcal{F}_{\mu(s)}\left((f_{JK})_{\mu(s)}{}^{\nu(n),\rho(m)}\, C^J_{\nu(m)}(x)C^K_{\rho(n)}(x)\right)}_{\in\,\bar{Q}_{m+n+2}}\,.\end{aligned} \tag{34}$$

The above statement is a key step in the analysis of conserved currents in $\bar{Q}_\infty$. It shows that there is *no* intrinsically pseudo-local conserved current which cannot be

[d]For massless fields in flat space the above does not apply since $\frac{1}{\Box} \notin Q_\infty$.

written as sum of *local* conserved currents. Hence, this result ensures the existence of a basis of local currents for the subspace $\mathcal{C}_\infty = \{j \in \bar{Q}_\infty \,|\, \mathcal{B}(j) \approx 0\}$. Furthermore, this property ensures that starting from $\mathcal{C}_k = \{j \in \bar{Q}_k \,|\, \mathcal{B}(j) \approx 0\}$ there exist continuous maps:

$$\pi_{\infty,k} : \mathcal{C}_\infty \to \mathcal{C}_k \,, \quad \text{s.t.} \quad \forall j \in \mathcal{C}_\infty \,, \quad \pi_k(j) \in \mathcal{C}_k \,. \tag{35}$$

The maps $\pi_{\infty,k}$ can be chosen such that:

$$\pi_{\infty,k}\left(C_{\mu(m)}C_{\nu(n)}\right) = \begin{cases} C_{\mu(m)}C_{\nu(n)} & \text{if} \quad n+m \leq k \\ 0 & \text{if} \quad n+m > k \end{cases}, \tag{36}$$

which is the natural definition induced by the corresponding one for the scalar jet vector C, upon taking its tensor product with itself.

A Functional Class Proposal: So far we have gathered the needed ingredients in order to define our functional class proposal. Note that while no non-trivial current exist in $\bar{Q}_\infty$ they do exist in $\bar{Q}_k$. In the following we will call canonical currents the representatives for the non-trivial currents in $\bar{Q}_k$. Metsaev classification and its extensions to AdS imply that there always exists a representative with at most $s + s_1 + s_2$ derivatives where s is the spin of the current and s_1 and s_2 are the spins of the fields involved. We will refer to the latter representatives as $\mathbf{j}$. Furthermore for each element $j \in \mathcal{C}_k$ the following decomposition is unique and well defined:

$$\forall j \in \mathcal{C}_k : \quad j = b_j \mathbf{j} + \mathcal{F}_{\mu(s)}(\mathcal{I}(C,C)) \,, \quad \mathcal{I}(C,C) \in \bar{Q}_{k-2} \,, \tag{37}$$

and allows to extract the projection of each local current on the non-trivial cohomology representatives. Therefore, given $j \in \mathcal{C}_\infty$ one can consider the following sequence of decompositions:

$$\pi_{\infty,k}(j) = b_k^{(j)} \mathbf{j} + \mathcal{F}(\mathcal{I}(C,C)) \,, \tag{38}$$

which is uniquely defined. Our construction then allows to define the above decomposition directly in $\mathcal{C}_\infty$ arriving at:

$$j = \lim_{k\to\infty} \pi_{\infty,k}(j) = \left(\lim_{k\to\infty} b_k^{(j)}\right) \mathbf{j} + \mathcal{F}(\lim_{k\to\infty} \mathcal{I}_k(C,C)) \,. \tag{39}$$

The inverse limit provides in this way a prescription to extract the coefficient of the local cohomology from any formal element $j \in \mathcal{C}_\infty$. It also provides with a criterion to study if the inverse limit of a formal element $j \in \mathcal{C}_\infty$ exists. We conclude this section by defining the space of *trivial* pseudo-local currents (pseudo-local improvements) as those elements $j \in \mathcal{C}_\infty$ such that:

$$\lim_{k\to\infty} b_k^{(j)} = 0 \,. \tag{40}$$

For each current $j \in \mathcal{C}_\infty$ with the above property we thus identify a subset of $\bar{Q}_\infty$ which is the functional space of allowed pseudo-local redefinitions.

3. Examples

This section is devoted to few tests and examples of the ideas described in this article. We first describe the string theory example decomposing the string vertex according to its canonical and improvement pieces and then move to AdS space. We also consider the implications of the functional class proposal introduced in the previous section on the simplest example of ϕ^4 coupling in AdS which has been subject of recent studies.[41,42]

3.1. *String theory*

The string theory case is one of the most instructive. String theory indeed naturally produces pseudo-local couplings involving infinitely many derivatives. This feature can be appreciated by computing the off-shell open string effective-action in the form:

$$S = \int \left[\tfrac{1}{2}\,\langle\Phi|L_0|\Phi\rangle + \ldots\right] + \langle 0|\,\mathcal{V}\,|\Phi\rangle\otimes|\Phi\rangle\otimes|\Phi\rangle + \mathcal{O}(\Phi^4)\,, \tag{41}$$

with $\mathcal{V}$ the string vertex which can be computed at tree-level by standard techniques.[11,13,14,43–47] Above the ... are the required terms which allow to recover the Virasoro lowest weight condition upon varying the quadratic action and which we will systematically omit in the following. The equations of motion following from the above action would then read:

$$(\,L_0|\Phi\rangle + \ldots) + \mathcal{V}\,|\Phi\rangle\otimes|\Phi\rangle + \mathcal{O}(\Phi^3) = 0\,. \tag{42}$$

The vertex $\mathcal{V}$ can be easily computed at cubic order and expressed in terms of Neumann coefficients by choosing a simple parametrisation for the coordinates around the vertex operators insertions: $\xi_i(y) = y - y_i$. In terms of the Neumann coefficients one arrives in full generality to the following well-known Gaussian type of expression:

$$\mathcal{V} = |y_{12}y_{13}y_{23}|\langle 0|\exp\left[-\frac{1}{2}\sum_{\substack{1\leq i<j\leq 3\\ 0\leq n,m\leq\infty}} \alpha_n^{\mu(i)}N^{ij}_{nm}(y_{ij})\alpha^{(j)}_{m\mu}\right]. \tag{43}$$

Upon computing explicitly the Neumann coefficients for the parametrisation $\xi_i(y) = y - y_i$, and restricting the attention for simplicity to the first Regge trajectory $|\Phi\rangle = \frac{1}{s!}\,\Phi_{\mu(s)}\,\alpha^{\mu}_{-1}\ldots\alpha^{\mu}_{-1}|0\rangle$, one finally arrives to:

$$\mathcal{V} = |y_{12}y_{13}y_{23}|\exp\left\{\sum_{i\neq j}^{3}\left[-\alpha'\partial_i\cdot\partial_j\ln|y_{ij}| - i\sqrt{2\alpha'}\,\frac{\alpha_1^{(j)}\cdot\partial_i}{y_{ij}} + \frac{\alpha_1^{(i)}\cdot\alpha_1^{(j)}}{y_{ij}^2}\right]\right\}. \tag{44}$$

For the first Regge trajectory the quadratic part also simplifies. Therefore, varying the above action one arrives to equations of motion of the type:

$$\mathcal{F}_{\mu(s)}(\Phi) = (\Box - \tfrac{s-1}{\alpha'})\Phi_{\mu(s)} + \ldots = J_{\mu(s)}(\Phi,\Phi)\,, \tag{45}$$

where we have dropped again unessential terms for our discussion proportional to traces and divergences of the fields. We have also considered a gauge fixing of the Stueckelberg gauge symmetries. The current $J_{\mu(s)}$ can be extracted from (44) and reads:

$$J(x,\alpha_{-1}|\Phi,\Phi) = e^{-\alpha'\Box \ln|\frac{y_{12}}{y_{31}y_{23}}|+i\sqrt{\frac{\alpha'}{2}}\frac{y_{12}}{y_{23}y_{31}}\alpha_{-1}\cdot\partial_{12}} \\ \times {}_{12}\langle 0|e^{+\frac{y_{12}}{y_{23}y_{31}}\alpha_{-1}\cdot(\alpha_1^{(1)}+\alpha_1^{(2)})+i\sqrt{2\alpha'}\left(\alpha_1^{(1)}\cdot\partial_2-\alpha_1^{(2)}\cdot\partial_1\right)+\alpha_1^{(1)}\cdot\alpha_1^{(2)}} \\ \times |\tfrac{y_{23}y_{31}}{y_{12}}|\,\Phi(x_1|\alpha_{-1}^{(1)})\Phi(x_2|\alpha_{-1}^{(2)})|0\rangle_{12}\Big|_{x_1=x_2=x}\,. \tag{46}$$

Notice that after varying the action and going on-shell we can use the equations of motion for the fields Φ_1 and Φ_2 which from now on will be considered on their linear mass-shell as in the discussion of Section 2. However, even after going on-shell at the EoMs level the current still remains pseudo-local. Only a pseudo-local redefinition would possibly remove the non-localities!

This makes manifest the pseudo-local nature of the string interactions. Importantly, one also recovers some off-shell ambiguities parametrised by the insertion positions[e] y_i.

In the following we are going to apply the functional class proposal discussed in this article to decompose the above currents and identify the corresponding improvement and canonical parts.

Decomposition in Canonical plus Improvements: As we have explained in Section 2 we can achieve a well-defined decomposition of any interaction by a limiting procedure. In the end this reduces the problem to study a basis of infinitely many local elements. On the other hand, improvement terms have the generic form:

$$\mathcal{F}_{\mu(s)}(\mathcal{I}(\Phi,\Phi)) = (\Box - \tfrac{s-1}{\alpha'})\mathcal{I}(\Phi,\Phi) + \ldots\,, \tag{47}$$

and by definition satisfy the Bianchi identities of the theory. The limiting procedure described in Section 2 is here further simplified by the fact that derivatives commute each other. The improvement decomposition for a spin-s current of the type:

$$(\Box - m^2)\Phi_{\mu(s)} = \Box^n \Phi\partial_{\mu(s)}\Phi\,, \tag{48}$$

is then given by factorising the equations of motion as:

$$(\Box - m^2)\Phi_{\mu(s)} = (m^2)^n \Phi\partial_{\mu(s)}\Phi \\ + (\Box - m^2)\sum_{i=1}^{n}\binom{n}{i}(\Box - m^2)^{i-1}(m^2)^{n-i}\,\Phi\partial_{\mu(s)}\Phi\,. \tag{49}$$

Here we have explicitly disentangled the improvement piece. One can then first expand the string current, perform the above decomposition and resum it. The

[e] Notice that if we had not fixed the coordinate parametrisation around each insertion position the string vertex would have explicitly depended on the choice of parametrisation.

final result is remarkably simple and reads:

$$J(x,\alpha_{-1}|\Phi,\Phi)=e^{-\alpha'(\Box-\frac{\hat{N}-1}{\alpha'})\ln|\frac{y_{12}}{y_{31}y_{23}}|}e^{i\sqrt{\frac{\alpha'}{2}}\,\alpha_{-1}\cdot\partial_{12}}$$
$$\times\,{}_{12}\langle 0|e^{\alpha_{-1}\cdot(\alpha_1^{(1)}+\alpha_1^{(2)})+i\sqrt{2\alpha'}\left(\alpha_1^{(1)}\cdot\partial_2-\alpha_1^{(2)}\cdot\partial_1\right)+\alpha_1^{(1)}\cdot\alpha_1^{(2)}}$$
$$\times\,\Phi(x_1|\alpha_{-1}^{(1)})\Phi(x_2|\alpha_{-1}^{(2)})|0\rangle_{12}\Big|_{x_1=x_2=x}\,. \tag{50}$$

The conclusion is then that the canonical current piece is given by:

$$J^{\text{can.}}(x,\alpha_{-1}|\Phi,\Phi)=e^{i\sqrt{\frac{\alpha'}{2}}\,\alpha_{-1}\cdot\partial_{12}+i\sqrt{2\alpha'}\left(\alpha_1^{(1)}\cdot\partial_2-\alpha_1^{(2)}\cdot\partial_1\right)+\alpha_1^{(1)}\cdot\alpha_1^{(2)}}$$
$$\times\,\Phi(x_1|\alpha_{-1}^{(1)})\Phi(x_2|\alpha_{-1}^{(2)})|0\rangle_{12}\Big|_{\substack{x_1=x_2=x\\ \alpha_{-1}^{(i)}=\alpha_{-1}}}\,,$$

and is local given any triple of spins matching Metsaev's classification. On the other hand the improvement current is pseudo-local even for fixed spin and is dressed by an exponential of $\Box$:

$$J^{\text{Impr.}}(x,\alpha_{-1}|\Phi,\Phi)=\left(e^{-\alpha'(\Box-\frac{\hat{N}-1}{\alpha'})\ln|\frac{y_{12}}{y_{31}y_{23}}|}-1\right)$$
$$e^{i\sqrt{\frac{\alpha'}{2}}\,\alpha_{-1}\cdot\partial_{12}+i\sqrt{2\alpha'}\left(\alpha_1^{(1)}\cdot\partial_2-\alpha_1^{(2)}\cdot\partial_1\right)+\alpha_1^{(1)}\cdot\alpha_1^{(2)}}\;\Phi(x_1|\alpha_{-1}^{(1)})\Phi(x_2|\alpha_{-1}^{(2)})|0\rangle_{12}\Big|_{\substack{x_1=x_2=x\\ \alpha_{-1}^{(i)}=\alpha_{-1}}}\,.$$

For instance, the spin-1 current built out of tachyons reads:

$$J^\mu\sim e^{-\alpha'\Box\ln|\frac{y_{23}}{y_{31}y_{12}}|}\Phi^\star\overleftrightarrow{\partial^\mu}\Phi=\Phi^\star\overleftrightarrow{\partial^\mu}\Phi-\alpha'\left(\ln|\frac{y_{23}}{y_{31}y_{12}}|\right)\Box\left(\Phi^\star\overleftrightarrow{\partial^\mu}\Phi\right)+\dots\,. \tag{51}$$

We conclude this section by stressing that all the dependence on the insertion points, as well as any dependence on the parametrisation around each insertion, just appears in the improvement and defines the corresponding field frame on the effective field theory side. Different choices of parametrisation or insertion positions then just correspond to different infinite derivative tails for each vertex. The infinite derivative improvement tail appears from this perspective as a key ingredient to achieve a string interpretation of the underlying (effective) field theory. We conclude this section by stressing that the improvement term is an improvement according to the definition (40). Notice also that any higher-derivative term in eq. (46) contributes to the canonical component! In particular all of them are needed to achieve the cancellation of any dependence on the y_i in the canonical current sector.

3.2. *Scalar currents in AdS_3 and AdS_4*

The limiting procedure described in Section 2 is more subtle in curved backgrounds because of the non-commutative nature of covariant derivatives. The first step is to fix a basis for the jets of the fields. A convenient choice, which is natural in Vasiliev's theory is to consider symmetrised and traceless derivatives:

$$C_{a(k)}(x)=i^k h_a^\mu\dots h_a^\mu\nabla_{\mu\{k\}}\Phi(x)\,. \tag{52}$$

The above jet components are precisely the field variables entering the unfolded equations. Furthermore, due to convenient isomorphisms for the isometry algebras $so(2,2) \sim sp(2) \oplus sp(2)$ and $so(3,2) \sim sp(4)$, the above totally symmetric tensors can be encoded into:

$$C_{\alpha(2k)}(x) \sim (ih^{\mu}_{\alpha\alpha}\nabla_\mu)^k\Phi(x)\,, \qquad C_{\alpha(k)\dot\alpha(k)}(x) \sim (ih^{\mu}_{\alpha\dot\alpha}\nabla_\mu)^k\Phi(x)\,. \tag{53}$$

Here we have used the standard conventions expressing the vielbein in terms of Pauli matrices as

$$h^{\alpha\alpha}_{\mu} = \frac{1}{2z}\,(1,\sigma_3,\sigma_1)^{\alpha\alpha}_{\mu}\,, \qquad h^{\alpha\dot\alpha}_{\mu} = \frac{1}{2z}(1,\sigma_3,\sigma_1,\sigma_2)^{\alpha\dot\alpha}_{\mu}\,. \tag{54}$$

The tracelessness condition becomes in spinorial language a simple consequence of symmetrisation of the spinorial indices. These are indeed contracted with the antisymmetric $sp(2)$ invariant tensor[f] $\epsilon_{\alpha\beta}$ ($\epsilon_{12} = 1$). The spinorial language is also convenient to express a basis for $\bar{Q}_\infty$ as:

$$j_{\alpha(2s)} = C_{\alpha(2s-2k)\nu(2l)}C^{\nu(2l)}{}_{\alpha(2k)}\,, \tag{55}$$

$$j_{\alpha(s)\dot\alpha(s)} = C_{\alpha(s-k)\nu(l)\dot\alpha(s-k)\dot\nu(l)}C_{\alpha(k)}{}^{\nu(l)}{}_{\dot\alpha(k)}{}^{\dot\nu(l)}\,. \tag{56}$$

In the following, for ease of notation and simplicity, we will concentrate on a conformally coupled scalar because in this case the corresponding HS currents can be improved to traceless currents:

$$d = 3\,, \qquad (\Box + \tfrac{3}{4})\Phi(x) = 0\,, \tag{57}$$

$$d = 4\,, \qquad (\Box + 2)\Phi(x) = 0\,. \tag{58}$$

The above mass shell conditions are equivalent to the following recursion relations for the jet variables:

$$\nabla C^{\alpha(2k)} - ih^{\alpha\alpha}C^{\alpha(2k-2)} + \tfrac{i}{2}\,h_{\gamma\gamma}C^{\gamma\gamma\alpha(2k)} = 0\,, \tag{59}$$

$$\nabla C^{\alpha(k)\dot\alpha(k)} - \tfrac{1}{2}\,h^{\alpha\dot\alpha}C^{\alpha(k-1)\dot\alpha(k-1)} - h_{\gamma\dot\gamma}C^{\gamma\alpha(k)\dot\gamma\dot\alpha(k)} = 0\,. \tag{60}$$

These allow to express covariant derivatives in terms of the other on-shell jet variables. In the spinorial language the corresponding quadratic equations read:

$$\mathcal{F}_{\alpha(2s)} = [\Box - s(s-2)]\phi_{\alpha(2s)} + \ldots = j_{\alpha(2s)}\,, \tag{61}$$

$$\mathcal{F}_{\alpha(s)\dot\alpha(s)} = [\Box - (s-2)(s+1)]\phi_{\alpha(s)\dot\alpha(s)} + \ldots = j_{\alpha(s)\dot\alpha(s)}\,. \tag{62}$$

The assumption of a conformally coupled scalar simplifies also the current conservation condition which usually involves both the current and its trace. In our case the conservation conditions for currents sourcing the Fronsdal tensor read as usual:

$$\nabla^{\beta\beta}j_{\beta\beta\alpha(2s-2)} \approx 0\,, \tag{63}$$

$$\nabla^{\beta\dot\beta}j_{\beta\alpha(s-1)\dot\beta\dot\alpha(s-1)} \approx 0\,, \tag{64}$$

respectively in 3 and 4 dimensions. The conservation condition can be solved in general[21,22,30,40,48–50] and admits the following solution in the space $\bar{Q}_\infty$ of pseudo-local currents.

[f] In the $sp(2)$ language we raise and lower indices as $A^\alpha = \epsilon^{\alpha\beta}A_\beta$ and $A_\alpha = A^\beta\epsilon_{\beta\alpha}$.

3d: In 3d the solution to current conservation reads:

$$j_{\alpha(2s)} = \sum_{l=0}^{\infty} \alpha_l \, j^{(l)}_{\alpha(2s)}\,, \tag{65}$$

expressed in terms of the following local basis of independently conserved currents:

$$j^{(2l)}_{\alpha(2s)} = \frac{(-1)^l}{(2l)!(2s)!} \sum_{k=0}^{s} (-1)^k \binom{2s}{2k} C_{\alpha(2s-2k)\nu(2l)}(x) C^{\nu(2l)}{}_{\alpha(2k)}(x)\,, \tag{66}$$

$$j^{(2l+1)}_{\alpha(2s)} = i\frac{(-1)^l}{(2l+1)!(2s)!} \sum_{k=0}^{s-1} (-1)^{k+1} \binom{2s}{2k+1} C_{\alpha(2s-2k-1)\nu(2l+1)}(x) C^{\nu(2l+1)}{}_{\alpha(2k+1)}(x)\,. \tag{67}$$

4d: In 4d the solution to current conservation reads:

$$j_{\alpha(s)\dot\alpha(s)} = \sum_{l=0}^{\infty} \alpha_l \, j^{(l)}_{\alpha(s)\dot\alpha(s)}\,, \tag{68}$$

expressed in terms of the following local basis of independently conserved currents:

$$j^{(l)}_{\alpha(s)\dot\alpha(s)} = (-1)^l \frac{1}{(l!)^2 (s!)^2} \sum_{k=0}^{s} (-1)^k \binom{s}{k}^2 C_{\alpha(s-k)\nu(l)\dot\alpha(s-k)\dot\nu(l)}(x) C_{\alpha(k)}{}^{\nu(l)}{}_{\dot\alpha(k)}{}^{\dot\nu(l)}(x)\,. \tag{69}$$

As anticipated, any pseudo-local conserved current $j \in Q_\infty$ can be written as (infinite) sum of local conserved currents. Therefore, the above result shows explicitly that for any current $j \in Q_\infty$ such that $\nabla \cdot j = 0$ then $j \in \mathcal{C}_\infty$. The above is just a convenient choice of basis for each $\mathcal{C}_k$ which is also convenient to study the action of the $\pi_{\infty,k}$-maps (36).

Decomposition in canonical plus improvements: The decomposition described in Section 2 amounts in the AdS case to a decomposition in improvement and canonical part for each element of the basis introduced above. Given any element of $\mathcal{C}_\infty$ one can indeed first decompose it in this basis. One then finds the corresponding decomposition of the associated local truncations simply from the decomposition of the basis elements.

By using the explicit form of the Fronsdal operator in spinorial notation it is possible to find the most general form of a local traceless improvement.[30] They read in terms of the basis introduced above:

$$\mathcal{I}^{(l)}_s = j^{(l-1)}_s - C^{(s)}_l \, j^{(0)}_s\,. \tag{70}$$

The coefficients $C^{(s)}_l$ read in 3d:

$$C^{(s)}_l = (-1)^l \frac{s\,(2s+l)!}{(2s)!(l+1)!} \Big[2\,(l+s)\, {}_2F_1(1, l+2s+1; l+2; -1) - l - 1\Big] + 4^{-s}(l+s)\,, \tag{71}$$

and in 4d:

$$C_l^{(s)} = -\frac{(-1)^l}{2(2s-1)} \frac{(l+s)!}{s!(s-1)!\Gamma(1+l-s)} \, {}_3F_2 \left(\begin{matrix} 1-s & 1-s & -2s \\ 2-2s & 1+l-s & \end{matrix} ; 1 \right) . \tag{72}$$

Their asymptotic behaviour for $l \to \infty$ is:

$$C_l^{(s)} \sim l^{2s-1} , \qquad d = 3 , \tag{73a}$$

$$C_l^{(s)} \sim l^{2s} , \qquad d = 4 . \tag{73b}$$

Notice that in the scalar case under consideration there is one local cohomology and one can choose a representative for it given precisely by $j_s^{(0)}$. The coefficient $C_l^{(s)}$ is precisely the cohomology coefficient of the decomposition of each $j_s^{(l)}$. The above spin-dependent behaviour shows a key difference among the flat and the AdS examples. In the latter the non-commutative nature of covariant derivatives shows a factorially growing behaviour for the contribution to the canonical current sector of the higher-derivative terms.

In this setting one can also compute the decomposition into canonical and improvement pieces of any element in $\mathcal{C}_\infty$ from:

$$\pi_{\infty,s+k}(j_s) = \sum_{l=0}^{k} \alpha_l \, j_s^{(l)} = \underbrace{\left(\sum_{l=0}^{k} \alpha_l \, C_{l+1}^{(s)} \right)}_{b_k} j_s^{(0)} + \text{Impr.} , \tag{74}$$

and taking the limit:

$$b = \lim_{k \to \infty} b_k = \sum_{l=0}^{\infty} \alpha_l \, C_{l+1}^{(s)} . \tag{75}$$

The existence and the result of the above limit allows to define accordingly well-defined pseudo-local currents (when the limit exists) and Improvements (when the limit exists and is zero).

Witten diagram computation in 3d and holographic checks of the functional class proposal: The simplest check of the functional class criterion proposed above is the holographic check.[30] One can indeed compute the Witten diagrams associated to the higher-derivative terms forming a basis for $\mathcal{C}_\infty$. The 3d computation can be done by standard techniques[30,51–55] and reads:

$$\int_{AdS_3} \phi^{\alpha(2s)} j_{\alpha(2s)}^{(l)} = \pi \, C_{l+1}^{(s)} \left(-\frac{1}{4} \right)^s \frac{(s-1)!}{|x_{12}|} \left(\frac{x_{12}^+}{x_{13}^+ x_{23}^+} \right)^s . \tag{76}$$

where the $C_l^{(s)}$ coincide with the coefficients given in eq. (71). Therefore, the Witten diagram computation leads to the same functional class criterion proposed in this note. This result also shows that our criterion for the functional class in HS theory gives a prescription for a proper set of non-local redefinitions that *do not* affect the computation of AdS/CFT observables (Witten diagrams).

Furthermore, from the AdS/CFT perspective a key aspect of the above computation that should be stressed is that the result of the Witten diagram for a pseudo-local interaction follows from (76) upon commuting the infinite sum over derivatives and the integral over AdS space. The locality criterion proposed in this paper turns out to be equivalent to the condition that the infinite sum over derivatives commutes with the integral over AdS-space.

3.3. ϕ^4 *couplings: an argument using cubic locality*

Before concluding we would like to try and push our analysis to the simplest cubic coupling at the EoMs level: the one involving three scalar fields as sources to the scalar EoMs. This coupling has been recently subject of investigations.[41,42] Let us stress that the analysis described here is preliminary and it is presented as an example of how similar issues as those described at cubic level are expected to manifest at higher orders.

In more details, in a theory of scalars there are infinitely many non-trivial cubic sources which can be written down. A simple way to perform the counting is to count inequivalent combinations of Mandelstam variables in flat space. One can then use the condition $s+t+u=$ const. to reduce the number of independent structures to:

$$s^n t^m\,, \qquad\qquad n\geq m\,, \tag{77}$$

where bose symmetry has been used to obtain the inequality. Back to coordinate space, the above basis can be conveniently recast into the following basis:

$$\mathcal{V}_4^{(s,l)}=C^{\alpha(s)\dot{\alpha}(s)}\,j^{(l)}_{\alpha(s)\dot{\alpha}(s)}\,,\qquad s=2n,\quad n\geq l\geq 0\,,\quad n,l\in\mathbb{N}\,. \tag{78}$$

We stress again that the above should be considered as a basis for the cubic scalar sources to the scalar field equations.

On the other hand, coupling the scalar to a tower of HS fields:

$$\mathcal{F}_{\alpha(s)\dot{\alpha}(s)}(\phi_s)=j^{(0)}_{\alpha(s)\dot{\alpha}(s)}\,, \tag{79a}$$

$$(\Box+2)\Phi(x)=\sum_{s=0}^{\infty}\phi_{\alpha(s)\dot{\alpha}(s)}C^{\alpha(s)\dot{\alpha}(s)}+\sum_{s,l}a_{s,l}\,j^{(l)}_{\alpha(s)\dot{\alpha}(s)}\,C^{\alpha(s)\dot{\alpha}(s)}\,, \tag{79b}$$

entangles the quartic self interaction of the scalar with the cubic interactions studied in the previous section[g]. At this order any redefinition of the type studied at the cubic level would induce a cubic scalar interaction of the type in eq. (78). We then conclude that at the quartic order the interactions in (78) can be removed up to

[g] Above we have also introduced an auxiliary scalar ϕ_0 with mass m^2 and corresponding interactions with the original scalar Φ on which j depends upon. Although this additional scalar is not present in Vasiliev's equation ϕ_0 can be decoupled in the limit $m^2\to\infty$. Introducing this auxiliary scalar is convenient at the perturbative level to argue about locality of quartic interactions.

a suitable choice of improvement interactions[h]. These would be generated by the redefinition removing the quartic contact terms and would still contribute to the quartic exchange amplitude.[41,56]

We can then reiterate the whole philosophy of this note applying it to the sum over s and l in (78). We first truncate both s and l to finite values and remove the corresponding local interactions by a corresponding redefinition. At the end we take the limit $s, l \to \infty$.

Since we are interested in the analysis of redundant basis for the quartic interaction we drop the inequality $n \geq l \geq 0$. In this way, starting from (79) we generate a pseudo-local cubic theory whose equations of motion read:

$$\mathcal{F}_{\alpha(s)\dot{\alpha}(s)}(\phi_s) = \tilde{j}_{\alpha(s)\dot{\alpha}(s)}(\Phi, \Phi)\,, \tag{80a}$$

$$(\Box + 2)\Phi(x) = \sum_{s=0}^{\infty} \phi_{\alpha(s)\dot{\alpha}(s)} C^{\alpha(s)\dot{\alpha}(s)}\,, \tag{80b}$$

$$\tilde{j}_{\alpha(s)\dot{\alpha}(s)}(\Phi, \Phi) \equiv j^{(0)}_{\alpha(s)\dot{\alpha}(s)}(\Phi, \Phi) - \sum_{l=0}^{\infty} \alpha_{s,l}\, \mathcal{F}_{\alpha(s)\dot{\alpha}(s)}\left(j^{(l)}_s(\Phi, \Phi)\right)\,, \tag{80c}$$

It is then natural to require that the latter pseudo-local expression satisfies our locality proposal in the limit $s, l \to \infty$[i]. Making use of the explicit form of the $\mathcal{F}_{\alpha(s)\dot{\alpha}(s)}$ operator one can express the current $\tilde{j}_s$ in the basis (68) obtaining for $d = 4$ the following coefficients:

$$\tilde{a}^{(s)}_l = \delta_{l,0} - \frac{1}{4(s-1)}\left[l^2 \alpha_{s,l-2} + \alpha_{s,l-1}\left(2ls + 2(l+1)^2 + s^2\right) + \alpha_{s,l}(l+s+2)^2\right]\,, \tag{81}$$

On can then argue that the coefficients in a cubic scalar source will be compatible with the definition of locality given in this note iff:

$$\sum_{l=0}^{\infty} \tilde{a}^{(s)}_l C^{(s)}_l = 1\,, \tag{82}$$

which is trivially satisfied for local interactions but quite non-trivial for pseudo-local ones. As an example of non-local interaction the following choice of coefficients:

$$\alpha_{s,l} = \frac{1}{l^n}\,, \qquad n \geq 1\,, \tag{83}$$

[h]This is not strictly true in the presence of quadratic spin-s sources to the scalar equations. However, we believe this subtlety to not invalidate our argument about locality since locality of a quartic vertex should not be influenced by the addition of other cubic interactions. Furthermore, considering a formal expansion parameter λ weighting the redefinition, the contribution to the quartic vertex generated by a quadratic spin-s source will be order λ^2 and hence subleading in the $\lambda \to 0$ limit.

[i]A subtlety of this reasoning is that one may as well conclude that it is not possible to move the quartic non-local interaction to the cubic order. However, since this is possible for any local truncation thereof and for any local coupling in general, this would indicate a problem for the existence of the corresponding pseudo-local limit. In this note we argue that this behavior may be considered as a sign of non-locality of the quartic interaction.

gives a non-local quartic interaction for any value of n. When the series in (82) is convergent but (82) is not satisfied, it is possible to "improve" the coefficients to make (82) satisfied by employing the exact representation for canonical currents:[30]

$$\alpha^{\Delta}_{s,l} = \Delta(-1)^l(2l+s+2)\left(\frac{l!}{(l+s+1)!}\right)^2, \tag{84}$$

which gives

$$\tilde{\alpha}_{s,l} = (\Delta+1)\delta_{l,0}\,. \tag{85}$$

However, this is not possible when the series is divergent which happens when $n \leq 2s+4$. We thus get the generic necessary but non sufficient condition:

$$\alpha_{s,l} \prec \frac{1}{l^{2s+4}}\,. \tag{86}$$

The above results are argued to extend our locality condition at quartic order[30,42] and show new subtleties which may arise in discussing quartic locality. Remarkably, it seems a generic feature of pseudo-local quartic interaction to not be compatible with (82) even when the series in (82) converges[j]. Notice that the basis (68) includes factorial damping coefficients which do not suffice to ensure locality of any (convergent) infinite combination thereof. In view of these results it would be desirable to have a better understanding of the locality properties of quartic vertices appearing in HS theory.

Sum over spins vs. sum over derivatives: An interesting interplay between sum over derivatives and sum over spins can be appreciated by expanding eq. (78) as:

$$\mathcal{V}_{s,l} = \sum_{k+q=s} \frac{(-1)^{l+q+k}}{(l!)^2(k)!^2(q!)^2}\, C^{\alpha(k)\alpha(q)\dot{\alpha}(k)\dot{\alpha}(q)} C_{\alpha(k)\nu(l)\dot{\alpha}(k)\dot{\nu}(l)} C_{\alpha(q)}{}^{\nu(l)}{}_{\dot{\alpha}(q)}{}^{\dot{\nu}(l)}\,. \tag{87}$$

One can then rewrite the same basis elements as:

$$\begin{aligned}\mathcal{V}_{s,l} &= \sum_{k+q=s} \frac{(-1)^k}{(k)!^2} C^{\alpha(q)\nu(l)\dot{\alpha}(q)\dot{\nu}(l)} \left[(-1)^{q+l}\frac{1}{q!^2 l!^2}\, C^{\alpha(k)}{}_{\alpha(q)}{}^{\dot{\alpha}(k)}{}_{\dot{\alpha}(q)} C_{\alpha(k)\nu(l)\dot{\alpha}(k)\dot{\nu}(l)}\right] \\ &= \sum_{k+q=s} C^{\alpha(q+l)\dot{\alpha}(q+l)} I^{(k)}_{\alpha(q+l)\dot{\alpha}(q+l)}\,. \end{aligned} \tag{88}$$

The latter cubic source can then be redefined by a redefinition of the spin-$(q+l)$ field analogously as in the discussion here-above. The only difference is that the corresponding redefinition will in this case generate also tracefull improvements at the cubic order. Dropping those,[30] one can still argue some condition for locality

[j] Similar features have been also appreciated in flat space.[56,57]

projecting everything on the canonical current sector. This time one can work at fixed $q+l$ obtaining a necessary condition for the $s\to\infty$ limit of the coefficients $\alpha_{s,l}$. In general, given the following redundant manifestly cyclic basis:

$$\mathcal{V}=\sum_{k,l,q}\frac{\alpha_{k,l,q}(-1)^{l+q+k}}{(l!)^2(k)!^2(q!)^2}\,C^{\alpha(k)\alpha(q)\dot{\alpha}(k)\dot{\alpha}(q)}C_{\alpha(k)\nu(l)\dot{\alpha}(k)\dot{\nu}(l)}C_{\alpha(q)}{}^{\nu(l)}{}_{\dot{\alpha}(q)}{}^{\dot{\nu}(l)}\,, \tag{89}$$

we obtain that the corresponding infinite set of coefficients can be improved to local coefficients as discussed above whenever the following convergence behaviour is verified:

$$\alpha_{k,l,q}\prec\frac{1}{k^{2(l+q)+4}}\,,\qquad k\to\infty\,,l+q=\text{const.}\,, \tag{90}$$

$$\alpha_{k,l,q}\prec\frac{1}{l^{2(q+k)+4}}\,,\qquad l\to\infty\,,q+k=\text{const.}\,, \tag{91}$$

$$\alpha_{k,l,q}\prec\frac{1}{q^{2(k+l)+4}}\,,\qquad q\to\infty\,,k+l=\text{const.}\,. \tag{92}$$

Otherwise the corresponding series will be divergent and there will be no way to improve the coefficients by employing the exact representative for canonical currents. The above basis gives a cyclic form of the convergence condition which shows some interplay between sum over derivatives and the sum over spin. In more physical terms, at the amplitude level, one can argue that what in the s-channel might appear as a non-locality in derivatives will appear in another channel as some divergent sum over exchanges (c.f. Section 4). On the other hand, something that appears local for that concerns the sum over l above may still be not admissible with respect to the sum over k or q.

4. Locality and AdS/CFT

Bulk locality at scales below the AdS length is a key concept which plays an important role within the holographic framework of the AdS/CFT correspondence.[2–5,58–61] In simple terms one deals with three length-scales: the AdS-length L, the string-length l_s and the Planck-length l_p. Standard AdS/CFT identifications can be then summarised as:

$$\left(\frac{L}{l_p}\right)^4=N\,,\qquad\qquad \left(\frac{L}{l_s}\right)^4=\lambda\,, \tag{93}$$

with $\lambda=Ng_{YM}^2$ the 't Hooft coupling and the identification of the string coupling constant as $g_s=g_{YM}^2$. The standard AdS/CFT lore then requires a large-N limit in order to recover a classical bulk description. Locality at sub-AdS distances is instead governed by the 't Hooft coupling constant. In particular, in the limit $\lambda\to\infty$ one would recover a local bulk-description. This limit requires strong coupling on the CFT side. In the HS case one is however interested into the opposite limit $\lambda\to0$ with $g_{YM}\to0$ so that $g_s\to0$ and both the boundary and the bulk side are weakly coupled. In this limit the string-length would satisfy $l_s\gg L$ and one naturally recovers the tensionless limit of string theory. HS theories are then conjectured to be an effective description of the bulk side of the correspondence in this regime.[18,19] However, due to the inequality $l_s \gg L$ one cannot expect to recover the same local bulk description which was available in the $\lambda \to \infty$ limit. We thus expect a generalisation of standard concepts of locality to be needed in order to consistently describe the bulk side in this regime. In this section we summarise some discussions carried at the conformal field theory level and their possible implications for bulk locality in the cases relevant for HS theories.

4.1. *Mellin amplitudes and higher-spin theories*

The Mellin representation[62] for conformal field theory correlator has shown to be very useful to study bulk-locality in the context of AdS/CFT. Indeed poles in the Mellin amplitude are associated to bulk exchanges in a remarkable analogy with flat space S-matrix.[63–65] In this language locality is translated into the analytic structure of Mellin amplitudes and in the corresponding singularities in analogy with S-matrix theory.[66,67]

The Mellin amplitude of the $O(N)$ vector model: In the free theory case CFT correlators can be easily computed by Wick-product. Computing the corresponding Mellin amplitude then amounts to invert the Mellin transform:[62]

$$\langle \mathcal{O}_\Delta(x_1)\mathcal{O}_\Delta(x_2)\mathcal{O}_\Delta(x_3)\mathcal{O}_\Delta(x_4)\rangle = \frac{\mathcal{N}}{(2\pi i)^2}\int d\delta_{ij}\,\tilde{M}(\delta_{ij})\prod_{i<j}^{4} x_{ij}^{-2\delta_{ij}}\,, \tag{94a}$$

$$\tilde{M}(\delta_{ij}) = M(\delta_{ij})\prod_{i<j}^{4}\Gamma(\delta_{ij})\,, \tag{94b}$$

where $\tilde{M}(\delta_{ij})$ is the Mellin transform of the CFT correlator, $M(\delta_{ij})$ is what is usually referred to as Mellin amplitude and $\mathcal{N}$ is some normalisation factor. Above we introduced the Mellin variables δ_{ij} playing a similar role to Mandelstam variables in flat space. These variables satisfy:

$$\delta_{ii} = 0\,, \qquad\qquad \sum_j \delta_{ij} = \Delta_j\,, \tag{95}$$

with Δ_j the operator dimensions. The suggestive nature of the above conditions can become more apparent by introducing auxiliary "momenta" k_i satisfying:

$$k_i^2 = -\Delta\,, \qquad\qquad \sum_i k_i = 0\,. \tag{96}$$

In the 4pt case and for equal conformal dimensions one can then define analogs of the Mandelstam variables as:

$$s = -(k_1+k_2)^2 = 2\Delta - 2\delta_{12}\,, \tag{97a}$$

$$t = -(k_1+k_3)^2 = 2\Delta - 2\delta_{13}\,, \tag{97b}$$

$$u = -(k_1+k_4)^2 = 2\Delta - 2\delta_{14}\,, \tag{97c}$$

satisfying the standard relation $s+t+u=4\Delta$. In the following we apply the above formalism to the 4pt correlator of four scalar operators $\mathcal{O}=\frac{1}{\sqrt{2N}}\phi^a\phi_a$ of dimension $\Delta=d-2$ which is given by:

$$\langle \mathcal{O}_\Delta(x_1)\mathcal{O}_\Delta(x_2)\mathcal{O}_\Delta(x_3)\mathcal{O}_\Delta(x_4)\rangle = \frac{1}{(x_{12}^2x_{34}^2)^\Delta}+\frac{1}{(x_{13}^2x_{24}^2)^\Delta}+\frac{1}{(x_{14}^2x_{23}^2)^\Delta}$$
$$+\frac{4}{N}\left(\frac{1}{(x_{12}^2x_{34}^2)^{\frac{\Delta}{2}}(x_{13}^2x_{24}^2)^{\frac{\Delta}{2}}}+\frac{1}{(x_{13}^2x_{24}^2)^{\frac{\Delta}{2}}(x_{14}^2x_{23}^2)^{\frac{\Delta}{2}}}+\frac{1}{(x_{14}^2x_{23}^2)^{\frac{\Delta}{2}}(x_{12}^2x_{34}^2)^{\frac{\Delta}{2}}}\right).$$

Remarkably, although the Mellin transform of a function:

$$f_M(\delta)\equiv\int_0^\infty dx\, x^{\delta-1}f(x)\,, \tag{98}$$

is ill defined when $f(x)=x^\Delta$, one can easily guess the corresponding result as a distributions by a simple inspection of (94a). It is indeed easy to show that the following distribution:

$$\tilde{M}(s,t,u)\sim\delta\left(\tfrac{s}{2}\right)\delta\left(\tfrac{t}{2}-\Delta\right)\delta\left(\tfrac{u}{2}-\Delta\right)$$
$$+\frac{4}{N}\delta\left(\tfrac{s}{2}-\tfrac{\Delta}{2}\right)\delta\left(\tfrac{t}{2}-\tfrac{\Delta}{2}\right)\delta\left(\tfrac{u}{2}-\Delta\right)+\text{cycl.}\,, \tag{99}$$

solves (94a) in the case of the free boson four-scalar correlator. Above we have introduced a third delta function and integration by using $s+u+t=4\Delta$ to arrive at a manifestly cyclic expression. One can recover a more standard expression depending upon two Mellin variables simply integrating on one among s,t,u.

Coming back to eq. (94a), let us stress that the Mellin amplitude is usually defined as the Mellin transform of the CFT correlator up to some Γ function factors. In the 4pt case this factor reads $M(\delta_{ij})=\tilde{M}(\delta_{ij})\left[\Gamma(\Delta-\frac{s}{2})\Gamma(\Delta-\frac{t}{2})\Gamma(\Delta-\frac{u}{2})\right]^{-2}$. These account for double-trace operators poles. Due to the singular structure (99) it is not clear to us how to treat the Γ-function factors whose arguments coincide with those of the δ-functions.

To summarise, this discussion shows that the Mellin transform of the free $O(N)$ model is naturally a distribution and does not display any pole in the Mellin-variables. Notice also that the above δ-functions are concentrated on points of the phase space which would not be allowed in the kinematical region for the generalised momenta k_i. This means that effectively the Mellin amplitude above *vanishes* for "physical" parameters, possibly as a consequence of the HS symmetry. Let us also stress that the above singular nature for the Mellin transform of the CFT correlator is a generic feature of CFTs with power-like dependence on the cross-ratios. This would remain true for instance for the critical $O(N)$ model at large N, conjectured[19] to be dual to Vasiliev's equations with scalar boundary conditions $\Delta=2$. On the other hand these features do not appear in strongly coupled CFTs.

Sum over exchanges and locality: The above result for the Mellin amplitude of the $O(N)$ model is expected, because the boundary theory is free, but it is also a bit puzzling in view of the HS dualities. From a bulk perspective one can indeed decompose the scattering process into an infinite number of HS exchanges which individually contribute with poles to the Mellin amplitude. This follows from the existence of non-trivial cubic couplings in the theory. The contribution of a spin-s exchange would indeed have the following generic structure:

$$M_n = \sum_m \frac{P_m^{(n)}(t)}{s-(\Delta_n+2m)}\,. \tag{100}$$

The sum over m parameterises the sum over descendants of the single trace operator exchanged and the numerators are some polynomial in the Mellin variables which closely resemble the analogous polynomials in the Mandelstam variables appearing at the S-matrix level.

In a theory including infinitely many HS particles the full exchange amplitude would then involve an infinite sum over primaries of arbitrary spin:

$$M = \sum_n g_n M_n = \sum_{n,m} \frac{g_n\, P_m^{(n)}(t)}{s-(\Delta_n+2m)}\,. \tag{101}$$

The full Mellin amplitude also involves contact terms[42] and the sum of those and the exchange[k] should then be equal to the distribution in (99). One possibility is to make an analogy with the case of conformal HS fields in flat space.[29] The corresponding exchanges are given in this case by Gegenbauer polynomial and the sum over spins in the conformal HS case is indeed a distribution, which we quote[29] here:

$$\sum_{s=0}^{\infty}(s+\tfrac{d-3}{2})\,C_s^{(\frac{d-3}{2})}(z) \sim \frac{(-1)^{d-4}}{(d-4)!}\delta^{[d-4]}(z)\,. \tag{102}$$

It is proportional to the $(d-4)$-th derivative of the δ-function.

Although the analysis carried out here is incomplete, one may draw similar conclusions for the bulk contact term as for the flat space case.[56,57,68] A pseudo-local behaviour may be expected in the regime under consideration where $\frac{L}{l_s} \ll 1$. In view of the above results and of those of Section 3.3, a better understanding of pseudo-locality in the bulk side at the quartic-order is a key step to improve our understanding of the HS-CFT dualities.

5. Conclusions

In this article we have studied locality in a field theory context with the aim of extending it to a quantum-gravity framework. In particular, we have formulated a functional class proposal to understand pseudo-local field redefinitions. Various examples have been considered and the functional class proposal has been tested in

[k] For exchanges involving external HS operators one can argue[56,57] the current exchange contribution not to be gauge invariant. In this case a non-vanishing contact term would be required to get a gauge invariant amplitude.

the string theory and in the AdS/CFT context. While the results of this article are encouraging and an extension of the locality proposal to the quartic order has been presented, more work is needed to achieve a better understanding of locality in the higher-spin framework both from a bulk and a boundary perspective. Hopefully, these investigation will shed some more light also on analogous questions in the context of Holography.

Acknowledgments

We would like to thank Glenn Barnich, Xavier Bekaert, Nicolas Boulanger, Slava Didenko, Maxim Grigoriev, Euihun Joung, Rakibur Rahman, Charlotte Sleight, Zhenya Skvortsov, Arkady Tseytlin and Mikhail Vasiliev for useful comments and discussions. We are also grateful to the organisers of the workshop "Higher Spin Gauge Theorie" at the Institute of Advanced Studies from Nanyang Technological University in Singapore for the opportunity to present this work. The research of M. Taronna is partially supported by the Fund for Scientific Research-FNRS Belgium, grant FC 6369 and by the Russian Science Foundation grant 14-42-00047 in association with Lebedev Physical Institute.

References

1. R. J. Eden, P. V. Landshoff, and D. I. Olive, *The analytic S-matrix*. 1966.
2. I. Heemskerk, J. Penedones, J. Polchinski, and J. Sully, Holography from Conformal Field Theory, *JHEP*. **10**, 079, (2009). doi: 10.1088/1126-6708/2009/10/079.
3. S. El-Showk and K. Papadodimas, Emergent Spacetime and Holographic CFTs, *JHEP*. **10**, 106, (2012). doi: 10.1007/JHEP10(2012)106.
4. A. L. Fitzpatrick and J. Kaplan, AdS Field Theory from Conformal Field Theory, *JHEP*. **02**, 054, (2013). doi: 10.1007/JHEP02(2013)054.
5. J. Maldacena, D. Simmons-Duffin, and A. Zhiboedov, Looking for a bulk point. (2015).
6. D. Amati, M. Ciafaloni, and G. Veneziano, Superstring Collisions at Planckian Energies, *Phys. Lett.* **B197**, 81, (1987). doi: 10.1016/0370-2693(87)90346-7.
7. D. J. Gross and P. F. Mende, The High-Energy Behavior of String Scattering Amplitudes, *Phys. Lett.* **B197**, 129, (1987). doi: 10.1016/0370-2693(87)90355-8.
8. D. J. Gross and P. F. Mende, String Theory Beyond the Planck Scale, *Nucl. Phys.* **B303**, 407, (1988). doi: 10.1016/0550-3213(88)90390-2.
9. D. Amati, M. Ciafaloni, and G. Veneziano, Classical and Quantum Gravity Effects from Planckian Energy Superstring Collisions, *Int. J. Mod. Phys.* **A3**, 1615–1661, (1988). doi: 10.1142/S0217751X88000710.
10. D. Amati, M. Ciafaloni, and G. Veneziano, Can Space-Time Be Probed Below the String Size?, *Phys. Lett.* **B216**, 41, (1989). doi: 10.1016/0370-2693(89)91366-X.
11. N. Moeller and P. C. West, Arbitrary four string scattering at high energy and fixed angle, *Nucl. Phys.* **B729**, 1–48, (2005). doi: 10.1016/j.nuclphysb.2005.09.036.
12. S. B. Giddings, D. J. Gross, and A. Maharana, Gravitational effects in ultrahigh-energy string scattering, *Phys. Rev.* **D77**, 046001, (2008). doi: 10.1103/PhysRevD.77.046001.
13. M. Taronna. *Higher Spins and String Interactions*. PhD thesis, Pisa U., (2010). URL `http://inspirehep.net/record/855395/files/arXiv:1005.3061.pdf`.

14. A. Sagnotti and M. Taronna, String Lessons for Higher-Spin Interactions, *Nucl. Phys.* **B842**, 299–361, (2011). doi: 10.1016/j.nuclphysb.2010.08.019.
15. A. Fotopoulos and M. Tsulaia, On the Tensionless Limit of String theory, Off - Shell Higher Spin Interaction Vertices and BCFW Recursion Relations, *JHEP.* **11**, 086, (2010). doi: 10.1007/JHEP11(2010)086.
16. D. Polyakov, Higher Spins and Open Strings: Quartic Interactions, *Phys. Rev.* **D83**, 046005, (2011). doi: 10.1103/PhysRevD.83.046005.
17. D. Polyakov, Higher Spins at the Quintic Order: Localization Effect and Simplifications, *Phys. Rev.* **D93**(4), 045001, (2016). doi: 10.1103/PhysRevD.93.045001.
18. E. Sezgin and P. Sundell, Massless higher spins and holography, *Nucl. Phys.* **B644**, 303–370, (2002). doi: 10.1016/S0550-3213(02)00739-3. [Erratum: Nucl. Phys.B660,403(2003)].
19. I. R. Klebanov and A. M. Polyakov, AdS dual of the critical O(N) vector model, *Phys. Lett.* **B550**, 213–219, (2002). doi: 10.1016/S0370-2693(02)02980-5.
20. N. Boulanger, S. Leclercq, and P. Sundell, On The Uniqueness of Minimal Coupling in Higher-Spin Gauge Theory, *JHEP.* **08**, 056, (2008). doi: 10.1088/1126-6708/2008/08/056.
21. P. Kessel, G. Lucena Gmez, E. Skvortsov, and M. Taronna, Higher Spins and Matter Interacting in Dimension Three, *JHEP.* **11**, 104, (2015). doi: 10.1007/JHEP11(2015)104.
22. N. Boulanger, P. Kessel, E. D. Skvortsov, and M. Taronna, Higher Spin Interactions in Four Dimensions: Vasiliev vs. Fronsdal. (2015).
23. R. R. Metsaev, Poincare invariant dynamics of massless higher spins: Fourth order analysis on mass shell, *Mod. Phys. Lett.* **A6**, 359–367, (1991). doi: 10.1142/S0217732391000348.
24. R. R. Metsaev, Cubic interaction vertices of massive and massless higher spin fields, *Nucl. Phys.* **B759**, 147–201, (2006). doi: 10.1016/j.nuclphysb.2006.10.002.
25. M. A. Vasiliev, Cubic Vertices for Symmetric Higher-Spin Gauge Fields in $(A)dS_d$, *Nucl. Phys.* **B862**, 341–408, (2012). doi: 10.1016/j.nuclphysb.2012.04.012.
26. E. Joung and M. Taronna, Cubic interactions of massless higher spins in (A)dS: metric-like approach, *Nucl. Phys.* **B861**, 145–174, (2012). doi: 10.1016/j.nuclphysb.2012.03.013.
27. N. Boulanger, D. Ponomarev, and E. D. Skvortsov, Non-abelian cubic vertices for higher-spin fields in anti-de Sitter space, *JHEP.* **05**, 008, (2013). doi: 10.1007/JHEP05(2013)008.
28. N. Boulanger, D. Ponomarev, E. D. Skvortsov, and M. Taronna, On the uniqueness of higher-spin symmetries in AdS and CFT, *Int. J. Mod. Phys.* **A28**, 1350162, (2013). doi: 10.1142/S0217751X13501625.
29. E. Joung, S. Nakach, and A. A. Tseytlin, Scalar scattering via conformal higher spin exchange. (2015).
30. E. D. Skvortsov and M. Taronna, On Locality, Holography and Unfolding, *JHEP.* **11**, 044, (2015). doi: 10.1007/JHEP11(2015)044.
31. J. Krasil'shchik and A. Verbovetsky, Geometry of jet spaces and integrable systems, *J. Geom. Phys.* **61**, 1633–1674, (2011). doi: 10.1016/j.geomphys.2010.10.012.
32. G. Barnich, M. Grigoriev, A. Semikhatov, and I. Tipunin, Parent field theory and unfolding in BRST first-quantized terms, *Commun. Math. Phys.* **260**, 147–181, (2005). doi: 10.1007/s00220-005-1408-4.
33. G. Barnich and M. Grigoriev, First order parent formulation for generic gauge field theories, *JHEP.* **01**, 122, (2011). doi: 10.1007/JHEP01(2011)122.
34. M. Grigoriev, Parent formulations, frame-like Lagrangians, and generalized auxiliary

fields, *JHEP.* **12**, 048, (2012). doi: 10.1007/JHEP12(2012)048.
35. R. Rahman and M. Taronna, From Higher Spins to Strings: A Primer. (2015).
36. V. E. Lopatin and M. A. Vasiliev, Free Massless Bosonic Fields of Arbitrary Spin in d-dimensional De Sitter Space, *Mod. Phys. Lett.* **A3**, 257, (1988). doi: 10.1142/S0217732388000313.
37. M. A. Vasiliev, Consistent Equations for Interacting Massless Fields of All Spins in the First Order in Curvatures, *Annals Phys.* **190**, 59–106, (1989). doi: 10.1016/0003-4916(89)90261-3.
38. M. A. Vasiliev, Higher spin gauge theories: Star product and AdS space. (1999).
39. E. Joung, M. Taronna, and A. Waldron, A Calculus for Higher Spin Interactions, *JHEP.* **07**, 186, (2013). doi: 10.1007/JHEP07(2013)186.
40. S. F. Prokushkin and M. A. Vasiliev, Cohomology of arbitrary spin currents in AdS(3), *Theor. Math. Phys.* **123**, 415–435, (2000). doi: 10.1007/BF02551048. [Teor. Mat. Fiz.123,3(2000)].
41. X. Bekaert, J. Erdmenger, D. Ponomarev, and C. Sleight, Towards holographic higher-spin interactions: Four-point functions and higher-spin exchange, *JHEP.* **03**, 170, (2015). doi: 10.1007/JHEP03(2015)170.
42. X. Bekaert, J. Erdmenger, D. Ponomarev, and C. Sleight, Quartic AdS Interactions in Higher-Spin Gravity from Conformal Field Theory, *JHEP.* **11**, 149, (2015). doi: 10.1007/JHEP11(2015)149.
43. S. Sciuto, The general vertex function in dual resonance models, *Lett. Nuovo Cim.* **2S1**, 411–418, (1969). doi: 10.1007/BF02755622.
44. P. Di Vecchia, R. Nakayama, J. L. Petersen, J. R. Sidenius, and S. Sciuto, Covariant N String Amplitude, *Nucl. Phys.* **B287**, 621, (1987). doi: 10.1016/0550-3213(87)90121-0.
45. A. Neveu and P. C. West, The Cyclic Symmetric Vertex for Three Arbitrary Neveu-schwarz Strings, *Phys. Lett.* **B180**, 34, (1986). doi: 10.1016/0370-2693(86)90129-2.
46. D. J. Gross and A. Jevicki, Operator Formulation of Interacting String Field Theory, *Nucl. Phys.* **B283**, 1, (1987). doi: 10.1016/0550-3213(87)90260-4.
47. P. Di Vecchia, K. Hornfeck, M. Frau, A. Lerda, and S. Sciuto, N String, g Loop Vertex for the Bosonic String, *Phys. Lett.* **B206**, 643–649, (1988). doi: 10.1016/0370-2693(88)90711-3.
48. O. A. Gelfond, E. D. Skvortsov, and M. A. Vasiliev, Higher spin conformal currents in Minkowski space, *Theor. Math. Phys.* **154**, 294–302, (2008). doi: 10.1007/s11232-008-0027-6.
49. O. A. Gelfond and M. A. Vasiliev, Unfolded Equations for Current Interactions of 4d Massless Fields as a Free System in Mixed Dimensions, *J. Exp. Theor. Phys.* **120**(3), 484–508, (2015). doi: 10.1134/S106377611503005X.
50. O. A. Gelfond and M. A. Vasiliev, Conserved Higher-Spin Charges in AdS_4. (2014).
51. C.-M. Chang and X. Yin, Higher Spin Gravity with Matter in AdS_3 and Its CFT Dual, *JHEP.* **10**, 024, (2012). doi: 10.1007/JHEP10(2012)024.
52. C.-M. Chang and X. Yin, Correlators in W_N Minimal Model Revisited, *JHEP.* **10**, 050, (2012). doi: 10.1007/JHEP10(2012)050.
53. M. Ammon, P. Kraus, and E. Perlmutter, Scalar fields and three-point functions in $D = 3$ higher spin gravity, *JHEP.* **07**, 113, (2012). doi: 10.1007/JHEP07(2012)113.
54. A. L. Fitzpatrick, J. Kaplan, and M. T. Walters, Universality of Long-Distance AdS Physics from the CFT Bootstrap, *JHEP.* **08**, 145, (2014). doi: 10.1007/JHEP08(2014)145.
55. K. B. Alkalaev and V. A. Belavin, Classical conformal blocks via AdS/CFT correspondence, *JHEP.* **08**, 049, (2015). doi: 10.1007/JHEP08(2015)049.

56. M. Taronna, Higher-Spin Interactions: four-point functions and beyond, *JHEP.* **04**, 029, (2012). doi: 10.1007/JHEP04(2012)029.
57. M. Taronna. *Higher-Spin Interactions: three-point functions and beyond.* PhD thesis, Pisa, Scuola Normale Superiore, (2012). URL `https://inspirehep.net/record/1188191/files/arXiv:1209.5755.pdf`.
58. J. Polchinski, S matrices from AdS space-time. (1999).
59. S. B. Giddings, Flat space scattering and bulk locality in the AdS / CFT correspondence, *Phys. Rev.* **D61**, 106008, (2000). doi: 10.1103/PhysRevD.61.106008.
60. M. Gary, S. B. Giddings, and J. Penedones, Local bulk S-matrix elements and CFT singularities, *Phys. Rev.* **D80**, 085005, (2009). doi: 10.1103/PhysRevD.80.085005.
61. A. L. Fitzpatrick, J. Kaplan, D. Poland, and D. Simmons-Duffin, The Analytic Bootstrap and AdS Superhorizon Locality, *JHEP.* **12**, 004, (2013). doi: 10.1007/JHEP12(2013)004.
62. G. Mack, D-independent representation of Conformal Field Theories in D dimensions via transformation to auxiliary Dual Resonance Models. Scalar amplitudes. (2009).
63. J. Penedones, Writing CFT correlation functions as AdS scattering amplitudes, *JHEP.* **03**, 025, (2011). doi: 10.1007/JHEP03(2011)025.
64. A. L. Fitzpatrick, J. Kaplan, J. Penedones, S. Raju, and B. C. van Rees, A Natural Language for AdS/CFT Correlators, *JHEP.* **11**, 095, (2011). doi: 10.1007/JHEP11(2011)095.
65. M. F. Paulos, Towards Feynman rules for Mellin amplitudes, *JHEP.* **10**, 074, (2011). doi: 10.1007/JHEP10(2011)074.
66. A. L. Fitzpatrick and J. Kaplan, Analyticity and the Holographic S-Matrix, *JHEP.* **10**, 127, (2012). doi: 10.1007/JHEP10(2012)127.
67. A. L. Fitzpatrick and J. Kaplan, Unitarity and the Holographic S-Matrix, *JHEP.* **10**, 032, (2012). doi: 10.1007/JHEP10(2012)032.
68. P. Dempster and M. Tsulaia, On the Structure of Quartic Vertices for Massless Higher Spin Fields on Minkowski Background, *Nucl. Phys.* **B865**, 353–375, (2012). doi: 10.1016/j.nuclphysb.2012.07.031.

Finite Temperature Maps in Vector/Higher Spin Duality

Antal Jevicki,* Kenta Suzuki† and Junggi Yoon‡

Department of Physics, Brown University, Providence, RI 02912, USA

**Antal_Jevicki@brown.edu*

†Kenta_Suzuki@brown.edu

‡Jung-Gi_Yoon@brown.edu

We discuss the question of reconstructing higher spin bulk fields from finite temperature N-component vector models. This is done in the framework of thermo-field quantum theory at Large N. A bi-local construction of connected dual space-times is accomplished, and issues related to the implementation of domain duality are discussed.

1. Introduction

The AdS/CFT correspondence based Large N describes Quantum Gravity in terms of standard QFT. This gives a framework for studying quantum properties of black holes through Quantum Field Theory at finite temperature. Various properties of black holes have been clarified in this way, in particular the origin and nature of black hole entropy, and the evaluation of free energy. The issues related to the information paradox, the reconstruction of physics behind the horizon are however still being vigorously debated [1].

A workable model of AdS/CFT for Higher Spins [2] is based on N-component vector type models [3, 4] (in d dimensions) which at Large N can be related to Vasiliev type Higher Spin Gravity theories [5–7] (in $(d+1)$ dimensions). This duality was confirmed in detailed studies at both tree [8] and one-loop level [9–11] (through $1/N$ expansion). A bi-local mechanism explaining holography was proposed in [12] and has been implemented in explicit terms in [13]. Here the bi-local composite operators of the $O(N)$ (or $U(N)$) model are seen to establish a bridge [14] for reconstruction of Higher Spin fields and bulk AdS space-time. Extensions to de Sitter case were considered [15]. In general the bi-local CFT observables provide a one-to-one map and a constructive representation of Holography [16–18].

The N-field vector model exhibits a phase transition at high temperature [19, 20] where the unconstrained bi-local structure is destroyed. This transition is analogous to the de-confinement [21] phase transition [22] of Matrix theories and QCD. It is then of definite interest to study the $O(N)$/Higher Spin Gravity duality at nonzero temperature. This can be done within the Thermo-field formalism with a doubling of the CFT Hamiltonians which is associated with the finite temperature Schwinger-Kelydish [23–25] time contour.

In earlier work [26, 27] and the present paper we study the Large N dynamics of the $O(N)$ field theory and its Higher Spin Gravity duality at finite temperature. One of main questions that one faces is the ability of recovering the high temperature phase of the theory which as described above corresponds to de-confinement and the emergence of partonic finite N structure. We discuss these questions in what follows and describe the relevant issue of the $O(N)$ singlet gauging which turns out to play a crucial role in the two different phases of the theory.

It is natural to expect that the Thermo-field scheme leads through Large N AdS/CFT correspondence to the mapping of Left/Right Penrose regions of typical AdS black holes [28]. The question has been raised of how two (decoupled CFTs) are capable of producing a connected [29] space-time, an issue which received contradictory answers in the literature [30]. The answer to this question concerns the reconstruction of local bulk fields [31] from two boundary CFT's (a domain AdS/CFT duality) [32]. Closely related is the issue of a possible reconstruction of behind the horizon [31, 33, 34] physics for black hole type space-times.

Our $O(N)$/Higher Spin Thermo-field duality construction offers some answers to the above questions which we highlight in this presentation. Through the associated bi-local representation we discuss the nature and properties of emergent bulk space-times steaming from finite (large) temperature. As we will emphasize central to the ability of reconstructing a connected space-time in the domain duality scheme is the nature of gauging (of the $O(N)$ symmetry) in the case of doubled CFT's. In particular the diagonal gauging leads to as we emphasize leads to recovery of bulk "evanescent" modes [35, 36] characteristic of black hole type space times. We also observe in this construction a mixing of spins at finite temperature. We have in this study considered mainly the (free) UV fixed point HS duality, this allowed for very explicit constructions and analysis we believe that certain important lessons can be learned from this simple case of domain duality. And while finite temperature CFT physics certainly contains implications on actual dual quantum black hole states, for the present case of Higher Spin Duality further understanding regarding the black hole geometries [37, 38] in Vasiliev's Higher Spin Gravity and their relationship with present thermal states is needed. Much like the better understood examples of AdS_3 Black hole [39, 40] these are most interesting topics of future investigations.

2. Phase Transitions in $O(N)$ Vector Models

We begin by elaborating on the phase transition characterizing $O(N)$ or $U(N)$ vector type models at Large N. The free energy of N-component vector models at large N in the singlet sector can be studied with the help of a $U(N)$ gauge field that couples to matter in fundamental representation [2, 19, 41] whose appearance is related to implementing the singlet Gauss's law constraint. The evaluation of free energy at Large N reduces to solving a saddle point equation for the density of eigenvalues of the $U(N)$ holonomy matrix. Positivity conditions of the density, which the solution of the saddle point equation should be subject to, lead to a Gross-Witten type phase transition in the vector model as performed by Shenker and Yin [20]. In the lower phase ($T \ll \sqrt{N}$), the density function is supported by all range of eigenvalues. The corresponding free energy is of order $\mathcal{O}(N^0)$

$$F_{\text{low}} \simeq 4\zeta(5)T^4. \tag{1}$$

The lower phase can be described by Hamiltonian formalism of collective field theory which captures the $O(N)$ singlet sector. In this collective field theory [10], the free energy of the lower phase is indeed reproduced

$$F_{\text{low}} = \sum_{\text{singlet states}} \log\left(1 - e^{-\beta H_{\text{col}}}\right) \tag{2}$$

as an unconstrained sum of composite bi-local degrees of freedom.

On the other hand, the upper phase ($T \gg \sqrt{N}$) is characterized by the density function supported by a finite range of interval. In this case, one has free energy of order $\mathcal{O}(N)$

$$F \simeq 4\zeta(3)NT^2 \tag{3}$$

which reflects that the partonic N dependent collection of degrees of freedom. In this case, it was observed that this can be found through the Lagrangian collective formalism of vector model [10, 12], which seemingly contains un-physical (two time) degrees of freedom. The result is simply gives the free energy of N-scalars:

$$F_{\text{high}} = N\text{tr}\,\log \Box_{\text{col}}. \tag{4}$$

This phenomenon was also observed in $U(N)$ fermi vector model [15] and fermion theories. One gets some insight into the phenomenon by considering the exact dimension of the Hilbert space of ($U(N)$) vector model. Through coherent state quantization this was evaluated by Berezin [42], and also in [15] to be given by

$$\dim \mathcal{H}_{N,K} = \prod_{j=0}^{K-1} \frac{j!(N+K+j)!}{(N+j)!(K+j)!} \tag{5}$$

where K is the volume of the momentum space. In the upper phase $(K \ll N)$[a], the dimension of the Hilbert space can be approximated by

$$\log \dim \mathcal{H}_{N,K} = 2NK \log 2 - \frac{N(N+1)}{2} \log K + \mathcal{O}(K^0) \tag{7}$$

[a]Since characteristic length in two dimension is $1/T^2$, the number of modes can be estimated by

$$K = \sum_{\vec{p}} 1 \sim T^2 V \tag{6}$$

where V is the area of the two-dimensional space.

The leading term corresponds to the dimension of Hilbert space of $2N$ fermi oscillator without $U(N)$ singlet constraint. Again, the $U(2N)$ constraint does not play a significant role in reducing the degree of freedom. The reduction can be found in the next leading correction, $-\frac{N(N+1)}{2}\log K$.

On the other hands, the dimension in the lower phase ($N \gg K$) becomes

$$\log \dim \mathcal{H}_{N,K} = K^2 \log N + \mathcal{O}(N^0). \tag{8}$$

The leading term represents the degree of freedom of K^2 bi-local creation operators. This is consistent with the fact that the canonical (unconstrained) collective field theory of the vector model captures the lower phase at finite temperature, while the upper phase requires implementation of finite N constraints, or as we will argue in the next section a more appropriate gauging of the $U(N)$ symmetry.

3. TFD of Collective $O(N)$ Vector Model

Thermo Field Dynamics(TFD), which is also known as the real time formalism for finite temperature theory by Schwinger and Keldyish [23, 24], is based on the closed time contour. Hilbert space of TFD is defined to be a direct product of the Hilbert space $\mathcal{H}$ of the original system and and identical fictitious one $\widetilde{\mathcal{H}}$. The associated Hamiltonian of TFD is given by

$$H_{\text{TFD}} \equiv H - \widetilde{H} \tag{9}$$

where the minus sign in front of $\widetilde{H}$ represents reverse time direction of the fictitious system. This turned out to be natural in gravity [28, 43].

At finite temperature, the thermal average of an observable $\mathcal{O}$ is given by

$$\langle \mathcal{O} \rangle_\beta = \frac{1}{Z(\beta)} \text{Tr}\left(e^{-\beta H}\mathcal{O}\right). \tag{10}$$

TFD is formulated in such a way that the thermal average is represented by the expectation value of the operator $\mathcal{O}$ with respect to a pure state in TFD which we call thermal vacuum $|0(\beta)\rangle$. Furthermore, it was also shown [44] that the thermal vacuum minimizes the free energy of the original system.

The TFD formalism of $O(N)$ vector model

$$H \equiv \sum_{i=1}^{N} \int d\vec{x} \left(\frac{1}{2}(\pi^i)^2 + \frac{1}{2}(\vec{\partial}\phi^i)^2\right) = \sum_{i=1}^{N} |\vec{p}| a^{i\dagger}(\vec{p}) a^i(\vec{p}) + E_0 \tag{11}$$

is given by introducing a identical fictitious system

$$\widetilde{H} \equiv \sum_{i=1}^{N} \int d\vec{x} \left(\frac{1}{2}(\widetilde{\pi}^i)^2 + \frac{1}{2}(\vec{\partial}\widetilde{\phi}^i)^2\right) = \sum_{i=1}^{N} |\vec{p}| \widetilde{a}^{i\dagger}(\vec{p}) \widetilde{a}^i(\vec{p}) + \widetilde{E}_0 \tag{12}$$

with the thermal vacuum being given by

$$|0(\beta)\rangle \equiv \exp\left[\sum_{j=1}^{N}\sum_{\vec{p}} \theta(\vec{p}) \left(a^{j\dagger}(\vec{p})\widetilde{a}^{j\dagger}(\vec{p}) - a^j(\vec{p})\widetilde{a}^j(\vec{p})\right)\right] |0\rangle \tag{13}$$

where $\theta(\vec{p}) \equiv \tanh^{-1} e^{-\beta|\vec{p}|/2}$. The thermal vacuum $|0(\beta)\rangle$ can be also seen as a Bogoliubov transformed vacuum. Consequently, it is not $a^j(\vec{p})$ or $\widetilde{a}^j(\vec{p})$ but the Bogoliubov transformed oscillators $a^j_\theta(\vec{p})$ and $\widetilde{a}^j_\theta(\vec{p})$ that annihilate the thermal vacuum:

$$a^j_\theta(\vec{p}) \equiv a^j(\vec{p}) \cosh\theta(\vec{p}) - \widetilde{a}^{j\dagger}(\vec{p}) \sinh\theta(\vec{p}) \tag{14}$$

and similarly for $a^{j\dagger}_\theta(\vec{p}), \widetilde{a}^j_\theta(\vec{p})$ and $\widetilde{a}^{j\dagger}_\theta(\vec{p})$. Furthermore, operators should be normal-ordered with respect to the thermal vacuum $|0(\beta)\rangle$, which leads to non-trivial consequence such as the appearance of extra modes [26].

In AdS/CFT, the higher spin theory in the bulk is dual to the single sector of the $O(N)$ vector model which can be successfully described by invariant collective field theory [13, 14]. The basic idea is to express the system in terms of $O(N)$ invariant variables,given by bi-local operators which then automatically implement the $O(N)$ singlet constraint.

In applying TFD to the $O(N)$ singlet sector, there arises a question of how to define a $O(N)$ singlet constraint in TFD. First, one can double the $O(N)$ singlet sector of the original zero temperature model. This is equivalent to imposing $O(N)\times O(N)$ singlet constraint on the doubled Hilbert space of $O(N)$ vector model. This collective (TFD) theory is described by the double set of $O(N)$ invariant bi-local fields $\Psi(\vec{x},\vec{y}) = \phi^i(\vec{x})\phi^i(\vec{y})$ and $\widetilde{\Psi}(\vec{x},\vec{y}) = \widetilde{\phi}^i(\vec{x})\widetilde{\phi}^i(\vec{y})$. Those two bi-local fields are certainly natural for this collective (TFD) theory because we just doubled the $O(N)$ singlet sector and, consequently, the collective $O(N)$ vector model (at $T=0$). It turns out that this collective (TFD) description is only appropriate for the lower phase of the $O(N)$ vector model, and is dual to the higher spin theory in the thermal AdS background.

To describe the upper phase of the theory, we find it necessary to enlarge the degree of freedom by relaxing the $O(N)\times O(N)$ singlet constraint on the doubled Hilbert space. It was observed earlier [45] that when original ($T=0$) Lagrangian poses a symmetry group G, the vacuum of TFD respects the diagonal subgroup of $G\times G$. For our case, gauging the diagonal $O(N)$ subgroup of $O(N)\times O(N)$ will be associated with the upper phase:

$$J^{ij} + \widetilde{J}^{ij}\,|\Phi\rangle = 0 \tag{15}$$

where J^{ij} and $\widetilde{J}^{ij}$ is a $O(N)$ generator of ϕ and $\widetilde{\phi}$, respectively. It follows that under the diagonal $O(N)$ symmetry invariant bi-local fields are

$$\phi^i(t,\vec{x})\phi^i(t,\vec{y})\ ,\ \phi^i(t,\vec{x})\widetilde{\phi}^i(t,\vec{y})\ ,\ \widetilde{\phi}^i(t,\vec{x})\widetilde{\phi}^i(t,\vec{y}). \tag{16}$$

The additional mixed bi-local field is a new degree of freedom in this TFD with relaxed constraint because it is non-singlet with respect to the $O(N)\times O(N)$ symmetry. Because this extra mode is consist of two vector fields with opposite time directions, it features spacelike momentum which can be seen from the bi-local (momentum) map, and therefore will turn out to correspond to so-called evanescent modes [32, 35, 36] of the bulk theory. These modes have been known to represent a signature associated of a horizon in the bulk.

Let us introduce a new notation $\phi(\vec{x}, j)$ $(j = 1, 2)$ for the $O(N)$ vector fields where $\phi_i(\vec{x}, 1)$ is the original $O(N)$ vector field $\phi_i(\vec{x})$ and $\phi_i(\vec{x}, 2)$ is the copied one $\widetilde{\phi}_i(\vec{x})$. This compact notation corresponds to doubling the space-time. Considering the time-direction of $\phi(\vec{x}, 2) = \widetilde{\phi}(\vec{x})$, we find it convenient to define $\Psi((\vec{x}, j), (\vec{y}, k))$

$$\Psi((\vec{x}, i), (\vec{y}, j)) \equiv \begin{pmatrix} \Psi((\vec{x},1),(\vec{y},1)) & \Psi((\vec{x},1),(\vec{y},2)) \\ \Psi((\vec{x},2),(\vec{y},1)) & \Psi((\vec{x},2),(\vec{y},2)) \end{pmatrix} \equiv \begin{pmatrix} \phi^j(\vec{x})\phi^j(\vec{y}) & i\phi^j(\vec{x})\widetilde{\phi}^j(\vec{y}) \\ i\widetilde{\phi}^j(\vec{x})\phi^j(\vec{y}) & -\widetilde{\phi}^j(\vec{x})\widetilde{\phi}^j(\vec{y}) \end{pmatrix}. \tag{17}$$

This new bi-local field covers all (diagonal) $O(N)$ invariant bi-locals in Eq. (16). In this notation, the collective TFD has the same form of ordinary collective $O(N)$ vector model (at $T = 0$) except for the fact that the bi-local space is enlarged. For example, $\Psi((\vec{x}, j), (\vec{y}, k))$ is symmetric in the enlarged bi-local space. Hence, defining the star product in the enlarged bi-local space, one can write the collective Hamiltonian for $\Psi((\vec{x}, j), (\vec{y}, k))$ as follows

$$H_{\mathrm{TFD}} = \frac{2}{N}\mathrm{Tr}\,[\Pi \star \Psi \star \Pi] + \frac{N}{8}\mathrm{Tr}\,\left[\Psi^{-1}\right] + \frac{N}{2}\mathrm{Tr}\,\left[-\nabla^2 \star \Psi\right] + \Delta V. \tag{18}$$

In Large N, the leading effective potential leads to the background field equation, which is solved by

$$\Psi_0((\vec{x}, i), (\vec{y}, j)) = \int \frac{d\vec{p}}{(2\pi)^2 2|\vec{p}|} \begin{pmatrix} \cosh F(\vec{p}) e^{i\vec{p}\cdot(\vec{x}-\vec{y})} & i \sinh F(\vec{p}) e^{i\vec{p}\cdot(\vec{x}-\vec{y})} \\ i \sinh F(\vec{p}) e^{i\vec{p}\cdot(\vec{x}-\vec{y})} & -\cosh F(\vec{p}) e^{i\vec{p}\cdot(\vec{x}-\vec{y})} \end{pmatrix} \tag{19}$$

where $F(\vec{k})$ is an arbitrary even function of $\vec{p}$. The arbitrariness of $F(\vec{p})$ implies that the above thermal ground state solution is not unique. This is not surprising because we have not taken the temperature dependence into account. Uniqueness could be accomplished by minimizing an operator representing free energy [44] in general. However, it is more practical to determine $F(\vec{p})$ from two-point function with respect to the thermal vacuum in Eq. (13) because the background field is equal to the (equal-time) two point function of the vector field with respect to $|0(\beta)\rangle$. In particular, the temperature dependence appears from the normal-ordering with respect to the thermal vacuum in the two-point function. Therefore, $F(\vec{p})$ is found to be

$$F(\vec{p}) = 2\theta(\vec{p}) = 2\tanh^{-1} e^{-\beta|\vec{p}|/2}. \tag{20}$$

The next step is to study fluctuation around the Large N thermal background. Expanding the bi-local field and its conjugate around the thermal background

$$\Psi(t; (\vec{x}, i), (\vec{y}, j)) = \Psi_0((\vec{x}, i), (\vec{y}, j)) + \frac{1}{\sqrt{N}}\eta(t; (\vec{x}, i), (\vec{y}, j)) \tag{21}$$

$$\Pi(t; (\vec{x}, i), (\vec{y}, j)) = \sqrt{N}\pi(t; (\vec{x}, i), (\vec{y}, j)) \tag{22}$$

the quadratic Hamiltonian reads

$$H^{(2)} = 2\mathrm{Tr}\,[\pi \star \Psi_0 \star \pi] + \frac{1}{8}\mathrm{Tr}\,\left[\Psi_0^{-1} \star \eta \star \Psi_0^{-1} \star \eta \star \Psi_0^{-1}\right]. \tag{23}$$

For spectrum of the linearized fluctuation, we found it convenient to utilize oscillator representation of the the fundamental vector fields instead of analyzing the quadratic Hamiltonian. The bi-local field $\Psi((\vec{x},i),(\vec{y},j))$ is quadratic in the vector fields $\phi^i(\vec{x},j)$ and, therefore, is a linear combination of $a^i_\theta(\vec{p}), a^{i\dagger}_\theta(\vec{p}), \widetilde{a}^i_\theta(\vec{p})$ and $\widetilde{a}^{i\dagger}_\theta(\vec{p})$ where θ represents the Bogoliubov transformed oscillators. Because the background field is a (equal-time) two-point function with respect to the thermal vacuum, by the normal-ordering of the bi-local field Ψ with respect to the thermal vacuum, the linearized fluctuation η can be expressed as a linear combination of normal-ordered quadratic composite operators of the (Bogoliubov transformed) oscillators in Eq. (14). The algebra of such quadratic composite operators and their Large N expansion was studied in [15, 42, 46] and [14, 26] by solving a Caimir constraint for the $O(N)$ singlet sector. For this, let us define bi-local creation operators in terms of quadratic creation operators as follows

$$\begin{pmatrix} \alpha^\dagger(\vec{p}_1,\vec{p}_2) & \gamma^\dagger(\vec{p}_1,\vec{p}_2) \\ \gamma^\dagger(\vec{p}_2,\vec{p}_1) & \widetilde{\alpha}^\dagger(\vec{p}_1,\vec{p}_2) \end{pmatrix} \equiv \frac{1}{\sqrt{2N|\vec{p}_1||\vec{p}_2|}} \begin{pmatrix} a^{i\dagger}_\theta(\vec{p}_1)a^{i\dagger}_\theta(\vec{p}_1) & a^{i\dagger}_\theta(\vec{p}_1)\widetilde{a}^{i\dagger}_\theta(\vec{p}_1) \\ \widetilde{a}^{i\dagger}_\theta(\vec{p}_1)a^{i\dagger}_\theta(\vec{p}_1) & \widetilde{a}^{i\dagger}_\theta(\vec{p}_1)\widetilde{a}^{i\dagger}_\theta(\vec{p}_1) \end{pmatrix}. \tag{24}$$

Together with their conjugate bi-local annihilation operators, they satisfy the canonical commutation relation in the enlarged bi-local momentum space:

$$\begin{aligned} \left[\alpha(\vec{p}_1,\vec{p}_2),\alpha^\dagger(\vec{p}_3,\vec{p}_4)\right] &= \frac{1}{2}\left\{\delta(\vec{p}_1-\vec{p}_3)\delta(\vec{p}_2,\vec{p}_4)+\delta(\vec{p}_1-\vec{p}_4)\delta(\vec{p}_2-\vec{p}_3)\right\} \\ \left[\widetilde{\alpha}(\vec{p}_1,\vec{p}_2),\widetilde{\alpha}^\dagger(\vec{p}_3,\vec{p}_4)\right] &= \frac{1}{2}\left\{\delta(\vec{p}_1-\vec{p}_3)\delta(\vec{p}_2-\vec{p}_4)+\delta(\vec{p}_1-\vec{p}_4)\delta(\vec{p}_2-\vec{p}_3)\right\} \\ \left[\gamma(\vec{p}_1,\vec{p}_2),\gamma^\dagger(\vec{p}_3,\vec{p}_4)\right] &= \frac{1}{2}\delta(\vec{p}_1-\vec{p}_3)\delta(\vec{p}_2-\vec{p}_4). \end{aligned} \tag{25}$$

Solving the Casimir constraint for the quadratic annihilation operators in Large N, we have

$$\frac{1}{\sqrt{2N|\vec{p}_1||\vec{p}_2|}} \begin{pmatrix} a^i_\theta(\vec{p}_1)a^i_\theta(\vec{p}_1) & a^i_\theta(\vec{p}_1)\widetilde{a}^i_\theta(\vec{p}_1) \\ \widetilde{a}^i_\theta(\vec{p}_1)a^i_\theta(\vec{p}_1) & \widetilde{a}^i_\theta(\vec{p}_1)\widetilde{a}^i_\theta(\vec{p}_1) \end{pmatrix} = \begin{pmatrix} \alpha(\vec{p}_1,\vec{p}_2) & \gamma(\vec{p}_1,\vec{p}_2) \\ \gamma(\vec{p}_2,\vec{p}_1) & \widetilde{\alpha}(\vec{p}_1,\vec{p}_2) \end{pmatrix} + \mathcal{O}\left(\frac{1}{N}\right). \tag{26}$$

Therefore, the linearized fluctuation can be expressed as a linear combination of the bi-local oscillators [26]. Note that the rest of quadratic composite operators such as $a^{i\dagger}a^i$ can be negligible because such operators are sub-leading in Large N expansion. From the mode expansion of the linearized fluctuation, one can read off the spectrum of the bi-local oscillators. This can be also achieved by (momentum) map from bi-local momentum $(\vec{p}_1,\vec{p}_2)$ to (p^μ,θ) where p^μ three-dimensional momentum for CFT and θ is an angle variable conjugate to the spin s.

First, the spectrum of the diagonal modes (See Eq. (24) and Eq. (26)) reads

$\alpha^\dagger(\vec{p}_1,\vec{p}_2)$	$p^0 = \lvert\vec{p}_1\rvert + \lvert\vec{p}_2\rvert$ $\vec{p} = \vec{p}_1 + \vec{p}_2$
$\widetilde{\alpha}^\dagger(\vec{p}_1,\vec{p}_2)$	$p^0 = -\lvert\vec{p}_1\rvert - \lvert\vec{p}_2\rvert$ $\vec{p} = -\vec{p}_1 - \vec{p}_2$

In fact, one can easily expect this spectrum from the momentum map because it map the three-dimensional momentum p^μ to the center of momentum of $(\vec{p}_1,\vec{p}_2)$.

From the spectrum, it is easy to see that these oscillators create a state with timelike momentum

$$(p^0)^2 - \vec{p}^2 \geqq 0. \tag{27}$$

On the other hand, the off-diagonal modes $\gamma^\dagger(\vec{p}_1, \vec{p}_2)$ has the following spectrum.

$\gamma^\dagger(\vec{p}_1, \vec{p}_2)$	$p^0 = \lvert\vec{p}_1\rvert - \lvert\vec{p}_2\rvert$ $\vec{p} = \vec{p}_1 - \vec{p}_2$

In this case, $\gamma^\dagger$ is associated with the mixed composite operator $\phi^i\widetilde{\phi}^i$, and the opposite time directions of them leads to such a spectrum. It is crucial to observe that $\gamma^\dagger$ is characterized by spacelike momentum

$$(p^0)^2 - \vec{p}^2 \leqq 0. \tag{28}$$

Those bi-local normal modes generate spin-s primary operators in CFT$_3$. This was explicitly shown in the previous work at zero temperature. For the case of the collective TFD of the $O(N)$ vector model, one can also construct analogous spin-s primary operators:

$$\begin{aligned} &\mathcal{O}_s^{ab}(x;\zeta) \\ &\equiv \sum_{n=0}^{[s/2]}\sum_{k=0}^{s-2n} \frac{(-4)^n(-1)^k}{\sqrt{N}(2n)!k!(s-2n-k)!} :(\zeta\cdot\partial)^{n+k}\phi^i_{(a)}(x)\;(\zeta\cdot\partial)^{s-n-k}\phi^i_{(b)}(x): \end{aligned} \tag{29}$$

where $\phi^i_{(1)}(x) \equiv \phi^i(x)$ and $\phi^i_{(2)}(x) \equiv \widetilde{\phi}^i(x)$ and ζ^μ is an auxiliary variable to package spacetime indices μ's of the spin-s operator. Here we have to stress that the definition of Left and Right CFT operators is given respectively by

$$\mathcal{O}_s \equiv \mathcal{O}_s^{11} \quad , \quad \widetilde{\mathcal{O}}_s \equiv \mathcal{O}_s^{22}. \tag{30}$$

We also emphasize that these CFT operators are normal-ordered not with respect to the Right and Left vacua $|0\rangle$ and $|\widetilde{0}\rangle$ but with respect to the thermal vacuum $|0(\beta)\rangle$. Because of the Bogoliubov transformation, the spin-s operators schematically become

$$\mathcal{O}_s \sim \underbrace{\alpha + \widetilde{\alpha}^\dagger}_{\text{timelike}} + \underbrace{\gamma + \gamma^\dagger}_{\text{spacelike}} \quad , \quad \widetilde{\mathcal{O}}_s \sim \underbrace{\widetilde{\alpha} + \alpha^\dagger}_{\text{timelike}} + \underbrace{\gamma + \gamma^\dagger}_{\text{spacelike}} \tag{31}$$

where the explicit form is given in the previous work [26]. Due to the non-trivial normal ordering with respect to the thermal vacuum, the spacelike momentum modes $\gamma^\dagger$ appears in the spin-s operators as spacelike momentum. To see this, one has to solve the conservation and the traceless conditions to get a physical spin-s operator. The solution for those condition depends whether the momentum of spin-s operator is timelike or spacelike. For timelike momentum, the spin-s operator $\mathcal{O}_s^{ab}$ with timelike momentum is expressed as a linear combination of α and $\widetilde{\alpha}$ (and their conjugates) which also have timelike momentum. The solution for spacelike momentum enables us to express the spin-s operator with spacelike momentum in terms of γ and $\gamma^\dagger$. It is important to note that a spin-s operator with timelike momentum commutes with one with spacelike momentum. Also, $\mathcal{O}$ with timelike momentum (spacelike momentum, respectively) commutes with $\widetilde{\mathcal{O}}$ (and the mixed operator $\mathcal{O}^{12}$). However, operators of the same type of momentum with different spins do not commute. For example,

$$\left[\mathcal{O}_s(p), \mathcal{O}^\dagger_{s'}(q)\right] = \mathbb{K}^{s,s'}_{\beta,p}\delta^{(3)}(p^\mu - q^\mu). \tag{32}$$

Since their commutation relation is a matrix in the spin space, one can diagonalize the commutation relation in spin space to define canonical operators $\mathcal{A}_s(p)$:

$$\mathcal{A}_s \sim \sum_{s'} (\mathbb{K}^{-\frac{1}{2}}_{\beta,p})^{s,s'}\mathcal{O}_{s'} \tag{33}$$

and similar for $\widetilde{\mathcal{A}}_s(p)$. Then, the canonical operators satisfy

$$\left[\mathcal{A}_s(k), \mathcal{A}^\dagger_{s'}(q)\right] = \delta_{s,s'}\delta^{(3)}(k-q) \quad , \quad \left[\widetilde{\mathcal{A}}_s(k), \widetilde{\mathcal{A}}^\dagger_{s'}(q)\right] = \delta_{s,s'}\delta^{(3)}(k-q)$$
$$\left[\mathcal{A}_s(k), \widetilde{\mathcal{A}}_{s'}(q)\right] = \left[\mathcal{A}_s(k), \widetilde{\mathcal{A}}^\dagger_{s'}(q)\right] = 0 \tag{34}$$

and others vanish. It is central that all the commutators between the dual pairs vanish. Note that we considered $\mathcal{O}_s$ and $\widetilde{\mathcal{O}}_s$ in defining the canonical operator. This is the second central features of the construction that it is sufficient to consider spin-s operator $\mathcal{O}_s$ and $\widetilde{\mathcal{O}}_s$ for the bi-local normal modes because α's and $\widetilde{\alpha}$'s are packaged into the timelike modes of $\mathcal{O}_s$ and $\widetilde{\mathcal{O}}_s$ while γ's are packaged into the spacelike modes. In the next section, we will give bulk interpretation. In the bulk interpretation, the timelike modes are associated with propagating modes in the bulk while the spacelike mode corresponds to evanescent modes which exponentially decay in the bulk.

4. Bulk Interpretation

A map from bi-local fields to the higher spin field in AdS background at zero temperature was fully constructed:

$$\vec{p} = \vec{p}_1 + \vec{p}_2 \tag{35}$$

$$p^z = 2\sqrt{|\vec{p}_1||\vec{p}_2|}\sin\left(\frac{\varphi_1 - \varphi_2}{2}\right) \tag{36}$$

$$\theta = \arctan\left(\frac{2\vec{p}_2 \times \vec{p}_1}{(|\vec{p}_2| - |\vec{p}_1|)p^z}\right) \tag{37}$$

where $\vec{p}_i = (|\vec{p}_i|\cos\varphi_i, |\vec{p}_i|\sin\varphi_i)$. Furthermore, this map also induce a explicit relation between spin-s operator on the boundary and the higher spin field in the bulk

$$\mathcal{H}^{(\pm)}_s(t;\vec{x},z) = C_{s,\pm}\int_{p^0>|\vec{p}|}\frac{d^2\vec{p}dp^0}{(2\pi)^3 2p^0}e^{-ip^0t+i\vec{p}\cdot\vec{x}}z^{\frac{1}{2}}\left[(p^0)^2 - \vec{p}^2\right]^{-\frac{s}{2}+\frac{1}{4}}$$
$$\times J_{\mp\frac{1}{2}}(\sqrt{(p^0)^2 - \vec{p}^2}\, z)\left[\mathcal{O}_s(p;\epsilon) \pm \mathcal{O}_s(p;\epsilon^*)\right] + \text{h.c} \tag{38}$$

where $(\pm)$ represents polarization of the bulk higher spin field.

At finite temperature, the TFD of CFT was suggested to generate a space-time similar to the eternal black hole which asymptotically approaches AdS_4. The fluctuation modes found in the linearized collective analysis are expected to fully reproduce the bulk modes of higher spin in the modified space-time. Using the knowledge of the AdS/bi-local CFT map, we can work out their properties.

In applying the bi-local map in Eq. (35)∼Eq. (37) to the collective TFD, a canonical spin-s CFT operator is required because the bi-local map as a canonical transformation guarantees the canonical structure of the bulk field. Therefore, we apply the map to the canonical operator $\mathcal{A}_s$ and $\widetilde{\mathcal{A}}_s$ rather than $\mathcal{O}_s$ and $\widetilde{\mathcal{O}}_s$. Also, in the zero temperature case, the integration in Eq. (38) runs over timelike momentum because the spin-s operator (equivalently, the bi-local normal mode) at zero temperature CFT has only timelike momentum. However, since the spin-s operator contains spacelike modes γ's at finite temperature, the integration should run over all momentum space. Finally, note that the map in Eq. (35)∼Eq. (37) induces the on-shell condition of bulk field:

$$(p^z)^2 = (p^0)^2 - \vec{p}^2 \tag{39}$$

where p^z is a canonical conjugate to the radial coordinate z. This implies that $\mathcal{A}_s$ and $\widetilde{\mathcal{A}}$ with timelike mode generates a propagating bulk mode (real p^z) while those with spacelike momentum produces an evanescent mode in the bulk which decays exponentially along the z direction (imaginary p^z). This is an extra mode compare to the zero temperature AdS/CFT. One could formally interpret the occurrence of spacelike modes of $\mathcal{A}_s$ and $\widetilde{\mathcal{A}}_s$ as an analytic continuation[b] of timelike modes characteristic of zero-temperature AdS/CFT. At finite temperature, the analytic continuation enlarges the momentum space of operators to include the evanescent modes. The existence of evanescent modes indicates that the effective potential $V(z)$ (in the language of [36]) is different from AdS case, and suggest horizon in the bulk. The information on the bulk geometry is require to derive the full bi-local map, which would define the wave functions for each mode in the bulk.

In the eternal black hole, one has four regions, I∼IV where region I and region III are right and left outside of the black hole and region II (or, region IV) is inside of the black hole containing future (or, past) horizon, respectively (see the earlier work [25, 47]). Based on the above observation, we are now construct higher spin fields from the canonical operators $\mathcal{A}_s$ and $\widetilde{\mathcal{A}}_s$ as follows. In the region I and region III, we have

$$\mathcal{H}_s^{\mathrm{I}}(t,\vec{x},z) \equiv \int_{k^0>0} \frac{dk^0 d\vec{p}}{(2\pi)^3} \left[f_{s,k}(t,\vec{x},z)\mathcal{A}_s(k) + f^*_{s,k}(t,\vec{x},z)\mathcal{A}^\dagger_s(k)\right] \tag{40}$$

$$\mathcal{H}_s^{\mathrm{III}}(t,\vec{x},z) \equiv \int_{k^0>0} \frac{dk^0 d\vec{p}}{(2\pi)^3} \left[f^*_{s,k}(t,\vec{x},z)\widetilde{\mathcal{A}}_s(k) + f_{s,k}(t,\vec{x},z)\widetilde{\mathcal{A}}^\dagger_s(k)\right] \tag{41}$$

where $\widetilde{\mathcal{A}}^\dagger_s(k)$ creates a mode $f_k(t,\vec{x},z)$ because the time direction in the region III

[b] We thank Soo-Jong Rey for this interpretation.

is opposite to that of the region I. Moreover, the higher spin field inside of the black hole is

$$\mathcal{H}_s^{\text{II}}(t,\vec{x},z) \equiv \int_{k^0>0} \frac{dk^0 d\vec{p}}{(2\pi)^3} \left[f_{s,k}^{(1)}(t,\vec{x},z)\mathcal{A}_s(k) + f_{s,k}^{(2)}(t,\vec{x},z)\widetilde{\mathcal{A}}_s^\dagger(k) + h.c. \right] \tag{42}$$

where $f_{s,k}^{(1)}(t,\vec{x},z)$ and $f_{s,k}^{(2)}(t,\vec{x},z)$ are two linearly independent solutions in the region II.

5. Rindler Higher Spin Gravity

Rindler-AdS spacetime serves as a particularly simple example of the "Thermofield double" construction of asymptotically AdS spacetime with an even horizon [48, 49]. In this section, as a concrete example of the Rindler-AdS/CFT correspondence, we employ the minimal bosonic higher spin and free $O(N)$ vector model duality [27].

Let us first give some basic aspects of Rindler-AdS spacetime and its relation to AdS black holes. Asymptotically AdS space admits three kinds of the static back hole solutions [50]. These three solutions are characterized by the curvature of the horizons (transverse spacial coordinates). For asymptotically AdS_{d+1} space, the geometry of the horizons of these black holes are (S^{d-1}, $\mathbb{R}^{d-1}$, H_{d-1}). Each of these black hole solutions is called "AdS-Schwarzschild", "AdS-Planar", and "AdS-Hyperbolic" black hole. Massless limit of these black hole solutions correspond to the global, Poincare, and Rindler coordinates of AdS spacetime, respectively. The geometry of the spatial infinity of these AdS black holes are given by $\mathbb{R} \times S^{d-1}$, $\mathbb{R}^{1,d-1}$, and $\mathbb{R} \times H^{d-1}$, where the corresponding pair of dual CFT's is defined [50].

In this section, we are particularly interested in 4-dimensional Rindler-AdS spacetime, whose metric is given by

$$ds^2 = \frac{1}{\rho^2}\left[-(1-\rho^2)d\tau^2 + \frac{d\rho^2}{1-\rho^2} + \frac{dx^2+d\sigma^2}{\sigma^2} \right]. \tag{43}$$

The range of each coordinate is

$$-\infty < \tau,\, x < \infty\,, \qquad 0 < \rho < 1 \qquad \sigma > 0\,. \tag{44}$$

Note that the boundary of this asymptotically AdS space is located at $\rho = 0$, while the Rindler horizon is at $\rho = 1$. One can immediately see from the metric that the boundary geometry is given by $\mathbb{R} \times H^2$.

One of the important characters of the black hole background is the presence of the so called "evanescent modes" [32, 36]. It was described that the evanescent modes exist in the AdS-Schwarzschild black hole background and in its infinite volume limit, AdS-Planar black hole background from the discussion of effective potentials [36]. The existence of the evanescent modes in Rindler-AdS background was suggested in the discussion of smearing function [32, 51], and we can also show this from the effective potential of scalar field in Rindler-AdS as follows [27]. The time and transverse coordinate dependence parts of the equation of motion for a conformally coupled scalar field Φ in 4d Rindler-AdS (Eq. (43)) can be solved by

$$\Phi(\tau,x,\sigma,\rho) = e^{-i\omega\tau - ik_x x}\,\sigma^{\frac{1}{2}} K_{i\nu}(|k_x|\sigma)\,\rho\, u(\rho)\,, \tag{45}$$

where $K_{i\nu}$ is the second kind of the modified Bessel function with a real-valued parameter ν. Here, we have also excluded the boundary scaling factor of the scalar field. Moving to the tortoise coordinate ρ_* defined by $d\rho_* = (1-\rho^2)^{-1}d\rho$, the equation of motion for $u(\rho_*)$ is reduced to a Schrödinger-like equation,

$$\frac{d^2u(\rho_*)}{d\rho_*^2} + \Big(\omega^2 - V(\rho_*)\Big)\, u(\rho_*) = 0\,, \tag{46}$$

with the effective potential

$$V(\rho_*) = \frac{\bar{\nu}^2}{\cosh^2(\rho_*)}\,, \tag{47}$$

where $\bar{\nu} = \sqrt{1/4+\nu^2}$. In the tortoise coordinate ρ_*, the AdS boundary is located at $\rho_* = 0$, while the Rindler horizon is at $\rho_* = \infty$. The effective potential becomes zero near horizon $\rho_* \to \infty$. Accordingly, evanescent modes ($\omega^2 \sim 0$) are allowed to exist.

Next, we move on to defining the dual CFT of Rindler-AdS higher spin theory. As we mentioned before, the boundary limit of the Rindler-AdS metric (Eq. (43)) leads to hyperbolic geometry $\mathbb{R}\times H^2$. Hence, the corresponding $O(N)$ vector model field ϕ_i^H is defined on this geometry. The index i of the field is the $O(N)$ index as $i = 1, \cdots, N$. This is absolutely correct; however, it is more convenient to move to the regular Rindler coordinates rather than the hyperbolic space. Namely, the theory is conformally invariant and the metric of $\mathbb{R}\times H^2$ is by the Weyl transformation related to the regular Rindler metric

$$ds^2 = -\sigma^2\, d\tau^2 + dx^2 + d\sigma^2\,. \tag{48}$$

Then, the dual CFT is $O(N)$ vector model defined in the regular-Rindler coordinates (Eq. (48)). The action is explicitly given by

$$S = \frac{1}{2}\int d\tau dx d\sigma \left[\frac{1}{\sigma}\,(\partial_\tau\phi_i^R)^2 - \sigma(\partial_x\phi_i^R)^2 - \sigma(\partial_\sigma\phi_i^R)^2\right]. \tag{49}$$

One can also show that the regular-Rindler $O(N)$ vector field ϕ_i^R is by field redefinition related to the original hyperbolic vector field as

$$\phi_i^H = \sqrt{\sigma}\,\phi_i^R\,. \tag{50}$$

The appropriately normalized solution for the equation of motion of the action, Eq. (49) is given

$$\phi_i^R(\tau,x,\sigma) = \int_0^\infty d\omega \int_{-\infty}^\infty dk\Big[\, b_i^R(\omega,k)\, g_{\omega,k}^R(\tau,x,\sigma) + \text{h.c.}\,\Big]\,, \tag{51}$$

where

$$g_{\omega,k}^R(\tau,x,\sigma) = \frac{1}{2\pi\sqrt{2\omega}}\,\frac{2K_{i\omega}(|k|\sigma)}{|\Gamma(i\omega)|}\, e^{ikx-i\omega\tau}\,. \tag{52}$$

After discussing Rindler-AdS spacetime and corresponding Rindler-CFT, we finally would like to construct a canonical transformation between the canonical variables of Rindler-AdS higher spin theory and the bi-local regular-Rindler $O(N)$ vector model. The singlet constraint Eq. (15) implies that we have three kinds of the bi-local collective oscillators

$$\begin{aligned}\alpha_{RR}(\vec{p}_1,\vec{p}_2) &= \sum_i b_i^R(\vec{p}_1)b_i^R(\vec{p}_2)\,,\\ \alpha_{LL}(\vec{p}_1,\vec{p}_2) &= \sum_i b_i^L(\vec{p}_1)b_i^L(\vec{p}_2)\,,\\ \gamma_{RL}(\vec{p}_1,\vec{p}_2) &= \sum_i b_i^R(\vec{p}_1)b_i^L(\vec{p}_2)\,,\end{aligned} \tag{53}$$

where $b_i^{R(L)}(\vec{p})$ are the oscillators of the right (left) boundary vector model.

In order to construct the $SO(2,3)$ AdS global generators and bi-local map in terms of Rindler-AdS coordinates, it is important to have an explicit canonical transformations between Poincare and Rindler coordinates of AdS_4 spacetime. This is because the global generators and bi-local map are already constructed in terms of the Poincare coordinates of AdS_4 [14]. The momentum part of the bi-local map is given in Eq. (35)$\sim$Eq. (37). Therefore, we expect that it is possible to construct these in Rindler-AdS just by using a canonical transformation from Poincare to Rindler coordinates of AdS. Indeed, this is the case.

We use the embedding space of AdS to derive a canonical transformation between Poincare and Rindler coordinates. It is well known that AdS_{d+1} space can be embedded into $(d+2)$-dimensional flat space $\mathbb{R}^{(2,d)}$. Equating parameterizations for the Poincare and Rindler coordinates of AdS_4, one can find explicit transformations from one to the other. The transformations from the Rindler-AdS coordinates (τ, x, σ, ρ) to the Poincare coordinates (t, x, y, z) are given by

$$t = \sigma\sqrt{1-\rho^2}\,\sinh\tau\,, \qquad x = x\,, \qquad y = \sigma\sqrt{1-\rho^2}\,\cosh\tau\,, \qquad z = \rho\sigma\,, \tag{54}$$

and also the inverse transformations can be found explicitly. Now that we know the explicit coordinate transformations between the Poincare and the Rindler coordinates, we can also write down the transformations in the momentum spaces. In this paper, we use the first-quantized particle picture. In that case, the momentum is just derivatives of its coordinates, so by chain rule, we can find an explicit transformations for the Poincare-coordinate momenta from Rindler-AdS canonical variables. Again, the inverse transformations are also explicitly found. One can check that the canonical commutation relations are preserved under this canonical transformation. Therefore, this transformation is indeed canonical transformation.

Now that we have obtained a canonical transformation from Poincare AdS canonical variables to Rindler-AdS canonical variables, we can transform the Poincare AdS $SO(2,3)$ generators and bi-local map to the Rindler-AdS counterparts of them. Without giving every formula of this bi-local map explicitly, we explain three important features of this bi-local map.

First, from the bi-local map one finds that the diagonal bi-local oscillators $\alpha_{RR}(\vec{p}_1,\vec{p}_2)$ and $\alpha_{LL}(\vec{p}_1,\vec{p}_2)$ can produce only real-valued AdS-radial direction momentum p^ρ.

$$\alpha^\dagger_{RR}(\vec{p}_1,\vec{p}_2): \quad p^\rho = 2\sigma\sqrt{|\vec{p}_1||\vec{p}_2|}\,\sin\left(\frac{\varphi_1-\varphi_2}{2}\right) - \frac{\sigma\rho}{\sqrt{1-\rho^2}}(p_1^\sigma + p_2^\sigma), \quad (55)$$

$$\alpha^\dagger_{LL}(\vec{p}_1,\vec{p}_2): \quad p^\rho = 2\sigma\sqrt{|\vec{p}_1||\vec{p}_2|}\,\sin\left(\frac{\varphi_1-\varphi_2}{2}\right) + \frac{\sigma\rho}{\sqrt{1-\rho^2}}(p_1^\sigma + p_2^\sigma), \quad (56)$$

where ρ and σ are bi-local parameters which written in terms of the canonical variables of the bi-local $O(N)$ Rindler vector model. They take real values for α_{RR} and α_{LL}. Here we have also defined $\vec{p}_1 = (p_1^x, p_1^\sigma)$, $\vec{p}_2 = (p_2^x, p_2^\sigma)$, $\varphi_1 = \arctan(p_1^\sigma/p_1^x)$, $\varphi_2 = \arctan(p_2^\sigma/p_2^x)$. From this observation, we conclude that these diagonal oscillators α_{RR}, α_{LL} create only propagating modes, and cannot create evanescent modes, which characterized by an imaginary radial-direction momentum, anywhere in the bulk Rindler-AdS space. On the other hand, the off-diagonal bi-local oscillator $\gamma_{RL}(\vec{p}_1,\vec{p}_2)$ in general produce a complex-valued radial direction momentum.

$$\gamma^\dagger_{RL}(\vec{p}_1,\vec{p}_2): \quad p^\rho = 2i\sigma\sqrt{|\vec{p}_1||\vec{p}_2|}\,\sin\left(\frac{\varphi_1-\varphi_2}{2}\right) - \frac{\sigma\rho}{\sqrt{1-\rho^2}}(p_1^\sigma - p_2^\sigma). \quad (57)$$

One should notice the pure imaginary factor i in the first term. Also for γ_{RL}, the bi-local parameter ρ and σ in general take complex values. Therefore, states created by this off-diagonal oscillator have exponentially growing amplitude, which we can identify as an evanescent mode.

Secondly, we discuss "quasinormal modes", which represents energy dissipation of a perturbed event horizon. It was shown that in four and higher dimensional Rindler-AdS space (massless topological black hole), quasinormal modes exist and the solutions can be exactly obtainable [52]. The quasinormal frequency takes complex value. In fact, the bi-local map is able to produce complex valued frequencies. In our notation, p^τ is the frequency of Rindler-AdS field. From the bi-local map, we find that the frequency produced by $\alpha^\dagger_{RR}$ or $\alpha^\dagger_{LL}$ is always real. On the other hand, states produced by $\gamma^\dagger_{RL}$ have in general complex valued frequencies which contain quasinormal modes, while $\alpha^\dagger_{RR}$ and $\alpha^\dagger_{LL}$ produce only real valued frequencies corresponding to propagating modes. The appearance of this complex frequency indeed comes from the coordinate dependence part of the bi-local map for p^τ. This coordinate mixing does not happen in the pure AdS case [13, 14], where momentum transformations for the pure AdS space are completely given in terms of bi-local $O(N)$ vector model momenta. See Eq. (35)~Eq. (37). Therefore, we obtain only real valued frequencies for the pure AdS background. The coordinate dependence part ($\sqrt{1-\rho^2}$) of the bi-local map for p^τ implies the presence of the event horizon ($\rho = 1$); this agrees with the physical meaning of the quasinormal modes.

Finally, we briefly note that, if we have only the diagonal oscillators $\alpha^\dagger_{RR}$, $\alpha^\dagger_{LL}$, the bi-local map for the Rindler-AdS coordinates can produce only the region outside of the Rindler horizon $0 < \rho < 1$. However, the entangled oscillator $\gamma^\dagger_{RL}$ can naturally produce the inside of the horizon $\rho > 1$. These three observations imply that the mixed bi-local field γ plays a central role in the construction: it secures the connectedness of Rindler-AdS spacetime.

In conclusion, we have emphasized in our construction the framework of "diagonal" gauging of $O(N)$ or $U(N)$. We have also seen that with the extra 'mixed' degrees of freedom one can reconstruct the modes characteristic of a space-time with horizon. Therefore even though the two (Left and Right) CFT's of the TFD are totally decoupled (and the Hamiltonian represents a sum of the two), there is an implicit "mixing" through the Gauss's law constraint. We find this to be needed for full reconstruction of the dual space-time. Namely, the diagonal gauging was central to the possibility of reconstructing a connected space-time. Since this feature does not seem to have been a part in the construction of earlier works, and in particular the construction of [53] we would like to provide one further observation. The 'extra' mixed degrees of freedom were seen to provide the additional space-like modes as featured in Eq. (32). One can formally therefore consider a decoupled set of Left-Right CFT operators and assume that through an analytic continuation one acquires the necessary additional degrees of freedom. Our construction shows however that the support for these comes from states associated with the 'mixed' Hilbert space.

Acknowledgments

We are grateful to Sumir Mathur and Soo-Jong Rey for usefull discussions and most relevant comments.

References

[1] A. Almheiri, D. Marolf, J. Polchinski and J. Sully, *Black Holes: Complementarity or Firewalls?*, *JHEP* **1302** (2013) 062 [arXiv:1207.3123 [hep-th]].

[2] B. Sundborg, *The Hagedorn transition, deconfinement and N=4 SYM theory*, *Nucl. Phys.* B **573** (2000) 349 [arXiv:hep-th/9908001].

[3] I. R. Klebanov and A. M. Polyakov, *AdS dual of the critical O(N) vector model*, *Phys. Lett.* B **550** (2002) 213 [arXiv:hep-th/0210114].

[4] E. Sezgin and P. Sundell, *Massless higher spins and holography*, *Nucl. Phys.* B **644** (2002) 303 [Erratum-ibid. B **660** (2003) 403] [arXiv:hep-th/0205131].

[5] M. A. Vasiliev, *Consistent equation for interacting gauge fields of all spins in (3+1)-dimensions*, *Phys. Lett.* B **243** (1990) 378.

[6] M. A. Vasiliev, *Higher spin gauge theories in four-dimensions, three-dimensions, and two-dimensions*, *Int. J. Mod. Phys.* D **5** (1996) 763 [arXiv:hep-th/9611024].

[7] X. Bekaert, S. Cnockaert, C. Iazeolla and M. A. Vasiliev, *Nonlinear higher spin theories in various dimensions*, [arXiv:hep-th/0503128].

[8] S. Giombi and X. Yin, *Higher Spin Gauge Theory and Holography: The Three-Point Functions*, *JHEP* **1009** (2010) 115 [arXiv:0912.3462 [hep-th]].

[9] S. Giombi and I. R. Klebanov, *One Loop Tests of Higher Spin AdS/CFT*, *JHEP* **1312** (2013) 068 [arXiv:1308.2337 [hep-th]].

[10] A. Jevicki, K. Jin and J. Yoon, *1/N and loop corrections in higher spin AdS_4/CFT_3 duality*, *Phys. Rev.* D **89** (2014) 085039 [arXiv:1401.3318 [hep-th]].

[11] S. Giombi, I. R. Klebanov and A. A. Tseytlin, *Partition Functions and Casimir Energies in Higher Spin AdS_{d+1}/CFT_d*, *Phys. Rev.* D **90** (2014) 024048 [arXiv:1402.5396 [hep-th]].
[12] S. R. Das and A. Jevicki, *Large N collective fields and holography*, *Phys. Rev. D* **68** (2003) 044011 [arXiv:hep-th/0304093].
[13] R. d. M. Koch, A. Jevicki, K. Jin and J. P. Rodrigues, *AdS_4/CFT_3 Construction from Collective Fields*, *Phys. Rev.* D **83** (2011) 025006 [arXiv:1008.0633 [hep-th]].
[14] R. d. M. Koch, A. Jevicki, J. P. Rodrigues and J. Yoon, *Canonical Formulation of O(N) Vector/Higher Spin Correspondence*, [arXiv:1408.4800 [hep-th]].
[15] D. Das, S. R. Das, A. Jevicki and Q. Ye, *Bi-local Construction of Sp(2N)/dS Higher Spin Correspondence*, *JHEP* **1301** (2013) 107 [arXiv:1205.5776 [hep-th]].
[16] E. Mintun and J. Polchinski, *Higher Spin Holography, RG, and the Light Cone* [arXiv:1411.3151 [hep-th]].
[17] R. G. Leigh, O. Parrikar and A. B. Weiss, *Exact renormalization group and higher-spin holography*, *Phys. Rev.* D **91** (2015) 026002 [arXiv:1407.4574 [hep-th]].
[18] R. de Mello Koch, A. Jevicki, J. P. Rodrigues and J. Yoon, *Holography as a Gauge Phenomenon in Higher Spin Duality*, *JHEP* **1501** (2015) 055 [arXiv:1408.1255 [hep-th]].
[19] H. J. Schnitzer, *Confinement/deconfinement transition of large N gauge theories with N(f) fundamentals: N(f)/N finite*, *Nucl. Phys.* B **695** (2004) 267 [arXiv:hep-th/0402219].
[20] S. H. Shenker and X. Yin, *Vector Models in the Singlet Sector at Finite Temperature*, (2011) [arXiv:1109.3519 [hep-th]].
[21] H. J. Schnitzer, *Confinement/deconfinement transition of large N gauge theories in perturbation theory with N(f) fundamentals: N(f)/N finite,* (2006) [arXiv:hep-th/0612099].
[22] O. Aharony, S. Giombi, G. Gur-Ari, J. Maldacena and R. Yacoby, *The Thermal Free Energy in Large N Chern-Simons-Matter Theories*, *JHEP* **1303** (2013) 121 [arXiv:1211.4843 [hep-th]].
[23] J. S. Schwinger, *Brownian motion of a quantum oscillator*, *J. Math. Phys.* **2** (1961) 407.
[24] L. V. Keldysh, *Diagram technique for nonequilibrium processes*, *Zh. Eksp. Teor. Fiz.* **47** (1964) 1515 [*Sov. Phys. JETP* **20** (1965) 1018].
[25] C. P. Herzog and D. T. Son, *Schwinger-Keldysh propagators from AdS/CFT correspondence*, emphJHEP **0303** (2003) 046 [arXiv:hep-th/0212072].
[26] A. Jevicki and J. Yoon, *Bulk from Bi-locals in Thermo Field CFT, JHEP* **1602** (2016) 090 [arXiv:1503.08484 [hep-th]].
[27] A. Jevicki and K. Suzuki, *Thermofield Duality for Higher Spin Rindler Gravity, JHEP* **1602** (2016) 094 [arXiv:1508.07956 [hep-th]].
[28] J. M. Maldacena, *Eternal black holes in anti-de Sitter*, *JHEP* **0304** (2003) 021 [arXiv:hep-th/0106112].
[29] S. D. Mathur, *What is the dual of two entangled CFTs?* (2014) [arXiv:1402.6378 [hep-th]].
[30] S. G. Avery and B. D. Chowdhury, *No Holography for Eternal AdS Black Holes* (2013) [arXiv:1312.3346 [hep-th]].
[31] A. Hamilton, D. N. Kabat, G. Lifschytz and D. A. Lowe, *Local bulk operators in AdS/CFT: A Holographic description of the black hole interior*, *Phys. Rev. D* **75** (2007) 106001 [Erratum-ibid. *D* **75** (2007) 129902] [arXiv:hep-th/0612053].
[32] R. Bousso, B. Freivogel, S. Leichenauer, V. Rosenhaus and C. Zukowski, *Null Geodesics, Local CFT Operators and AdS/CFT for Subregions*, *Phys. Rev.* D **88**

(2013) 064057 [arXiv:1209.4641 [hep-th]].
[33] A. Hamilton, D. N. Kabat, G. Lifschytz and D. A. Lowe, *Local bulk operators in AdS/CFT: A Boundary view of horizons and locality, Phys. Rev. D* **73** (2006) 086003 [arXiv:hep-th/0506118].
[34] K. Papadodimas and S. Raju, *State-Dependent Bulk-Boundary Maps and Black Hole Complementarity, Phys. Rev. D* **89** (2014) 086010 [arXiv:1310.6335 [hep-th]].
[35] S. Leichenauer and V. Rosenhaus,*AdS black holes, the bulk-boundary dictionary, and smearing functions, Phys. Rev.* D **88** (2013) 026003 [arXiv:1304.6821 [hep-th]].
[36] S. J. Rey and V. Rosenhaus, *Scanning Tunneling Macroscopy, Black Holes, and AdS/CFT Bulk Locality, JHEP* **1407** (2014) 050 [arXiv:1403.3943 [hep-th]].
[37] V. E. Didenko and M. A. Vasiliev, *Static BPS black hole in 4d higher-spin gauge theory, Phys. Lett.* B **682** (2009) 305 [*Phys. Lett.* B **722** (2013) 389] [arXiv:0906.3898 [hep-th]].
[38] C. Iazeolla and P. Sundell, *Families of exact solutions to Vasiliev's 4D equations with spherical, cylindrical and biaxial symmetry, JHEP* **1112** (2011) 084 [arXiv:1107.1217 [hep-th]].
[39] P. Kraus and E. Perlmutter, *Partition functions of higher spin black holes and their CFT duals, JHEP* **1111** (2011) 061 [arXiv:1108.2567 [hep-th]].
[40] M. R. Gaberdiel, T. Hartman and K. Jin, *Higher Spin Black Holes from CFT, JHEP* **1204** (2012) 103 [arXiv:1203.0015 [hep-th]].
[41] O. Aharony, J. Marsano, S. Minwalla, K. Papadodimas and M. Van Raamsdonk, *The Hagedorn - deconfinement phase transition in weakly coupled large N gauge theories, Adv. Theor. Math. Phys.* **8** (2004) 603 [arXiv:hep-th/0310285].
[42] F. A. Berezin, *Models of Gross-Neveu Type as Quantization of Classical Mechanics With Nonlinear Phase Space, Commun. Math. Phys.* **63** (1978) 131.
[43] W. Israel, *Thermo field dynamics of black holes, Phys. Lett.* A **57** (1976) 107.
[44] Y. Takahashi and H. Umezawa, *Thermo field dynamics, Int. J. Mod. Phys.* B **10** (1996) 1755 [*Collective Phenomena* **2** (1975) 55].
[45] A. J. Niemi and G. W. Semenoff, *Finite Temperature Quantum Field Theory in Minkowski Space, Annals Phys.* **152** (1984) 105.
[46] F. A. Berezin, *Quantization in complex symmetric spaces, Math.USSR Izv.*, 9 (1975) 341.
[47] P. Kraus, H. Ooguri and S. Shenker, *Inside the horizon with AdS / CFT, Phys. Rev.* D **67** (2003) 124022 [arXiv:hep-th/0212277].
[48] B. Czech, J. L. Karczmarek, F. Nogueira and M. Van Raamsdonk, *Rindler Quantum Gravity, Class. Quant. Grav.* **29**, 235025 (2012) [arXiv:1206.1323 [hep-th]].
[49] M. Parikh and P. Samantray, *Rindler-AdS/CFT,* (2012) [arXiv:1211.7370 [hep-th]].
[50] R. Emparan, *AdS / CFT duals of topological black holes and the entropy of zero energy states, JHEP* **9906** (1999) 036 [arXiv:9906040[hep-th]].
[51] A. Hamilton, D. N. Kabat, G. Lifschytz and D. A. Lowe, *Holographic representation of local bulk operators, Phys. Rev.* D **74** (2006) 066009 [arXive:hep-th/0606141].
[52] R. Aros, C. Martinez, R. Troncoso and J. Zanelli, *Quasinormal modes for massless topological black holes, Phys. Rev.* D **67** (2003) 044014 [arXive:hep-th/0211024].
[53] K. Papadodimas and S. Raju, *An Infalling Observer in AdS/CFT, JHEP* **1310** (2013) 212 [arXiv:1211.6767 [hep-th]].
[54] S. D. Mathur, *The information paradox: conflicts and resolutions, Pramana* **79** (2012) 1059 [arXiv:1201.2079 [hep-th]].
[55] D. Marolf and J. Polchinski, *Gauge/Gravity Duality and the Black Hole Interior, Phys. Rev. Lett.* **111** (2013) 171301 [arXiv:1307.4706 [hep-th]].

[56] M. Van Raamsdonk, *Building up spacetime with quantum entanglement*, *Gen. Rel. Grav.* **42** (2010) 2323 [*Int. J. Mod. Phys. D* **19** (2010) 2429] [arXiv:1005.3035 [hep-th]].
[57] J. Maldacena and L. Susskind, *Cool horizons for entangled black holes*, *Fortsch. Phys.* **61** (2013) 781 [arXiv:1306.0533 [hep-th]].
[58] M. Beccaria and A. A. Tseytlin, *Vectorial AdS_5/CFT_4 duality for spin-one boundary theory*, *J. Phys.* A **47** (2014) 492001 [arXiv:1410.4457 [hep-th]].
[59] G. W. Semenoff and H. Umezawa, *Functional Methods in Thermo Field Dynamics: A Real Time Perturbation Theory for Quantum Statistical Mechanics*, *Nucl. Phys.* B **220** (1983) 196.
[60] R. de Mello Koch, A. Jevicki, K. Jin, J. P. Rodrigues and Q. Ye, *S=1 in O(N)/HS duality*, *Class. Quant. Grav.* **30** (2013) 104005 [arXiv:1205.4117 [hep-th]].
[61] E. Mintun, J. Polchinski and V. Rosenhaus, *Bulk-Boundary Duality, Gauge Invariance, and Quantum Error Correction*, [arXiv:1501.06577 [hep-th]].

On (Un)Broken Higher-Spin Symmetry in Vector Models

E. D. Skvortsov

Arnold Sommerfeld Center for Theoretical Physics
Ludwig-Maximilians University Munich
Theresienstr. 37, D-80333 Munich, Germany

Lebedev Institute of Physics
Leninsky ave. 53, 119991 Moscow, Russia
eugene.d.skvortsov@gmail.com

The simplest consequences of exact and broken higher-spin symmetry are studied. The one-loop anomalous dimensions of higher-spin currents are determined from the multiplet recombination in the spirit of the modern bootstrap programme: the Wilson-Fisher CFT is studied both in the $4-\epsilon$-expansion and in the large-N. The bulk implications are briefly addressed: part of the higher-spin theory cubic action is reconstructed; one-loop corrections to the AdS masses of higher-spin fields are discussed.

1. Introduction

Wilson-Fisher CFT's describe many second-order phase transitions in the real world: vapor-water critical point, super-fluid λ-point, Ising model at the Curie temperature and many others. Wilson-Fisher CFT's remain unsolved, though they can consistently be approached by various expansion schemes: $4-\epsilon$[1] and large-N.[2–4] Also, the numerical bootstrap programme[5] allowed to determine some of the anomalous dimensions to a very high precision.[6]

With the dawn of the AdS/CFT correspondence[7–9] there had been attempts to find models that are simpler than full string theory vs. $\mathcal{N}=4$ SYM.[10,11] According to the Klebanov-Polyakov conjecture,[12] see also Refs. 13, 14, the large-N Wilson-Fisher CFT should be dual to the four-dimensional higher-spin theory.[15] With the different choice of the boundary conditions the same higher-spin (HS) theory should be dual to just a free scalar — free $O(N)$ model. First remarkable tests of this HS AdS/CFT duality have been performed in Ref. 16 with a precursor in Ref. 13. Recently, the HS AdS/CFT duality has successfully passed the one-loop tests in Refs. 17–22 and references therein/thereon.

The higher-spin theories are almost uniquely fixed by the infinite-dimensional gauge symmetries thereof,[23,24] which come from gauging the higher-spin algebras.[25] Therefore, through the AdS/CFT duality the Wilson-Fisher CFT should inherit higher-spin symmetries as global ones, which can then be viewed as an infinite-dimensional extension of the conformal symmetry.

It turns out that the higher-spin (HS) symmetry works in a way that differs from its two-dimensional cousin Virasoro. When some HS algebra is a symmetry of a CFT, which means that there are conserved tensors $\partial^m J_{ma_2\ldots a_s} = 0$ on top of the stress-tensor, one can prove that the CFT is a free one.[26–30a] Moreover, all the correlation functions are given by the HS algebra invariants.[32,33] This is the case for the duality with the free $O(N)$ model.

In Wilson-Fisher CFT the HS symmetry is broken or deformed, with the non-conservation of the higher-spin tensors having the specific form $\partial \cdot J = gJJ$ of double-trace operators, see e.g. Refs. 34–37. In particular, in Ref. 37 it was shown that the broken higher-spin symmetry is powerful enough as to fix all the three-point functions up to few numbers to the leading order in g.

In the present paper we extract the anomalous dimensions from the non-conservation of the higher-spin currents to the first order in the coupling constant. To do so we need to work out the non-conservation operators JJ and compute the correlation function thereof. There is a number of analytical studies appeared recently[37–41] that pursue new and closely related methods. In particular,[40] suggested to use the multiplet recombination, i.e. quantum equations of motion. In the case of HS currents the non-conservation operator JJ recombines with the short multiplet of the HS current $J_{a_1\ldots a_s}$ to form a long one, which is a non-conserved current. While phenomenologically HS currents may not be of much interest except for the few lower ones, it is the presence of an infinity of such 'almost conserved' HS tensors that is the signature of Wilson-Fisher CFT's and it is these currents that are dual to the HS fields of the HS theory in AdS. Therefore, it is important to zoom in on this subsector of the CFT.

The outline is as follows. In Section 2 we review the case of unbroken HS symmetries and reveal some of the simplest constraints that conserved higher-spin tensors impose on a CFT. Free boson and fermion are considered as examples and the relevant correlation functions of the higher-spin currents are computed. Next, in Section 3 we review the two mechanisms of HS symmetry breaking: classical and quantum; derive the non-conservation operators for $4-\epsilon$ and large-N Wilson-Fisher CFT and extract the anomalous dimensions of the HS currents. In the last Section 4 we address several application for the AdS/CFT and the dual higher-spin theory: fix a part of the cubic action and make a conjecture for the full answer, suggesting that the duals of the free boson and fermion are the same in AdS_4/CFT^3; discuss one-loop corrections to the $AdS_{5-\epsilon}$-masses of higher-spin fields. In Appendices A and B we review basic facts about CFT correlators and Wilson-Fisher CFT's. In Appendix C we show how to extend the Vasiliev-type theories with the duals of

[a]There is a number of earlier works addressing the same issue[31] from a different perspective.[25]

higher-trace operators to the lowest order.

2. Unbroken HS Symmetry

In the present section we consider unbroken HS symmetry whose signature is the existence in a CFT of at least one HS conserved current on top of the stress-tensor[b,c]

$$\partial^b J_{ba(s-1)}(x) = 0\,, \qquad\qquad s > 2\,. \tag{1}$$

First, we review the implications of unbroken HS symmetries — the constraints imposed on a CFT by HS currents, simplest HS Ward identities. Then, we consider free boson and free fermion as examples of CFT's with HS symmetries, which is aimed at collecting two- and three-point functions needed for the study of HS breaking and for the reconstruction of the AdS/CFT duals.

2.1. *Constraints by unbroken HS symmetry*

Our main assumption is the existence of at least one HS current (1) in a CFT, where by CFT we mean all the usual conditions, of which the most important for us is the presence of the unique stress-tensor.

Simplest implications of HS symmetry. For the case of the $2d$ minimal models the decoupling of a Virasoro singular vector imposed on the three-point function $\langle O_{\Delta_1} O_{\Delta_2} O_{\Delta_3}\rangle$ constrains Δ_i. Likewise, the decoupling of the divergence of the HS current imposed on the simplest three-point functions results in important hints. First of all, HS currents can relate scalar operators of the same dimension only:[d]

$$\mathbb{D}_1 \langle J_s O_{\Delta_1} O_{\Delta_2}\rangle = 0 \qquad \Longrightarrow \qquad \Delta_1 = \Delta_2\,. \tag{2}$$

This fact is already true for the global symmetry current $s = 1$ and the stress-tensor $s = 2$, see e.g. Ref. 42. Secondly, the three-point function of two HS currents and one scalar operator leads to[e]

$$\mathbb{D}_{1,2} \langle J_s J_{s'} O_{\Delta}\rangle = 0 \qquad \Longrightarrow \qquad \Delta = d - 2\,. \tag{3}$$

This dramatic restriction requires a genuine HS current and is not true for the $s = 1, 2$ currents that can have scalars of arbitrary dimension in the OPE. The value of $d - 2$ suggests that O is nothing but ϕ^2 for a free scalar ϕ and the presence of a free field is not what one would expect from a nontrivial CFT.

[b] The term HS current is a bit unfortunate in this context, because what we have is a HS conserved tensor, which can be used to construct HS currents by contracting it with Killing tensors. We will continue loosely refer to all of them as HS currents.

[c] The indices $a, b, \ldots$ are those of the d-dimensional flat space where a CFT lives. A group of s totally-symmetric indices $a_1 \ldots a_s$ is abbreviated as $a(s)$.

[d] The structure of the two- and three-point functions in CFT and the useful technique of Thomas derivative to deal with tensorial primaries are reviewed in Appendix A. In practice the conservation is imposed with the help of the Thomas derivative and we denote the operator that checks the conservation of the i-th operator as $\mathbb{D}_i$ in the index free notation, which in components reduces to (1).

[e] Here we assumed parity and we are in generic dimension. If all possible conformal structures are taken into account it is possible to see free fermion, etc. See Ref. 43 for some $3d$ examples.

From OPE to HS algebra. The OPE is the basic tool to study CFT's. For operators carrying nontrivial tensor representations, e.g. HS currents, the OPE can be quite complicated. In general in the OPE of two HS currents we expect to find

$$J_{s_1} J_{s_2} = \delta_{s_1,s_2} \langle J_{s_1} J_{s_2} \rangle + \sum J_k + O_2 \,, \tag{4}$$

where there is a two-point function, a sum over HS currents of different spins and, possibly, the spin-zero J_0 operator may also be present. Some other operators, collectively denoted as O_2 must also be present as the N-counting shows.[12] For example, the normal product $: J_{s_1} J_{s_2} :$ should be there. For the purpose of studying the HS currents content the structure of the OPE can be simplified by integrating over the insertion of J_{s_1} to form a HS charge. First, one builds a set of currents by contracting a HS conserved tensor with a conformal Killing tensor $v^{a(s-1)}$:

$$j_m(v) = J_{ma(s-1)} v^{a(s-1)} \,, \qquad\qquad \partial^a v^{a(s-1)} - \text{traces} = 0 \,, \tag{5}$$

which applies to the stress-tensor as well, for which case Killing tensor becomes Killing vector. Such currents are conserved in the usual sense of $\partial^m j_m(v) = 0$ and can be integrated to form a charge

$$Q(v) = \int dS^m j_m(v) \,. \tag{6}$$

When applied to the OPE one gets the action of the HS charge on the HS current:

$$[Q_{s_1}, J_{s_2}] = \sum_s J_s \,, \tag{7}$$

which is nothing but the action of the HS algebra formed by Q_s for all spins s present in the spectrum on the HS currents themselves. This was the main starting point in Ref. 26. One can make one step further in abstracting the HS algebra — to integrate the second time, which leads to the Lie bracket of two HS charges:

$$[Q_{s_1}, Q_{s_2}] = \sum Q_k \,. \tag{8}$$

The commutator has to obey the Jacobi identity and the algebra formed by Q_s contains the conformal algebra realized by the stress-tensor via Q_2 and at least one genuine HS charge Q_s. Then the Jacobi identity can be solved. This was studied in Ref. 28. The Jacobi identity is purely algebraic and does not appeal to any local field realization of the HS currents.

The main result of studying HS algebras via the Lie bracket (8) or the representation theory via (7) is that:[26–30] (i) there are infinitely many of HS currents/charges once at least one is present; (ii) the correlators are those of one of the free CFT's.

Let us note that the unitarity does not matter much for our considerations. In particular, the presence of at least one HS current still requires all of them to be present. As an example, one can consider free fields obeying higher order equations $\Box^k \phi = 0$, $k > 1$,[44] the HS currents are still there. In the free theory the OPE is just Wick's theorem. The full OPE structure for some of the free CFT's was explicitly worked out in Ref. 45, where one can see how the action of the HS charges leads to HS algebras via Wick's theorem.

Simplest HS Ward Identities. Many of the results known for the case of $s = 1$ global symmetry currents and $s = 2$ stress-tensor can be generalized to HS currents. As was discussed above the OPE contains the complete information about HS symmetries. Let us study the simplest implications of the HS Ward identities. The basic Ward identity in the integral form for a symmetry δ^v generated by current $j_m(v)$ is

$$-\int dS^m \langle j_m(v)O(x_1)\ldots\rangle = \sum_i \langle\ldots\delta_i^v O(x_i)\ldots\rangle\,. \tag{9}$$

A useful example is provided by the free scalar field $\phi(x)$ in d-dimensions. There is a lot of HS transformations generated by various Killing tensors contracted with the same HS current. One can consider the simplest instance given by hyper-translations $\delta_v = v^{a(s-1)}\partial_{a(s-1)}$ for $v^{a(s-1)}$ constant. The action of the spin-s charge on ϕ and ϕ^2 are

$$[Q_s,\phi] = v^{a(s-1)}\partial_{a(s-1)}\phi \tag{10}$$

$$[Q_s,\phi^2] = \partial^{s-1}\phi^2 + \partial^{s-3}J_2 + \ldots + J_{s-1} \tag{11}$$

where in the last line we just sketched the general structure, see e.g. Ref. 26. The first formula is self-evident since $\delta\phi = v^{a(s-1)}\partial_{a(s-1)}\phi$ is a symmetry of the Klein-Gordon equation $\Box\phi = 0$ for $v^{a(s-1)}$ constant. It is also the symmetry of the action $\frac{1}{2}\int(\partial\phi)^2$.

Contrary to the spin-two and spin-one charges the action of the HS charges may involve an infinite number of fields. Indeed, the global symmetry or the conformal symmetry slices the totality of fields into irreducible representations of a given symmetry. The HS algebra is infinite-dimensional. Its smallest representations are the identity and the free field ϕ itself, which plays the role of the fundamental one. Other representations are given by tensor powers of ϕ and decompose into infinite number of conformal fields. The simplest case is the tensor square $\phi\otimes\phi$ and the Flato-Fronsdal theorem[46] tells that it decomposes into all HS currents $J_0\oplus J_2\oplus J_4\oplus\ldots$ with $J_0 = \phi^2$ being the first degenerate member.[f]

When applied to three-point functions $\langle J_s OO\rangle$ the Ward identity (9) relates them to two-point functions $\langle OO\rangle$, which allows to fix the coupling constants. Indeed, the three-point function of a HS current and a weight-Δ scalar operator O_Δ is fixed by conformal symmetry up to a number g_{s00}, the coupling constant:

$$\langle J_s O_\Delta O_\Delta\rangle = g_{s00}\overline{\langle J_s O_\Delta O_\Delta\rangle}\,, \tag{12}$$

where we introduce the standard conformal structure:

$$\overline{\langle J_s O_\Delta O_\Delta\rangle} = (Q\cdot\xi)^s\left(\frac{x_{23}^2}{x_{12}^2 x_{13}^2}\right)^{\frac{d-2}{2}}\frac{1}{(x_{23}^2)^\Delta}\,. \tag{13}$$

Here, ξ^a is light-like auxiliary vector that projects out traces, see Appendix A, and the correlator is factorized via the conformally-covariant vector

$$Q^a = \left(\frac{x_{21}^a}{x_{21}^2} - \frac{x_{31}^a}{x_{31}^2}\right)\,. \tag{14}$$

[f]The Flato-Fronsdal theorem was generalized to any d in Ref. 47.

Likewise, the two-point function $\langle O_\Delta O_\Delta\rangle$ is fixed up to a number C_{OO} to be the standard conformal structure:

$$\langle O_\Delta(x_1)O_\Delta(x_2)\rangle = C_{OO}\overline{\langle O_\Delta O_\Delta\rangle}\,, \qquad \overline{\langle O_\Delta O_\Delta\rangle} = \frac{1}{(x_{12}^2)^\Delta}\,. \tag{15}$$

Couplings g_{s00} are related in a simple way to the OPE coefficients of J_s in the OO OPE. In order to fix the coupling constants g_{s00} we can consider the simplest Ward identity for the hyper-translations. By integrating over a small ball around the insertion of the first operator, which is conveniently placed at $y=0$, we should get:[g]

$$\xi^a\dots\xi^a\int\partial^m\langle J_{ma(s-1)}O_\Delta(0)O_\Delta(z)\rangle = \xi^a\dots\xi^a\partial^y_a\dots\partial^y_a\langle O_\Delta(y)O_\Delta(z)\rangle\big|_{y=0}\,. \tag{16}$$

The right-hand side is simple while the integral can be done by replacing the correlation function by its actual value with a unit coefficient in front of the standard structure:

$$\xi^a\dots\xi^a\int\partial^m\overline{\langle J_{ma(s-1)}O_\Delta(0)O_\Delta(z)\rangle} = S_d f_s(\xi\cdot z)^{s-1}\frac{1}{(z^2)^{\Delta+s-1}}\,,$$

$$\xi^a\dots\xi^a\partial^y_a\dots\partial^y_a\langle O_\Delta(y)O_\Delta(z)\rangle\big|_{y=0} = \frac{2^{s-1}\Gamma[\Delta+s-1]}{\Gamma[\Delta]}(\xi\cdot z)^{s-1}\frac{1}{(z^2)^{\Delta+s-1}}\,,$$

$$f_s = \frac{2^{d+2s-4}\Gamma\left(\frac{d}{2}\right)\Gamma\left(\frac{d-3}{2}+s\right)}{\sqrt{\pi}(d+2s-4)\Gamma(s)\Gamma(d+s-3)}\,,$$

where S_d is the area of the d-dimensional sphere, $S_d = 2\pi^{\frac{d}{2}}/\Gamma[\frac{d}{2}]$. Now we can express all the cubic couplings g_{s00} in terms of the two-point normalization factor C_{OO}:

$$g_{s00} = \frac{C_{OO}}{S_d F_s}\,, \qquad F_s = \frac{f_s\Gamma[\Delta]}{2^{s-1}\Gamma[\Delta+s-1]}\,. \tag{17}$$

In the case of $s=1,2$ the above formula boils down to the well-known, see e.g. Ref. 48, relations

$$g_{100} = \frac{C_{OO}}{S_d}\,, \qquad g_{200} = \frac{C_{OO}d\Delta}{(d-1)S_d}\,. \tag{18}$$

For example, for the free $3d$ scalar and fermion — weight-one ϕ^2 and weight-two $\bar\psi\psi$ scalar operators coupled to HS currents of even spins — we find

$$\Delta = 1: \qquad g_{s00} = C_{OO}2^{1-s}(2s-1)\Gamma(s)^2\,, \tag{19}$$

$$\Delta = 2: \qquad g_{s00} = C_{OO}2^{1-s}(2s-1)\Gamma(s)^2 s\,, \tag{20}$$

the ratio being $1/s$. Let us note that the Ward identity for the HS symmetry we studied above is the simplest one. There are Ward identities with more than one HS currents, which allows to uncover the full HS algebra, as was done in Ref. 26 for the $3d$ case.

[g]The integral form of the Ward identity is easier to work with since the integrals are well-defined. In the differential form of the Ward identities one has to regularize the otherwise ill-defined distributions. For example, one can allow one of the fields to have an infinitesimal anomalous dimension or directly apply differential regularization.

HS Invariants. Since many of the coefficients depend on the way the two- and three-point functions are normalized it is important to pass to invariants. For example, the normalization of the two-point functions is arbitrary unless it is a global symmetry current or the stress-tensor, whose normalization is fixed by the Ward identities and contain important information, which is called central charges. There is only one invariant that is relevant for $\langle J_s J_0 J_0 \rangle$:

$$I_{s00} = \frac{\langle J_s J_0 J_0 \rangle^2}{\langle J_s J_s \rangle \langle J_0 J_0 \rangle^2}\,, \tag{21}$$

where it is assumed that some standard representatives of the correlation functions are chosen, e.g. (15), (13) and we take the ratio of the factors multiplying those. In terms of the HS algebra structure constants f_{s_1,s_2,s_3} and the invariant metric $g_{s_1,s_2} \sim \delta_{s_1,s_2}$ defined by the three- and two-point functions, respectively, the ratio above corresponds to $f^2_{s00} g^{ss} g^{00} g^{00}$. In case of the free $O(N)$ boson/fermion this invariant scales as N^{-1} with N, as will be shown below.

HS central charges. In a given CFT one might have to rescale the operators as to have the Ward identities satisfied with the canonical normalization. This can be done by replacing J_s with $J_s \lambda_s$ where $\lambda_s = C_{OO}/(F_s S_d g_{s00})$. As a result the Ward identities look canonical, but the two-point function of the HS currents gets rescaled and is expressed in terms of the invariant:

$$C_{ss} = \frac{1}{I_{s00} F_s^2 S_d^2}\,. \tag{22}$$

The HS central charges[49] can be defined by analogy with the global symmetry and the stress-tensor central charges as:

$$\langle J_s J_s \rangle = \frac{C_{J,s}}{S_d^2} \frac{(P_{12})^s}{(x_{12}^2)^{d+s-2}}\,, \qquad\qquad C_{J,s} = \frac{1}{I_{s00}^2 F_s^2}\,, \tag{23}$$

where P_{12} is the unique two-point conformally-covariant structure

$$P_{12} = \xi^1_a \xi^2_b \left(\delta^{ab} - 2\frac{x_{12}^a x_{12}^b}{x_{12}^2} \right). \tag{24}$$

$C_{J,s}$ scales as N in the free vector models.

2.2. *Free boson*

In $O(N)$ vector models the fundamental field is an $O(N)$ vector, the spin field $\phi^i(x)$, $i = 1, \ldots, N$. If the model is the free one or the interacting one taken at the strict $N \to \infty$ limit, there is a set of conserved HS currents. The expressions for the HS currents have been obtained a countable number of times in the literature, see e.g. Ref. 50, but let us construct them once again by reiterating the method of Ref. 50. The most general distribution of derivatives over the two fields can be obtained via

$$J^{ij}(x,\xi) = F(\xi \cdot \partial_1, \xi \cdot \partial_2) \; : \phi^i(x_1) \phi^j(x_2) : \Big|_{x_1 = x_2 = x}\,, \tag{25}$$

where we again prefer to hide all indices away by contracting them with an auxiliary vector ξ^a. It can be rewritten by stripping off the center of mass coordinates as

$$F(u,v) = \sum_s (u+v)^s F_s\left(\frac{u-v}{u+v}\right). \tag{26}$$

The HS currents are primaries and hence are traceless, which is taken into account by the light-like polarization vector ξ^a, $\xi\cdot\xi = 0$. As a consequence, the conservation has to be imposed with the help of the Thomas operator, see Appendix A. The solution to the conservation constraint is rather simple ($\nu = (d-3)/2$):

$$F(t,w) = (1-2wt+t^2)^{-\nu}\,, \qquad\qquad F(t,w) = \sum_s F_s(w)t^s\,, \tag{27}$$

and, as is well-known, yields Gegenbauer polynomials, $F_s(w) = C^\nu_s(w)$. Therefore, the generating function of all the HS currents is

$$J^{ij}(x,\xi) = \left(1 - 2(\partial_1 - \partial_2)(\partial_1+\partial_2) + (\partial_1+\partial_2)^2\right)^{-\nu} :\phi^i(x_1)\phi^j(x_2): \Big|_{x_1=x_2=x}\,,$$

where we abbreviated $\partial_i = \xi\cdot\partial_i$.

In general HS currents carry a rank-two reducible representation of $O(N)$ and can be decomposed into three irreducible components:

$$J^{ij} = J^{\square\square} + J^{\begin{smallmatrix}\square\\\square\end{smallmatrix}} + \frac{1}{N}J^{\bullet}\,, \qquad \begin{aligned} J^{\square\square} &= \frac{1}{2}(J^{ij}+J^{ji}) - \frac{1}{N}\delta^{ij}\delta_{kn}J^{kn}\,,\\ J^{\begin{smallmatrix}\square\\\square\end{smallmatrix}} &= \frac{1}{2}(J^{ij}-J^{ji})\,,\\ J^{\bullet} &= \delta_{kn}J^{kn}\,. \end{aligned} \tag{28}$$

The behavior of the three components at the quantum level is quite different. The singlet sector, which the stress-tensor belongs to, is important for AdS/CFT.

In what follows we will need two- and three-point functions, which can be computed by simple Wick contractions of the spin field ϕ^i. The canonically normalized two-point function of ϕ^i is

$$\langle\phi^i(x_1)\phi^j(x_2)\rangle = \delta^{ij}\frac{\Gamma[d/2-1]}{4\pi^{d/2}(x_{12}^2)^{d/2-1}} = \frac{\delta^{ij}}{4\pi^{d/2}}\int d\alpha\,\alpha^{d/2-2}e^{-\alpha x_{12}^2}\,, \tag{29}$$

where the latter expression is the most useful one as it allows to replace complicated tensor structures resulting from derivatives of the spin field with polynomials in the Schwinger parameters.

The two-point function $\langle J_s J_s\rangle$ can be straightforwardly evaluated with the orthogonality relation $\langle J_{s_1}J_{s_2}\rangle \sim \delta_{s_1,s_2}$ boiling down to that of Gegenbauer polynomials:

$$\langle J^{ij}_s J^{kl}_s\rangle = [\delta^{ik}\delta^{jl} + (-)^s\delta^{il}\delta^{jk}]C_s \times \overline{\langle J_s J_s\rangle}\,, \tag{30}$$

$$C_j = \frac{\pi 2^{-2d+j+8}\Gamma(d+j-3)\Gamma(d+2j-2)}{\left(4\pi^{d/2}\right)^2 j!(d+2j-3)\Gamma\left(\frac{d}{2}-\frac{3}{2}\right)^2}\,, \tag{31}$$

where we singled out the two-point structure with a unit coefficient

$$\overline{\langle J_s J_s\rangle} = \frac{(P_{12})^s}{(x_{12}^2)^{d+s-2}}\,, \tag{32}$$

which plays the role of the standard one. The two-point function can be projected onto the three irreducible $O(N)$-structures as:

$$\langle J_s^{\bullet} J_s^{\bullet}\rangle = N[1+(-)^s]C_s\overline{\langle J_s J_s\rangle}\,, \tag{33}$$

$$\langle J_s^{\substack{\square\\\square}} J_s^{\substack{\square\\\square}}\rangle = \frac{1}{2}(\delta^{ik}\delta^{jl}-\delta^{il}\delta^{jk})[1-(-)^s]C_s\overline{\langle J_s J_s\rangle}\,, \tag{34}$$

$$\langle J_s^{\square\square} J_s^{\square\square}\rangle = \frac{1}{2}(\delta^{ik}\delta^{jl}+\delta^{il}\delta^{jk}-\frac{2}{N}\delta^{ij}\delta^{kl})C_s[1+(-)^s]\overline{\langle J_s J_s\rangle}\,. \tag{35}$$

The two-point function of singlets scales as N and the anti-symmetric currents are nontrivial for odd spins while the singlet and symmetric currents are so for even spins. Analogously, the three-point functions can be found to be normalized as

$$\langle J_s^{ij} J_0 J_0\rangle = 4\delta^{ij}[1+(-)^s]C_{sOO}\times\overline{\langle J_s O_{d-2} O_{d-2}\rangle}\,, \tag{36}$$

$$C_{sOO} = \pi^{-\frac{3d}{2}}2^{s-6}\Gamma\left(\frac{d}{2}-1\right)^2\Gamma\left(\frac{d}{2}+s-1\right)\,, \tag{37}$$

where 4 is because of the Wick contractions and we split off the standard conformal structure (13). The three-point function of the spin field with the HS current is

$$\langle J_s^{ij}\phi^k\phi^l\rangle = [\delta^{ik}\delta^{jl}+(-)^s\delta^{il}\delta^{jk}]C_{s\phi\phi}\times\overline{\langle J_s O_{\frac{d-2}{2}} O_{\frac{d-2}{2}}\rangle}\,, \tag{38}$$

$$C_{s\phi\phi} = \frac{\pi^{-d}2^{s-4}\Gamma\left(\frac{d}{2}-1\right)\Gamma\left(\frac{d}{2}+s-1\right)\Gamma(d+s-3)}{s!\Gamma(d-3)}\,. \tag{39}$$

The invariant (21) built of the single-trace correlation functions (33) is

$$I_{s00} = \frac{\sqrt{\pi}2^{-d-s+7}\Gamma\left(\frac{d}{2}+s-1\right)\Gamma(d+s-3)}{N\Gamma\left(\frac{d}{2}-1\right)^2\Gamma(s+1)\Gamma\left(\frac{d-3}{2}+s\right)} \tag{40}$$

and can also be extracted from Ref. 51. It is I_{s00} that has to be reproduced by the dual HS theory upon plugging boundary-to-bulk propagators into the cubic vertices as will be discussed in Section 4. Combining the invariant with the discussion above we see that the HS central charge is

$$C_{J,s} = \frac{\sqrt{\pi}N2^{-d-s+3}\Gamma(s)^2\Gamma(s+1)\Gamma\left(\frac{d}{2}+s-1\right)\Gamma(d+s-3)}{\Gamma\left(\frac{d}{2}\right)^2\Gamma\left(\frac{d-3}{2}+s\right)}\,, \tag{41}$$

where the formula works by construction for even spins. This formula agrees with the $s=2$ case, for which it gives the well-known[48]

$$C_T = \frac{dN}{d-1}\,. \tag{42}$$

2.3. *Free fermion*

As in any free CFT there are HS currents in the free fermion model too. The same idea as in the scalar case leads to

$$J^i{}_j(x,\xi) = F(\xi\cdot\partial_1,\xi\cdot\partial_2)\ :\bar{\psi}^i(x_1)(\gamma\cdot\xi)\psi_j(x_2):\Big|_{x_1=x_2=x}, \tag{43}$$

where $i,j=1,\ldots\tilde{N}$ are the $U(\tilde{N})$ indices and

$$F(u,v)=\sum_s (u+v)^{s-1}F_s\left(\frac{u-v}{u+v}\right). \tag{44}$$

The solution is again given by Gegenbauer polynomials with a bit different parameters ($\nu=\frac{d-3}{2}$):[h]

$$F(t,w)=t(1-2wt+t^2)^{-\nu-1}, \qquad F(t,w)=\sum_s F_s(w)t^s. \tag{45}$$

In other words, $F_s(w)=C^{\nu+1}_{s-1}(w)$ and

$$\begin{aligned} J^i{}_j(x,\xi) &= (\partial_1+\partial_2)\Big(1-2(\partial_1-\partial_2)(\partial_1+\partial_2)+(\partial_1+\partial_2)^2\Big)^{-\nu-1}\times\\ &\quad\times :\bar{\psi}^i(x_1)(\gamma\cdot\xi)\psi_j(x_2):\Big|_{x_1=x_2=x}. \end{aligned} \tag{46}$$

The two-point function of ψ^i differs from that of the scalar field by an extra $\not{\partial}=\gamma\cdot\partial$:

$$\langle\psi_j(x_1)\bar{\psi}^i(x_2)\rangle = -\delta^i_j\frac{1}{S_d}\frac{\not{x}}{(x^2_{12})^{d/2}} = \tag{47}$$

$$= \delta^i_j\not{\partial}\frac{\Gamma[\frac{d}{2}-1]}{4\pi^{d/2}(x^2_{12})^{d/2-1}} = \not{\partial}\frac{\delta^i_j}{4\pi^{d/2}}\int d\alpha\,\alpha^{d/2-2}e^{-\alpha x^2_{12}}. \tag{48}$$

A new interesting feature is that the HS currents can have more complicated symmetry types.[47,53,54] Namely, we can replace $(\gamma\cdot\xi)$ with $\gamma^{u[q-1]v}\xi_v$ where $\gamma^{u[q]}$ is the anti-symmetrized product of γ-matrices and an additional projector is needed to make the currents Young-irreducible, i.e. quasi-primary. We will not consider such currents, restricting to the totally-symmetric ones, which are present in any free CFT.

The two-point function of the HS currents can be computed in full analogy to the free boson case. The $U(\tilde{N})$-decomposition can easily be worked out and we record for the future just the normalization of the singlet HS currents:

$$\langle J^\bullet_s J^\bullet_s\rangle = NC_s\overline{\langle J_sJ_s\rangle}, \tag{49}$$

$$C_j = \frac{(-)^j\pi^{1-d}2^{-2d+j+1}\Gamma(d+j-2)\Gamma(d+2j-3)}{\Gamma\left(\frac{d-1}{2}\right)^2\Gamma(j)} \tag{50}$$

where $N=\tilde{N}\mathrm{tr}\mathbf{1}$ is the total number of fermions. The non-singlet currents have an obvious tensor structure and $\mathrm{tr}\mathbf{1}$ instead of N. Note that $(-)^j$ is exactly what

[h] In Ref. 52 another presentation for the HS currents was found, which up to an overall coefficient can be obtained by Taylor expanding the Gegenbauer polynomial.

one needs for the reflection positivity, but it would be better to introduce i^s into the currents. A word of warning is that the lowest singlet $O = J_0 = \bar{\psi}\psi$ cannot be obtained from J_s at $s = 0$, but since the three-point function $\langle OOO \rangle = 0$ we can ignore this case. After rescaling the generating function appropriately, this result matches the one in Ref. 52, which was obtained by difficult resummation. Also, the three-point function is of interest

$$\langle J_s J_0 J_0 \rangle = 2NC_{sOO} \times \overline{\langle J_s O_{d-1} O_{d-1} \rangle}\,, \tag{51}$$

$$C_{jOO} = \frac{\pi^{-\frac{3d}{2}} 2^{j-4} \Gamma\left(\frac{d}{2}\right)^2 \Gamma\left(\frac{d}{2}+j-1\right) \Gamma(d+j-2)}{\Gamma(d-1)\Gamma(j)}\,, \tag{52}$$

since it determines the coupling constants. It vanishes for odd spins, which is implicit in the formula. In AdS_4/CFT_3 duality the free boson and fermion have the same dual, in particular the HS algebras are the same.[28,55] The invariant I_{s00} is

$$I_{s00} = \frac{\sqrt{\pi}(-)^s 2^{-d-s+5} \Gamma\left(\frac{d}{2}+s-1\right) \Gamma(d+s-2)}{N\Gamma\left(\frac{d}{2}\right)^2 \Gamma(s) \Gamma\left(\frac{d-3}{2}+s\right)} \tag{53}$$

and differs by $s(d+s-3)/(d-2)^2$ from that of the free boson. We will see in Section 4 how this difference is compensated by changing the boundary conditions from free boson $\Delta = 1$ to free fermion $\Delta = 2$ for the AdS duals of $O = \phi^2, \bar{\psi}\psi$.

2.4. *HS singletons*

Lastly, we note that the generalization of the above formulae to the conformal fields with higher spin is obvious, see also Ref. 50. For example, in the case of the $n = (d-2)/2$-forms $A_{a[n]}$ in even dimension, of which $4d$ Maxwell field A_m is a particular case, we find that the totally-symmetric HS currents

$$J = F(\xi \cdot \partial_1, \xi \cdot \partial_2)\, \xi^a \xi^a : G_{ab[n]}(x_1) G_a{}^{b[n]}(x_2) : \Big|_{x_1 = x_2 = x} \tag{54}$$

are generated by $C^{\nu+2}_{j-2}(w)$. Here $G_{a[n+1]} = \partial_a A_{a[n]} +$ permutations is the totally anti-symmetric field-strength. Similarly, the HS singletons[56] can be taken into account, the currents being constructed by $F = C^{\nu+2s}_{j-2s}(w)$

$$J = F(\xi \cdot \partial_1, \xi \cdot \partial_2)\, \xi^{a(2s)} : G_{a(s),b[n](s)}(x_1) G_{a(s)}{}^{b[n](s)}(x_2) : \Big|_{x_1 = x_2 = x}\,, \tag{55}$$

where $G_{a(s)[n+1]}$ is the field-strength that has the symmetry of the rectangular $s \times n+1$ Young diagram. As is known, see Ref. 57 for the explicit construction in $4d$, the lowest spin for the conserved tensor built out of a spin-s conformal field is $2s$, which is also clear from the index of the Gegenbauer polynomial here-above.

3. HS Symmetry Breaking

In this section we discuss two possible ways to break HS symmetries: classical and quantum, which are realized in the theories of Yang-Mills and Wilson-Fisher type.[i] Then, we scrutinize the quantum breaking and study rather general implications of the conformal symmetry and quantum equations of motion for HS currents. The non-conservation operators for Wilson-Fisher CFT in $4-\epsilon$ and large-N are derived and used to extract the anomalous dimensions of the HS currents at the first non-trivial order in the coupling constant. In Appendix B we collected basic facts about the models studied in the main part.

3.1. *Classical and Quantum Breaking*

There are at least two mechanisms to break HS symmetries known. Due to AdS/CFT each way of breaking has both CFT and AdS interpretation. When the HS symmetry is exact the conservation of a HS current in a free CFT_d is dual to the gauge transformations for the Fronsdal field[58] in AdS_{d+1}:[j]

$$\partial^m J_{ma(s-1)} = 0 \qquad \Longleftrightarrow \qquad \delta\Phi_{\underline{m}(s)} = \nabla_{\underline{m}}\xi_{\underline{m}(s-1)} + O(z, \Phi\xi)\,. \tag{56}$$

This is true near the boundary. When the bulk field departs from the AdS boundary at $z=0$ interactions switch on and the free gauge symmetries get deformed by the terms of the form $\Phi\xi$ (like in Yang-Mills and gravity) and more complicated and nonlinear ones (as different from Yang-Mills and gravity). The duals of the free CFT's are HS theories as was discussed in Refs. 10–12, 59. HS theories should gauge the HS algebra of a given free CFT, but at present only the dual of the free boson is known in any AdS_{d+1}[60] and the dual of the free fermion is available in AdS_4 thanks to the already mentioned coincidence that the HS algebra of the free boson is identical to that of the free fermion in $3d$. The same HS theories can be the duals of interacting CFT's by changing the boundary conditions within the unitarity window.[12–14]

When HS symmetry is broken the conservation of the HS currents is replaced by non-conservation, the structure being dependent on the mechanism of breaking. Accordingly, the dual bulk fields have to acquire a mass, which can be done classically[61] and then followed by radiative corrections or immediately via loops.[34]

Classical, BEH. One can think of a HS generalization of the usual Brout-Engler-Higgs mechanism — a massless vector boson swallows the Goldstone scalar which makes the right number of degrees of freedom to become a massive vector. In the case of the original particle being a massless spin $s>1$ field in AdS_{d+1} the correct partner is a massive spin-$(s-1)$ field of mass $M^2=(d+s-2)(s-1)$ in the units of the cosmological constant.

[i]This is how we distinguish between classical and quantum. It will be clear that the definitions are well-defined in a more broad sense when a CFT is weakly-coupled.
[j]The indices $\underline{m}, \underline{n}, \ldots = 0, \ldots, d$ are the indices of AdS_{d+1} tensors.

The field realization of the HS BEH mechanism is nicely incorporated into the Lagrangian description of massive HS fields proposed by Zinoviev.[62] The idea is to take the sum of the Fronsdal Lagrangians for $j = 0, 1, \ldots, s$, which has the right total number of degrees of freedom to describe a single massive spin-s field. All possible mixings and mass-like terms are then added to make the Lagrangian irreducible, as well as the gauge transformations $\delta\Phi_j = \nabla\xi_j$ are appended with the gauge parameters of fields with neighboring spins:

$$\mathcal{L}_{s,m^2} = \sum_{j=0}^{j=s} \mathcal{L}_j + \sum_j \left[\Phi_j \nabla \Phi_{j-1} + m_j^2 \Phi_j^2\right], \qquad \delta\Phi_j = \nabla\xi_j + \xi_{j+1} + \eta\xi_{j-1}, \tag{57}$$

where η is the AdS metric. The gauge invariance fixes all the free coefficients in terms of a single constant, which can be identified with the mass. At the special point where the AdS-mass is that of the spin-s massless field the Lagrangian splits into two parts: the Fronsdal one for a spin-s massless field and the Lagrangian for a massive spin-$(s-1)$ field with mass M^2:

$$\mathcal{L}_{s,m^2}\Big|_{m^2=(d+s-2)(s-2)-s} = \mathcal{L}_s + \mathcal{L}_{s-1,M^2}. \tag{58}$$

On the CFT side, the HS current non-conservation is determined by some single-trace operator present in the theory:

$$\mathbb{D}J = gO_1. \tag{59}$$

A good example is Yang-Mills theory. At the zero coupling it has a HS symmetry manifested by $J_s = \mathrm{tr}[A\partial^s A] = \mathrm{tr}[F\partial^{s-2}F]$. When the interactions are switched on, the currents are no longer conserved $\mathbb{D}J_s = g\,\mathrm{tr}[A\partial^{s-1}A^2] + O(g^2) = g\,\mathrm{tr}[F\partial^{s-2}FA] + O(g^2)$. This type of breaking was studied in the case of $\mathcal{N} = 4$ SYM in Refs. 31, 63–65.

Quantum, GPZ. The second mechanism can be called the quantum breaking.[34,37] On the CFT side the non-conservation is governed by double- and/or triple-trace operators built out of the HS currents J themselves:

$$\mathbb{D}J = gJJ \text{ possibly plus } g^2JJJ. \tag{60}$$

This equation is true not only in perturbation theory where the operators renormalize order by order, but at the full quantum level. The simplest consequences of this non-conservation condition we will explore below. It is not clear what should happen in the bulk. In Ref. 34 it was proposed that the HS fields acquire the mass via loops, which results in a jump in the number of physical degrees of freedom. A delicate UV/IR interplay is required and to see how this is realized on the AdS side is an open problem.

3.2. *Insertions of equations of motion*

In this Section we insert the equations of motion into various correlation functions to read off the relations between the CFT data. The main idea is to take advantage of the conformal symmetry when extracting information from the non-conservation of the HS currents. While the HS currents are exactly conserved in the free CFT or at $N = \infty$ in Wilson-Fisher CFT, apart from $N = \infty$ the short multiplets of HS currents recombine with the long multiplets formed by double- and triple-trace operators as to form long multiplets, which is what (60) describes. This phenomenon originates from the recombination of ϕ itself, whose structure depends on the quantum equations of motion.[40] We do not consider all possible correlation functions that can be of interest, but only those that talk to the anomalous HS currents.

3.2.1. *Anomalous Two-Point Function of Scalar Field*

The simplest case is to consider the two-point function of the spin field:[40]

$$\langle \phi^i \phi^j \rangle = \delta^{ij} \frac{C_{\phi\phi}}{(x_{12}^2)^{\Delta_\phi}}\,, \tag{61}$$

where $\Delta_\phi = d/2 - 1 + \gamma_\phi$. Then we impose the equations of motion abstractly, without having to specify the composite operator that becomes the descendant $\Box\phi$:

$$\Box_1\Box_2 \langle \phi\phi \rangle = 4\gamma_\phi(\gamma_\phi + 1)(2\gamma_\phi + d - 2)(2\gamma_\phi + d)\frac{C_{\phi\phi}}{(x_{12}^2)^{\Delta_\phi+2}}\,. \tag{62}$$

Another way to compute the same is to drag $\Box$ inside the correlator and replace it with the composite operator it is supposed to recombine with. This depends on the model. In the large-N Wilson-Fisher $\Box\phi^i = g_\star \sigma\phi^i$ and via the Wick theorem we can get to the lowest order:

$$\Box_1\Box_2 \langle \phi^i \phi^j \rangle = g_\star^2 \langle (\sigma\phi^i)(\sigma\phi^j) \rangle = \delta^{ij} g_\star^2 \frac{C_{\sigma\sigma}C_{\phi\phi}}{(x_{12}^2)^{\Delta_\phi+2}}\,, \tag{63}$$

where there is a critical coupling $g_\star$ and the σ-field two-point function normalization $C_{\sigma\sigma}$. Therefore, to the lowest order in perturbation theory we find a relation between the three quantities:

$$C_{\sigma\sigma} g_\star^2 = 4d(d-2)\gamma_\phi\,. \tag{64}$$

This relation is indeed true as can be seen from Appendix B. One more example is the ϕ_6^3-theory where $\Box\phi = g_\star\phi^2$. It is formally the same as (64) where $d = 6$ and $\sigma = \phi$. Analogously, in the $4-\epsilon$ Wilson-Fisher case we have $\Box\phi^i = g_\star\phi^i(\phi)^2$ and

$$\Box_1\Box_2 \langle \phi^i \phi^j \rangle = g_\star^2 \langle (\phi^i\phi^2)(\phi^j\phi^2)) \rangle = \delta^{ij} 2(N+2) g_\star^2 \frac{C_{\phi\phi}^2}{(x_{12}^2)^{3\Delta_\phi}}\,, \tag{65}$$

which results in

$$\gamma_\phi = \frac{(N+2)g_\star^2}{256\pi^4}\,. \tag{66}$$

In practice it is convenient to measure the anomalous dimensions of composite operators in terms of that of ϕ rather than less physically interesting quantities like $C_{\sigma\sigma}$ and $g_\star$. Eq. (64), (66) allows us to do so since it is $C_{\sigma\sigma}g_\star^2$ that shows up in the formulas. The insertions of equations of motion was successfully applied recently in Ref. 40,66,67.

3.2.2. *Anomalous Two-Point Function of HS Currents*

Interactions can break the HS symmetry and, as a result, the HS currents are no longer conserved and develop an anomalous dimension $\gamma = \gamma_s$

$$\langle J_s(x_1,\xi_1)J_s(x_2,\xi_2)\rangle = \frac{C_{J,s}}{\mu^{2\gamma}(x_{12}^2)^{d+s-2+\gamma}}(P_{12})^s\,, \tag{67}$$

where $C_{J,s}$ is the spin-s central charge and μ is the renormalization scale to compensate for the anomalous dimension γ, which we drop in what follows.

The main idea of the Anselmi's trick[31] is that one can check the conservation of the anomalous HS current inside the two-point function in two different ways. First of all, one can take double-divergence of the two-point function as a whole, i.e. to check the conservation of (67), the result being proportional to the anomalous dimension. Secondly, one can drag the double-divergence inside the two-point function, i.e. replace $\mathbb{D}J$ with its expression in a given theory. This can be written as

$$\langle \mathbb{D}_1 J_s(x_1)\mathbb{D}_2 J_s(x_2)\rangle = \mathbb{D}_1\mathbb{D}_2\langle J_s(x_1)J_s(x_2)\rangle\,. \tag{68}$$

The identity above becomes a source of nontrivial information when the two sides can be computed independently. For example, the non-conservation is governed by some small parameter $g_\star$:

$$\mathbb{D}J_s = g_\star K_s\,. \tag{69}$$

Then, the left-hand side of (68) is proportional to $(g_\star)^2$, while the right-hand side can be explicitly evaluated from (67). One observes that the anomalous dimension is by $(g_\star)^2$ closer than before — one needs to evaluate $\langle KK\rangle$ and $\langle JJ\rangle$ to n loops in order to determine the anomalous dimension at loop order $n+1$. In particular, it is enough to work with an almost 'classical' theory in order to find anomalous dimensions to the one-loop order. One more advantage of the method is that it allows to treat by the same means different theories in diverse dimensions that are originally studied by distinct methods. Let us compute the right-hand side of (68) by taking the divergence at both the insertions ($h = d/2$):

$$\mathbb{D}_1\mathbb{D}_2\langle J_s(x_1)J_s(x_2)\rangle = \frac{(P_{12})^{s-2}}{(x_{12}^2)^{d+s-1+\gamma}}\left[P_{12}a(\gamma) + \tilde{P}_{12}b(\gamma)\right], \tag{70}$$

$$a(\gamma) = 2\gamma s(h+s-2)\left(2h^2 + 2\gamma(h+s-2) + 3hs - 5h + s^2 - 4s + 3\right)\,, \tag{71}$$

$$b(\gamma) = -2\gamma^2(s-1)s(h+s-3)(h+s-2)\,, \tag{72}$$

where there is an additional conformal structure

$$\tilde{P}_{12} = 2\frac{(x_{12}\cdot\xi_1)(x_{12}\cdot\xi_2)}{x_{12}^2}\,. \tag{73}$$

The HS current is a primary operator by definition, but this is not the case for its divergence, which is a typical descendant. The new structure $\tilde{P}_{12}$ appears since the two-point functions of descendants are more complicated. When the anomalous dimension γ vanishes the divergence becomes a primary operator that decouples and in this case the HS current is exactly conserved, which explains the factors of γ and γ^2 in front of the usual conformal structure P_{12} and the unusual one $\tilde{P}_{12}$.

Since the space-time dependence of (70) is fixed, it is convenient to take it at $x_1 = x$, $x_2 = 0$, $\xi_1 = \xi_2 = \xi$, which results in Ref. 65

$$\mathbb{D}_1\mathbb{D}_2\langle J_s(x_1)J_s(0)\rangle = \left(\frac{-2(\xi\cdot x)^2}{x^2}\right)^{s-1}\frac{1}{(x^2)^{d+s-1+\gamma}}c(\gamma)\,, \tag{74}$$

$$c(\gamma) = 2\gamma s(h+s-2)(\gamma(hs+h+(s-2)s-1)+(h+s-1)(2h+s-3))\,. \tag{75}$$

Therefore, to the lowest order we find

$$c(\gamma) = \gamma c_1 + O(\gamma^2)\,, \qquad c_1 = \frac{1}{2}s(d+s-3)(d+2s-4)(d+2s-2)\,. \tag{76}$$

The two-point function of K has the same space-time dependence and is defined by a number $C_{K,s}$

$$\langle K_s(x_1)K_s(x_2)\rangle = C_{K,s}\left(\frac{-2(\xi\cdot x)^2}{x^2}\right)^{s-1}\frac{1}{(x^2)^{d+s-1+\gamma}}\,. \tag{77}$$

Finally, the equation (68) for the anomalous dimension to the lowest order is

$$g_\star^2\frac{C_{K,s}}{C_{J,s}} = c(\gamma) = g_\star^2 c_1\gamma_1 + \text{higher orders}\,, \tag{78}$$

where in the last expression we assume that the anomalous dimension is expanded as $\gamma = g_\star^2\gamma_1 + \ldots$. Next, we will derive K for a number of models and compute γ_1.

3.3. *Non-conservation of HS currents*

Below, we give the explicit form of the composite operators responsible for the non-conservation of HS currents in some bosonic models that are described in Appendix B in more detail.

General Formulae. The HS currents are no longer conserved in the interacting vector model, the non-conservation having the form of double-trace operators for the bosonic models:

$$\mathbb{D}J = gJJ\,. \tag{79}$$

Explicitly, using the generating function (25) we find:

$$\mathbb{D}J_s(x,\xi) = (\xi\cdot\partial_1 + \xi\cdot\partial_2)^{s-1}\times$$
$$\times\left[f_1(u)\Box_1\phi(x_1)^i\phi^j(x_2) + f_2(u)\phi(x_1)^i\Box_2\phi^j(x_2)\right]\Big|_{x_1=x_2=x},$$

where $u = (\xi\cdot\partial_1 - \xi\cdot\partial_2)/(\xi\cdot\partial_1+\xi\cdot\partial_2)$ and the $\partial_1\cdot\partial_2$-terms vanish because J was designed as to ensure $\mathbb{D}J(x,\xi)=0$ whenever $\Box_{1,2}=0$. The expression in the square bracket can be worked out without using the explicit form of the quantum equations of motion. It can be simplified with the help of the recurrence relations that the Gegenbauer polynomials obey with the result:[k]

$$K = f_1(x) + f_2(y)\,,\qquad \begin{aligned} f_1(x) &= +4\nu(\nu+1)(x-1)C^{(\nu+2)}_{s-2}(x) + \nu(2\nu+1)C^{(\nu+1)}_{s-1}(x)\,,\\ f_2(y) &= -4\nu(\nu+1)(y+1)C^{(\nu+2)}_{s-2}(y) - \nu(2\nu+1)C^{(\nu+1)}_{s-1}(y)\,,\end{aligned}$$

and the variables x and y point split the equations of motion of interest. Note that $K(x,y)$ inherits the symmetry of the original currents under the permutation of the arguments, i.e. $K(-y,-x) = (-)^j K(x,y)$ and $f_2(-x)$ is $(-)^s f_1(x)$.

Large-N. For the sigma model the source is bilinear:

$$\Box\phi^i = g_\star\phi^i\sigma\,. \tag{80}$$

Here the double-trace operator K_s measuring the non-conservation is:

$$K_s = \mathbb{D}J_s(x,\xi) = g_\star(\xi\cdot\partial_1+\xi\cdot\partial_2+\xi\cdot\partial_3)^{s-1}\,K_s(x,y)\,\phi^i(x_1)\phi^j(x_2)\sigma(x_3)\Big|_{x_i=x},$$

where

$$x = \frac{\xi\cdot\partial_1 - \xi\cdot\partial_2 + \xi\cdot\partial_3}{\xi\cdot\partial_1 + \xi\cdot\partial_2 + \xi\cdot\partial_3}\,,\qquad y = \frac{\xi\cdot\partial_1 - \xi\cdot\partial_2 - \xi\cdot\partial_3}{\xi\cdot\partial_1 + \xi\cdot\partial_2 + \xi\cdot\partial_3}\,. \tag{81}$$

The same expression works fine for the $6d$ ϕ^3-theory with the understanding that σ is one of the ϕ fields whose two-point function is that of the free scalar field to the lowest order, while for the large-N vector model $G_\sigma^{-1} = -\frac{1}{2}G_\phi^2$.

The form of the non-conservation operator just obtained is sufficient to do computations, but it is also instructive to derive the decomposition in terms of composite operators JJ built of the single-trace operators:

$$\mathbb{D}J_s = g_\star\sum_{a+c<s} f_{a,c}\,\partial^a J_{s-1-a-c}\,\partial^c\sigma\,. \tag{82}$$

Directly re-expanding the generating function K we find (recall that $\nu=(d-3)/2$)

$$f_{a,c} = -\frac{2(a+c+1)!(a+c-\nu-s+1)(a+2c+1-2(s+\nu))!}{a!c!(c+1)!(a+c-2(s+\nu))!}\,. \tag{83}$$

Eq. (82) is helpful in evaluating the two-point function $\langle KK\rangle$ as it diagonalizes J and σ. For the case of the spin-four non-conservation operator we find $f_{1,0}/f_{0,1}|_{s=4} = -2/5$ as in Refs. 35, 37:

$$\mathbb{D}J_4 \sim J_2\,\partial\sigma - \frac{2}{5}\partial J_2\,\sigma\,. \tag{84}$$

[k]These formulae agree for $d=6$ with Ref. 65.

The main source of the HS symmetry breaking in Refs. 35, 37 was the violation of parity coming from the Chern-Simons coupling. Note that the formulae derived above apply strictly speaking only to the case of non-singlet currents as for the singlet ones it yields the terms of the form $J_0\sigma$ and there is no ϕ^2 operator, but there is $\phi^i\phi^j$, $i \neq j$.

Wilson-Fisher $4-\epsilon$. For the Wilson-Fisher model in $4-\epsilon$ the source is trilinear:

$$\Box\phi^i = g_\star \phi^i \phi^2 \,, \tag{85}$$

which leads to

$$K_s = g_\star \left(\sum_{k=1}^{4} \xi\cdot\partial_k\right)^{s-1} K_s(x,y)\, \phi^i(x_1)\phi^j(x_2)\phi^k(x_3)\phi_k(x_4)\Big|_{x_i=x}\,, \tag{86}$$

where $K(x,y)$ is the same function as before, but the arguments have to be changed to

$$x = \frac{\xi\cdot\partial_1 - \xi\cdot\partial_2 + \xi\cdot\partial_3 + \xi\cdot\partial_4}{\xi\cdot\partial_1 + \xi\cdot\partial_2 + \xi\cdot\partial_3 + \xi\cdot\partial_4}\,, \qquad y = \frac{\xi\cdot\partial_1 - \xi\cdot\partial_2 - \xi\cdot\partial_3 - \xi\cdot\partial_4}{\xi\cdot\partial_1 + \xi\cdot\partial_2 + \xi\cdot\partial_3 + \xi\cdot\partial_4}\,. \tag{87}$$

Again one can diagonalize the non-conservation operator and represent it as JJ:

$$\mathbb{D}J^{ij}_s = \sum_{a+c+s'=s-1} A^{s'}_{a,c}\, \partial^a J^{ij}_{s'}\, \partial^c J_0\,, \tag{88}$$

where the second factor is always $J_0 = \phi^2$ and the coefficients are

$$A^s_{a,c} = \frac{2(-1)^a(\nu+s)(a+c+1)!(a+c+2\nu+2s+1)!}{a!c!(c+1)!(a+2(\nu+s))!}\,. \tag{89}$$

$A^s_{a,c}$ vanishes whenever $a+c$ is even. For example, the non-conservation operator for the spin-two current

$$\mathbb{D}J^{ij}_2 = a_1\partial J^{ij}_0 J_0 + a_2 J^{ij}_0 \partial J_0\,, \quad a_1 = -8\nu(\nu+1)\,, \quad a_2 = 4\nu(\nu+1)(2\nu+1)\,, \tag{90}$$

is non-trivial for the non-singlet sector, but vanishes for the singlet one at $d=4$ since $a_2 = -a_1 = 6$ for $\nu = \frac{1}{2}$, i.e. the traceless stress-tensor is conserved as expected.[1]

3.4. *Anomalous dimensions*

Below we combine the formulae for the non-conservation of the HS currents derived in the previous Section with the Wick theorem as to get the anomalous dimensions γ_s of the HS currents to the one-loop order. The main formula is (78), where $C_{J,s}$ was computed in Section 2. In addition (64) and (66) allows us to express γ_s in the units of γ_ϕ. The only missing ingredient is $C_{K,s}$, i.e. the two-point functions $\langle KK\rangle$.

[1]There is, of course, a conserved stress-tensor for any d in the ϕ^4 theory, but it is not traceless and the traceless one is conserved only in $d=4$.

3.4.1. $4 - \epsilon$

To determine the anomalous dimension of the non-singlet currents we need to compute two contributions — two distinct Wick contractions:

$$\langle K^{12} K^{12} \rangle = 2N\langle 1234\rangle + 8\langle 1324\rangle\,, \tag{91}$$

where we indicated the N factors explicitly as well as the permutation of the ϕ's in (86). Also, we took only the K^{12} component as a generic non-singlet. The correlator $\langle K_j K_j \rangle$ is expected to have the form

$$I_\sigma = C_{K,j}\frac{[-2(\xi\cdot x)^2]^{j-1}}{(x^2)^{2d+2j-6}}\,, \tag{92}$$

which should be applied to the $d = 4$ case. The prefactor splits as

$$C_{K,j} = \frac{C^4_{\phi\phi}\nu^2 2^{j-1}\Gamma[2d+2j-6]}{\Gamma[d/2-1]^4}\times I\,, \tag{93}$$

where $C_{\phi\phi}$ is the ϕ two-point function normalization (29) and I is the integral over the Schwinger parameters that depends on the permutation. We only need:

$$I_{1234} = \frac{(j-1)j(j+1)(j+2)}{2j+1}\,, \qquad I_{2314} = \frac{j(j+1)\left((-)^j(j^2+j-4)-2\right)}{2(2j+1)}\,, \tag{94}$$

$$I_{1324} = I_{3214} = (-)^j I_{2314}\,, \qquad I_{3412} = -\frac{2\left((-)^j+1\right)j(j+1)}{2j+1}\,. \tag{95}$$

One has to distinguish between odd and even spins. The relation (66) between the coupling constant $g_\star^2$ and anomalous dimension of the spin field γ_ϕ allows us to express everything in the units of γ_ϕ:

$$\gamma^{\square\square} = 2\gamma_\phi\left(1 - \frac{2(N+6)}{(N+2)s(s+1)}\right)\,, \tag{96}$$

$$\gamma^{\begin{smallmatrix}\square\\ \square\end{smallmatrix}} = 2\gamma_\phi\left(1 - \frac{2}{s(s+1)}\right)\,, \tag{97}$$

where the only difference between the two cases comes from whether the current has even or odd spin. This is the correct formula that dates back to Wilson and Kogut,[68] see also Ref. 69 for the more clean decomposition into $O(N)$ irreducible structures and higher orders. Note that (97) vanishes for $s = 1$ since it is the global $O(N)$-symmetry current. For the singlet HS currents we find the similar permutations but with different prefactors:

$$\langle K^\bullet K^\bullet \rangle \sim 4N^2\langle 1234\rangle + 16N\langle 2314\rangle + 4N^2\langle 3412\rangle\,, \tag{98}$$

which again leads to Wilson and Kogut:

$$\gamma = 2\gamma_\phi\left(1 - \frac{6}{s(s+1)}\right)\,. \tag{99}$$

The anomalous dimension vanishes for the stress-tensor $s = 2$, as is expected.

3.4.2. *Large-N*

In the case of the large-N expansion first of all we strip off the trivial factors from $\langle KK\rangle$:

$$\langle K_j K_j\rangle = C_{K,j}\frac{[-2(\xi\cdot x)^2]^{j-1}}{(x^2)^{d+2j-2}}\,. \tag{100}$$

After simple manipulations we find ($\Delta_\phi = d/2-1$, $\Delta_\sigma = 2$)

$$C_{K,j} = \nu^2\frac{C^2_{\phi\phi}C_{\sigma\sigma}}{\Gamma[\Delta_\phi]^2\Gamma[\Delta_\sigma]}\Gamma[d+2j+\Delta_\sigma-4]2^{j-d-\Delta_\sigma+2}\times I\,, \tag{101}$$

where I is the integral over the Schwinger parameters:

$$I\Big|_{\Delta_\sigma=2} = \frac{32\sqrt{\pi}(d-1)\Gamma\left(\frac{d}{2}-1\right)\Gamma(d+s-1)}{(d+2s-3)\Gamma\left(\frac{d-1}{2}\right)\Gamma(d+1)\Gamma(s-1)}\,. \tag{102}$$

Using (64) we can write the anomalous dimensions as a multiple of γ_ϕ:

$$\gamma_s = \gamma_\phi\frac{8(s-1)(d+s-2)}{(d+2s-4)(d+2s-2)}\,, \tag{103}$$

which is in accordance with Lang and Ruhl.[4] This formula gives $\gamma_s^{\begin{smallmatrix}\square\\\square\end{smallmatrix}}$ for odd s and $\gamma_s^{\square\square}$ for even s. The same formula is applicable to the large-N expansion of the recently studied $O(N)$ vector-model in $4<d<6$.[70–73] It vanishes for $s=1$ as before.

3.4.3. *Six Dimensions and Nearby*

Another useful toy models are the six-dimensional ϕ^3 theories

$$S = \int d^dx\left(\frac{1}{2}(\partial\phi)^2 + \frac{g}{2}d^{IJK}\phi_I\phi_J\phi_K\right)\,, \tag{104}$$

for various d^{IJK}, see e.g. Refs. 69, 74, of which we consider two particular cases: the scalar QCD[75,76]

$$S = \int d^dx\left(\frac{1}{2}(\partial\phi)^2 + \frac{1}{2}(\partial\sigma)^2 + \frac{g}{2}\phi_i\phi^i\sigma\right)\,, \tag{105}$$

whose β-function can be made to vanish up to g^5 by adjusting the field content; also we include the recently studied UV completion of the $O(N)$ vector-model in $4<d<6$[71–73]

$$S = \int d^dx\left(\frac{1}{2}(\partial\phi)^2 + \frac{1}{2}(\partial\sigma)^2 + \frac{g_1}{2}\phi_i\phi^i\sigma + \frac{g_2}{6}\sigma^3\right)\,. \tag{106}$$

For simplicity we restrict ourselves to the non-singlet currents built out of ϕ^i. In six dimensions we have $C_{\sigma\sigma} = C_{\phi\phi} = \Gamma[d/2-1]/(4\pi^{d/2})$, $\Delta_\sigma = \Delta_\phi = d/2-1 = 2$. From the computational point of view the only difference between this model and the large-N vector-model is in that the normalization of the σ-field is that of the free field. Therefore, we can directly use (102) at $d = 6$. For the anti-symmetric $O(N)$-representation (the same for the symmetric) we find

$$\gamma_s = 2\gamma_\phi \frac{(s-1)(s+4)}{(s+1)(s+2)} = 12\gamma_\phi \left(\frac{1}{6} - \frac{1}{(s+1)(s+2)}\right), \tag{107}$$

which is in accordance with Ref. 41, 65, 75, 76. The anomalous dimension vanishes at $s = 1$ for the current is that of the global $O(N)$-symmetry. The same formula applies to the $6 - \epsilon$ expansion of (106) and to (105) at $6d$.

3.5. *HS algebra interpretation*

Below we vaguely argue how the anomalous dimension of the HS currents at one-loop order are related to the HS algebra representation theory. Despite not having anything to do with loop integrals the computations above reveal that the anomalous dimensions are simpler than some of the ingredients used to derive them: the two-point functions of the HS currents almost cancel those of the non-conservation operators yielding a simple rational expression.

Let us recall that the HS currents multiplet $\mathbb{J} = \sum_s J_s$ forms an irreducible representation $\mathbb{J}$ of the HS algebra corresponding to a free field they are built from. Therefore, the non-conservation operator K belongs to its tensor square for the case of the $4 - \epsilon$ expansion, where we depart from the free theory,

$$K - \mathbb{D}J \in \mathbb{J} \otimes \mathbb{J} \tag{108}$$

In fact, the tensor product we need is $\mathbb{J} \otimes J_0$. By definition, the generating function we found for K can be re-expanded in terms of the Clebsh-Gordon coefficients of the HS algebra, which is (89). For the case of the $6 - \epsilon$-expansion we need instead $\mathbb{J} \otimes \phi$.

Another important ingredient is the kinematical factor c_1 (76) that comes from decoupling the divergence of the HS current inside the two-point function. This tells us that we cannot express everything in terms of the HS algebra associated with the free boson, for which the HS currents are strictly conserved. To get this factor right one needs the HS algebra for generalized free fields near the unitarity bound. Such HS algebra is defined[77] as a quotient of the universal enveloping algebra $U(so(d,2))$ of the conformal algebra by certain ideal:

$$\mathrm{hs}(\lambda) = U\left(\begin{smallmatrix}\square\\\square\end{smallmatrix}\right)/I_\lambda\,, \qquad I_\lambda = \begin{smallmatrix}\square\\\square\\\square\\\square\end{smallmatrix} \oplus (C_2 - C_2(\Delta))\,, \tag{109}$$

where $\begin{smallmatrix}\square\\\square\end{smallmatrix}$ stands for the $so(d,2)$ generators $T_{AB} = -T_{BA}$ and the first part of the ideal is generated by the totally anti-symmetric bilinear $T_{[AB}T_{CD]}$. The first factor of the ideal implies that the field is a scalar one, while the second factor fixes the value of the Casimir operator C_2 to be that of the generalized free field of weight-Δ, i.e. $C_2(\Delta) = \Delta(\Delta - d)$. We are interested in $\Delta = (d-2)/2 + \epsilon$. At $\epsilon = 0$ one finds an additional ideal in $\mathrm{hs}(\lambda)$ and the quotient is the Eastwood-Vasiliev HS algebra[60,78] that corresponds to the free boson. In all the cases the correlation functions of J's can be computed as HS invariants,[32,33] the same for $J \otimes J$ etc. as we argue in Appendix C.

The interpretation of the large-N limit is less straightforward. The auxiliary weight-two field is not a free one, except for $d = 3$ where one can think of σ as $\bar{\psi}\psi$ for the purpose of deriving the Clebsh-Gordon coefficients. In generic dimension one has to take the representation that is dual to $\mathbb{J}$, which will give $\tilde{J}_0$ of weight-two.

It would be important to reformulate the entire procedure in terms of the higher-spin symmetries with the goal to reveal the right algebraic structure governing the HS currents subsector of Wilson-Fisher CFT's.

4. AdS

In this section we discuss various implications for the AdS dual HS theories that can be extracted from the CFT results obtained above. Firstly, we reconstruct the part of the dual HS theory that produces the required three-point functions $\langle J_s OO \rangle$ or better to say the invariants I_{s00} (21). Secondly, we discuss the expected one-loop corrections to the AdS-masses of HS fields. In Appendix C we show how to enlarge HS theories with the duals of the multi-trace operators.

4.1. *Cubic couplings with two scalar legs*

The knowledge of the correlation functions of some CFT can be used to reconstruct the dual AdS theory, see e.g. Ref. 79 for the recent study. The reconstruction approach should work at least at the level of trees since the kinematical volume is the same on the CFT and AdS sides. For example, the number of independent cubic vertices in AdS_{d+1}[80] is equal to the number of independent three-point structures in CFT$_d$.[81] The four-point functions of the scalar operators are fixed up to a function of two conformally-invariant cross-ratios, which is in accordance with the quartic vertices in AdS having a doubly infinite expansion in derivatives.[82] The explicit check up to the quartic order on the example of the free boson vs. a subsector of the $4d$ HS theory was done in Ref. 82, where the crucial observation was that the AdS exchange diagrams reproduce the contribution of the single-trace operators to the four-point function, with the remnant corresponding to the double-trace operators accounted for by the quartic contact vertices.

One can start with a given free CFT, compute the three-point invariant I_{s00} (21) and manufacture the interaction vertex in AdS as to reproduce the correlation functions of that CFT. In the case of the $\langle J_s O_\Delta O_\Delta \rangle$ correlator the relevant part of the bulk action is

$$S = \sum_s \int \frac{1}{2}[\Phi_{\underline{m}(s)}(\Box - m_s^2)\Phi^{\underline{m}(s)} + \ldots] + \sum_s g_s \int \Phi_{\underline{m}(s)} J^{\underline{m}(s)} = S_0 + S_1\,, \quad (110)$$

$$m_{s>0}^2 = \Lambda[(d+s-2)(s-2)-s]\,, \qquad m_0^2 = \Delta(\Delta - d)\,, \quad (111)$$

where we truncated the Fronsdal action to its Klein-Gordon part since it is well-known. The first part makes the free action S_0, while the second one makes the interactions S_1. The interaction is due to the currents $J^{\underline{m}(s)}$ that the HS fields couple to. The currents are built of the scalar field Φ_0 that is a part of the HS multiplet and is included in (110) as $s = 0$. The form of the current can be changed by adding improvements

$$J_{\underline{m}(s)} = \Phi_0(x) \left(\overleftrightarrow{\nabla}_{\underline{m}}\right)^s \Phi_0(x) + O(g_{\underline{mm}}, \Lambda)\,, \tag{112}$$

where we drop the terms with the AdS_{d+1} metric $g_{\underline{mm}}$ and cosmological constant Λ since the boundary-to-bulk propagator for the HS fields[83] is naturally traceless. For the purpose of computing the simplest Witten diagram ⅄ we can pass to the simplified interaction:

$$\int \Phi_{\underline{m}(s)} J^{\underline{m}(s)} \sim \int \Phi_{\underline{m}(s)} \nabla^{\underline{m}(s)} \Phi_0 \Phi_0\,. \tag{113}$$

Therefore, the part of the cubic action that is responsible for $\langle J_s O_\Delta O_\Delta \rangle$ is:

$$S = \sum_s \int \frac{1}{2} [\Phi_{\underline{m}(s)} (\Box - m_s^2) \Phi^{\underline{m}(s)} + \ldots] + \sum_s g_s \int \Phi_{\underline{m}(s)} \nabla^{\underline{m}(s)} \Phi_0 \Phi_0\,. \tag{114}$$

The only information needed from the AdS side is the coefficient of the standard three-point function $\overline{\langle J_s O_\Delta O_\Delta \rangle}$ (13) that is produced from the standard bulk vertex, which was done in Ref. 84 with the result

$$\int \Phi_{\underline{m}(s)} \nabla^{\underline{m}(s)} \Phi_0 \Phi_0 = \tilde{b}_{s00} \times \overline{\langle J_s O_\Delta O_\Delta \rangle}\,, \tag{115}$$

$$\tilde{b}_{s00} = \frac{2^{-5+2s} \pi^{-d/2} (-3+d+2s) \Gamma\left[-1+\frac{d}{2}+s\right]^3 \Gamma[-3+d+s] \Gamma[-1+s+\Delta]^2}{\Gamma[-2+d+2s]^2 \Gamma[\Delta]^2}\,, \tag{116}$$

which corresponds to the unit normalization of the two-point functions.[84] Therefore, $g_s \tilde{b}_{s00}$ should exactly match the CFT invariant I_{s00} (21). For the cases of the free boson and fermion we find:

$$\text{boson}: \qquad g_s^B = \frac{1}{\sqrt{N}} \frac{\pi^{\frac{d-3}{4}} 2^{\frac{1}{2}(3d+s-1)} \Gamma\left(\frac{d-1}{2}\right)}{\Gamma(d+s-3)} \sqrt{\frac{\Gamma\left(\frac{d-1}{2}+s\right)}{\Gamma(s+1)}}\,, \tag{117}$$

$$\text{fermion}: \qquad g_s^F = g_s^B \frac{s}{(d+s-3)}\,. \tag{118}$$

Because of the relation between g^B and g^F displayed in the last line the duals of the free boson and free fermion are different for $d \neq 3$ and should be one and the same HS theory for $d = 3$. It is quite remarkable that the difference between the correlation functions in the two CFT's is compensated by the difference of the same bulk integral with different boundary conditions. The case of the free boson was considered in Ref. 82. At $d = 3$ the coupling has an especially simple form:

$$g_s^B\Big|_{d=3} = g_s^F\Big|_{d=3} = \frac{2^{\frac{s+8}{2}}}{\sqrt{N}\Gamma(s)}\,. \tag{119}$$

There are at least two alternative ways to achieve the same result. Firstly, one can carefully fix the normalization of the two-point functions from the quadratic part of the bulk action, i.e. from the Fronsdal action, as was computed by Metsaev in Ref. 85. Secondly, the Ward identities fix the coupling constants for the CFT three-point functions. Likewise, we can use them to fix the part of the cubic action by studying the bulk counter-part of the Ward identities, as in Ref. 86 for $s = 1$. Moreover, using the fact that the divergence of the HS field boundary-to-bulk propagator with respect to the boundary data is a pure gauge transformation in the bulk,[83] one can see that the CFT HS Ward identities lead to the AdS HS Ward identities.

At this point it is sensible to ask what is the first test of the tree-level HS AdS/CFT that does not immediately follow from the bulk/boundary Ward identities. Indeed, the Fronsdal gauge transformations $\delta\Phi_s = \nabla\xi_{s-1}$ receive corrections as to make the interaction part S_1 of the action gauge invariant. For example, (110) is gauge invariant provided the scalar field transformations, which are trivial at the free level $\delta_0\Phi_0 = 0$, get corrected as $\delta_1\Phi_0 = \sum g_s\nabla\ldots\nabla\xi_s\Phi_0$. In general, the condition for the action to be gauge-invariant to the cubic order is

$$\delta_1 S_0 + \delta_0 S_1 + O(g^2) = 0 \qquad \Longleftrightarrow \qquad \mathbb{D}\langle JJJ\rangle = \langle JJ\rangle\,. \tag{120}$$

The left-hand side can be seen to be related to the CFT Ward identities on the right. When the Ward identities are applied to the holographic three-point functions $\mathbb{D}\langle JJJ\rangle = \mathbb{D}S_1$, the property of the boundary-to-bulk propagators[83] turns $\mathbb{D}S_1$ into $\delta_0 S_1$. If $\delta_0 S_1 \neq 0$ the correction to the gauge transformations δ_1 is needed to ensure the gauge invariance of the action. Therefore, if $\delta_0 S_1 \neq 0$ there is a relation between S_1 and S_0. This is how the gauge invariance in the bulk is related to the CFT Ward identities.

However, if $\delta_0 S_1 = 0$, which is the case for the abelian vertices of the type Weyl tensor cubed, no relation between S_1 and S_0 follows. The number of abelian vertices should be in accordance with the number of the CFT structures that are well-defined as distributions and obey the trivial Ward identities.[87] Therefore, the abelian part of S_1 cannot be determined this way and makes a nontrivial prediction. The HS Ward identities (120.right) capture the full structure of the HS algebra, which is in accordance with the non-abelian vertices being fixed in terms of the HS algebra structure constants as in Refs. 88–95.

It seems that one can fix the complete cubic action of the $4d$ HS theory by employing the old result by Metsaev.[96,97] Namely, the most general ansatz for the $4d$ cubic vertices[98] in flat space was taken in Refs. 96, 97 and the closure of the Poincare algebra at the quartic order was studied. It was found that all the cubic couplings are fixed in terms of a single coupling constant with the result:

$$g_{s_1,s_2,s_3} \sim \frac{1}{\Gamma[s_1+s_2+s_3]}\,, \tag{121}$$

where we display the important part of the spin dependence. Remarkably, g_{s_1,s_2,s_3} agrees with g_s obtained above for $s_2 = s_3 = 0$. The rationale for why the same formula is expected to work in AdS_4 is as follows. The interactions of HS fields contain higher derivatives which are accompanied by negative powers of the cosmological constant — the crucial part of the Fradkin-Vasiliev mechanism,[88,99] which is at the heart of the common belief that no sensible flat limit exists for HS theories. However, the flat limit exists for cubic vertices[100] — in such a limit only the highest derivative terms survive and go over into the vertices classified by Metsaev in flat space.[80,101] The flat limit of the quartic vertex should reveal some non-localities and indeed those are present in Ref. 96,97. Therefore, not only the cubic formula (121) should work but there can be a meaningful flat limit of the HS theory with some mild non-localities present.

4.2. *AdS masses at one loop*

Despite some mild pathologies present in $4-\epsilon$ expansion of the Wilson-Fisher (WF) CFT whenever the space dimension is non-integer,[102] the physical observables are well-defined. The very existence of the $4-\epsilon$ approach suggests that the dual HS theory might be defined in $AdS_{5-\epsilon}$. In principle, there exist d-dimensional Vasiliev equations[60] at any integer d. They are difficult to define at fractional space-time dimension due to the HS algebra that is built in. One can expect that whenever some observable can be scalarized it then can easily be extended to any d. The duality between $WF_{4-\epsilon}$ and the HS theory in $AdS_{5-\epsilon}$ is complementary to the Klebanov-Polyakov conjecture that deals with the large-N expansion.

In case the bulk counterpart of the quantum HS symmetry breaking is in accordance with Ref. 34 the one-loop anomalous dimensions re-derived above with the help of the broken HS symmetry can be used to predict the corrections to the AdS_5 masses of HS fields:

$$\delta m_s^2 = -2(s-2)\epsilon + 2\epsilon^2\gamma_\phi\left(1-\frac{6}{s(s+1)}\right)\,, \tag{122}$$

where the first term is just the ϵ-expansion of the Fronsdal field's mass (111). Similar prediction within the $1/N$-duality was given by Ruhl in Ref. 103 for the AdS_4/CFT_3 duality:

$$\delta m_s^2 = 4\gamma_\phi(s-2)\,, \qquad \gamma_s\Big|_{d=3} = 4\gamma_\phi\frac{(s-2)}{(2s-1)}\,, \qquad \gamma_\phi\Big|_{d=3} = \frac{4}{3\pi^2}\,. \tag{123}$$

Borrowing the result by Ruhl and Lang[4,104] on anomalous dimensions of the HS currents in any dimension d, which can be simplified to

$$\gamma_s = \frac{8\gamma_\phi}{(d+2s-4)(d+2s-2)}\left((d+s-2)(s-1)-\frac{\Gamma[d+1]\Gamma[s+1]}{2(d-1)\Gamma[d+s-3]}\right)\,,$$

one expects to find at one-loop for the mass shift of AdS_{d+1} HS fields:

$$\delta m_s^2 = (d+2s-4)\gamma_s\,. \tag{124}$$

Note that there are two types of contributions to the anomalous dimensions: one is exactly the same as for the non-singlet currents (103) and another one comes in the

Feynman diagrams language from the additional diagrams where the indices can form closed loops, which are absent for the non-singlet case. In other words[105] first term comes from the σ-exchange while the second receives contributions from the sea of the HS currents. It would be interesting to see how such contributions can be manufactured in AdS.

5. Conclusions

Our main conclusions are as follows. It is very likely that the presence of an at least one HS conserved tensor in a $d > 2$ CFT implies that the CFT is a free one in disguise. Therefore, what makes the HS symmetry interesting is its breaking.[31,37]

There are at least two different ways to break HS symmetries: classical and quantum, the former is realized in Yang-Mills type theories and the latter in Wilson-Fisher CFT's. The bulk realization of the classical breaking is via the usual BEH mechanism and requires HS theories to be extended with an appropriate 'matter content'. On contrary, the quantum breaking does not seem to require any new fields to be coupled, but it has not been yet observed in detail. Knowing the one-loop anomalous dimensions of the HS currents allows to make predictions for the corrections to the AdS-masses of HS fields. In particular it may be instructive to extend the duality to fractional dimensions where the $4-\epsilon$ Wilson-Fisher CFT is dual to HS theory in $AdS_{5-\epsilon}$.

The quantum breaking is realized in the CFT's with less operators at disposal. In both the cases, there is an exact quantum equation of motion that describes the recombination of HS currents with a non-conservation operator. The difference is in the structure of such an operator. We have assumed that the non-conservation equation is $\partial\cdot J = gJJ$ and have used this equation to extract anomalous dimensions of the HS currents to the lowest nontrivial order.

There is a number of obvious extensions of the present work. First of all, one can extend the treatment of the large-N to the singlet sector as well. One can study the fermionic vector-models and more complicated models of Yukawa type. Also, one can allow for parity-violating non-conservation operators as in Refs. 35, 37, which occurs in the Chern-Simons matter theories.

It would be important to understand what is the algebraic structure behind the quantum HS symmetry breaking, which should allow to avoid any Feynman diagrams computation at higher orders.

Given a CFT one can attempt to manufacture the AdS vertices as to reproduce the given correlation functions. The success story in the case of the free $3d$ scalar vs. a subsector of the $4d$ HS theory up to some of the quartic vertices is in Refs. 82, 106. The reconstruction approach raises at least two questions. First, does it work for any CFT, even such as free CFT's? Second, more specific, is the dual of the free boson the same HS theory that solves the Noether procedure for HS fields? The independent check from the bulk side of the reconstruction approach would be important. It is possible to fix certain parts of the HS theory action directly on the AdS side.[88–94] In all known cases the coupling constants are related to the HS algebra in a simple way. Moreover, the bulk/boundary Ward identities arguments show that such results are not independent from the CFT. The nontrivial test at the cubic level should come from the abelian part of the vertices.

Acknowledgments

I would like to thank Alexander Manashov and John Gracey for correspondence. I am indebted to Alexander Manashov, Ruslan Metsaev, Dmitry Ponomarev, Charlotte Sleight, Arkady Tseytlin and Sasha Zhiboedov for the very useful discussions and comments and to Dmitry Ponomarev for proofreading. I also would like to thank the organizers of the 3rd Higher-Spins and Holography conference, Moscow, Russia and International Workshop on Higher Spin Gauge Theories, Singapore, where some of these results were reported. I am also grateful to the organizers of the School and Workshop on Higher Spins, Strings and Dualities at Quintay, December 7-18, 2015, Chile for the warm hospitality during the final stage of this work, which was supported by Conicyt grant DPI 20140115. The research was supported in parts by the RFBR Grant No 14-02-01172. This work was supported by the DFG Transregional Collaborative Research Centre TRR 33 and the DFG cluster of excellence Origin and Structure of the Universe.

A. Two- and Three-Point Functions, Thomas Derivative

As is well-known, see e.g. Ref. 42, the two- and three-point functions are fixed by conformal symmetry up to few numbers. For the $\langle 00\rangle$, $\langle ss\rangle$, $\langle 000\rangle$ and $\langle s00\rangle$ cases there is a unique conformally-invariant structure and the ambiguity is in overall factor only, which we omit:

$$\begin{aligned}
\langle O_\Delta(x_1)O_\Delta(x_2)\rangle &= \frac{1}{(x_{12}^2)^\Delta}\,,\\
\langle O_\Delta^{a(s)}(x_1)O_\Delta^{b(s)}(x_2)\rangle &= \frac{1}{(x_{12}^2)^\Delta}\big(P^{ab}\dots P^{ab}-\text{traces}\big)\,,\\
\langle O_{\Delta_1}(x_1)O_{\Delta_2}(x_2)O_{\Delta_3}(x_3)\rangle &= \frac{1}{(x_{12}^2)^{\frac{\Delta_1+\Delta_2-\Delta_3}{2}}(x_{23}^2)^{\frac{\Delta_2+\Delta_3-\Delta_1}{2}}(x_{13}^2)^{\frac{\Delta_1+\Delta_3-\Delta_2}{2}}}\,,\\
\langle O_{\Delta_1}^{a(s)}(x_1)O_{\Delta_2}(x_2)O_{\Delta_3}(x_3)\rangle &= \big(Q^a\dots Q^a-\text{traces}\big)\times\\
&\quad\times\langle O_{\Delta_1}(x_1)O_{\Delta_2}(x_2)O_{\Delta_3}(x_3)\rangle\left(\frac{x_{12}^2x_{13}^2}{x_{23}^2}\right)^{\frac{s}{2}}\,,\\
P^{ab} &= I^{ab}(x_{12})\,, \qquad\qquad I_{ab}(x)=\delta_{ab}-\frac{2x_ax_b}{x^2}\,,\\
Q^a &= \left(\frac{x_{21}^a}{x_{21}^2}-\frac{x_{31}^a}{x_{31}^2}\right)\,.
\end{aligned}$$

Throughout the text we use the structures above as the standard ones. Since the primary operators have to be traceless tensors we need either to impose the trace projector explicitly or consistently work on the space of all tensors modulo pure traces. The second option is more appealing and can be implemented by contracting all tensor indices with a light-like polarization vectors ξ^a, $\xi^a\xi_a = 0$, which takes away the traces:

$$P = P^{ab}\xi^1_a\xi^2_b\,, \qquad\qquad Q = Q^a\xi^1_a\,, \tag{A.1}$$

and in the most of the cases we can employ a single polarization vector, $\xi_{1,2} = \xi$.

The price to pay is that in order to release an index one needs to use the Thomas derivative, which is an operator that preserves the equivalence relation $f(\xi) \sim f(\xi) + \xi^2 g(\xi)$, i.e. $\mathrm{d}_a(\xi^2 f(\xi)) = O(\xi^2)$. It was obtained in Ref. 50 and by Thomas,[107] see also Ref. 81:

$$\mathrm{d}_a = \left(\frac{d}{2} - 1 + \xi^m \partial^\xi_m\right)\partial^\xi_a - \frac{1}{2}\xi_a \Delta^\xi\,. \tag{A.2}$$

In order to check the conservation of a HS current we need to combine the usual derivative with the Thomas operator:

$$\mathbb{D}O(x;\xi) = \frac{\partial}{\partial x_a}\mathrm{d}_a O(x;\xi)\,. \tag{A.3}$$

B. Models

Below we briefly summarize the definitions and the lowest order results for the models that we consider in the main text.

B.1. *Wilson-Fisher* $4-\epsilon$

The famous Wilson-Fisher model[1] is defined by the following action:

$$S = \int d^dx\,\left[(\partial\phi)^2 + \frac{g\mu^\epsilon}{4}(\phi^2)^2\right]. \tag{B.1}$$

The one-loop results[68] for the β-function and anomalous dimensions of the operators ϕ and ϕ^2 are:

$$\beta = -\epsilon g + (N+8)\frac{g^2}{8\pi^2}\,, \qquad g_* = \frac{8\pi^2}{N+8}\epsilon\,, \tag{B.2}$$

$$\gamma_\phi = \frac{N+2}{4(N+8)^2}\epsilon^2\,, \qquad \Delta_\phi = \frac{d}{2} - 1 + \gamma_\phi\,, \tag{B.3}$$

$$\gamma_{\phi^2} = \frac{N+2}{N+8}\epsilon\,, \qquad \Delta_\phi = d - 2 + \gamma_{\phi^2}\,, \tag{B.4}$$

and one has to replace d with $d-\epsilon$. The anomalous dimensions of the HS currents have been computed in Ref. 68,69 and we quote them in the main text.

B.2. *Large-N vector-model*

The large-N expansion is obtained by introducing the Hubbard-Stratanovich field into (B.1):

$$S = \int d^d x \left(\frac{1}{2}(\partial\phi)^2 + \frac{1}{2}\sigma\phi^2 - \frac{3N}{2g}\sigma^2 \right) . \tag{B.5}$$

The two-point function of the spin field ϕ^i is that of the free field, while the two-point function of the auxiliary field σ to the leading order is

$$\langle \sigma(x_1)\sigma(x_2) \rangle = \frac{C_{\sigma\sigma}}{(x_{12}^2)^2} , \qquad C_{\sigma\sigma} = \frac{2^{d+2} \sin\left(\frac{\pi d}{2}\right) \Gamma\left(\frac{d-1}{2}\right)}{\pi^{3/2}\Gamma\left(\frac{d}{2}-2\right)} . \tag{B.6}$$

The leading $1/N$ corrections to the dimension of ϕ and σ[2–4,104,108,109] are

$$\Delta_\phi = \frac{d}{2} - 1 + \frac{1}{N}\gamma_\phi + O\left(\frac{1}{N^2}\right) , \qquad \gamma_\phi = \frac{2\sin\left(\frac{\pi d}{2}\right)\Gamma(d-2)}{\pi\Gamma\left(\frac{d}{2}+1\right)\Gamma\left(\frac{d}{2}-2\right)} , \tag{B.7}$$

$$\Delta_\sigma = 2 + \frac{1}{N}\gamma_\sigma + O\left(\frac{1}{N^2}\right) , \qquad \gamma_\sigma = \frac{4(d-2)(d-2)}{(d-4)}\gamma_\phi . \tag{B.8}$$

The anomalous dimensions for the HS currents are known to the leading order only for the singlet sector[4,104] and to $1/N^2$ for the non-singlets.[110] In the main text we use $\Box\phi^i = g_\star \sigma\phi^i$, where $g_\star = 1$.

C. Duals of the Multi-Trace Operators

The Vasiliev HS theories consists of the gauge fields plus a number of scalar/fermions that belong to the same HS algebra multiplet as the graviton and genuine HS gauge fields. There are several reasons to discuss how to extend these HS theories with matter-like fields: making contact with string theory, whose spectrum is by far larger than the spectrum of any of the HS theories — constant vs. exponential growth of the number of states; understanding the classical breaking of HS symmetries.

It turns out that one can easily extend the Vasiliev HS theories with the fields that are dual to multi-trace operators. This works at least to the lowest order. HS theories can always be thought of as the duals of some free CFT's. In the latter case there is a free field ϕ that corresponds to a representation, say S, of the conformal algebra. Then, the HS algebra is the algebra of all endomorphisms of S, i.e. $S \otimes S^*$. The single-trace operators correspond to $S \otimes S$. As was illustrated above $S \otimes S$ contains an infinite number of HS currents. In the AdS_4/CFT_3 case the HS algebra is the algebra of functions in two pairs of operators obeying canonical commutation relations[111]

$$[Y^A, Y^B] = 2iC^{AB} , \qquad A, B, \ldots = 1, \ldots, 4 . \tag{C.1}$$

Formally, $S \otimes S$ and $S \otimes S^*$ are isomorphic as vector spaces and one can embed the duals of the single-trace operators into the HS algebra, which leads to the twisted-adjoint action[23]

$$dC = \omega \star C - C \star \pi(\omega) , \tag{C.2}$$

where ω is a connection of the HS algebra and π is induced by the auto-morphism that flips the sign of the AdS-translation (in the CFT it exchanges translations and inversions).

Multi-trace operators are given by higher products $(S \otimes S)^k$ and it is possible to embed them into $(S \otimes S^*)^k$. To add the duals of multi-trace operators one just needs to take several copies of the same oscillators

$$[Y^A_m, Y^B_n] = 2iC^{AB}\delta_{mn} \qquad n, m = 1, \ldots, k\,. \tag{C.3}$$

Multiple copies of the HS algebra oscillators naturally show up when studying the conformal HS fields in generalized space-times as in Ref. 112,113. The equations for the duals of the multi-trace operators are

$$dC_k = \omega \star C_k - C_k \star \pi(\omega)\,, \tag{C.4}$$

where ω acts on C_k as on the tensor product. Such equation describes the coupling between gauge HS fields and the duals of multi-trace operators to the lowest order. Higher-trace operators are dual to massive fields as the higher tensor products decompose into long representations.[47]

The lowest component in the order-k multi-trace tower is the dual of $(\phi^2)^k$-type operators, which has to have the conformal weight $\Delta_k = k(d-2)$, i.e. k in $d = 3$. This can be easily seen by repeating the derivation of the boundary-to-bulk propagators in Ref. 33. In particular, the correlation functions in the free theory can be reproduced as HS invariants, which immediately follows from Ref. 33. Such invariants are needed to reproduce the correlation functions of double-trace operators JJ from the HS algebra perspective.

References

1. K. G. Wilson and M. E. Fisher, Critical exponents in 3.99 dimensions, *Phys. Rev. Lett.* **28**, 240–243, (1972). doi: 10.1103/PhysRevLett.28.240.
2. R. Abe, Expansion of a Critical Exponent in Inverse Powers of Spin Dimensionality, *Prog. Theor. Phys.* **49(1)**, 113–128, (1973). doi: doi:10.1143/PTP.49.113.
3. A. N. Vasiliev, M. Pismak, Yu, and Yu. R. Khonkonen, Simple Method of Calculating the Critical Indices in the $1/N$ Expansion, *Theor. Math. Phys.* **46**, 104–113, (1981). doi: 10.1007/BF01030844. [Teor. Mat. Fiz.46,157(1981)].
4. K. Lang and W. Ruhl, Anomalous dimensions of tensor fields of arbitrary rank for critical nonlinear O(N) sigma models at $2 < d < 4$ to first order in 1/N, *Z. Phys.* **C51**, 127–132, (1991). doi: 10.1007/BF01579569.
5. S. El-Showk, M. F. Paulos, D. Poland, S. Rychkov, D. Simmons-Duffin, and A. Vichi, Solving the 3D Ising Model with the Conformal Bootstrap, *Phys. Rev.* **D86**, 025022, (2012). doi: 10.1103/PhysRevD.86.025022.
6. S. El-Showk, M. F. Paulos, D. Poland, S. Rychkov, D. Simmons-Duffin, and A. Vichi, Solving the 3d Ising Model with the Conformal Bootstrap II. c-Minimization and Precise Critical Exponents, *J. Stat. Phys.* **157**, 869, (2014). doi: 10.1007/s10955-014-1042-7.
7. J. M. Maldacena, The large N limit of superconformal field theories and supergravity, *Adv. Theor. Math. Phys.* **2**, 231–252, (1998).

8. S. S. Gubser, I. R. Klebanov, and A. M. Polyakov, Gauge theory correlators from non-critical string theory, *Phys. Lett.* **B428**, 105–114, (1998). doi: 10.1016/S0370-2693(98)00377-3.
9. E. Witten, Anti-de Sitter space and holography, *Adv. Theor. Math. Phys.* **2**, 253–291, (1998).
10. B. Sundborg, Stringy gravity, interacting tensionless strings and massless higher spins, *Nucl. Phys. Proc. Suppl.* **102**, 113–119, (2001). doi: 10.1016/S0920-5632(01)01545-6.
11. E. Sezgin and P. Sundell, Massless higher spins and holography, *Nucl.Phys.* **B644**, 303–370, (2002). doi: 10.1016/S0550-3213(02)00739-3.
12. I. R. Klebanov and A. M. Polyakov, AdS dual of the critical O(N) vector model, *Phys. Lett.* **B550**, 213–219, (2002). doi: 10.1016/S0370-2693(02)02980-5.
13. E. Sezgin and P. Sundell, Holography in 4D (super) higher spin theories and a test via cubic scalar couplings, *JHEP.* **0507**, 044, (2005). doi: 10.1088/1126-6708/2005/07/044.
14. R. G. Leigh and A. C. Petkou, Holography of the N=1 higher spin theory on AdS(4), *JHEP.* **0306**, 011, (2003).
15. M. A. Vasiliev, Consistent equation for interacting gauge fields of all spins in (3+1)-dimensions, *Phys. Lett.* **B243**, 378–382, (1990).
16. S. Giombi and X. Yin, Higher Spin Gauge Theory and Holography: The Three-Point Functions, *JHEP.* **1009**, 115, (2010). doi: 10.1007/JHEP09(2010)115.
17. S. Giombi, I. R. Klebanov, S. S. Pufu, B. R. Safdi, and G. Tarnopolsky, AdS Description of Induced Higher-Spin Gauge Theory, *JHEP.* **1310**, 016, (2013). doi: 10.1007/JHEP10(2013)016.
18. A. Tseytlin, On partition function and Weyl anomaly of conformal higher spin fields, *Nucl.Phys.* **B877**, 598–631, (2013). doi: 10.1016/j.nuclphysb.2013.10.009.
19. S. Giombi and I. R. Klebanov, One Loop Tests of Higher Spin AdS/CFT, *JHEP.* **1312**, 068, (2013). doi: 10.1007/JHEP12(2013)068.
20. S. Giombi, I. R. Klebanov, and A. A. Tseytlin, Partition Functions and Casimir Energies in Higher Spin AdS_{d+1}/CFT_d, *Phys. Rev.* **D90**(2), 024048, (2014). doi: 10.1103/PhysRevD.90.024048.
21. M. Beccaria, X. Bekaert, and A. A. Tseytlin, Partition function of free conformal higher spin theory, *JHEP.* **1408**, 113, (2014). doi: 10.1007/JHEP08(2014)113.
22. M. Beccaria and A. Tseytlin, On higher spin partition functions. (2015).
23. M. A. Vasiliev, Consistent equations for interacting massless fields of all spins in the first order in curvatures, *Annals Phys.* **190**, 59–106, (1989).
24. M. A. Vasiliev, Dynamics of massless higher spins in the second order in curvatures, *Phys. Lett.* **B238**, 305–314, (1990).
25. E. S. Fradkin and M. A. Vasiliev, Candidate to the role of higher spin symmetry, *Ann. Phys.* **177**, 63, (1987).
26. J. Maldacena and A. Zhiboedov, Constraining Conformal Field Theories with A Higher Spin Symmetry. (2011).
27. V. Alba and K. Diab, Constraining conformal field theories with a higher spin symmetry in d=4. (2013).
28. N. Boulanger, D. Ponomarev, E. Skvortsov, and M. Taronna, On the uniqueness of higher-spin symmetries in AdS and CFT. (2013). doi: 10.1142/S0217751X13501625.
29. Y. S. Stanev, Constraining conformal field theory with higher spin symmetry in four dimensions, *Nucl. Phys.* **B876**, 651–666, (2013). doi: 10.1016/j.nuclphysb.2013.09.002.
30. V. Alba and K. Diab, Constraining conformal field theories with a higher spin sym-

metry in d¿ 3 dimensions. (2015).
31. D. Anselmi, The N=4 quantum conformal algebra, *Nucl. Phys.* **B541**, 369–385, (1999). doi: 10.1016/S0550-3213(98)00848-7.
32. N. Colombo and P. Sundell, Higher Spin Gravity Amplitudes From Zero-form Charges. (2012).
33. V. Didenko and E. Skvortsov, Exact higher-spin symmetry in CFT: all correlators in unbroken Vasiliev theory, *JHEP.* **1304**, 158, (2013). doi: 10.1007/JHEP04(2013)158.
34. L. Girardello, M. Porrati, and A. Zaffaroni, 3-D interacting CFTs and generalized Higgs phenomenon in higher spin theories on AdS, *Phys. Lett.* **B561**, 289–293, (2003). doi: 10.1016/S0370-2693(03)00492-1.
35. S. Giombi, S. Minwalla, S. Prakash, S. P. Trivedi, S. R. Wadia, and X. Yin, Chern-Simons Theory with Vector Fermion Matter, *Eur. Phys. J.* **C72**, 2112, (2012). doi: 10.1140/epjc/s10052-012-2112-0.
36. S. Giombi and X. Yin, The Higher Spin/Vector Model Duality, *J.Phys.* **A46**, 214003, (2013). doi: 10.1088/1751-8113/46/21/214003.
37. J. Maldacena and A. Zhiboedov, Constraining conformal field theories with a slightly broken higher spin symmetry, *Class. Quant. Grav.* **30**, 104003, (2013). doi: 10.1088/0264-9381/30/10/104003.
38. Z. Komargodski and A. Zhiboedov, Convexity and Liberation at Large Spin, *JHEP.* **11**, 140, (2013). doi: 10.1007/JHEP11(2013)140.
39. L. F. Alday and A. Zhiboedov, Conformal Bootstrap With Slightly Broken Higher Spin Symmetry. (2015).
40. S. Rychkov and Z. M. Tan, The ϵ-expansion from conformal field theory, *J. Phys.* **A48**(29), 29FT01, (2015). doi: 10.1088/1751-8113/48/29/29FT01.
41. A. N. Manashov and M. Strohmaier, Conformal constraints for anomalous dimensions of leading twist operators, *Eur. Phys. J.* **C75**(8), 363, (2015). doi: 10.1140/epjc/s10052-015-3595-2.
42. E. S. Fradkin and M. Ya. Palchik, *Conformal quantum field theory in D-dimensions.* 1996.
43. S. Giombi, S. Prakash, and X. Yin, A Note on CFT Correlators in Three Dimensions. (2011).
44. X. Bekaert and M. Grigoriev, Higher order singletons, partially massless fields and their boundary values in the ambient approach, *Nucl. Phys.* **B876**, 667–714, (2013). doi: 10.1016/j.nuclphysb.2013.08.015.
45. O. A. Gelfond and M. A. Vasiliev, Operator algebra of free conformal currents via twistors, *Nucl. Phys.* **B876**, 871–917, (2013). doi: 10.1016/j.nuclphysb.2013.09.001.
46. M. Flato and C. Fronsdal, One Massless Particle Equals Two Dirac Singletons: Elementary Particles in a Curved Space. 6., *Lett.Math.Phys.* **2**, 421–426, (1978). doi: 10.1007/BF00400170.
47. M. A. Vasiliev, Higher spin superalgebras in any dimension and their representations, *JHEP.* **12**, 046, (2004).
48. H. Osborn and A. C. Petkou, Implications of conformal invariance in field theories for general dimensions, *Annals Phys.* **231**, 311–362, (1994). doi: 10.1006/aphy.1994.1045.
49. D. Anselmi, Theory of higher spin tensor currents and central charges, *Nucl. Phys.* **B541**, 323–368, (1999). doi: 10.1016/S0550-3213(98)00783-4.
50. N. S. Craigie, V. K. Dobrev, and I. T. Todorov, Conformally Covariant Composite Operators in Quantum Chromodynamics, *Annals Phys.* **159**, 411–444, (1985). doi: 10.1016/0003-4916(85)90118-6.
51. D. E. Diaz and H. Dorn, On the AdS higher spin / O(N) vector model corre-

spondence: Degeneracy of the holographic image, *JHEP.* **07**, 022, (2006). doi: 10.1088/1126-6708/2006/07/022.
52. D. Anselmi, Higher spin current multiplets in operator product expansions, *Class. Quant. Grav.* **17**, 1383–1400, (2000). doi: 10.1088/0264-9381/17/6/305.
53. K. Alkalaev, Mixed-symmetry tensor conserved currents and AdS/CFT correspondence, *J. Phys.* **A46**, 214007, (2013). doi: 10.1088/1751-8113/46/21/214007.
54. K. Alkalaev, Massless hook field in AdS(d+1) from the holographic perspective, *JHEP.* **01**, 018, (2013). doi: 10.1007/JHEP01(2013)018.
55. S. E. Konstein and M. A. Vasiliev, Extended higher spin superalgebras and their massless representations, *Nucl. Phys.* **B331**, 475–499, (1990).
56. X. Bekaert and M. Grigoriev, Manifestly conformal descriptions and higher symmetries of bosonic singletons, *SIGMA.* **6**, 038, (2010).
57. O. Gelfond, E. Skvortsov, and M. Vasiliev, Higher spin conformal currents in Minkowski space, *Theor.Math.Phys.* **154**, 294–302, (2008). doi: 10.1007/s11232-008-0027-6.
58. C. Fronsdal, Massless fields with integer spin, *Phys. Rev.* **D18**, 3624, (1978).
59. V. E. Didenko and E. D. Skvortsov, Towards higher-spin holography in ambient space of any dimension, *J. Phys.* **A46**, 214010, (2013). doi: 10.1088/1751-8113/46/21/214010.
60. M. A. Vasiliev, Nonlinear equations for symmetric massless higher spin fields in (a)ds(d), *Phys. Lett.* **B567**, 139–151, (2003).
61. M. Bianchi, P. J. Heslop, and F. Riccioni, More on La Grande Bouffe, *JHEP.* **08**, 088, (2005). doi: 10.1088/1126-6708/2005/08/088.
62. Y. M. Zinoviev, On massive high spin particles in (a)ds. (2001).
63. B. Eden, C. Jarczak, and E. Sokatchev, A Three-loop test of the dilatation operator in N = 4 SYM, *Nucl. Phys.* **B712**, 157–195, (2005). doi: 10.1016/j.nuclphysb.2005.01.036.
64. J. Henn, C. Jarczak, and E. Sokatchev, On twist-two operators in N=4 SYM, *Nucl. Phys.* **B730**, 191–209, (2005). doi: 10.1016/j.nuclphysb.2005.09.043.
65. A. V. Belitsky, J. Henn, C. Jarczak, D. Mueller, and E. Sokatchev, Anomalous dimensions of leading twist conformal operators, *Phys. Rev.* **D77**, 045029, (2008). doi: 10.1103/PhysRevD.77.045029.
66. A. Raju, ϵ-Expansion in the Gross-Neveu CFT. (2015).
67. S. Ghosh, R. K. Gupta, K. Jaswin, and A. A. Nizami, ϵ-Expansion in the Gross-Neveu Model from Conformal Field Theory. (2015).
68. K. G. Wilson and J. B. Kogut, The Renormalization group and the epsilon expansion, *Phys. Rept.* **12**, 75–200, (1974). doi: 10.1016/0370-1573(74)90023-4.
69. V. M. Braun and A. N. Manashov, Evolution equations beyond one loop from conformal symmetry, *Eur. Phys. J.* **C73**, 2544, (2013). doi: 10.1140/epjc/s10052-013-2544-1.
70. X. Bekaert, E. Meunier, and S. Moroz, Towards a gravity dual of the unitary Fermi gas, *Phys. Rev.* **D85**, 106001, (2012). doi: 10.1103/PhysRevD.85.106001.
71. L. Fei, S. Giombi, and I. R. Klebanov, Critical $O(N)$ models in $6-\epsilon$ dimensions, *Phys. Rev.* **D90**(2), 025018, (2014). doi: 10.1103/PhysRevD.90.025018.
72. L. Fei, S. Giombi, I. R. Klebanov, and G. Tarnopolsky, Three loop analysis of the critical O(N) models in $6-\epsilon$ dimensions, *Phys. Rev.* **D91**(4), 045011, (2015). doi: 10.1103/PhysRevD.91.045011.
73. J. A. Gracey, Four loop renormalization of ϕ^3 theory in six dimensions, *Phys. Rev.* **D92**(2), 025012, (2015). doi: 10.1103/PhysRevD.92.025012.
74. O. F. de Alcantara Bonfim, J. E. Kirkham, and A. J. McKane, Critical Exponents

to Order ϵ^3 for ϕ^3 Models of Critical Phenomena in Six ϵ-dimensions, *J. Phys.* **A13**, L247, (1980). doi: 10.1088/0305-4470/13/7/006. [Erratum: J. Phys.A13,3785(1980)].
75. L. Baulieu, E. G. Floratos, and C. Kounnas, Parton Model Interpretation of the Cut Vertex Formalism, *Nucl. Phys.* **B166**, 321, (1980). doi: 10.1016/0550-3213(80) 90230-8.
76. S. V. Mikhailov and A. V. Radyushkin, Evolution Kernel for the Pion Wave Function: Two Loop Calculation in Scalar ϕ^3 in Six-dimensions Model, *Theor. Math. Phys.* **65**, 999, (1986). doi: 10.1007/BF01028633. [Teor. Mat. Fiz.65,44(1985)].
77. K. B. Alkalaev, M. Grigoriev, and E. D. Skvortsov, Uniformizing higher-spin equations, *J. Phys.* **A48**(1), 015401, (2015). doi: 10.1088/1751-8113/48/1/015401.
78. M. G. Eastwood, Higher symmetries of the Laplacian, *Annals Math.* **161**, 1645–1665, (2005). doi: 10.4007/annals.2005.161.1645.
79. S. El-Showk and K. Papadodimas, Emergent Spacetime and Holographic CFTs, *JHEP.* **10**, 106, (2012). doi: 10.1007/JHEP10(2012)106.
80. R. R. Metsaev, Cubic interaction vertices for massive and massless higher spin fields, *Nucl. Phys.* **B759**, 147–201, (2006). doi: 10.1016/j.nuclphysb.2006.10.002.
81. M. S. Costa, J. Penedones, D. Poland, and S. Rychkov, Spinning Conformal Correlators, *JHEP.* **11**, 071, (2011). doi: 10.1007/JHEP11(2011)071.
82. X. Bekaert, J. Erdmenger, D. Ponomarev, and C. Sleight, Quartic AdS Interactions in Higher-Spin Gravity from Conformal Field Theory, *JHEP.* **11**, 149, (2015). doi: 10.1007/JHEP11(2015)149.
83. A. Mikhailov, Notes on higher spin symmetries. (2002).
84. M. S. Costa, V. Gonçalves, and J. a. Penedones, Spinning AdS Propagators, *JHEP.* **09**, 064, (2014). doi: 10.1007/JHEP09(2014)064.
85. R. R. Metsaev, Gauge invariant two-point vertices of shadow fields, AdS/CFT, and conformal fields, *Phys. Rev.* **D81**, 106002, (2010). doi: 10.1103/PhysRevD.81.106002.
86. D. Z. Freedman, S. D. Mathur, A. Matusis, and L. Rastelli, Correlation functions in the CFT(d) / AdS(d+1) correspondence, *Nucl. Phys.* **B546**, 96–118, (1999). doi: 10.1016/S0550-3213(99)00053-X.
87. J. Erdmenger and H. Osborn, Conserved currents and the energy momentum tensor in conformally invariant theories for general dimensions, *Nucl. Phys.* **B483**, 431–474, (1997). doi: 10.1016/S0550-3213(96)00545-7.
88. E. S. Fradkin and M. A. Vasiliev, Cubic Interaction in Extended Theories of Massless Higher Spin Fields, *Nucl. Phys.* **B291**, 141, (1987). doi: 10.1016/0550-3213(87) 90469-X.
89. M. A. Vasiliev, Cubic interactions of bosonic higher spin gauge fields in ads(5), *Nucl. Phys.* **B616**, 106–162, (2001).
90. K. B. Alkalaev and M. A. Vasiliev, N = 1 supersymmetric theory of higher spin gauge fields in ads(5) at the cubic level, *Nucl. Phys.* **B655**, 57–92, (2003).
91. M. A. Vasiliev, Cubic Vertices for Symmetric Higher-Spin Gauge Fields in $(A)dS_d$, *Nucl. Phys.* **B862**, 341–408, (2012). doi: 10.1016/j.nuclphysb.2012.04.012.
92. N. Boulanger, D. Ponomarev, and E. Skvortsov, Non-abelian cubic vertices for higher-spin fields in anti-de Sitter space, *JHEP.* **1305**, 008, (2013). doi: 10.1007/JHEP05(2013)008.
93. E. D. Skvortsov and M. Taronna, On Locality, Holography and Unfolding, *JHEP.* **11**, 044, (2015). doi: 10.1007/JHEP11(2015)044.
94. P. Kessel, G. Lucena Gómez, E. Skvortsov, and M. Taronna, Higher Spins and Matter Interacting in Dimension Three, *JHEP.* **11**, 104, (2015). doi: 10.1007/JHEP11(2015) 104.
95. P. Kessel, C. Sleight, and M. Taronna, *to appear.* (2015).

96. R. R. Metsaev, S matrix approach to massless higher spins theory. 2: The Case of internal symmetry, *Mod. Phys. Lett.* **A6**, 2411–2421, (1991). doi: 10.1142/S0217732391002839.
97. R. R. Metsaev, Poincare invariant dynamics of massless higher spins: Fourth order analysis on mass shell, *Mod. Phys. Lett.* **A6**, 359–367, (1991). doi: 10.1142/S0217732391000348.
98. A. K. H. Bengtsson, I. Bengtsson, and N. Linden, Interacting Higher Spin Gauge Fields on the Light Front, *Class. Quant. Grav.* **4**, 1333, (1987). doi: 10.1088/0264-9381/4/5/028.
99. E. S. Fradkin and M. A. Vasiliev, On the Gravitational Interaction of Massless Higher Spin Fields, *Phys. Lett.* **B189**, 89–95, (1987). doi: 10.1016/0370-2693(87)91275-5.
100. X. Bekaert, N. Boulanger, and P. Sundell, How higher-spin gravity surpasses the spin two barrier: no-go theorems versus yes-go examples, *Rev.Mod.Phys.* **84**, 987–1009, (2012). doi: 10.1103/RevModPhys.84.987.
101. R. R. Metsaev, Cubic interaction vertices for higher spin fields. (1996).
102. M. Hogervorst, S. Rychkov, and B. C. van Rees, Unitarity violation at the Wilson-Fisher fixed point in 4-epsilon dimensions. (2015).
103. W. Ruhl, The Masses of gauge fields in higher spin field theory on AdS(4), *Phys. Lett.* **B605**, 413–418, (2005). doi: 10.1016/j.physletb.2004.11.050.
104. K. Lang and W. Ruhl, The Critical O(N) sigma model at dimensions $2 < d < 4$: Fusion coefficients and anomalous dimensions, *Nucl. Phys.* **B400**, 597–623, (1993). doi: 10.1016/0550-3213(93)90417-N.
105. L. F. Alday, A. Bissi, and T. Lukowski, Large spin systematics in CFT, *JHEP.* **11**, 101, (2015). doi: 10.1007/JHEP11(2015)101.
106. X. Bekaert, J. Erdmenger, D. Ponomarev, and C. Sleight, Towards holographic higher-spin interactions: Four-point functions and higher-spin exchange, *JHEP.* **03**, 170, (2015). doi: 10.1007/JHEP03(2015)170.
107. T. Y. Thomas, On conformal geometry, *Proceedings of the National Academy of Sciences.* **12**(5), 352–359, (1926).
108. A. N. Vasiliev, Yu. M. Pismak, and Yu. R. Khonkonen, 1/N Expansion: calculation of the exponent η in the order $1/n^3$ by the conformal bootstrap method, *Theor. Math. Phys.* **50**, 127–134, (1982). doi: 10.1007/BF01015292. [Teor. Mat. Fiz.50,195(1982)].
109. A. Petkou, Conserved currents, consistency relations and operator product expansions in the conformally invariant O(N) vector model, *Annals Phys.* **249**, 180–221, (1996). doi: 10.1006/aphy.1996.0068.
110. S. E. Derkachov and A. N. Manashov, The Simple scheme for the calculation of the anomalous dimensions of composite operators in the 1/N expansion, *Nucl. Phys.* **B522**, 301–320, (1998). doi: 10.1016/S0550-3213(98)00103-5.
111. M. A. Vasiliev, Extended higher spin superalgebras and their realizations in terms of quantum operators, *Fortsch. Phys.* **36**, 33–62, (1988).
112. O. Gelfond and M. Vasiliev, Higher rank conformal fields in the Sp(2M) symmetric generalized space-time, *Theor.Math.Phys.* **145**, 1400–1424, (2005). doi: 10.1007/s11232-005-0168-9.
113. O. Gelfond and M. Vasiliev, Unfolded Equations for Current Interactions of 4d Massless Fields as a Free System in Mixed Dimensions. (2010).

Extended Anti-de Sitter Hypergravity in 2 + 1 Dimensions and Hypersymmetry Bounds

Marc Henneaux[a], Alfredo Pérez[b], David Tempo[a,b] and Ricardo Troncoso[b]

[a] *Université Libre de Bruxelles and International Solvay Institutes, ULB Campus Plaine C.P.231, B-1050 Bruxelles, Belgium*

[b] *Centro de Estudios Científicos (CECs), Av. Arturo Prat 514, Valdivia, Chile*

In a recent paper (JHEP **1508** (2015) 021), we have investigated hypersymmetry bounds in the context of simple anti-de Sitter hypergravity in 2+1 dimensions. We showed that these bounds involved non linearly the spin-2 and spin-4 charges, and were saturated by a class of extremal black holes, which are $\frac{1}{4}$-hypersymmetric. We continue the analysis here by considering (M, N)-extended anti-de Sitter hypergravity models, based on the superalgebra $osp(M|4) \oplus osp(N|4)$. The asymptotic symmetry superalgebra is then the direct sum of two-copies of a W-superalgebra that contains $so(M)$ (or $so(N)$) Kac-Moody currents of conformal weight 1, fermionic generators of conformal weight $5/2$ and bosonic generators of conformal weight 4 in addition to the Virasoro generators. The nonlinear hypersymmetry bounds on the conserved charges are derived and shown to be saturated by a class of extreme hypersymmetric black holes which we explicitly construct.

1. Introduction

Simple anti-de Sitter hypergravity in three dimensions is a consistent higher spin field theory involving fields of spins 2, 4 and $\frac{5}{2}$ that is invariant under hypersymmetry, a higher spin fermionic symmetry with spin-$\frac{3}{2}$ parameter. In the limit of zero cosmological constant, the spin-4 field decouples and the theory of the remaining fields reduces to the hypergravity model of [1] that has been recently reformulated as a Chern-Simons theory in [2]. The theory has no local degrees of freedom, but possesses the rich asymptotics of higher spin gauge fields in $2+1$ dimensions described by W-(super)algebras [3–5], in this case $W_{(2,\frac{5}{2},4)}$ [6].

The asymptotic symmetry algebra has interesting consequences since it implies "hypersymmetry bounds", in much the same way as supersymmetry implies supersymmetry bounds. In [6] we explicitly derived the hypersymmetry bounds for simple AdS hypergravity and analyzed them for different types of solutions, in particular, for black holes. We showed that the hypersymmetric black holes saturate the bounds and are extremal, in the sense that they lie on the border of the region within which a sensible thermodynamics (real entropy) can be defined.

The purpose of this work is to extend the analysis to hypergravity models with more hypersymmetries ("extended hypergravity"). This is done along the following lines. First, in the next section (Section 2), we describe the extended hypergravity models: we discuss the underlying superalgebras and write down the action. Then, in Section 3, we study the asymptotics using well-established methods and show that the asymptotic superalgebra is an extension of the algebra of [6] by $so(M)$ (respectively $so(N)$) Kac-Moody currents under which the fermionic hypercharges transform in the $\underline{\mathbf{M}}$ (respectively, in the $\underline{\mathbf{N}}$). We derive the form of the nonlinear hypersymmetry bounds in 4. Next, in Section 5, we construct the black hole solutions and discuss their thermodynamics. Finally, we show that the hypersymmetric black holes are extremal and saturate the hypersymmetry bounds (Section 6). Section 7 collects our concluding remarks.

2. Extended Anti-de Sitter Hypergravities in 2 + 1 Dimensions

(M, N)-extended AdS hypergravities in three-dimensional spacetimes are hypersymmetric extensions of $sp(4)$ higher spin gravity, described by a Chern-Simons theory with gauge algebra $sp(4) \oplus sp(4)$ where the gravitational subalgebra $sp(2) \simeq sl(2, \mathbb{R})$ is principally embedded on each side. The $sp(4)$ higher spin gravity contains the graviton and a spin-4 field. Here, the word "spin" refers to the conformal weight of the corresponding asymptotic generators in the conformal algebra at infinity, see below.

One may "hypersymmetrize" the $sp(4)$ higher spin gravity and construct (M, N)-extended AdS hypergravities, with M hypersymmetries in one chiral sector and N hypersymmetries in the other, by embedding $sp(4) \oplus sp(4)$ in an appropriate superalgebra. We consider hypersymmetric extensions such that the resulting superalgebra has the following properties:

- The bosonic subalgebra is the direct sum $sp(4) \oplus \mathcal{G}$ where $\mathcal{G}$ is a compact algebra.
- The fermionic generators transform in the $\underline{\mathbf{4}}$ of $sp(4)$ (and in some representation of $\mathcal{G}$ on which we do not impose any a priori requirement).

The first condition implies that the extra bosonic fields in the theory, coming in addition to the graviton and its spin-4 companion, have all lower spin 1. These extra fields are just the gauge fields associated with the internal R-symmetry described by $\mathcal{G}$. The second condition guarantees that the fermionic fields have all spin $\frac{5}{2}$ ("hypergravitini").

The algebra $sp(4)$ is the algebra underlying anti-de Sitter hypergravity in three dimensions, but it is also the anti-de Sitter algebra in 4 dimensions. As such, its graded extensions have been systematically studied in the early days of supergravity. It turns out that there is only one class of graded extensions, given by $osp(M|4)$ [7, 8]. Thus, while there are seven distinct types of extended supergravity models in three dimensions [7–11], there is only one type of extended hypergravity models. The underlying superalgebras are $osp(M|4) \oplus osp(N|4)$.

The $(M, N) = (1, 1)$ case of [6] is described by the superalgebra $osp(1|4) \oplus osp(1|4)$. It contains, in addition to the graviton and its spin-4 companion, a spin $\frac{5}{2}$ field on each chiral side, called the "hypergravitino". In the extended case, there are more "hypergravitini" and these transform in the $\underline{\mathbf{M}}$ (respectively, in the $\underline{\mathbf{N}}$) of $so(M)$ (respectively, $so(N)$). There are also extra gauge fields transforming in the adjoint of $so(M)$ (respectively, $so(N)$).

For definiteness, we shall focus from now on the chiral sector with superalgebra $osp(M|4)$. Similar considerations apply to the other sector. The (anti)commutation relations of $osp(M|4)$ are explicitly:

$$
\begin{aligned}
\left[L_i, L_j\right] &= (i-j)\, L_{i+j}\,, \\
\left[L_i, U_m\right] &= (3i-m)\, U_{i+m}\,, \\
\left[L_i, T^{IJ}\right] &= 0\,, \\
\left[L_i, \mathcal{S}^I_p\right] &= \left(\frac{3}{2}i - p\right) \mathcal{S}^I_{i+p}\,, \\
\left[U_m, U_n\right] &= \\
&\frac{1}{12}(m-n)\left(\left(m^2+n^2-4\right)\left(m^2+n^2-\frac{2}{3}mn-9\right)-\frac{2}{3}(mn-6)\,mn\right) L_{m+n} \\
&\quad + \frac{1}{6}(m-n)\left(m^2-mn+n^2-7\right) U_{m+n}\,, \qquad (1)\\
\left[T^{IJ}, T^{KL}\right] &= \delta^{IK}T^{JL} - \delta^{IL}T^{JK} - \delta^{JK}T^{IL} + \delta^{JL}T^{IK}\,, \\
\left[U_m, T^{IJ}\right] &= 0\,, \\
\left[U_m, \mathcal{S}^I_p\right] &= \frac{1}{24}\left(2m^3 - 8m^2p + 20mp^2 + 82p - 23m - 40p^3\right) \mathcal{S}^I_{i+p}\,, \\
\left[T^{IJ}, \mathcal{S}^K_p\right] &= \delta^{IK}\mathcal{S}^J_p - \delta^{JK}\mathcal{S}^I_p\,, \\
\left\{\mathcal{S}^I_p, \mathcal{S}^J_q\right\} &= \delta^{IJ}\left(U_{p+q} + \frac{1}{12}\left(6p^2 - 8pq + 6q^2 - 9\right) L_{p+q}\right) \\
&\quad - \frac{5}{12}(p-q)(2p^2+2q^2-5)T^{IJ}\,.
\end{aligned}
$$

Here L_i, with $i = 0, \pm 1$, stand for the generators that span the gravitational $sl\,(2,\mathbb{R})$ subalgebra, while $T^{IJ} = -T^{JI}$, with $I, J = 1, \cdots, M$, are the spin-0 $so(M)$ generators, which will yield spin-1 fields in the Chern-Simons theory. The U_m and $\mathcal{S}^I_p$, with $m = 0, \pm 1, \pm 2, \pm 3$ and $p = \pm\frac{1}{2}, \pm\frac{3}{2}$, will yield the spin-4 and spin-$\frac{5}{2}$ fields, respectively.

The dynamics of (M, N)-extended hypergravity follows from the difference of two Chern-Simons actions, $I = I_{CS}\left[A^+\right] - I_{CS}\left[A^-\right]$, with

$$
I_{CS}\left[A\right] = \frac{k_4}{4\pi}\int str\left[AdA + \frac{2}{3}A^3\right], \qquad (2)
$$

where the level, $k_4 = k/10$, is expressed in terms of the Newton constant and the AdS radius according to $k = \ell/4G$. In eq. (2) $str\left[\cdots\right]$ stands for the supertrace of the fundamental ($(4+M)\times(4+M)$ or $(4+N)\times(4+N)$) matrix representation of $osp(M|4)$ and the gauge fields $A^{\pm}$ correspond to the two independent copies $osp(M|4)$ and $osp(N|4)$. A convenient matrix representation of the generators T^{IJ} is such that the lower diagonal block is given by

$$\left(T^{IJ}\right)^{K}{}_{L} = -2\delta^{K[I}\delta^{J]}_{L} \,,$$

and hence

$$str\left(T^{IJ}T^{KL}\right) = 4\delta^{K[I}\delta^{J]L} \,.$$

3. Asymptotic Structure of Extended Hypergravities

3.1. *Boundary conditions*

In order to discuss the boundary conditions, we perform – as it has now become standard – the gauge transformation of [12] that eliminates asymptotically the radial dependence of the connections, so that $A^{\pm} = g_{\pm}^{-1}a^{\pm}g_{\pm} + g_{\pm}^{-1}dg_{\pm}$, with

$$a^{\pm} = a^{\pm}_{\varphi}(t,\varphi)\,d\varphi + a^{\pm}_{t}(t,\varphi)\,dt \tag{3}$$

(to leading order). Then, following the lines of [3–5, 13], we impose that at any fixed time slice $t = t_0$, the deviations with respect to the reference background go asymptotically along the lowest (highest) $sl(2)$-weight vectors for each $sl(2)$-representation occurring in the theory, i.e.,

$$a^{\pm}_{\varphi} = L_{\pm 1} - \frac{2\pi}{k}\tilde{\mathcal{L}}^{\pm}(\varphi)\,L_{\mp 1} + \frac{\pi}{5k}\mathcal{U}^{\pm}(\varphi)\,U_{\mp 3} - \frac{2\pi}{k}\psi^{\pm}_{I}(\varphi)\,\mathcal{S}^{I}_{\mp\frac{3}{2}} - \frac{5\pi}{k}\mathcal{J}^{\pm}_{IJ}(\varphi)T^{IJ} \,. \tag{4}$$

All components $\mathcal{J}^{\pm}_{IJ}$ along the internal symmetry generators T^{IJ}, which are $sl(2)$-scalars, are allowed. In (4), $\tilde{\mathcal{L}}^{\pm}(\varphi)$ is defined in terms of what will become the Virasoro generators $\mathcal{L}^{\pm}$ through

$$\tilde{\mathcal{L}}^{\pm} = \mathcal{L}^{\pm} - \frac{5\pi}{2k}\mathcal{J}^{\pm}_{IJ}\mathcal{J}^{\pm IJ} \,. \tag{5}$$

The two expressions differ by the familiar Sugawara term quadratic in the currents.

3.2. *Asymptotic symmetries*

Exactly as in [3–5, 13], one then finds that the fall-off conditions (4) are maintained under a restricted set of gauge transformations, $\delta a^{\pm} = d\Omega^{\pm} + [a^{\pm},\Omega^{\pm}]$, where, on each slice, the Lie-algebra-valued parameters

$$\Omega^{\pm} = \Omega^{\pm}\left[\epsilon_{\pm},\chi_{\pm},\zeta_{IJ\pm},\vartheta^{I}_{\pm}\right] \,, \tag{6}$$

depend on $(2 + \frac{M^{\pm}(M^{\pm}-1)}{2})$ bosonic and $M^{\pm}$ fermionic functions of φ, given by $\epsilon_{\pm},\chi_{\pm},\zeta_{IJ\pm}$, and $\vartheta_{\pm}$, respectively. Here, we have set $M^{+} = M$ and $M^{-} = N$. They take the form

$$\Omega^{\pm}\left[\epsilon_{\pm},\chi_{\pm},\zeta_{IJ\pm},\vartheta^{I}_{\pm}\right] = \epsilon_{\pm}(\varphi)\,L_{\pm 1} - \chi_{\pm}(\varphi)\,U_{\pm 3} \mp \vartheta_{\pm[I]}(\varphi)\,\mathcal{S}^{[I]}_{\pm\frac{3}{2}}$$
$$+\left(\zeta_{IJ\pm}(\varphi) - \frac{5\pi}{k}\epsilon_{\pm}(\varphi)\,\mathcal{J}^{\pm}_{IJ}\right)T^{IJ} + \eta^{\pm}\left[\epsilon_{\pm},\chi_{\pm},\nu_{IJ\pm},\vartheta^{I}_{\pm}\right] \,, \tag{7}$$

where the $\eta^{\pm}$'s, and the precise way in which the fields $\mathcal{L}^{\pm}, \mathcal{U}^{\pm}$, $\mathcal{J}^{\pm}_{IJ}$, $\psi^{\pm I}$ transform, are explicitly given in Appendix **A**. These expressions involve the fields $\mathcal{L}^{\pm}$, $\mathcal{U}^{\pm}$, $\mathcal{J}^{\pm}_{IJ}$, $\psi^{\pm I}$ and the independent gauge parameters $\epsilon_{\pm}, \chi_{\pm}, \zeta_{IJ\pm}$, and $\vartheta_{\pm}$, as well as their derivatives with respect to φ.

The boundary conditions (4) define phase space at a given instant of time. Phase space histories fulfill (4) at all times, i.e., take the form (4) with the functions $\tilde{\mathcal{L}}^{\pm}$, $\mathcal{U}^{\pm}$, $\psi^{\pm}_{I}$ and $\mathcal{J}^{\pm}_{IJ}$ now depending also on t. These boundary conditions are of course preserved by gauge transformations of the form (7) with parameters $\epsilon_{\pm}(t,\varphi)$, $\chi_{\pm}(t,\varphi)$, $\vartheta_{\pm[I]}(t,\varphi)$, $\zeta_{IJ\pm}(t,\varphi)$ that are time-dependent too. In particular, the motion in time is a gauge transformation with gauge parameter $a^{\pm}_{t}$. This implies that the asymptotic behaviour of $a^{\pm}_{t}$ has to be given by [14, 15]

$$a^{\pm}_{t} = \pm\Omega^{\pm}\left[\xi_{\pm}, \mu_{\pm}, \nu_{IJ\pm}, \varrho^{I}_{\pm}\right] , \tag{8}$$

where $\Omega^{\pm}$ is defined through (7), and $\xi_{\pm}$, $\mu_{\pm}$, $\nu_{IJ\pm}$, $\varrho^{I}_{\pm}$ can be identified with the "chemical potentials" when one goes to the thermodynamical formulation. Once the temporal components of the vector potential have been chosen, the parameters $\epsilon_{\pm}$, $\chi_{\pm}$, $\zeta_{IJ\pm}$, $\vartheta^{I}_{\pm}$ of the residual gauge transformations must fulfill certain differential equations of first order in time expressing that the $a^{\pm}_{t}$'s are left invariant by the transformations, which may be regarded as "deformed chirality conditions".

3.3. *Generators of asymptotic symmetries*

Following the canonical approach [16], one finds that the generators of the asymptotic symmetries are

$$\mathcal{Q}^{\pm}\left[\epsilon_{\pm}, \chi_{\pm}, \vartheta_{I\pm}\right] = -\int d\varphi\left(\epsilon_{\pm}\mathcal{L}^{\pm} + \chi_{\pm}\mathcal{U}^{\pm} - \zeta^{IJ}_{\pm}\mathcal{J}^{\pm}_{IJ} - i\vartheta^{I}_{\pm}\psi^{\pm}_{I}\right) . \tag{9}$$

(modulo bulk terms proportional to the constraints that we will not write explicitly and that can be taken strongly equal to zero if one uses the Dirac bracket - which coincides with the Poisson bracket for gauge invariant functions).

Since the Poisson brackets fulfill $\left[\mathcal{Q}[\eta_1], \mathcal{Q}[\eta_2]\right]_{PB} = -\delta_{\eta_1}\mathcal{Q}[\eta_2]$, the algebra of the canonical generators can be easily found from the transformation law of the fields, and it is explicitly written down in Appendix **B**.

Expanding in Fourier modes, $X = \frac{1}{2\pi}\sum_m X_m e^{im\varphi}$, the asymptotic symmetry algebra reads

$$\begin{aligned}
i\left[L_m, L_n\right]_{PB} &= (m-n)\, L_{m+n} + \frac{k}{2}m^3\delta^0_{m+n} \,,\\
i\left[L_m, U_n\right]_{PB} &= (3m-n)\, U_{m+n} \,,\\
i\left[L_m, J^{IJ}_n\right]_{PB} &= -nJ^{IJ}_{m+n} \,,\\
i\left[L_m, \psi^{I}_n\right]_{PB} &= \left(\frac{3}{2}m - n\right)\psi^{I}_{m+n} \,,
\end{aligned}$$

$$
\begin{aligned}
i\left[U_m, J_n^{IJ}\right]_{PB} &= 0\,,\\
i\left[J_m^{IJ}, (J_{KL})_n\right]_{PB} &= -4i\delta^{[I}_{[K}\left(\mathcal{J}^{J]}{}_{L]}\right)_{m+n} - \frac{2k}{5}n\delta^{[I}{}_K\delta^{J]}{}_L\delta^0_{m+n}\,,\\
\left[J_m^{IJ}, \psi_n^K\right]_{PB} &= 2\delta^{K[I}\psi^{J]}_{m+n}\,,
\end{aligned}
$$

$$
\begin{aligned}
i\left[U_m, U_n\right]_{PB} &= \frac{1}{2^2 3^2}(m-n)\left(3m^4 - 2m^3 n + 4m^2 n^2 - 2mn^3 + 3n^4\right)\left(L_{m+n} - \frac{5\pi}{2k}\Lambda^{(2)}_{m+n}\right)\\
&\quad + \frac{1}{6}(m-n)\left(m^2 - mn + n^2\right)\mathcal{U}_{m+n} - \frac{2^3 3\pi}{k}(m-n)\Lambda^{(6)}_{m+n}\\
&\quad - \frac{7^2\pi}{3^2 k}(m-n)\left(m^2 + 4mn + n^2\right)\tilde{\Lambda}^{(4)}_{m+n} + \frac{k}{2^3 3^2}m^7\delta^0_{m+n}\,,\\
i\left[U_m, \psi_n^I\right]_{PB} &= \frac{1}{2^2 3}\left(m^3 - 4m^2 n + 10mn^2 - 20n^3\right)\psi^I_{m+n} - \frac{23\pi}{3k}i\Lambda^{(11/2)I}_{m+n}\\
&\quad + \frac{70\pi}{k}im^2\Lambda^{(7/2)I}_{m+n} + \frac{\pi}{3k}(23m - 82n)\Lambda^{(9/2)I}_{m+n}\,,\\
i\left[\psi_m^I, \psi_n^J\right]_{PB} &= \delta^{IJ}\left[U_{m+n} + \tfrac{1}{2}\left(m^2 - \tfrac{4}{3}mn + n^2\right)\left(L_{m+n} - \tfrac{5\pi}{2k}\Lambda^{(2)}_{m+n}\right) + \tfrac{k}{6}m^4\delta^0_{m+n}\right]\\
&\quad - \frac{5}{6}i(m-n)\left(m^2 + n^2\right)J^{IJ}_{m+n}\\
&\quad - \frac{25\pi}{k}\left(m^2 + n^2\right)\Lambda^{(2)IJ}_{m+n} + \frac{50\pi}{3k}i\left(m\Lambda^{(3)JI}_{m+n} + n\Lambda^{(3)IJ}_{m+n}\right) - \frac{3\pi}{k}\Lambda^{(4)(IJ)}_{m+n}\,,
\end{aligned}
\tag{10}
$$

where $\Lambda_m^{(l)}$ stand for the Fourier modes of the corresponding nonlinear terms (see Appendix **B**). The central charge in the Virasoro subalgebra is the same as that for pure gravity [17].

The modes fulfill the following reality conditions: $(L_m)^* = L_{-m}$, $(U_m)^* = U_{-m}$, $\left(\psi_m^I\right)^* = \psi^I_{-m}$, $\left(\mathcal{J}_m^{IJ}\right)^* = \mathcal{J}^{IJ}_{-m}$ so that the functions $\mathcal{L}^\pm$, $\mathcal{U}^\pm$, $\psi^\pm$ and $\mathcal{J}^{IJ}$ are real.

The asymptotic symmetry algebra given above is the classical limit of an extension of the superalgebra $W_{\left(2,\frac{5}{2},4\right)}$ of [18, 19]. This extension involves M spin-$\frac{5}{2}$ fermionic generators ψ_m^I and $\frac{M(M-1)}{2}$ Kac-Moody currents $\mathcal{J}_n^{IJ}$. The spin-$\frac{5}{2}$ fermionic generators ψ_m^I are the "hypersymmetry" generators, and transform in the $\underline{\mathbf{M}}$ of $so(M)$. The algebra $W_{\left(2,\frac{5}{2},4\right)}$ corresponds to $M = 1$.

4. Hypersymmetry Bounds from the Asymptotic Symmetry Algebra

4.1. *Boundary conditions and spectral flow*

We focus for definiteness on the + copy and drop the subscript "+". Similar considerations apply to the − sector.

The fermions are subject to boundary conditions of the form

$$\psi_I(\varphi + 2\pi) = R_{IJ}\psi_J(\varphi) \tag{1}$$

where the matrix $R = (R_{IJ})$ is an element of $O(M)$, which we can take to be either the identity, or a fixed element of $O(M)$ with determinant -1 discussed below.

Different boundary conditions are related to these ones by spectral flow [20] (see also [13] for a discussion in the similar AdS_3 extended supergravity context).

When M is odd, one may assume $R = \mathbb{1}$ (periodic boundary conditions) or $R = -\mathbb{1}$ (antiperiodic boundary conditions). In both cases, the affine generators $\mathcal{J}_{IJ}$ are periodic and the corresponding affine algebra is untwisted. When M is even, $M = 2r$, one may assume $R = \mathbb{1}$ (periodic boundary conditions) or, if $R \neq \mathbb{1}$, that it defines an outer automorphism of $SO(2r)$. In that latter case, the affine generators are not periodic and the affine algebra is twisted.

We shall restrict the analysis to periodic boundary conditions ("Ramond case"). This is motivated by the fact that we are interested in black holes. The situations found in $(1,1)$ hypergravity and supergravity indicate that black hole solutions naturally admit in both cases the periodic spin structure [6, 21]. Note that the antiperiodic case ("Neveu-Schwarz" case) is automatically included when M is even since then, as mentioned above, it can be related to the periodic case by spectral flow [20] ($-\mathbb{1} \in SO(2m)$). With periodic boundary conditions, the Fourier labels m in $X = \frac{1}{2\pi}\sum_m X_m e^{im\varphi}$ are integers for all fields X.

4.2. *Hypersymmetry bounds*

The Poisson Bracket of the fermionic generator of the asymptotic symmetry hyperalgebra in (10), implies interesting hypersymmetry bounds. These were discussed in great generality in [6]. Here, we focus on the bounds that hold in the context of periodic boundary conditions.

We consider bosonic configurations carrying global charges with only zero modes, given by $L_0 = 2\pi\mathcal{L}$, $U_0 = 2\pi\mathcal{U}$ and $J_0^{IJ} = 2\pi\mathcal{J}^{IJ}$, for each copy. Furthermore, we assume without loss of generality that the affine Kac-Moody currents have been brought to the Cartan subalgebra by conjugation, so that $\mathcal{J}^{IJ}$ has only non-vanishing components $\mathcal{J}^{2i-1\,2i}$ ($i = 1, 2, \cdots, \text{rank}SO(M) = \left[\frac{M}{2}\right] = r$). We set $M = 2r$ when M is even, or $M = 2r+1$ when M is odd, and

$$\mathcal{J}^{2i-1\,2i} = j_i \epsilon^{2i-1\,2i} \tag{2}$$

The anticommutators of the hypersymmetry generators with $m = -n = p \geq 0$ are then found to reduce to

$$(2\pi)^{-1}\left(\hat{\psi}_p^I \hat{\psi}_{-p}^J + \hat{\psi}_{-p}^J \hat{\psi}_p^I\right) = B_p^{IJ}\ , \tag{3}$$

with

$$\begin{aligned} B_p^{IJ} &= \left(\mathcal{U} + \frac{3\pi}{k}\tilde{\mathcal{L}}^2\right)\delta^{IJ} + \frac{500\pi^2}{3k^2}\mathcal{J}^I{}_K\left(\tilde{\mathcal{L}}\mathcal{J}^{JK} + \frac{5\pi}{k}\mathcal{J}_{MN}\mathcal{J}^{JM}\mathcal{J}^{KN}\right) \\ &\quad - \frac{100 i\pi}{3k}\left(\tilde{\mathcal{L}}\mathcal{J}^{IJ} + \frac{10\pi}{k}\mathcal{J}^I{}_M\mathcal{J}^{ML}\mathcal{J}^J{}_L\right)p + \frac{5}{3}\left(\delta^{IJ}\tilde{\mathcal{L}} + \frac{30\pi}{k}\mathcal{J}^I{}_K\mathcal{J}^{JK}\right)p^2 \\ &\quad - \frac{10}{3}i\mathcal{J}^{IJ}p^3 + \frac{k}{12\pi}\delta^{IJ}p^4\ . \end{aligned} \tag{4}$$

Note that

$$\left(B_p^{IJ}\right)^\dagger = B_p^{JI}\,, \tag{5}$$

and as one sees explicitly from (4), B_p^{IJ} has both a real symmetric part and a pure imaginary, antisymmetric part.

Now, the hermitian operator $\hat{\psi}_p^I \hat{\psi}_{-p}^I + \hat{\psi}_{-p}^I \hat{\psi}_p^I$ is positive definite for each I and p. This implies, in the classical limit, that the global charges fulfill the bounds

$$B_p^{II} \geq 0\,, \tag{6}$$

(no summation over I). The bound $B_0^{II} \geq 0$ for $p = 0$ reads

$$B_0^{II} \equiv \left(\mathcal{U} + \frac{3\pi}{k}\tilde{\mathcal{L}}^2\right) + \frac{500\pi^2}{3k^2}\left(\tilde{\mathcal{L}}(j_i)^2 + \frac{5\pi}{k}(j_i)^4\right) \geq 0\,, \tag{7}$$

with $I = 1, 2, \cdots 2r$. Note that when M is odd, there is an additional bound corresponding to $I = 2r + 1$,

$$B_0^{2r+1\,2r+1} \equiv \left(\mathcal{U} + \frac{3\pi}{k}\tilde{\mathcal{L}}^2\right) \geq 0\,. \tag{8}$$

These bounds are manifestly nonlinear.

One can express the bounds for $p > 0$ in terms of the bounds for $p = 0$ as

$$B_p^{II} = B_0^{II} + \frac{5}{3}\left(\tilde{\mathcal{L}} + \frac{30\pi}{k}(j_i)^2\right)p^2 + \frac{k}{12\pi}p^4 \geq 0\,.$$

Now, in the black hole case, one must have $\tilde{\mathcal{L}} \geq 0$ (see below) and so one finds that the bounds with $p > 0$ are automatic consequences of bounds with $p = 0$, which are thus the strongest.

One can derive further bounds involving the mixed terms B_p^{IJ} with $J \neq I$. To illustrate the procedure, consider for definitess $I = 1$ and $J = 2$, for which B_p^{12} does not identically vanish. Form the complex fields $\chi_p = \psi_p^1 + i\psi_p^2$ and $\omega_p = \psi_p^1 - i\psi_p^2$. From the conditions $\chi_p(\chi_p)^\dagger + (\chi_p)^\dagger\chi_p \geq 0$ and $\omega_p(\omega_p)^\dagger + (\omega_p)^\dagger\omega_p \geq 0$, one gets

$$B_p^{11} + B_p^{22} \geq \pm i(B_p^{12} - B_p^{21}), \tag{9}$$

i.e., given that $B_p^{11} = B_p^{22}$ and $B_p^{12} = -B_p^{21}$,

$$B_p^{11} \geq \pm iB_p^{12}\,. \tag{10}$$

In general, this bound is independent from the previous ones, but it is not so in the black hole case. Indeed, the condition (10) can be conveniently factorized as

$$\left[\left(p - \frac{10\pi}{k}j_1\right)^2 + \lambda_{[+]}^2\right]\left[\left(p - \frac{10\pi}{k}j_1\right)^2 + \lambda_{[-]}^2\right] \geq 0\,, \tag{11}$$

where $\pm\lambda_{[+]}$ and $\pm\lambda_{[-]}$ correspond to the eigenvalues of the $sp\,(4)$ dynamical gauge fields introduced in the next section. In the black hole case, these eigenvalues are necessarily real (see below), so that for this class of solutions the bounds in (11) are clearly fulfilled. One refers for this reason to the bounds (7) with $p = 0$ as the "strongest bounds" in the black hole context.

5. Black Holes

5.1. *Black hole connection and regularity conditions*

Higher spin black holes generalizing the pure gravity black hole [22, 23] have been investigated first in the pioneering work [24–26], reviewed in [27]. A different class of black hole solutions differing in their asymptotics was subsequently derived in [14, 15, 28]. We follow this approach as it is clearly compatible with the asymptotic W-symmetry algebra exhibited above.

In the absence of a well-defined spacetime geometry, higher spin black holes are defined through the Euclidean continuation [24, 25], as regular flat connections on the solid torus with well-defined thermodynamics (real entropy). We follow this point of view but, however, as in [15], we impose the above boundary conditions on the connection and not ones that would modify the asymptotic behaviour of a_{φ}^{pm}. For the (M, M)-extended AdS hypergravity theory, the Euclidean connection that describes the black holes is a direct generalization of the simple hypergravity black hole of [6] and can be written as

$$\begin{aligned} a = &\left(L_1 - \frac{2\pi}{k}\tilde{\mathcal{L}}L_{-1} + \frac{\pi}{5k}\mathcal{U}U_{-3} - \frac{5\pi}{k}\mathcal{J}_{IJ}T^{IJ}\right)d\varphi - \left\{i\xi\left(L_1 - \frac{2\pi}{k}\tilde{\mathcal{L}}L_{-1} + \frac{\pi}{5k}\mathcal{U}U_{-3}\right.\right. \\ &\left. + \frac{5\pi}{k}\mathcal{J}_{IJ}T^{IJ}\right) - i\mu\left[U_3 - \frac{6\pi}{k}\tilde{\mathcal{L}}U_1 - \frac{22\,\pi^2}{15\,k^2}\left(\mathcal{U} + \frac{60\pi}{11k}\tilde{\mathcal{L}}^2\right)\tilde{\mathcal{L}}U_{-3}\right. \\ &\left.\left. + \frac{6\pi}{k}\mathcal{U}L_{-1} + \frac{\pi}{k}\left(\mathcal{U} + \frac{12\pi}{k}\tilde{\mathcal{L}}^2\right)U_{-1}\right] - i\nu_{IJ}T^{IJ}\right\}d\tau\,, \end{aligned} \tag{12}$$

where $\tilde{\mathcal{L}}$ is given by

$$\tilde{\mathcal{L}} = \mathcal{L} - \frac{5\pi}{2k}\mathcal{J}_{IJ}\mathcal{J}^{IJ}\,, \tag{13}$$

while the components of the zero modes of the $so(M)$ Kac-Moody currents, and their corresponding chemical potentials are constrained to commute by the field equations, and hence

$$\mathcal{J}_{[I}^{\ K}\nu_{J]K} = 0\,. \tag{14}$$

As above, if we assume that $\mathcal{J}_{IJ}T^{IJ}$ belong to the Cartan subalgebra of $so(M)$, this condition implies that the chemical potentials $\nu_{IJ}T^{IJ}$ also do. Therefore,

$$\nu_{2i-1\,2i} = \nu_i\epsilon_{2i-1\,2i}\,.$$

Due to the fact that the thermal cycles are contractible, the holonomy of the gauge fields along them has to be trivial. These are the so-called "regularity conditions". For the branch of solutions that is continuously connected to the BTZ black hole [22, 23], possibly endowed with an $so(M)$ field, the regularity conditions read

$$e^{a_\tau^{sp(4)}}.e^{i\left(\nu_{IJ} + \frac{5\pi}{k}\xi\mathcal{J}_{IJ}\right)T^{IJ}} = \begin{pmatrix} -\mathbb{1}_{4\times 4} & 0 \\ 0 & \mathbb{1}_{M\times M}\end{pmatrix}. \tag{15}$$

Hence, the chemical potentials fulfill

$$\nu_i + \frac{5\pi}{k}\xi j_i = 2\pi n_i \,, \tag{16}$$

where n_i stands for a set of integers, and

$$\xi = \frac{\pi}{5^2 2}\left[\frac{3^3\lambda_{[-]}^3 - 41\left(3\lambda_{[+]} - \lambda_{[-]}\right)\lambda_{[-]}\lambda_{[+]} - 3^2\lambda_{[+]}^3}{\left(\lambda_{[-]}^2 - \lambda_{[+]}^2\right)\lambda_{[-]}\lambda_{[+]}}\right] , \tag{17}$$

$$\mu = \frac{3\pi}{5}\left[\frac{3\lambda_{[-]} - \lambda_{[+]}}{\left(\lambda_{[-]}^2 - \lambda_{[+]}^2\right)\lambda_{[-]}\lambda_{[+]}}\right] , \tag{18}$$

with $\lambda_{[\pm]}$ given by

$$\lambda_{[\pm]}^2 = \frac{10\pi}{k}\left(\tilde{\mathcal{L}} \pm \frac{4}{5}\sqrt{\tilde{\mathcal{L}}^2 - \frac{3k}{16\pi}\mathcal{U}}\right) . \tag{19}$$

One gets exactly the same regularity condition in the $sp(4)$ sector (in terms of $\tilde{\mathcal{L}}$) as in the simple hypergravity case considered in [6] (or, for that matter, as in the case of pure $sp(4)$ gravity). We also note that the natural value for the integers n_i characterizing the holonomy of the internal $SO(M)$ symmetry is $n_i = 0$ since otherwise there might appear to be a δ-function source of quantized strength in the non-gravitational, internal, sector, but we shall temporarily allow for more general n_i's to see how these integers enter the entropy. One could similarly allow for more general solutions of the regularity conditions involving different integers in the $sp(4)$ sector; the above choice corresponds to the BTZ branch.

5.2. *Entropy*

We use the correct canonical expression for the black hole entropy adapted to the above boundary conditions derived first in [29, 30], which can also be alternatively written according to [15, 31, 32] as

$$S = -2k_4 \mathrm{Im}\left(\mathrm{str}\left[a_\tau a_\varphi\right]\right) \tag{20}$$

(in the conventions of [15]). Once evaluated for the solution in (12), the black hole entropy becomes

$$S = 8\pi \mathrm{Re}\left[\xi\mathcal{L} + 2\mu\mathcal{U} + \frac{1}{2}\nu^{IJ}\mathcal{J}_{IJ}\right] . \tag{21}$$

Plugging then the expressions for the chemical potentials into (21) allows one to express the black hole entropy in terms of the (extensive) global charges. One gets

$$S = \frac{2\pi k}{5}\mathrm{Re}\left(3\lambda_{[+]} + \lambda_{[-]} + \frac{5}{2k}n_i j_i\right) . \tag{22}$$

For the natural $so(M)$ holonomy $n_i = 0$, this expression becomes

$$S = \frac{2\pi k}{5}\mathrm{Re}\left(3\lambda_{[+]} + \lambda_{[-]}\right) . \tag{23}$$

It is only for this branch that the black hole entropy reduces to the horizon area over $4G$ when the spin-4 field is turned off.

Two points are worth being pointed out: (i) as can already be seen for the coupled pure gauge-gravitational fields without higher spin gauge fields described by the gauge algebra $sl(2,\mathbb{R}) \oplus so(M)$, the $so(M)$ gauge fields are "gravitationally stealth" in the sense of [33], i.e., they do not generate a back reaction on the metric because their contribution to the stress energy vanishes; they only contribute to a redefinition of the asymptotic Virasoro generators; (ii) the black hole entropy of the $n_i = 0$ branch (23) is also blind to them if one expresses it in terms of the tilted Virasoro generators but not so if one uses the Virasoro generators fulfilling the above asymptotic algebra and directly related to the mass M and the angular momentum J. The black hole entropy can also detect non-vanishing n_i, see (22).

For the $n_i = 0$ branch that we consider from now on, the Lorentzian continuation of the entropy reads

$$S = \pi\sqrt{\frac{2}{5}\pi k}\left[\sqrt{\tilde{\mathcal{L}}^+}\left(\sqrt{1-\frac{4}{5}\sqrt{1-\frac{3k\mathcal{U}^+}{16\pi\left(\tilde{\mathcal{L}}^+\right)^2}}}+3\sqrt{1+\frac{4}{5}\sqrt{1-\frac{3k\mathcal{U}^+}{16\pi\left(\tilde{\mathcal{L}}^+\right)^2}}}\right)\right.$$
$$\left.+\sqrt{\tilde{\mathcal{L}}^-}\left(\sqrt{1-\frac{4}{5}\sqrt{1-\frac{3k\mathcal{U}^-}{16\pi\left(\tilde{\mathcal{L}}^-\right)^2}}}+3\sqrt{1+\frac{4}{5}\sqrt{1-\frac{3k\mathcal{U}^-}{16\pi\left(\tilde{\mathcal{L}}^-\right)^2}}}\right)\right]. \quad (24)$$

Requiring the entropy to be well-defined, i.e., being real and positive, implies that the eigenvalues $\lambda_{[\pm]}$ should be real. This forces then the spin-4 charges to be bounded according to

$$-\left(\tilde{\mathcal{L}}^\pm\right)^2 \leq \frac{k}{3\pi}\mathcal{U}^\pm \leq \frac{2^4}{3^2}\left(\tilde{\mathcal{L}}^\pm\right)^2, \quad (25)$$

in addition to $\tilde{\mathcal{L}}^\pm \geq 0$. The bounds are saturated in the extremal cases, and only the lower one in (25) corresponds to the hypersymmetry bound aforementioned. Note that the range of positive spin-4 charges is larger than that of the negative ones.

6. Killing Vector-Spinors

Bosonic configurations that admit unbroken hypersymmetries have to fulfill the following Killing vector-spinor equation

$$\delta a = d\theta + [a, \theta] = 0\,, \quad (26)$$

where the parameter θ is purely fermionic, given by $\theta = \theta_I^p \mathcal{S}_p^I$ for both copies, and globally well-defined.

Equivalently, the Killing vector-spinor equation can be obtained from promoting the corresponding asymptotic symmetries to hold everywhere and not just asymptotically. Therefore, in the case of the plus copy ($a_\varphi = a_\varphi^+$), the fermionic parameter is of the form $\theta = \Omega^+ \left[0,0,0,\vartheta^I\right]$, which explicitly reads

$$\theta = -\vartheta_I \mathcal{S}^I_{\frac{3}{2}} + \left(\vartheta'_I + \frac{10}{k}\mathcal{J}_I{}^K\vartheta_K\right)\mathcal{S}^I_{\frac{1}{2}} - \frac{1}{2}\left(\vartheta''_I - \frac{6\pi}{k}\tilde{\mathcal{L}}\vartheta_I - \frac{100\pi^2}{k^2}\mathcal{J}_{IK}\mathcal{J}^{JK}\vartheta_J \right.$$
$$\left. + \frac{20\pi}{k}\mathcal{J}_I{}^K\vartheta'_K\right)\mathcal{S}^I_{-\frac{1}{2}} + \frac{1}{6}\left[\left(\vartheta''_I + \frac{30\pi}{\kappa}\mathcal{J}_I{}^K\vartheta'_K - \frac{14\pi}{3k}\tilde{\mathcal{L}}\vartheta_I - \frac{300\pi^2}{k^2}\mathcal{J}_{IK}\mathcal{J}^{JK}\vartheta_J\right)' \right.$$
$$\left. - \frac{140\pi^2}{k^2}\left(\tilde{\mathcal{L}}\mathcal{J}_I{}^J + \frac{50\pi}{7k}\mathcal{J}_{IM}\mathcal{J}^{MK}\mathcal{J}^J_K\right)\vartheta_J\right]\mathcal{S}^I_{-\frac{3}{2}}\,. \tag{27}$$

The condition that a_φ should be left strictly unchanged then implies that the parameters ϑ_I should satisfy the following differential equations:

$$\left[\left(\mathcal{U} + \frac{3\pi}{k}\tilde{\mathcal{L}}^2\right)\delta^J_I + \frac{500\pi^2}{3k^2}\mathcal{J}_{IK}\left(\tilde{\mathcal{L}}\mathcal{J}^{JK} + \frac{5\pi}{k}\mathcal{J}_{MN}\mathcal{J}^{JM}\mathcal{J}^{KN}\right)\right]\vartheta_J$$
$$- \frac{100\pi}{3k}\left(\mathcal{J}_I{}^K\tilde{\mathcal{L}} + \frac{10\pi}{k}\mathcal{J}_{IM}\mathcal{J}^{ML}\mathcal{J}^K_L\right)\vartheta'_K \tag{28}$$
$$- \frac{5}{3}\left(\delta^K_I\tilde{\mathcal{L}} + \frac{30\pi}{k}\mathcal{J}_{IJ}\mathcal{J}^{KJ}\right)\vartheta''_K + \frac{10}{3}\mathcal{J}_I{}^K\vartheta'''_K + \frac{k}{12\pi}\vartheta''''_I = 0\,.$$

From the experience gathered with black holes within $(1,1)$ hypergravity or supergravity [6, 21], it is reasonable to assume that the fermionic parameters are constant, given by $\vartheta^I = \vartheta^I_0$. The Killing vector-spinor equations (28) then reduce to

$$\left[\left(\mathcal{U} + \frac{3\pi}{k}\tilde{\mathcal{L}}^2\right)\delta^J_I + \frac{500\pi^2}{3k^2}\mathcal{J}_{IK}\left(\tilde{\mathcal{L}}\mathcal{J}^{JK} + \frac{5\pi}{k}\mathcal{J}_{MN}\mathcal{J}^{JM}\mathcal{J}^{KN}\right)\right]\vartheta_J = B^{IK}_0\vartheta_K = 0\,, \tag{29}$$

which clearly admit non trivial solutions if the matrix B^{IK}_0 has zero eigenvalues. Since black holes are well-defined provided $\tilde{\mathcal{L}} \geq 0$, and the spin-4 charges fulfill eq. (25), the Killing vector-spinor equations (28) possess non trivial solutions only when the lower bound in (25) is saturated, i.e., only for negative spin-4 charges given by

$$\mathcal{U} = -\frac{3\pi}{k}(\tilde{\mathcal{L}})^2\,.$$

When this condition is fulfilled,

- there is at least one Killing vector-spinor when M is odd (corresponding to $B^{2r+1\,2r+1}_0 = 0$) and more if some currents j_i vanish;
- there are Killing vector-spinors when M is even only if some currents j_i vanish.
- The maximum number of hypersymmetries is thus M. It is attained when all the j_i's vanish and correspond to $\frac{M}{4}$-hypersymmetry, in agreement with the $\frac{1}{4}$-hypersymmetry found in [6] for $M = 1$.

It is straightforward to verify that the remaining Killing vector-spinor equation, that come from preserving the form of the Lagrange multiplier a_t globally, is also fulfilled.

7. Conclusions

In this note, we have extended the analysis of hypersymmetry bounds of [6] to extended AdS_3 hypergravity. These bounds follow from the asymptotic symmetry superalgebra and involve the charges nonlinearly. Just as in [6], we have found that the bounds are saturated by a class of extremal black holes, which are hypersymmetric (i.e., possess Killing vector-spinors). However, not all extremal black holes are hypersymmetric. The fact that extremality and super/hypersymmetry do not coincide in the context of higher spin black holes has been discussed recently in the thorough work [34], which focuses on (an appropriate real form of) the superalgebra $sl(3|2)$.

Hypersymmetric solutions of a different types (solitons) have been also explored in [6]. The extension of that analysis to extended hypergravity is left for future study.

Finally, we note that nonlinear bounds have also been found in the context of asymptotically flat solutions of hypergravity in the case of fermionic fields of spin $s = n + \frac{3}{2}$, with $n > 0$ (which, in the case of $n = 0$, i.e., supergravity, turn out to be linear) [35].

Acknowledgments

D.T. holds a Marina Solvay fellowship. The work of M.H. and D.T. is partially supported by the ERC through the "SyDuGraM" Advanced Grant, by FNRS-Belgium (convention FRFC PDR T.1025.14 and convention IISN 4.4503.15) and by the "Communaut Franaise de Belgique" through the ARC program. The work of A.P., D.T. and R.T. is partially funded by the Fondecyt grants N° 11130262, 11130260, 1130658, 1121031. The Centro de Estudios Científicos (CECs) is funded by the Chilean Government through the Centers of Excellence Base Financing Program of Conicyt.

Appendix A. Explicit form of the asymptotic symmetries

The Lie-algebra-valued parameter $\eta^{\pm}$ that appears in the asymptotic gauge symmetries spanned by $\Omega^{\pm}$ in eq. (6) is given by

$$
\begin{aligned}
\eta^{\pm}\left[\epsilon_{\pm},\chi_{\pm},\vartheta^{I}_{\pm}\right] &= -\frac{3\pi}{k}\left(i\psi^{\pm}_{I}\vartheta^{I}_{\pm}+\frac{2}{3}\epsilon_{\pm}\tilde{\mathcal{L}}^{\pm}+2\chi_{\pm}\mathcal{U}^{\pm}-\frac{k}{6\pi}\epsilon''_{\pm}\right)L_{\mp 1}\mp\epsilon'_{\pm}L_{0}\\
&+\frac{6\pi}{k}\left(\chi_{\pm}\tilde{\mathcal{L}}^{\pm}-\frac{k}{12\pi}\chi''_{\pm}\right)U_{\pm 1}\mp\frac{2\pi}{k}\left(\chi_{\pm}\tilde{\mathcal{L}}^{\pm\prime}+\frac{8}{3}\chi'_{\pm}\tilde{\mathcal{L}}^{\pm}-\frac{k}{12\pi}\chi'''_{\pm}\right)U_{0}\\
&-\frac{\pi}{2k}\left[i\psi^{\pm}_{I}\vartheta^{I}_{\pm}+2\left(\mathcal{U}^{\pm}-\frac{1}{2}\tilde{\mathcal{L}}^{\pm\prime\prime}+\frac{12\pi}{k}\left(\tilde{\mathcal{L}}^{\pm}\right)^{2}\right)\chi-\frac{11}{3}\chi'_{\pm}\tilde{\mathcal{L}}^{\pm\prime}\right.\\
&\left.-\frac{14}{3}\chi''_{\pm}\tilde{\mathcal{L}}^{\pm}+\frac{k}{12\pi}\chi^{(4)}_{\pm}\right]U_{\mp 1}\pm\chi'_{\pm}U_{\pm 2}\pm\frac{\pi}{2k}\left[i\psi^{\pm}_{I}\vartheta^{I\prime}_{\pm}+\frac{1}{5}i\psi^{\pm\prime}_{I}\vartheta^{I}_{\pm}\right.\\
&+\frac{8\pi}{k}i\mathcal{J}^{I\pm}_{\ \ J}\psi^{\pm}_{I}\vartheta^{J\prime}_{\pm}-\frac{5}{3}\chi''_{\pm}\tilde{\mathcal{L}}^{\pm\prime}-\frac{4}{3}\tilde{\mathcal{L}}^{\pm}\chi'''_{\pm}+\frac{2}{5}\left(\mathcal{U}^{\pm}-\frac{1}{2}\tilde{\mathcal{L}}^{\pm\prime\prime}+\frac{18\pi}{k}\left(\tilde{\mathcal{L}}^{\pm}\right)^{2}\right)'\chi_{\pm}\\
&\left.+\frac{6}{5}\left(\mathcal{U}^{\pm}-\frac{7}{9}\tilde{\mathcal{L}}^{\pm\prime\prime}+\frac{44\pi}{3k}\left(\tilde{\mathcal{L}}^{\pm}\right)^{2}\right)\chi'_{\pm}+\frac{k}{60\pi}\chi^{(5)}_{\pm}\right]U_{\mp 2}\\
&-\frac{\pi}{4k}\left\{i\psi^{\pm}_{I}\vartheta^{I\prime\prime}_{\pm}+\frac{1}{15}i\left(\psi^{\pm\prime\prime}_{I}-2^{4}\frac{5\pi}{k}\tilde{\mathcal{L}}^{\pm}\psi^{\pm}_{I}+2^{3}\frac{5\pi}{k}\mathcal{J}^{J\pm}_{\ \ I}\psi^{\pm\prime}_{J}+2^{2}\frac{35\pi}{k}\mathcal{J}^{J\pm\prime}_{\ \ I}\psi^{\pm}_{J}\right.\right.\\
&\left.-\frac{10^{3}\pi^{2}}{k^{2}}\mathcal{J}^{J\pm}_{\ \ K}\mathcal{J}_{I}^{\ \ K\pm}\psi^{\pm}_{J}\right)\vartheta^{I}_{\pm}+\frac{2}{5}i\left(\psi^{\pm\prime}_{I}+\frac{40\pi}{k}\mathcal{J}^{J\pm}_{\ \ I}\psi^{\pm\prime}_{J}\right)\vartheta^{I\prime}_{\pm}-\chi'''_{\pm}\tilde{\mathcal{L}}^{\pm\prime}-\frac{4}{5}\epsilon_{\pm}\mathcal{U}^{\pm}\\
&+\frac{2}{3}\left(\mathcal{U}^{\pm}-\frac{13}{10}\mathcal{L}^{\pm\prime\prime}+\frac{272\pi}{15k}\left(\tilde{\mathcal{L}}^{\pm}\right)^{2}\right)\chi''_{\pm}+\frac{8}{15}\left(\mathcal{U}^{\pm}-\frac{17}{24}\tilde{\mathcal{L}}^{\pm\prime\prime}+\frac{241\pi}{12k}\left(\tilde{\mathcal{L}}^{\pm}\right)^{2}\right)'\chi'_{\pm}\\
&+\frac{40\pi}{3k}\left[i\psi^{\pm}_{I}\psi^{\pm I\prime}-\frac{12\pi}{5k}\left(\tilde{\mathcal{L}}^{\pm}\right)^{3}-\frac{11}{5^{2}}\mathcal{U}^{\pm}\tilde{\mathcal{L}}^{\pm}+\frac{3^{2}}{5^{2}}\left(\tilde{\mathcal{L}}^{\pm\prime}\right)^{2}+\frac{23}{50}\tilde{\mathcal{L}}^{\pm\prime\prime}\tilde{\mathcal{L}}^{\pm}\right.\\
&\left.\left.+\frac{k}{10^{2}\pi}\left(\mathcal{U}^{\pm}-\frac{1}{2}\tilde{\mathcal{L}}^{\pm\prime\prime}\right)''+\frac{10\pi}{k}i\mathcal{J}^{IJ\pm}\psi^{\pm}_{I}\psi^{\pm}_{J}\right]\chi_{\pm}-\frac{5}{9}\chi^{(4)}_{\pm}\mathcal{L}^{\pm}+\frac{k}{180\pi}\chi^{(6)}_{\pm}\right\}U_{\mp 3}\\
&-\frac{2\pi}{k}\left\{\epsilon_{\pm}\psi^{\pm}_{I}+\frac{1}{2}\vartheta_{\pm I}\mathcal{L}^{\pm\prime}+\frac{7}{6}\left[\delta^{J}_{\ I}\mathcal{L}^{\pm}-\frac{15}{7}\left(\mathcal{J}_{I}^{\ \ J\pm\prime}-\frac{10}{k}\mathcal{J}^{JK\pm}\mathcal{J}^{\pm}_{IK}\right)\right]\vartheta'_{\pm J}\right.\\
&-\frac{5}{2}\mathcal{J}_{I}^{\ \ J\pm}\vartheta''_{\pm J}-\frac{5}{3}\left[\psi^{\pm\prime\prime}_{I}-\frac{52\pi}{5k}\left(\tilde{\mathcal{L}}^{\pm}\psi^{\pm}_{I}-\frac{25}{13}\mathcal{J}_{I}^{\ \ J\pm}\psi^{\pm\prime}_{J}-\frac{25}{26}\mathcal{J}_{I}^{\ \ J\pm\prime}\psi^{\pm}_{J}\right.\right.\\
&\left.\left.+\frac{125}{13}\psi^{\pm}_{P}\mathcal{J}^{KP\pm}\mathcal{J}^{\pm}_{KI}\right)\right]\chi_{\pm}-\frac{25}{6}\left(\psi^{\pm\prime}_{I}+\frac{10\pi}{k}\psi^{\pm}_{J}\mathcal{J}_{I}^{\ \ J\pm}\right)\chi'_{\pm}\\
&-\frac{5}{6}\left[\mathcal{J}^{\pm\prime\prime}_{IJ}-\frac{10\pi}{k}\left(\mathcal{J}_{I}^{\ \ K\pm\prime}\mathcal{J}^{\pm}_{JK}+\frac{7}{5}\tilde{\mathcal{L}}\mathcal{J}^{\pm}_{IJ}+2\mathcal{J}_{I}^{\ \ K\pm}\mathcal{J}^{\pm\prime}_{JK}+\frac{10\pi}{k}\mathcal{J}^{MN\pm}\mathcal{J}^{\pm}_{IM}\mathcal{J}^{\pm}_{JN}\right)\right]\vartheta^{J}_{\pm}\\
&\left.-\frac{17}{6}\chi''_{\pm}\psi^{\pm}_{I}-\frac{k}{12\pi}\vartheta'''_{\pm I}\right\}\mathcal{S}^{I}_{\mp\frac{3}{2}}\pm\frac{3\pi}{k}\left[\left(\delta^{J}_{\ I}\tilde{\mathcal{L}}^{\pm}-\frac{5}{3}\mathcal{J}_{I}^{\ \ J\pm\prime}+\frac{50}{3k}\mathcal{J}^{JK\pm}\mathcal{J}^{\pm}_{IK}\right)\vartheta_{\pm J}\right.\\
&\left.-5\chi'_{\pm}\psi^{\pm}_{I}-\frac{10}{3}\mathcal{J}_{I}^{\ \ J\pm}\vartheta'_{\pm J}-\frac{10}{3}\left(\psi^{\pm\prime}_{I}+\frac{10\pi}{k}\psi^{\pm}_{J}\mathcal{J}_{I}^{\ \ J\pm}\right)\chi_{\pm}-\frac{k}{6\pi}\vartheta''_{\pm I}\right]\mathcal{S}^{I}_{\mp\frac{1}{2}}\\
&+\frac{20\pi}{k}\left(\chi_{\pm}\psi^{\pm}_{I}+\frac{1}{2}\mathcal{J}^{\pm}_{IJ}\vartheta^{J}_{\pm}+\frac{k}{20\pi}\vartheta'_{\pm I}\right)\mathcal{S}^{I}_{\pm\frac{1}{2}}\,,
\end{aligned}
$$

while the transformation laws of the fields $\mathcal{L}^{\pm}$, $\mathcal{U}^{\pm}$, $\mathcal{J}^{\pm}_{IJ}$, $\psi^{\pm}$, explicitly reads

$$\delta\mathcal{L}^{\pm} = 2\epsilon'_{\pm}\mathcal{L}^{\pm} + \epsilon_{\pm}\mathcal{L}^{\pm\prime} - \frac{k}{4\pi}\epsilon'''_{\pm} + 3\mathcal{U}^{\pm\prime}\chi_{\pm} + 4\mathcal{U}^{\pm}\chi'_{\pm} + \frac{5}{2}i\psi^{\pm}_{I}\left(\vartheta^{I}_{\pm}\right)' + \frac{3}{2}i\psi^{\pm\prime}_{I}\vartheta^{I}_{\pm} - \mathcal{J}^{\pm}_{IJ}\left(\zeta^{IJ}_{\pm}\right)' ,$$

$$\delta j^{\pm}_{IJ} = \epsilon'_{\pm}j^{\pm}_{IJ} + \epsilon_{\pm}j^{\pm\prime}_{IJ} + 4\mathcal{J}^{\pm K}_{[I|}\zeta^{\pm}_{J]K} + \frac{k}{5\pi}\zeta^{\pm\prime}_{IJ} - 2i\psi^{\pm}_{[I|}\vartheta_{\pm|J]} ,$$

$$\begin{aligned}
\delta\psi^{\pm}_{I} &= \frac{5}{2}\epsilon'_{\pm}\psi^{\pm}_{I} + \epsilon_{\pm}\psi^{\pm\prime}_{I} + 2\zeta_{I}{}^{K}\psi^{\pm}_{K} - \left[\left(\mathcal{U}^{\pm} - \frac{1}{2}\tilde{\mathcal{L}}^{\pm\prime\prime}\right)\delta^{K}_{I} + \frac{5}{6}\left(\mathcal{J}^{\pm K}_{I}\right)''' + \frac{3\pi}{k}\Lambda^{(4)K}_{\pm I}\right]\vartheta_{\pm K} \\
&+ \frac{5}{3}\left[\left(\tilde{\mathcal{L}}^{\pm}\delta^{K}_{I} - 2\left(\mathcal{J}^{\pm K}_{I}\right)'\right)\vartheta'_{\pm K} - \frac{k}{20\pi}\vartheta'''_{\pm I}\right]' + \frac{82\pi}{3k}\left(\Lambda^{(9/2)\prime}_{\pm} - \frac{23}{82}\Lambda^{(11/2)}_{\pm} - \frac{5k}{82\pi}\psi^{\pm\prime\prime\prime}_{I}\right)\chi_{\pm} \\
&- 7\left(\psi^{\pm\prime}_{I} + \frac{10\pi}{k}\Lambda^{(7/2)}_{\pm I}\right)\chi''_{\pm} + \frac{35\pi}{k}\left(\Lambda^{(9/2)}_{\pm I} - \frac{k}{6\pi}\psi^{\pm\prime\prime}_{I}\right)\chi'_{\pm} - \frac{35}{12}\chi'''_{\pm}\psi^{\pm}_{I} \\
&- \frac{5}{3}\left[\left(\mathcal{J}^{\pm K}_{I}\right)' + \frac{30\pi}{k}\Lambda^{(2)K}_{\pm I}\right]\vartheta''_{\pm K} - \frac{10}{3}\mathcal{J}_{\pm I}{}^{K}\vartheta'''_{\pm K} + \frac{10^{2}}{3k}\Lambda^{(3)K}_{\pm I}\vartheta'_{\pm K} ,
\end{aligned}$$

$$\begin{aligned}
\delta\mathcal{U}^{\pm} &= 4\epsilon'_{\pm}\mathcal{U}^{\pm} + \epsilon_{\pm}\mathcal{U}^{\pm\prime} + \frac{23\pi}{3k}i\left(\Lambda^{(11/2)}_{\pm I} + \Lambda^{(9/2)\prime}_{\pm I} + \frac{210}{23}\Lambda^{(7/2)\prime\prime}_{\pm I} - \frac{k}{92\pi}\psi^{\pm\prime\prime\prime}_{I}\right)\vartheta^{I}_{\pm} \\
&+ \frac{35\pi}{k}i\left(\Lambda^{(9/2)}_{\pm I} + 4\Lambda^{(7/2)\prime}_{\pm I} - \frac{k}{60\pi}\psi^{\pm\prime\prime}_{I}\right)\left(\vartheta^{I}_{\pm}\right)' - \frac{7}{4}i\left(\psi^{\pm\prime}_{I} - \frac{40\pi}{k}\Lambda^{(7/2)}_{\pm I}\right)\left(\vartheta^{I}_{\pm}\right)'' \\
&- \frac{35}{12}i\psi^{\pm}_{I}\left(\vartheta^{I}_{\pm}\right)''' - \frac{1}{6}\left[\left(\mathcal{U}^{\pm} - \frac{1}{2}\tilde{\mathcal{L}}^{\pm\prime\prime}\right)'' + \frac{144}{k}\left(\Lambda^{(6)}_{\pm} - \frac{49}{216}\tilde{\Lambda}^{(4)\prime\prime}_{\pm}\right)\right]'\chi_{\pm} \\
&- \frac{5}{6}\left[\left(\mathcal{U}^{\pm} - \frac{2}{3}\tilde{\mathcal{L}}^{\pm\prime\prime}\right)'' + \frac{288}{5k}\Lambda^{(6)}_{\pm}\right]\chi'_{\pm} + \frac{14}{9}\left[\tilde{\mathcal{L}}^{\pm\prime\prime} - \frac{27}{28}\mathcal{U}^{\pm} - \frac{21\pi}{k}\tilde{\Lambda}^{(4)}_{\pm}\right]'\chi''_{\pm} \\
&+ \frac{7}{3}\left[\tilde{\mathcal{L}}^{\pm\prime\prime} - \frac{3}{7}\mathcal{U}^{\pm} - \frac{28\pi}{3k}\tilde{\Lambda}^{(4)}_{\pm}\right]\chi''' + \frac{35}{18}\tilde{\mathcal{L}}^{\pm\prime}\chi^{(4)}_{\pm} + \frac{7}{9}\tilde{\mathcal{L}}^{\pm}\chi^{(5)}_{\pm} - \frac{k}{2^{4}3^{2}\pi}\chi^{(7)}_{\pm} ,
\end{aligned}$$

where

$$\begin{aligned}
\Lambda^{(2)}_{\pm IJ} &= \mathcal{J}^{\pm}_{IK}\mathcal{J}^{\pm K}_{J} , \\
\Lambda^{(3)}_{\pm IJ} &= \tilde{\mathcal{L}}^{\pm}\mathcal{J}^{\pm}_{IJ} - \Lambda^{(2)\prime}_{\pm IJ} - \mathcal{J}^{\pm}_{IK}\left(\mathcal{J}^{\pm K}_{J}\right)' - \frac{10\pi}{k}\Lambda^{(2)K}_{\pm I}\mathcal{J}^{\pm}_{KJ} , \\
\Lambda^{(4)}_{\pm IJ} &= \delta_{IJ}\left(\tilde{\mathcal{L}}^{\pm}\right)^{2} - \frac{50}{9}\left[\mathcal{J}^{\pm}_{IJ}\tilde{\mathcal{L}}^{\pm} - \frac{3}{2}\Lambda^{(2)\prime}_{\pm IJ} + \mathcal{J}^{\pm\prime}_{IK}\mathcal{J}^{\pm K}_{J}\right]' - \frac{25}{9}\mathcal{J}^{\pm\prime}_{IK}\mathcal{J}^{\pm K\prime}_{J} \\
&\quad - \frac{500\pi}{9k}\left[\tilde{\mathcal{L}}^{\pm}\Lambda^{(2)}_{\pm IJ} - \frac{1}{2}\mathcal{J}^{\pm MN}\left(\mathcal{J}^{\pm}_{IM}\mathcal{J}^{\pm}_{NJ}\right)' - \mathcal{J}^{\pm}_{IM}\left(\mathcal{J}^{\pm MN}\mathcal{J}^{\pm}_{NJ}\right)'\right] \\
&\quad + \frac{2500\pi}{9k^{2}}\Lambda^{(2)K}_{\pm I}\Lambda^{(2)}_{\pm KJ} , \\
\tilde{\Lambda}^{(4)}_{\pm} &= \left(\tilde{\mathcal{L}}^{\pm}\right)^{2} ,
\end{aligned}$$

$$
\begin{aligned}
\Lambda_{\pm}^{(6)} &= -\frac{7}{18}\mathcal{U}^{\pm}\tilde{\mathcal{L}}^{\pm} + \frac{295}{432}\left(\tilde{\mathcal{L}}^{\pm\prime}\right)^{2} + \frac{22}{27}\tilde{\mathcal{L}}^{\pm\prime\prime}\tilde{\mathcal{L}}^{\pm} + \frac{25}{12}i\psi_{K}^{\pm}\psi^{\pm K\prime} \\
&\quad - \frac{8\pi}{3k}\left[\left(\tilde{\mathcal{L}}^{\pm}\right)^{3} - \frac{125}{16}i\mathcal{J}^{\pm KL}\psi_{K}^{\pm}\psi_{L}^{\pm}\right] . \\
\Lambda_{\pm I}^{(7/2)} &= \psi_{K}^{\pm}\mathcal{J}_{I}^{\pm K} , \\
\Lambda_{\pm I}^{(9/2)} &= \tilde{\mathcal{L}}^{\pm}\psi_{I}^{\pm} + \frac{5}{3}\psi_{K}^{\pm}\mathcal{J}_{I}^{\pm K\prime} - \frac{10}{3}\Lambda_{\pm I}^{(7/2)\prime} - \frac{50\pi}{3k}\psi_{K}^{\pm}\Lambda_{\pm I}^{(2)K} , \\
\Lambda_{\pm I}^{(11/2)} &= \frac{27}{23}\left(\tilde{\mathcal{L}}^{\pm\prime}\psi_{I}^{\pm} - \frac{260}{27}\mathcal{J}_{I}^{\pm K\prime}\psi_{K}^{\pm\prime} - \frac{370}{81}\mathcal{J}_{I}^{\pm K\prime}\psi_{K}^{\pm\prime\prime} - \frac{260}{81}\mathcal{J}_{I}^{\pm K\prime\prime}\psi_{K}^{\pm}\right) \\
&\quad + \frac{820\pi}{23k}\left(\tilde{\mathcal{L}}^{\pm}\Lambda_{\pm I}^{(7/2)} - \frac{25}{41}\Lambda_{\pm I}^{(2)K}\psi_{K}^{\pm\prime} + \frac{130}{123}\Lambda_{\pm K}^{(7/2)}\mathcal{J}_{I}^{\pm K\prime} + \frac{55}{123}\mathcal{J}_{I}^{\pm K}\Lambda_{\pm K}^{(7/2)\prime}\right) \\
&\quad + \frac{5000\pi^{2}}{23k^{2}}\Lambda_{\pm K}^{(7/2)}\Lambda_{\pm I}^{(2)K} .
\end{aligned}
$$

Here the prime denotes derivative with respect to φ, and $\chi_{\pm}^{(n)}$ denotes the n-th derivative of $\chi_{\pm}$.

Appendix B. Poisson brackets of the canonical generators

The Poisson brackets of the asymptotic symmetry generators are given by

$$
\begin{aligned}
\left[\mathcal{L}(\varphi),\mathcal{L}(\phi)\right]_{PB} &= -2\delta'(\varphi-\phi)\mathcal{L}(\varphi) - \delta(\varphi-\phi)\mathcal{L}'(\varphi) + \frac{k}{4\pi}\delta'''(\varphi-\phi) , \\
\left[\mathcal{L}(\varphi),\mathcal{U}(\phi)\right]_{PB} &= -4\delta'(\varphi-\phi)\mathcal{U}(\varphi) - 3\delta(\varphi-\phi)\mathcal{U}'(\varphi) , \\
\left[\mathcal{L}(\varphi),\mathcal{J}^{IJ}(\phi)\right]_{PB} &= -\mathcal{J}^{IJ}(\varphi)\delta'(\varphi-\phi) , \\
\left[\mathcal{L}(\varphi),\psi^{[I]}(\phi)\right]_{PB} &= -\frac{5}{2}\delta'(\varphi-\phi)\psi^{[I]}(\varphi) - \frac{3}{2}\delta(\varphi-\phi)\psi^{[I]\prime}(\varphi) , \\
\left[\mathcal{U}(\varphi),\mathcal{J}^{IJ}(\phi)\right]_{PB} &= 0 , \\
\left[\mathcal{J}^{IJ}(\varphi),\mathcal{J}_{KL}(\phi)\right]_{PB} &= -4\delta^{[I}_{[K}\mathcal{J}^{J]}_{L]}(\varphi)\delta(\varphi-\phi) - \frac{k}{5\pi}\delta^{[I}_{[K}\delta^{J]}_{L]}\delta'(\varphi-\phi) , \qquad \text{(B.1)} \\
\left[\mathcal{J}^{IJ}(\varphi),\psi^{K}(\phi)\right]_{PB} &= 2\delta^{K[I}\psi^{J]}(\varphi)\delta(\varphi-\phi) , \\
\left[\mathcal{U}(\varphi),\psi^{I}(\phi)\right]_{PB} &= \frac{1}{12}\left\{\left[\psi^{I\prime\prime}(\varphi) - \frac{92\pi}{k}\left(\Lambda^{(9/2)I}(\varphi) + \frac{210}{23}\Lambda^{(7/2)I\prime}(\varphi)\right)\right]'\right. \\
&\quad \left. - \frac{92\pi}{k}\Lambda^{(11/2)I}(\varphi)\right\}\delta(\varphi-\phi) + \frac{7}{12}\left(\psi^{I\prime\prime}(\varphi) - \frac{60\pi}{k}\Lambda^{(9/2)I}(\varphi)\right. \\
&\quad \left. - \frac{240\pi}{k}\Lambda^{(7/2)I\prime}(\varphi)\right)\delta'(\varphi-\phi) + \frac{35}{12}\psi^{I}(\varphi)\delta'''(\varphi-\phi) \\
&\quad + \frac{7}{4}\left(\psi^{I\prime} - \frac{40\pi}{k}\Lambda^{(7/2)I}(\varphi)\right)\delta''(\varphi-\phi) ,
\end{aligned}
$$

$$
\begin{aligned}
[\mathcal{U}(\varphi), \mathcal{U}(\phi)]_{PB} = & \frac{5}{6}\left[\left(\mathcal{U}(\varphi) - \frac{2}{3}\tilde{\mathcal{L}}''(\varphi)\right)'' + \frac{288\pi}{5k}\Lambda^{(6)}(\varphi)\right]\delta'(\varphi-\phi) \\
& + \frac{1}{6}\left[\left(\mathcal{U}(\varphi) - \frac{1}{2}\tilde{\mathcal{L}}''(\varphi) - \frac{98\pi}{3k}\tilde{\Lambda}^{(4)}(\varphi)\right)'' + \frac{144\pi}{k}\Lambda^{(6)}(\varphi)\right]'\delta(\varphi-\phi) \\
& + \frac{3}{2}\left(\mathcal{U}(\varphi) - \frac{28}{27}\tilde{\mathcal{L}}''(\varphi) + \frac{196\pi}{9k}\tilde{\Lambda}^{(4)}(\varphi)\right)'\delta''(\varphi-\phi) \\
& + \left(\mathcal{U}(\varphi) - \frac{7}{3}\tilde{\mathcal{L}}''(\varphi) + \frac{196\pi}{9k}\tilde{\Lambda}^{(4)}(\varphi)\right)\delta'''(\varphi-\phi) \\
& - \frac{35}{18}\tilde{\mathcal{L}}'(\varphi)\delta^{(4)}(\varphi-\phi) - \frac{7}{9}\tilde{\mathcal{L}}(\varphi)\delta^{(5)}(\varphi-\phi) + \frac{k}{144\pi}\delta^{(7)}(\varphi-\phi), \\
i\left[\psi^I(\varphi), \psi^J(\phi)\right]_{PB} = & \delta(\phi-\varphi)\left[\delta^{IJ}\mathcal{U}(\varphi) - \frac{1}{2}\delta^{IJ}\tilde{\mathcal{L}}''(\varphi) + \frac{5}{6}\mathcal{J}^{IJ''}(\varphi) + \frac{3\pi}{k}\Lambda^{(4)(IJ)}(\varphi)\right. \\
& \left. + \frac{25\pi}{k}\Lambda^{(2)JI'}(\varphi) + \frac{50\pi}{3k}\Lambda^{(3)JI'}(\varphi)\right] + \frac{10}{3}\mathcal{J}^{IJ}(\varphi)\delta'''(\phi-\varphi) \\
& + \delta'(\phi-\varphi)\left[\frac{10}{3}\mathcal{J}^{IJ''}(\varphi) - \frac{5}{3}\delta^{IJ}\tilde{\mathcal{L}}'(\varphi) - \frac{100\pi}{3k}\Lambda^{(3)[IJ]}(\varphi)\right. \\
& \left. + \frac{50\pi}{k}\Lambda^{(2)JI'}(\varphi)\right] + \frac{k}{12\pi}\delta^{IJ}\delta''''(\phi-\varphi) \\
& + \delta''(\phi-\varphi)\left[5\mathcal{J}^{IJ'}(\varphi) - \frac{5}{3}\delta^{IJ}\tilde{\mathcal{L}}(\varphi) + \frac{50\pi}{k}\Lambda^{(2)IJ}(\varphi)\right], \qquad \text{(B.2)}
\end{aligned}
$$

so that once expanded in Fourier modes, the algebra corresponds to the one in eqs. (10)

References

[1] C. Aragone and S. Deser, "Hypersymmetry in $D = 3$ of Coupled Gravity Massless Spin 5/2 System," *Class. Quant. Grav.* **1** (1984) L9.

[2] O. Fuentealba, J. Matulich and R. Troncoso, "Extension of the Poincaré group with half-integer spin generators: hypergravity and beyond," arXiv:1505.06173 [hep-th].

[3] M. Henneaux and S. -J. Rey, "Nonlinear W_∞ as Asymptotic Symmetry of Three-Dimensional Higher Spin Anti-de Sitter Gravity," *JHEP* **1012**, 007 (2010) [arXiv:1008.4579 [hep-th]].

[4] A. Campoleoni, S. Fredenhagen, S. Pfenninger and S. Theisen, "Asymptotic symmetries of three-dimensional gravity coupled to higher-spin fields," *JHEP* **1011**, 007 (2010) [arXiv:1008.4744 [hep-th]].

[5] M. Henneaux, G. Lucena Gómez, J. Park and S. J. Rey, "Super-W(infinity) Asymptotic Symmetry of Higher-Spin AdS_3 Supergravity," *JHEP* **1206**, 037 (2012) [arXiv:1203.5152 [hep-th]].

[6] M. Henneaux, A. Pérez, D. Tempo and R. Troncoso, "Hypersymmetry bounds and three-dimensional higher-spin black holes," *JHEP* **1508** (2015) 021 doi:10.1007/JHEP08(2015)021 [arXiv:1506.01847 [hep-th]].

[7] V. G. Kac, "A Sketch of Lie Superalgebra Theory," *Commun. Math. Phys.* **53** (1977) 31; "Lie Superalgebras," *Adv. Math.* **26** (1977) 8.

[8] W. Nahm, "Supersymmetries and their representations," *Nucl. Phys. B* **135** (1978) 149.

[9] M. Gunaydin, G. Sierra and P. K. Townsend, "The Unitary Supermultiplets of D = 3 Anti-De Sitter and D = 2 Conformal Superalgebras," *Nucl. Phys. B* **274** (1986) 429.

[10] A. Achucarro and P. K. Townsend, "A Chern-Simons Action for Three-Dimensional anti-De Sitter Supergravity Theories," *Phys. Lett. B* **180** (1986) 89. doi:10.1016/0370-2693(86)90140-1

[11] A. Achucarro and P. K. Townsend, "Extended Supergravities in $d = (2+1)$ as Chern-Simons Theories," *Phys. Lett. B* **229** (1989) 383. doi:10.1016/0370-2693(89)90423-1

[12] O. Coussaert, M. Henneaux and P. van Driel, "The Asymptotic dynamics of three-dimensional Einstein gravity with a negative cosmological constant," *Class. Quant. Grav.* **12**, 2961 (1995) [gr-qc/9506019].

[13] M. Henneaux, L. Maoz and A. Schwimmer, "Asymptotic dynamics and asymptotic symmetries of three-dimensional extended AdS supergravity," *Annals Phys.* **282**, 31 (2000) [hep-th/9910013].

[14] M. Henneaux, A. Pérez, D. Tempo and R. Troncoso, "Chemical potentials in three-dimensional higher spin anti-de Sitter gravity," *JHEP* **1312**, 048 (2013) [arXiv:1309.4362 [hep-th]].

[15] C. Bunster, M. Henneaux, A. Pérez, D. Tempo and R. Troncoso, "Generalized Black Holes in Three-dimensional Spacetime," *JHEP* **1405**, 031 (2014) [arXiv:1404.3305 [hep-th]].

[16] T. Regge and C. Teitelboim, "Role of Surface Integrals in the Hamiltonian Formulation of General Relativity," *Annals Phys.* **88**, 286 (1974).

[17] J. D. Brown and M. Henneaux, "Central Charges in the Canonical Realization of Asymptotic Symmetries: An Example from Three-Dimensional Gravity," *Commun. Math. Phys.* **104**, 207 (1986).

[18] J. M. Figueroa-O'Farrill, S. Schrans and K. Thielemans, "On the Casimir algebra of B(2)," *Phys. Lett. B* **263**, 378 (1991).

[19] S. Bellucci, S. Krivonos and A. S. Sorin, "Linearizing W(2,4) and WB(2) algebras," *Phys. Lett. B* **347**, 260 (1995) [hep-th/9411168].

[20] A. Schwimmer and N. Seiberg, "Comments on the N=2, N=3, N=4 Superconformal Algebras in Two-Dimensions," *Phys. Lett. B* **184** (1987) 191. doi:10.1016/0370-2693(87)90566-1

[21] O. Coussaert and M. Henneaux, "Supersymmetry of the (2+1) black holes," *Phys. Rev. Lett.* **72**, 183 (1994) [hep-th/9310194].

[22] M. Bañados, C. Teitelboim and J. Zanelli, "The Black hole in three-dimensional space-time," *Phys. Rev. Lett.* **69**, 1849 (1992) [hep-th/9204099].

[23] M. Bañados, M. Henneaux, C. Teitelboim and J. Zanelli, "Geometry of the (2+1) black hole," *Phys. Rev. D* **48**, 1506 (1993) [gr-qc/9302012].

[24] M. Gutperle and P. Kraus, "Higher Spin Black Holes," *JHEP* **1105**, 022 (2011) [arXiv:1103.4304 [hep-th]].

[25] M. Ammon, M. Gutperle, P. Kraus and E. Perlmutter, "Spacetime Geometry in Higher Spin Gravity," *JHEP* **1110**, 053 (2011) [arXiv:1106.4788 [hep-th]].

[26] A. Castro, E. Hijano, A. Lepage-Jutier and A. Maloney, "Black Holes and Singularity Resolution in Higher Spin Gravity," *JHEP* **1201**, 031 (2012) [arXiv:1110.4117 [hep-th]].

[27] M. Ammon, M. Gutperle, P. Kraus and E. Perlmutter, "Black holes in three dimensional higher spin gravity: A review," *J. Phys. A* **46**, 214001 (2013) [arXiv:1208.5182 [hep-th]].

[28] A. Pérez, D. Tempo and R. Troncoso, "Higher Spin Black Holes," *Lect. Notes Phys.* **892**, 265 (2015) [arXiv:1402.1465 [hep-th]].

[29] A. Pérez, D. Tempo and R. Troncoso, "Higher spin gravity in 3D: Black holes, global charges and thermodynamics," *Phys. Lett. B* **726**, 444 (2013) [arXiv:1207.2844 [hep-th]].

[30] A. Pérez, D. Tempo and R. Troncoso, "Higher spin black hole entropy in three dimensions," *JHEP* **1304**, 143 (2013) [arXiv:1301.0847 [hep-th]].
[31] M. Bañados, R. Canto and S. Theisen, "The Action for higher spin black holes in three dimensions," *JHEP* **1207** (2012) 147 doi:10.1007/JHEP07(2012)147 [arXiv:1204.5105 [hep-th]].
[32] J. de Boer and J. I. Jottar, "Thermodynamics of higher spin black holes in AdS_3," *JHEP* **1401**, 023 (2014) [arXiv:1302.0816 [hep-th]].
[33] E. Ayon-Beato, C. Martinez, R. Troncoso and J. Zanelli, "Gravitational Cheshire effect: Nonminimally coupled scalar fields may not curve spacetime," *Phys. Rev. D* **71**, 104037 (2005) doi:10.1103/PhysRevD.71.104037 [hep-th/0505086].
[34] M. Bañados, A. Castro, A. Faraggi and J. I. Jottar, "Extremal Higher Spin Black Holes," arXiv:1512.00073 [hep-th].
[35] O. Fuentealba, J. Matulich and R. Troncoso, "Asymptotically flat structure of hypergravity in three spacetime dimensions," *JHEP* **1510**, 009 (2015) doi:10.1007/JHEP10(2015)009 [arXiv:1508.04663 [hep-th]].

Quasiconformal Group Approach to Higher Spin Algebras, Their Deformations and Supersymmetric Extensions

Murat Günaydin

Institute for Gravitation and the Cosmos
Physics Department, Pennsylvania State University
University Park, PA 16802, USA
mgunaydin@psu.edu

The quasiconformal method provides us with a unified approach to the construction of minimal unitary representations (minrep) of noncompact groups, their deformations as well as their supersymmetric extensions. We review the quasiconformal construction of the minrep of $SO(d,2)$, its deformations and their applications to unitary realizations of $AdS_{(d+1)}/CFT_d$ higher spin algebras and their deformations for arbitrary d and supersymmetric extensions for $d \leq 6$. $AdS_{(d+1)}/CFT_d$ higher spin algebras, their deformations and supersymmetric extensions are given by the enveloping algebras of the quasiconformal realizations of the minrep, its deformations and supersymmetric extensions, respectively, and are in one-to-one correspondence with massless conformal fields for arbitrary d and massless conformal supermultiplets for $d \leq 6$.

1. Introduction

The minimal unitary representation (minrep) of a noncompact Lie group, as defined by Joseph,[1] is realized over an Hilbert space of functions that depend on the smallest number of variables possible. The mathematics literature on minreps of noncompact groups is quite extensive. We refer to Refs. 2–14 and the references therein. After the discovery of the novel geometric quasiconformal realizations of non-compact groups[15,16] a unified approach to their minreps was developed. The minimal unitary realization is obtained by quantization of the geometric quasiconformal action. This was first carried out explicitly for the largest exceptional group $E_{8(8)}$[17] which is the U-duality group of maximal supergravity in three dimensions. The minrep of the $3d$ U-duality group $E_{8(-24)}$ of the exceptional supergravity[18] via the quasiconformal approach was constructed in Ref. 19. A unified approach to the construction of the minreps of noncompact groups by quantizing their quasiconformal actions was formulated in Ref. 20. The results of Ref. 20 extend to the minreps of noncompact superalgebras $\mathfrak{g}$ whose even subalgebras are of the form $\mathfrak{h} \oplus \mathfrak{sl}(2,\mathbb{R})$ where $\mathfrak{h}$ is compact. The quasiconformal construction of the minreps of non-compact Lie algebras $SU(n,m|p+q)$ and $OSp(2N^*|2M)$ were subsequently developed in Refs. 21–23.[a]

The quasiconformal construction of the minrep of AdS_5/CFT_4 group $SU(2,2)$ and its supersymmetric extensions $SU(2,2|N)$ were studied in Ref. 21. The minrep of $SU(2,2)$ describes a massless scalar field in $4d$. It admits a one parameter (ζ) family of deformations labelled by helicity, which can be continuous[b]. For a positive (negative) integer value of ζ, the deformed minrep describes a massless conformal field transforming in $\left(0\,,\,\frac{\zeta}{2}\right)$ $\left(\left(-\frac{\zeta}{2},0\right)\right)$ representation of the Lorentz group. The minrep and its deformations for integer ζ are isomorphic to the doubleton representations of $SU(2,2)$ that were constructed using covariant twistorial oscillators.[24–26] Similarly, the minrep of $SU(2,2\,|\,N)$ and its deformations for integer ζ are isomorphic to the doubleton supermultiplets that were studied in Refs. 24–26. The minimal unitary supermultiplet of $PSU(2,2|4)$ is the $N=4$ Yang-Mills supermultiplet in $d=4$ that was first constructed as a CPT self-conjugate . doubleton supermuliplet in Ref. 24.

The minrep of AdS_7/CFT_6 group $SO(6,2)=SO^*(8)$ and its deformations were constructed in Ref. 22. The deformations of the minrep of $SO(6,2)$ are labelled by the "spin" t of an $SU(2)$ subgroup . The $SU(2)$ spin t is the analog of helicity in six dimensions. The minrep of $SO^*(8)$ and its deformations describe massless conformal fields in six dimensions. These results were then extended to $6d$ superconformal algebras $OSp(8^*|2N)$.[22,23] The minimal unitary supermultiplet of $OSp(8^*|4)$ describes the massless conformal $(2,0)$ supermultiplet in $6d$, which was first constructed as a doubleton supermultiplet in Ref. 27.

The quasiconformal realizations of minreps of noncompact groups are in general nonlinear involving operators that are cubic and quartic in the coordinates and momenta. However for symplectic groups quasiconformal construction of the minrep coincides with the oscillator construction involving bilinears of twistorial oscillators.[20] Hence the minrep of AdS_4/CFT_3 group $Sp(4,\mathbb{R})$ is simply the scalar singleton of Dirac and it admits a single deformation which is the spinor singleton. They were referred to as the remarkable representations of AdS_4 group by Dirac.[28] The singleton supermultiplets of AdS_4 superalgebras $OSp(2n|4,\mathbb{R})$, in particular the $N=8$ superalgebra $OSp(8|4,\mathbb{R})$, were first constructed in Refs. 29, 30 using the oscillator methods developed earlier.[31,32] Oscillator construction of general supermultiplets of $OSp(N|4,\mathbb{R})$ was later given in Refs. 27, 33.

All the massless higher spin representations of AdS_4 group $SO(3,2)$ occur in the tensor product of two singleton representations.[34] The higher spin theories in AdS_4 were studied extensively by Fronsdal and collaborators.[35–38] In the 1980s Fradkin and Vasiliev pioneered the study of higher spin theories involving fields

[a]We will use the notation for labelling Lie (super)groups and Lie (super)algebras interchangeably throughout the paper.

[b]The parameter ζ is twice the helicity.

containing infinitely many spins $0 \leq s < \infty$.[39,40] Much work has been done on higher spin theories in the intervening years, in particular after the work of Ref. 41 who conjectured that Vasiliev's higher spin theory[42] in AdS_4 is dual to $O(N)$ vector model in $3d$. The conjecture of Refs. 41 was checked by Giombi and Yi by calculating some of the three point functions of higher spin currents in the bulk and matching them with those of free and critical $O(N)$ vector models in $3d$.[43,44] For reviews on higher spin theories we refer to Refs. 45–51 and references therein.

That the Fradkin-Vasiliev higher spin algebra in AdS_4[39] is the enveloping algebra of the singletonic realization of $Sp(4,\mathbb{R})$ was first pointed out in Ref. 52. Again in Ref. 52 it was suggested that the higher spin algebras of Fradkin-Vasiliev type in AdS_5 and AdS_7 can similarly be obtained from the doubletonic realizations of $SU(2,2)$ and $SO(6,2)$, respectively. Higher spin superalgebras in AdS_4, AdS_5 and AdS_7 could then be realized as enveloping algebras of the singletonic realization of $OSp(N/4,\mathbb{R})$ and doubletonic realizations of $SU(2,2|N)$ and $OSp(8^*|2N)$.[52] Conformal higher spin superalgebras in four dimensions were studied in Ref. 53. Higher spin theories and supersymmetric extensions in AdS_5 and AdS_7 were studied later in Refs. 54–57 using the doubletonic realizations of Refs. 25–27, 58. Higher spin superalgebras in higher dimensions were also studied in Ref. 59.

Vasiliev pointed out that to obtain the standart bosonic higher spin algebra in $AdS_{(d+1)}$, one has to quotient the enveloping algebra of $SO(d,2)$ by the ideal that annihilates the scalar "singleton" representation.[46] Later Eastwood identified this ideal to be the Joseph ideal.[60] This is consistent with the observation that AdS_4/CFT_3 higher spin algebra is given by the enveloping algebra of the singletonic realization of $SO(3,2)$[52] since the Joseph ideal vanishes identically for the singleton. However the Joseph ideal does not vanish identically as operators for the doubletonic realizations of $SO(4,2)$ and $SO(6,2)$ in terms of covariant twistorial oscillators. This is where the importance of quasiconformal approach to the construction of higher spin algebras and superalgebras and their deformations becomes manifest. The Joseph ideal vanishes identically as operators in the quasiconformal construction of the minrep of $SO(d,2)$ and hence its enveloping algebra leads directly to the higher spin algebra without the need for quotienting. This approach also allows one to define deformations and supersymmetric extensions of higher spin algebras.

The vanishing of the Joseph ideal for the quasiconformal realization of minreps of $SO(4,2)$ and $SO(6,2)$ was shown in Refs. 61, 62. Hence their enveloping algebras lead directly to unitary realizations of the bosonic AdS_5/CFT_4 and AdS_7/CFT_6 higher spin algebras, respectively. The enveloping algebras of the deformed minreps of $SU(2,2)$ and of $SO^*(8)$ and their supersymmetric extensions yield infinite families of AdS_5/CFT_4 and AdS_7/CFT_6 higher spin algebras and superalgebras, respectively.[61,62]

These results were extended to AdS_6/CFT_5 higher spin algebra in Ref. 63. The minrep of $SO(5,2)$ admits a unique deformation which describes a conformally massless spinor field in $5d$. In five dimensions there exists a unique simple conformal superalgebra , namely $F(4)$ with the even subalgebra $SO(5,2) \oplus SU(2)$. The minimal unitary supermultiplet of $F(4)$ decomposes into the deformed minrep and two copies of the minrep. Its enveloping algebra defines the unique higher spin superalgebra in AdS_6.

Extension of the above results to higher dimensions was given in Ref. 64 where it was shown that the minrep of $SO(d,2)$ and its deformations are in one-to-one correspondence with massless conformal fields in d dimensional spacetimes. For the quasiconformal realization of the minrep of $SO(d,2)$ generators of the Joseph ideal vanish and its enveloping algebra yields directly the $AdS_{(d+1)}/CFT_d$ higher spin algebra. The enveloping algebra of a deformation of the minrep leads to a deformed higher spin algebra. In odd dimensions there exists a unique deformation. In even dimensions there exist infinitely many deformations whose enveloping algebras define an infinite family of higher spin algebras.

Below we will review the quasiconformal realization of $SO(d,2)$ and its quantization that leads directly to the minimal unitary representation of $SO(d,2)$. We shall then discuss the deformations of the minrep and show that there is a one-to-one correspondence between the minrep and its deformations and massless conformal fields in d dimensional Minkowskian spacetimes. This will be followed by a review of the application of these results to higher spin algebras, their deformations and supersymmetric extensions.

2. Geometric Realization of $SO(d,2)$ as a Quasiconformal Group

The geometric quasiconformal realization of the anti-de Sitter (conformal) group $SO(d,2)$ in $(d+1)$ (d) dimensions that was given in[16] is based on the 5-grading of the Lie algebra $\mathfrak{so}(d,2)$ with respect to its maximal rank subalgebra $\mathfrak{so}(1,1) \oplus \mathfrak{so}(d-2) \oplus \mathfrak{so}(2,1)$:

$$\begin{aligned}\mathfrak{so}(d,2) = \mathbf{1}^{(-2)} \oplus (\mathbf{d-2,2})^{(-1)} \oplus [\,\Delta \oplus \mathfrak{so}(d-2) \oplus \mathfrak{su}(1,1)\,]^{0} \\ \oplus (\mathbf{d-2,2})^{(+1)} \oplus \mathbf{1}^{(+2)}\end{aligned}$$

where the generator that determines the five grading is denoted as Δ. We should stress that $SO(d-2)$ is the simple part of the little group of massless particles in d-dimensional Minkowski space-time. The generators of quasiconformal group action of $SO(d,2)$ are realized as nonlinear differential operators acting on a $(2d-3)$-dimensional space $\mathcal{T}$ with coordinates $\mathcal{X} = \left(X^{i,a}, x\right)$, where $X^{i,a}$ transform in the $(d-2,2)$ representation of $\mathfrak{so}(d-2) \oplus \mathfrak{su}(1,1)$ subalgebra, with $i = 1, \ldots, d-2$ and $a = 1, 2$, and x is a singlet coordinate.

Labelling the generators with grade $-2, -1, 0, +1$ and $+2$ as $K_-, U_{i,a}$, $[\Delta \oplus \mathcal{L}_{ij} \oplus$

$\mathcal{M}_{ab}]$, $\widetilde{U}_{i,a}$ and K_+, respectively, their quasiconformal realizations have the form:

$$\begin{aligned}
K_+ &= \frac{1}{2}\left(2x^2 - \mathcal{I}_4\right)\frac{\partial}{\partial x} - \frac{1}{4}\frac{\partial \mathcal{I}_4}{\partial X^{i,a}}\eta^{ij}\epsilon^{ab}\frac{\partial}{\partial X^{j,b}} + x\,X^{i,a}\frac{\partial}{\partial X^{i,a}} \\
U_{i,a} &= \frac{\partial}{\partial X^{i,a}} - \eta_{ij}\epsilon_{ab}\,X^{j,b}\frac{\partial}{\partial x} \\
\mathcal{L}_{ij} &= \eta_{ik}X^{k,a}\frac{\partial}{\partial X^{j,a}} - \eta_{jk}X^{k,a}\frac{\partial}{\partial X^{i,a}} \\
\mathcal{M}_{ab} &= \epsilon_{ac}X^{i,c}\frac{\partial}{\partial X^{i,b}} + \epsilon_{bc}X^{i,c}\frac{\partial}{\partial X^{i,a}} \\
K_- &= \frac{\partial}{\partial x} \quad , \qquad \Delta = 2\,x\frac{\partial}{\partial x} + X^{i,a}\frac{\partial}{\partial X^{i,a}} \quad , \qquad \widetilde{U}_{i,a} = [U_{i,a}\,,\,K_+]
\end{aligned} \tag{2.1}$$

where $\mathcal{L}_{ij}$ and $\mathcal{M}_{ab}$ are the generators of $SO(d-2)$ and $SU(1,1)$ subgroups, respectively,($i,j,k,l = 1,\ldots,d-2$; $a,b,c,d = 1,2$) and ϵ^{ab} is the inverse symplectic tensor, such that $\epsilon^{ab}\epsilon_{bc} = \delta^a{}_c$. $\mathcal{I}_4(X)$ denotes the quartic polynomial of the coordinates $X^{i,a}$

$$\mathcal{I}_4(X) = \eta_{ij}\eta_{kl}\epsilon_{ac}\epsilon_{bd}X^{i,a}X^{j,b}X^{k,c}X^{l,d} \tag{2.2}$$

which is an invariant of $SO(d-2)\times SU(1,1)$ subgroup[c]. In the above expression, ϵ_{ab} is the symplectic invariant tensor of $SU(1,1)$ and η_{ij} is the invariant metric of $SO(d-2)$ in the fundamental representation, which we choose as $\eta_{ij} = -\delta_{ij}$ to be consistent with the conventions of Ref. 16.

The grade +1 generators $\widetilde{U}_{i,a}$ are obtained by substituting the expression for the quartic invariant in the above equations:

$$\begin{aligned}
\widetilde{U}_{i,a} &= [U_{i,a}\,,\,K_+] \\
&= \eta_{ij}\epsilon_{ad}\left(\eta_{kl}\epsilon_{bc}X^{j,b}X^{k,c}X^{l,d} - x\,X^{j,d}\right)\frac{\partial}{\partial x} + x\frac{\partial}{\partial X^{i,a}} \\
&\quad - \eta_{ij}\epsilon_{ab}\,X^{j,b}X^{l,c}\frac{\partial}{\partial X^{l,c}} - \epsilon_{ad}\eta_{kl}\,X^{l,d}X^{k,c}\frac{\partial}{\partial X^{i,c}} \\
&\quad + \epsilon_{ad}\eta_{ij}\,X^{l,d}X^{j,b}\frac{\partial}{\partial X^{l,b}} + \eta_{ij}\epsilon_{bc}X^{j,b}X^{l,c}\frac{\partial}{\partial X^{l,a}}.
\end{aligned} \tag{2.3}$$

The generators of $SO(d,2)$ satisfy the following commutation relations:

$$\begin{aligned}
[\mathcal{L}_{ij}\,,\,\mathcal{L}_{kl}] &= \eta_{jk}\mathcal{L}_{il} - \eta_{ik}\mathcal{L}_{jl} - \eta_{jl}\mathcal{L}_{ik} + \eta_{il}\mathcal{L}_{jk} \\
[\mathcal{M}_{ab}\,,\,\mathcal{M}_{cd}] &= \epsilon_{cb}\mathcal{M}_{ad} + \epsilon_{ca}\mathcal{M}_{bd} + \epsilon_{db}\mathcal{M}_{ac} + \epsilon_{da}\mathcal{M}_{bc}
\end{aligned} \tag{2.4a}$$

$$\begin{aligned}
[\Delta\,,\,K_\pm] &= \pm 2\,K_\pm & [K_-\,,\,K_+] &= \Delta \\
[\Delta\,,\,U_{i,a}] &= -U_{i,a} & \left[\Delta\,,\,\widetilde{U}_{i,a}\right] &= \widetilde{U}_{i,a} \\
[U_{i,a}\,,\,K_+] &= \widetilde{U}_{i,a} & \left[\widetilde{U}_{i,a}\,,\,K_-\right] &= -U_{i,a} \\
[U_{i,a}\,,\,U_{j,b}] &= 2\,\eta_{ij}\epsilon_{ab}\,K_- & \left[\widetilde{U}_{i,a}\,,\,\widetilde{U}_{j,b}\right] &= 2\,\eta_{ij}\epsilon_{ab}\,K_+
\end{aligned} \tag{2.4b}$$

[c] Note that we have the isomorphisms $SU(1,1) \equiv SO(2,1) \equiv Sp(2,\mathbb{R})$.

$$[\mathcal{L}_{ij}\,,\,U_{k,a}] = \eta_{jk}U_{i,a} - \eta_{ik}U_{j,a} \qquad \left[\mathcal{L}_{ij}\,,\,\widetilde{U}_{k,a}\right] = \eta_{jk}\widetilde{U}_{i,a} - \eta_{ik}\widetilde{U}_{j,a}$$
$$[\mathcal{M}_{ab}\,,\,U_{i,c}] = \epsilon_{cb}U_{i,a} + \epsilon_{ca}U_{i,b} \qquad \left[\mathcal{M}_{ab}\,,\,\widetilde{U}_{i,c}\right] = \epsilon_{cb}\widetilde{U}_{i,a} + \epsilon_{ca}\widetilde{U}_{i,b} \tag{2.4c}$$

$$\left[U_{i,a}\,,\,\widetilde{U}_{j,b}\right] = \eta_{ij}\epsilon_{ab}\,\Delta - 2\,\epsilon_{ab}\mathcal{L}_{ij} - \eta_{ij}\mathcal{M}_{ab}\,. \tag{2.4d}$$

The above nonlinear realization has a geometric interpretation as the invariance group of a light-cone with respect to a quartic distance function in the space $\mathcal{T}$. The quartic norm of a vector $\mathcal{X} = (X, x)$ in the $(2d-3)$-dimensional space $\mathcal{T}$ is given as

$$\mathcal{N}_4\,(\mathcal{X}) = \mathcal{I}_4\,(X) + 2\,x^2 \tag{2.5}$$

and the quartic distance function between any two points $\mathcal{X}$ and $\mathcal{Y}$ in $\mathcal{T}$ is then defined as follows:[17,20]

$$d\,(\mathcal{X},\mathcal{Y}) = \mathcal{N}_4\,(\delta\,(\mathcal{X},\mathcal{Y})) \tag{2.6}$$

where the "symplectic" difference $\delta\,(\mathcal{X},\mathcal{Y})$ is defined as

$$\delta\,(\mathcal{X},\mathcal{Y}) = \left(X^{i,a} - Y^{i,a}\,,\, x - y - \eta_{ij}\epsilon_{ab}\,X^{i,a}Y^{j,b}\right) = -\delta\,(\mathcal{Y},\mathcal{X})\,. \tag{2.7}$$

The lightlike separations between any two points with respect to this quartic distance function are left invariant under the quasiconformal group action of $SO(d,2)$ on $\mathcal{T}$.

3. Minimal Unitary Representation of $SO(d,2)$ from its Quasiconformal Realization

The "quantization" of the geometric quasiconformal realization of a noncompact group leads directly to its minimal unitary representation.[16,17,19,20,64] To "quantize" the quasiconformal realization of $SO(d,2)$, we split the $2(d-2)$ variables $X^{i,a}$ defined in Section 2 into $(d-2)$ coordinates X^i and $(d-2)$ conjugate momenta P_i as

$$X^i = X^{i,1} \qquad P_i = \eta_{ij}\,X^{j,2} \tag{3.1}$$

and introduce a momentum p conjugate to the singlet coordinate x as well. Furthermore we treat them as quantum mechanical operators satisfying the canonical commutation relations

$$\left[X^i\,,\,P_j\right] = i\delta^i_j \qquad [x\,,\,p] = i\,. \tag{3.2}$$

Instead of the coordinates X^i and momenta P_i it will be convenient to work with bosonic oscillator annihilation operators a_i and creation operators $a_i^\dagger$, defined as

$$a_i = \frac{1}{\sqrt{2}}\left(X^i + i\,P_i\right) \qquad a_i^\dagger = \frac{1}{\sqrt{2}}\left(X^i - i\,P_i\right). \tag{3.3}$$

They satisfy the commutation relations

$$\left[a_i\,,\,a_j^\dagger\right] = \delta_{ij} \qquad [a_i\,,\,a_j] = \left[a_i^\dagger\,,\,a_j^\dagger\right] = 0\,. \tag{3.4}$$

For the minimal unitary realization of $\mathfrak{so}(d,2)$ the 5-grading

$$\mathfrak{so}(d,2) = \mathfrak{g}^{(-2)} \oplus \mathfrak{g}^{(-1)} \oplus [\,\Delta \oplus \mathfrak{so}(d-2)_L \oplus \mathfrak{su}(1,1)_M\,] \oplus \mathfrak{g}^{(+1)} \oplus \mathfrak{g}^{+2)}$$

is determined by the $SO(1,1)$ generator

$$\Delta = \frac{1}{2}\,(xp + px)\ . \tag{3.5}$$

The generators of $\mathfrak{su}(1,1)_M \subset \mathfrak{g}^{(0)}$ are bilinears of the bosonic oscillators:

$$M_+ = \frac{1}{2} a_i^\dagger a_i^\dagger \qquad M_- = \frac{1}{2} a_i a_i \qquad M_0 = \frac{1}{4}\left(a_i^\dagger a_i + a_i a_i^\dagger\right) \tag{3.6}$$

and satisfy

$$[M_-\,,\,M_+] = 2\,M_0 \qquad\qquad [M_0\,,\,M_\pm] = \pm\,M_\pm\,. \tag{3.7}$$

Its quadratic Casimir $\mathcal{M}^2$ is given by:

$$\mathcal{C}_2\left[\mathfrak{su}(1,1)_M\right] = \mathcal{M}^2 = {M_0}^2 - \frac{1}{2}\,(M_+M_- + M_-M_+). \tag{3.8}$$

The subalgebra $\mathfrak{so}(d-2)_L \subset \mathfrak{g}^{(0)}$ is also realized as bilinears of the bosonic oscillators:

$$L_{ij} = i\left(a_i^\dagger a_j - a_j^\dagger a_i\right) \tag{3.9}$$

and satisfy the commutation relations

$$[L_{ij}\,,\,L_{kl}] = i\,(\delta_{jk}L_{il} - \delta_{ik}L_{jl} - \delta_{jl}L_{ik} + \delta_{il}L_{jk})\ . \tag{3.10}$$

The Casimir $\mathcal{L}^2$ of $\mathfrak{so}(d-2)_L$

$$\mathcal{C}_2\left[\mathfrak{so}(d-2)_L\right] = \mathcal{L}^2 = L_{ij}L_{ij} \tag{3.11}$$

is related to the Casimir of $\mathfrak{su}(1,1)_M$ as follows:

$$\mathcal{L}^2 = 8\,\mathcal{M}^2 - \frac{1}{2}\,(d-2)\,(d-6)\ . \tag{3.12}$$

The generator in $\mathfrak{g}^{(-2)}$ is defined as

$$K_- = \frac{1}{2} x^2\,. \tag{3.13}$$

The generators $(U_i, U_i^\dagger)$ in grade -1 subspace are realized as bilinears of x and the bosonic oscillators

$$U_i = x\,a_i \qquad\qquad U_i^\dagger = x\,a_i^\dagger\,. \tag{3.14}$$

They close into K_- under commutation and form a Heisenberg subalgebra

$$\left[U_i\,,\,U_j^\dagger\right] = 2\,\delta_{ij}\,K_- \qquad\qquad [U_i\,,\,U_j] = \left[U_i^\dagger\,,\,U_j^\dagger\right] = 0 \tag{3.15}$$

with $K_- = \frac{1}{2}x^2$ playing the role of "central charge".

The quartic invariant $\mathcal{I}_4$ of $SO(d-2)_L \times SU(1,1)_M$ subgroup that enters the quasiconformal realization goes over to a linear function of the quadratic Casimir of

$SO(d-2)_L \times SU(1,1)_M$ after "quantization". In particular the grade +2 generator K_+ becomes:

$$K_+ = \frac{1}{2}p^2 + \frac{1}{x^2}\mathcal{G} \tag{3.16}$$

with $\mathcal{G}$ given by the Casimir $\mathcal{L}^2$:

$$\mathcal{G} = \frac{1}{4}\mathcal{L}^2 + \frac{1}{8}(d-3)(d-5) = 2\mathcal{M}^2 + \frac{3}{8}. \tag{3.17}$$

The generators $W_i, W_i^\dagger$ in grade +1 subspace are given by the commutators of the grade −1 generators with K_+:

$$\begin{aligned} W_i &= -i\,[U_i\,,\,K_+] = p\,a_i - \frac{i}{x}\left[\frac{1}{2}(d-3)\,a_i + i\,L_{ij}\,a_j\right] \\ W_i^\dagger &= -i\left[U_i^\dagger\,,\,K_+\right] = p\,a_i^\dagger - \frac{i}{x}\left[\frac{1}{2}(d-3)\,a_i^\dagger + i\,L_{ij}\,a_j^\dagger\right]. \end{aligned} \tag{3.18}$$

The positive grade generators form a Heisenberg algebra as well:

$$\left[W_i\,,\,W_j^\dagger\right] = 2\,\delta_{ij}\,K_+ \qquad [W_i\,,\,W_j] = \left[W_i^\dagger\,,\,W_j^\dagger\right] = 0 \tag{3.19}$$

with the generator K_+ playing the role of central charge. The commutators of grade −2 and grade +1 generators close into grade −1 subspace:

$$[W_i\,,\,K_-] = -i\,U_i \qquad \left[W_i^\dagger\,,\,K_-\right] = -i\,U_i^\dagger. \tag{3.20}$$

The generators $K_\pm$ and Δ form a distinguished $\mathfrak{su}(1,1)$ subalgebra labelled as $\mathfrak{su}(1,1)_K$:

$$[K_-\,,\,K_+] = i\,\Delta \qquad [\Delta\,,\,K_\pm] = \pm 2i\,K_\pm. \tag{3.21}$$

Its quadratic Casimir

$$\mathcal{C}_2\left[\mathfrak{su}(1,1)_K\right] = \mathcal{K}^2 = \Delta^2 - 2\,(K_+K_- + K_-K_+) \tag{3.22}$$

is related to the quadratic Casimir of $\mathfrak{so}(d-2)_L$ (and that of $\mathfrak{su}(1,1)_M$) as follows

$$\mathcal{K}^2 = -\frac{1}{2}\mathcal{L}^2 - \frac{1}{4}(d-2)(d-6) = -4\,\mathcal{M}^2\,. \tag{3.23}$$

Grade ±1 generators transform in the $(d-2,2)$ representation $SO(d-2)_L \times SU(1,1)_M$:

$$\begin{aligned} &[M_0\,,\,U_i] = -\frac{1}{2}U_i && [M_0\,,\,W_i] = -\frac{1}{2}W_i \\ &[M_+\,,\,U_i] = -U_i^\dagger && [M_+\,,\,W_i] = -W_i^\dagger \\ &[M_-\,,\,U_i] = 0 && [M_-\,,\,W_i] = 0 \\ &[L_{ij}\,,\,U_k] = i\,(\delta_{jk}U_i - \delta_{ik}U_j) && [L_{ij}\,,\,W_k] = i\,(\delta_{jk}W_i - \delta_{ik}W_j) \end{aligned} \tag{3.24}$$

$$\begin{array}{ccccc} & & K_- & & \\ & U_i & | & U_i^\dagger & \\ M_- & --- & (\Delta \oplus L_{ij} \oplus M_0) & --- & M_+ \\ & W_i & | & W_i^\dagger & \\ & & K_+ & & \end{array}$$

Table 1: Above we give the 5×5 grading of the Lie algebra of $\mathfrak{so}(d,2)$. The vertical 5-grading is determined by Δ and the horizontal 5-grading is determined by M_0.

$$\begin{aligned} \left[M_0 \,,\, U_i^\dagger\right] &= \frac{1}{2} U_i^\dagger & \left[M_0 \,,\, W_i^\dagger\right] &= \frac{1}{2} W_i^\dagger \\ \left[M_+ \,,\, U_i^\dagger\right] &= 0 & \left[M_+ \,,\, W_i^\dagger\right] &= 0 \\ \left[M_- \,,\, U_i^\dagger\right] &= U_i & \left[M_- \,,\, W_i^\dagger\right] &= W_i \\ \left[L_{ij} \,,\, U_k^\dagger\right] &= i\left(\delta_{jk} U_i^\dagger - \delta_{ik} U_j^\dagger\right) & \left[L_{ij} \,,\, W_k^\dagger\right] &= i\left(\delta_{jk} W_i^\dagger - \delta_{ik} W_j^\dagger\right) \end{aligned} \tag{3.25}$$

and close into the generators of $SO(d-2)_L \times SU(1,1)_M$:

$$\begin{aligned} [U_i \,,\, W_j] &= 2i\, \delta_{ij}\, M_- & \left[U_i^\dagger \,,\, W_j\right] &= \delta_{ij}\, (2i\, M_0 - \Delta) + 2\, L_{ij} \\ \left[U_i^\dagger \,,\, W_j^\dagger\right] &= 2i\, \delta_{ij}\, M_+ & \left[U_i \,,\, W_j^\dagger\right] &= \delta_{ij}\, (2i\, M_0 + \Delta) - 2\, L_{ij}. \end{aligned} \tag{3.26}$$

For the minimal unitary realization given above the quadratic Casimir of $\mathfrak{so}(d,2)$ turns out to be

$$\mathcal{C}_2\left[\mathfrak{so}(d,2)\right] = -\frac{1}{2}\left(d^2 - 4\right) . \tag{3.27}$$

The Lie algebra of $SO(d,2)$ admits a different 5-grading determined by the compact generator M_0 in addition to the 5-grading determined by the non-compact generator Δ. The resulting 5×5 grading is given in Table 1. We should note that the generators of $SO(d,2)$ given above are not all Hermitian. Since we are interested in unitary realizations one must go to a basis in which all the generators at Hermitian (anti-Hermitian) with pure imaginary (real) structure constants and determine the real form corresponding to the unitary realization , which in our case turns out to be $SO(d,2)$.

The little group of massless particles in d dimensional Minkowskian spacetime is the Euclidean group $E_{(d-2)}$ in $(d-2)$ dimensions acting on the transverse coordinates whose Lie algebra is the semi-direct sum

$$\mathfrak{e}_{(d-2)} = \mathfrak{so}(d-2) \circledS\, t_{(d-2)}$$

where $t_{(d-2)}$ denotes the generators of translations. There are two different such embeddings of the Euclidean group $E_{(d-2)}$, namely one acting on the transverse coordinates X_i and the other acting on the transverse momenta P_i inside the quasiconformal realization of $SO(d,2)$ which we shall denote as $E^X_{(d-2)}$ and $E^P_{(d-2)}$, respectively. Their generators are

$$E^X_{(d-2)} \Longrightarrow [L_{ij} \oplus T^X_i] \tag{3.28}$$

$$E^P_{(d-2)} \Longrightarrow [L_{ij} \oplus T^P_i] \tag{3.29}$$

where

$$T^X_i = \frac{i}{\sqrt{2}}(U^\dagger_i - U_i) = xP_i$$

and

$$T^P_i = \frac{1}{\sqrt{2}}(U^\dagger_i + U_i) = xX_i.$$

Both of these Euclidean subgroups have extensions to conformal groups $SO(d-1,1)$ in $(d-2)$ Euclidean dimensions as subgroups of $SO(d,2)$, which we denote as $SO^X(d-1,1)$ and $SO^P(d-1,1)$. The Lie algebras of these Euclidean conformal subgroups in $(d-2)$ dimensions have the standard three graded decompositions:

$$\mathfrak{so}^X(d-1,1) = K^X_i \oplus (L_{ij} + D^X) \oplus T^X_i \tag{3.30}$$

$$\mathfrak{so}^P(d-1,1) = K^P_i \oplus (L_{ij} + D^P) \oplus T^P_i \tag{3.31}$$

where D^X and D^P are the the generators of respective scale transormations that determine the 3-grading given by

$$D^X = \tfrac{1}{2}[\Delta - i(M_+ - M_-)] = \tfrac{1}{2}(xp + px) + \tfrac{1}{2}(X_iP_i + P_iX_i) \tag{3.32}$$

$$D^P = \tfrac{1}{2}[\Delta + i(M_+ - M_-)] = \tfrac{1}{2}(xp + px) - \tfrac{1}{2}(X_iP_i + P_iX_i). \tag{3.33}$$

The special conformal generators K^X_i and K^P_i acting on transverse coordinates X_i and momenta P_i are given by

$$K^X_i = \tfrac{1}{\sqrt{2}}(W_i + W^\dagger_i) = p\,X_i - \tfrac{i}{x}\left(\tfrac{(d-3)}{2}X_i + iP_iX_jX_j - iP_jX_jX_i\right) \tag{3.34}$$

$$K^P_i = \tfrac{i}{\sqrt{2}}(W^\dagger_i - W_i) = p\,P_i - \tfrac{i}{x}\left(\tfrac{(d-3)}{2}P_i - iX_iP_jP_j + iX_jP_jP_i\right). \tag{3.35}$$

They satisfy the commutation relations

$$[T^X_i, K^X_j] = -2i\delta_{ij}D^X + 2iL_{ij} \tag{3.36}$$

$$[T^P_i, K^P_j] = 2i\delta_{ij}D^P - 2iL_{ij} \tag{3.37}$$

$$[T^X_i, K^P_j] = iL_+ \tag{3.38}$$

$$[T^P_i, K^X_j] = iL_- \tag{3.39}$$

where $L_+ = P_iP_i$ and $L_- = X_iX_i$. We have a 5-grading of the Lie algebra $\mathfrak{so}(d,2)$ with respect to the generator $L_0 = \frac{1}{2}(D^X - D^P) = X_iP_i + P_iX_i$ which together with Δ defines a 5×5 grading as indicated in Table 2.

Therefore the quasiconformal realization of $SO(d,2)$ can be interpreted as the minimal Lie algebra containing the Euclidean conformal Lie algebra acting on transverse coordinates and the dual Euclidean conformal Lie algebra acting on the corresponding transverse momenta. The common subgroup of these two Euclidean conformal groups is $SO(d-2)$.

$$\begin{array}{ccccc} & & K_- & & \\ & T_i^P & | & T_i^X & \\ L_- & --- & (\Delta \oplus L_{ij} \oplus L_0) & --- & L_+ \\ & K_i^X & | & K_i^P & \\ & & K_+ & & \end{array}$$

Table 2: Above we give the 5×5 grading of the Lie algebra of $\mathfrak{so}(d,2)$ in an Hermitian basis. The vertical 5-grading is determined by Δ and the horizontal 5-grading is determined by L_0.

4. 3-grading of $\mathfrak{so}(d,2)$ as a Conformal Algebra in d-Dimensions

The conformal group $SO(d,2)$ in d-dimensional Minkowskian space-time are generated by translations ($\mathcal{P}_\mu$), special conformal generators ($\mathcal{K}_\mu$) and Lorentz ($\mathcal{M}_{\mu\nu}$) and scale tranformations ($\mathcal{D}$). Its Lie algebra $\mathfrak{so}(d,2)$ has a 3-grading determined by the generator $\mathcal{D}$ of scale transformations (dilatations):

$$\mathfrak{so}(d,2) = \mathcal{K}_\mu \oplus (\mathcal{M}_{\mu\nu} + \mathcal{D}) \oplus \mathcal{P}_\mu . \tag{4.1}$$

In terms of the generators given in the previous section the dilatation generator is given by

$$\mathcal{D} = \frac{1}{2}\left[\Delta - i\left(M_+ - M_-\right)\right] \tag{4.2}$$

and the Lorentz generators $\mathcal{M}_{\mu\nu}$ $(\mu, \nu = 0, \ldots, d-1)$ are

$$\begin{aligned} \mathcal{M}_{0i} &= \frac{1}{2\sqrt{2}}\left(U_i + U_i^\dagger\right) + \frac{i}{2\sqrt{2}}\left(W_i - W_i^\dagger\right), \quad \mathcal{M}_{ij} = L_{ij} \\ \mathcal{M}_{i,d-1} &= \frac{1}{2\sqrt{2}}\left(U_i + U_i^\dagger\right) - \frac{i}{2\sqrt{2}}\left(W_i - W_i^\dagger\right), \quad \mathcal{M}_{0,d-1} = \frac{1}{2}\left[\Delta + i\left(M_+ - M_-\right)\right] \end{aligned} \tag{4.3}$$

where $i, j, ... = 1, 2, ..., (d-2)$.

The translation $\mathcal{P}_\mu$ and special conformal generators $\mathcal{K}_\mu$ $(\mu = 0, \ldots, d-1)$ are given by

$$\begin{aligned} \mathcal{P}_0 &= K_+ + M_0 + \frac{1}{2}\left(M_+ + M_-\right) \\ \mathcal{P}_i &= \frac{1}{\sqrt{2}}\left(W_i + W_i^\dagger\right) \qquad (i = 1, \ldots, d-2) \\ \mathcal{P}_{d-1} &= K_+ - M_0 - \frac{1}{2}\left(M_+ + M_-\right) \end{aligned} \tag{4.4}$$

$$\begin{aligned} \mathcal{K}_0 &= K_- + M_0 - \frac{1}{2}\left(M_+ + M_-\right) \\ \mathcal{K}_i &= -\frac{i}{\sqrt{2}}\left(U_i - U_i^\dagger\right) \qquad (i = 1, \ldots, d-2) \\ \mathcal{K}_{d-1} &= -K_- + M_0 - \frac{1}{2}\left(M_+ + M_-\right) . \end{aligned} \tag{4.5}$$

They satisfy the commutation relations:

$$\begin{aligned}
[\mathcal{M}_{\mu\nu}\,,\,\mathcal{M}_{\rho\tau}] &= i\,(\eta_{\nu\rho}\mathcal{M}_{\mu\tau} - \eta_{\mu\rho}\mathcal{M}_{\nu\tau} - \eta_{\nu\tau}\mathcal{M}_{\mu\rho} + \eta_{\mu\tau}\mathcal{M}_{\nu\rho}) \\
[\mathcal{P}_{\mu}\,,\,\mathcal{M}_{\nu\rho}] &= i\,(\eta_{\mu\nu}\,\mathcal{P}_{\rho} - \eta_{\mu\rho}\,\mathcal{P}_{\nu}) \\
[\mathcal{K}_{\mu}\,,\,\mathcal{M}_{\nu\rho}] &= i\,(\eta_{\mu\nu}\,\mathcal{K}_{\rho} - \eta_{\mu\rho}\,\mathcal{K}_{\nu}) \\
[\mathcal{D}\,,\,\mathcal{M}_{\mu\nu}] &= [\mathcal{P}_{\mu}\,,\,\mathcal{P}_{\nu}] = [\mathcal{K}_{\mu}\,,\,\mathcal{K}_{\nu}] = 0 \\
[\mathcal{D}\,,\,\mathcal{P}_{\mu}] &= +i\,\mathcal{P}_{\mu} \qquad\qquad [\mathcal{D}\,,\,\mathcal{K}_{\mu}] = -i\,\mathcal{K}_{\mu} \\
[\mathcal{P}_{\mu}\,,\,\mathcal{K}_{\nu}] &= 2i\,(\eta_{\mu\nu}\,\mathcal{D} + \mathcal{M}_{\mu\nu})
\end{aligned} \tag{4.6}$$

where $\eta_{\mu\nu} = \mathrm{diag}(-,+,\ldots,+)$ is the Minkoswki metric in d dimensions. The Poincaré mass operator in d dimensions vanishes identically

$$\mathscr{M}^2 = \eta_{\mu\nu}\mathcal{P}^{\mu}\mathcal{P}^{\nu} = 0. \tag{4.7}$$

Similarly one finds that the square of special conformal generator $\mathcal{K}_{\mu}$ vanishes identically as well

$$\eta^{\mu\nu}\mathcal{K}_{\mu}\mathcal{K}_{\nu} = 0\,. \tag{4.8}$$

5. 3-grading of $\mathfrak{so}(d,2)$ with Respect to its Maximal Compact Subalgebra $\mathfrak{so}(d) \oplus \mathfrak{so}(2)$

The generators $C_{MN} \oplus H$ $(M, N = 1, \ldots, d)$ of the maximal compact subalgebra $\mathfrak{so}(d) \oplus \mathfrak{so}(2)$ of $\mathfrak{so}(d,2)$ can be expressed in terms of the Lorentz covariant generators $\mathcal{M}_{\mu\nu}, \mathcal{D}, \mathcal{K}_{\mu}$ and $\mathcal{P}_{\mu}$ as follows

$$H = \frac{1}{2}(\mathcal{P}_0 + \mathcal{K}_0) = \frac{1}{2}\,(K_+ + K_-) + M_0 \tag{5.1}$$

$$\begin{aligned}
C_{ij} &= L_{ij} & C_{i,d-1} &= \mathcal{M}_{i,d-1} \\
C_{d-1,d} &= \frac{1}{2}\,(\mathcal{P}_{d-1} - \mathcal{K}_{d-1}) & C_{i,d} &= \frac{1}{2}\,(\mathcal{P}_i - \mathcal{K}_i).
\end{aligned} \tag{5.2}$$

They satisfy the commutation relations

$$[C_{MN}\,,\,C_{PQ}] = i\,(\delta_{NP}C_{MQ} - \delta_{MP}C_{NQ} - \delta_{NQ}C_{MP} + \delta_{MQ}C_{NP})\,. \tag{5.3}$$

The Lie algebra $\mathfrak{so}(d,2)$ admits a "compact" 3-grading determined by the $SO(2)$ generator H (conformal hamiltonian):

$$\mathfrak{so}(d,2) = C_M^- \oplus (C_{MN} + H) \oplus C_M^+ \tag{5.4}$$

where the grade $+1$ generators C_M^+ are given by

$$\begin{aligned}
C_i^+ &= \mathcal{M}_{0i} - \frac{i}{2}(\mathcal{K}_i + \mathcal{P}_i) = \frac{1}{\sqrt{2}}\left(U_i^{\dagger} - i\,W_i^{\dagger}\right) \qquad (i = 1, \ldots, d-2) \\
C_{d-1}^+ &= \mathcal{M}_{0,(d-1)} - \frac{i}{2}(\mathcal{P}_{(d-1)} + \mathcal{K}_{(d-1)}) = \frac{1}{2}\,[\Delta - i\,(K_+ - K_-)] + i\,M_+ \\
C_d^+ &= \frac{1}{2}(\mathcal{P}_0 - \mathcal{K}_0) + i\mathcal{D} = \frac{i}{2}\,[\Delta - i\,(K_+ - K_-)] + M_+\,.
\end{aligned} \tag{5.5}$$

The grade -1 generators C^-_M $(M, N, \cdots = 1, \ldots, d)$ are Hermitian conjugates of the grade $+1$ generators C^+_M :

$$
\begin{aligned}
C^-_i &= \mathcal{M}_{0i} + \frac{i}{2}(\mathcal{K}_i + \mathcal{P}_i) = \frac{1}{\sqrt{2}}(U_i + i\,W_i) \qquad\qquad (i = 1, \ldots, d-2) \\
C^-_{d-1} &= \mathcal{M}_{0,(d-1)} + \frac{i}{2}(\mathcal{P}_{(d-1)} + \mathcal{K}_{(d-1)}) = \frac{1}{2}\left[\Delta + i\,(K_+ - K_-)\right] - i\,M_- \\
C^-_d &= \frac{1}{2}(\mathcal{P}_0 - \mathcal{K}_0) - i\mathcal{D} = -\frac{i}{2}\left[\Delta + i\,(K_+ - K_-)\right] + M_- \,.
\end{aligned}
\tag{5.6}
$$

They satisfy

$$
\begin{aligned}
\left[C^+_M \,, C_{NP}\right] &= i\left(\delta_{MN}\, C^+_P - \delta_{MP}\, C^+_N\right) \\
\left[C^-_M \,, C_{NP}\right] &= i\left(\delta_{MN}\, C^-_P - \delta_{MP}\, C^-_N\right) \\
\left[H \,, C_{MN}\right] &= \left[C^+_M \,, C^+_N\right] = \left[C^-_M \,, C^-_N\right] = 0 \\
\left[H \,, C^+_M\right] &= +C^+_M \qquad\qquad \left[H \,, C^-_M\right] = -C^-_M \\
\left[C^+_M \,, C^-_N\right] &= 2\left(-\delta_{MN}\, H + i\, C_{MN}\right).
\end{aligned}
\tag{5.7}
$$

6. Hilbert Space of the Minimal Unitary Representation

The conformal Hamiltonian (or AdS energy operator) H, given in equation (5.1), can be written as the sum of Hamiltonians H_i of $(d-2)$ bosonic oscillators (a_i) and the Hamiltonian $H_\odot$ of a singular oscillator:

$$
\begin{aligned}
H &= \frac{1}{2}\left(K_+ + K_-\right) + M_0 \\
&= \frac{1}{4}\left(x^2 + p^2\right) + \frac{1}{2\,x^2}\mathcal{G} + \frac{1}{4}\sum_i \left(a_i^\dagger a_i + a_i a_i^\dagger\right) \\
&= H_\odot + \sum_i H_i
\end{aligned}
\tag{6.1}
$$

where

$$
H_\odot = \frac{1}{2}\left(K_+ + K_-\right) = \frac{1}{4}\left(x^2 + p^2\right) + \frac{1}{2\,x^2}\mathcal{G} \quad , \qquad H_i = \frac{1}{2}\, a_i^\dagger a_i + \frac{1}{4} \quad , \text{(no sum)}
\tag{6.2}
$$

$H_\odot$ is the Hamiltonian of a singular harmonic oscillator. The role of the coupling constant for the singular potential

$$
V(x) = \frac{\mathcal{G}}{x^2}
\tag{6.3}
$$

is played by the operator $\mathcal{G}$ given in equation (3.17). The Hamiltonian has the same form as in conformal quantum mechanics of Ref. 65 as well as in the Calogero models.[66,67] It generates the compact $U(1)$ subgroup of the one-dimensional conformal group $SO(2,1) = SU(1,1)_K$ acting on the singlet coordinate x. The other generators of $SU(1,1)_K$ in the compact three grading with respect to $H_\odot$ are given by the following linear combinations

$$B_{\odot}^{+} = -\frac{i}{2}\,[\Delta - i\,(K_{+} - K_{-})]$$
$$B_{\odot}^{-} = \frac{i}{2}\,[\Delta + i\,(K_{+} - K_{-})]. \tag{6.4}$$

They satisfy the commutation relations:

$$\left[B_{\odot}^{-}\,,\,B_{\odot}^{+}\right] = 2\,H_{\odot} \qquad \left[H_{\odot}\,,\,B_{\odot}^{+}\right] = B_{\odot}^{+} \qquad \left[H_{\odot}\,,\,B_{\odot}^{-}\right] = -\,B_{\odot}^{-}. \tag{6.5}$$

As a basis for the Hilbert space of the minimal unitary representation of $SO(d,2)$ we shall consider states which are "twisted tensor products" of the states of the Fock space $\mathcal{F}$ of $(d-2)$ bosonic oscillators a_i with the states of the singular oscillator that furnish a unitary representation of $SU(1,1)$[d] . The Fock space of the a-type oscillators is spanned by the states of the form

$$|n_1, n_2, \ldots, n_{d-2}\rangle = \prod_i \frac{1}{\sqrt{n_i!}}\,(\,a_i^{\dagger}\,)^{n_i}\,|0\rangle \tag{6.6}$$

where n_i are non-negative integers and $|0\rangle$ is the corresponding Fock vacuum.

In the coordinate representation the state(s) with the lowest eigenvalue of the conformal Hamiltonian $H_{\odot}$ are wave-functions of the form

$$\psi_0^{\alpha_{g_0}}\,(x) = C_0\,x^{\alpha_{g_0}}\,e^{-x^2/2} \tag{6.7}$$

where C_0 is a normalization constant ,g_0 is the lowest eigenvalue of $\mathcal{G}$ and

$$\alpha_{g_0} = \frac{1}{2} \pm \sqrt{2g_0 + \frac{1}{4}}\,. \tag{6.8}$$

The Fock vacuum $|0\rangle$ is the eigenstate of $\mathcal{G}$ with the lowest eigenvalue g_0. For the minrep of $SO(d,2)$ given above, the lowest possible value of g_0 is

$$g_0 = \frac{1}{8}\,(d-3)\,(d-5) \tag{6.9}$$

Denoting the state corresponding to the wave function $\psi_0^{\alpha_{g_0}}\,(x)$ with the Fock vacuum $|0\rangle$ as $|\psi_0^{\alpha_{g_0}}, 0\rangle$ we have

$$B_{\odot}^{-}\,|\psi_0^{\alpha_{g_0}}, 0\rangle = 0\,. \tag{6.10}$$

The Hermiticity of $H_{\odot}$ implies

$$g_0 \geq -\frac{1}{8} \tag{6.11}$$

and the normalizability of $|\psi_0^{\alpha_{g_0}}, 0\rangle$ requires

$$\alpha_{g_0} > -\frac{1}{2}\,. \tag{6.12}$$

[d]We use the term twisted tensor product since the eigenvalues of the Hamiltonian of the singular oscillator depend on the states of the Fock space $\mathcal{F}$.

The lowest eigenvalue g_0 leads to two possible values for α_{g_0} namely $(5-d)/2$ and $(d-3)/2$. However, the normalizability of the states in representation space requires one to choose:[64]

$$\alpha_{g_0} = \frac{(d-3)}{2} \,. \tag{6.13}$$

The corresponding tensor product state $|\psi_0^{(d-3)/2}, 0\rangle$ is an eigenstate of $H_\odot$ with the lowest eigenvalue $E_\odot^{\alpha_{g_0}} = \frac{1}{4}(d-2)$:

$$H_\odot |\psi_0^{(d-3)/2}, 0\rangle = \frac{1}{4}(d-2)\,|\psi_0^{(d-3)/2}, 0\rangle \,. \tag{6.14}$$

The state $|\psi_0^{(d-3)/2}(x)\,, 0\rangle$ is the unique state annihilated by all the grade -1 generators C_M^- and transforms as a singlet of $SO(d)$ subgroup. This shows that the minrep is a unitary lowest weight (positive energy) representation. Furthermore it is annihilated by all the translation generators U_i of the little group $SO(d-2)ⓈT_{(d-2)}$ of massless particles in d dimensions. Therefore the minrep of $SO(d,2)$ describes a massless conformal scalar field in d dimensional Minkowskian spacetime. We refer to[64] for the K-type decomposition of the minrep of $SO(d,2)$ with respect to its maximal compact subgroup.

7. Deformations of the Minimal Unitary Representation of $SO(d,2)$ and Massless Conformal Fields

The minrep of $SO(d,2)$ describes a conformally massless scalar field in d dimensions. In this section we shall review the "deformations" of the minrep and establish their one-to-one correspondence with conformally massless fields in d dimensional Minkoswkian spacetimes.[64] By a deformation we mean adding spin terms S_{ij} of the orbital generators L_{ij} of the little group $SO(d-2)$ of massless particles such that all the Jacobi identities are satisfied[e]:

$$L_{ij} \longrightarrow J_{ij} = L_{ij} + S_{ij} \tag{7.1}$$

This requirement leads to following expression for the generator K_+ of the deformed minrep:

$$K_+ = \frac{1}{2}p^2 + \frac{1}{x^2}\left(\frac{1}{2}\mathcal{J}^2 - \frac{1}{4}\mathcal{L}^2 - \frac{(d-6)}{2(d-2)}\mathcal{S}^2 + \frac{1}{8}(d-3)(d-5)\right) \tag{7.2}$$

where $\mathcal{J}^2 = J_{ij}J_{ij}$ and $\mathcal{S}^2 = S_{ij}S_{ij}$. Under the deformation the generators $M_{\pm,0}$ and Δ in grade 0 subspace, U_i and $U_i^\dagger$ in grade -1 subspace, and K_- in grade -2 subspace of $\mathfrak{so}(d,2)$ remain unchanged. The grade $+1$ generators W_i and $W_i^\dagger$ get modified as follows:

$$\begin{aligned} W_i &= p\,a_i - \frac{i}{x}\left[\frac{1}{2}(d-3)\,a_i + i\,(L_{ij} + 2\,S_{ij})\,a_j\right] \\ W_i^\dagger &= p\,a_i^\dagger - \frac{i}{x}\left[\frac{1}{2}(d-3)\,a_i^\dagger + i\,(L_{ij} + 2\,S_{ij})\,a_j^\dagger\right]. \end{aligned} \tag{7.3}$$

[e]For $d=4$ the little group is $U(1)$ and deformations are parametrized by helicity which can take on continuous values.[21] This explains the origin of the term "deformation" which has been used in all dimensions later.

Furthermore the spin generators S_{ij} are required to satisfy the constraint:

$$\Delta_{ij} = S_{ik}S_{jk} + S_{jk}S_{ik} - \frac{2}{(d-2)}\,\mathcal{S}^2\,\delta_{ij} = 0 \tag{7.4}$$

which turn out to be the same constraint satisfied by little group generators of massless representations of the Poincaré group that extend to unitary representations of the conformal group in d dimensions.[68,69] This correspondence between massless conformal fields in d dimensions and the minrep of $SO(d,2)$ and its deformations were established earlier for dimensions $d = 4, 5$ and 6 in Refs. 21–23, 63. More recently this one-to-one correspondence was extended to all spacetime dimensions in Ref. 64. We should note that each deformation leads to a different unitary realization of the $SU(1,1)_K$ subgroup and thus establishes a correspondence between massless conformal fields in any dimension and the Calogero models.

For the deformed minrep one finds the following expressions for the Casimirs of various subalgebras

$$\begin{aligned}
\mathcal{C}_2\left[\mathfrak{so}(d-2)_J\right] &= J_{ij}J_{ij} = \mathcal{J}^2 \\
\mathcal{C}_2\left[\mathfrak{su}(1,1)_M\right] &= {M_0}^2 - \frac{1}{2}\left(M_+M_- + M_-M_+\right) = \mathcal{M}^2 = \frac{1}{8}\,\mathcal{L}^2 + \frac{1}{16}\,(d-2)\,(d-6) \\
\mathcal{C}_2\left[\mathfrak{su}(1,1)_K\right] &= \Delta^2 - 2\left(K_+K_- + K_-K_+\right) = \mathcal{K}^2 \\
&= -\mathcal{J}^2 + \frac{1}{2}\,\mathcal{L}^2 + \frac{(d-6)}{(d-2)}\,\mathcal{S}^2 - \frac{1}{4}\,(d-2)\,(d-6).
\end{aligned} \tag{7.5}$$

The coset $SO(d,2)/SO(d-2) \times SO(2,2)$ generators satisfy

$$[UW] = U_iW_i^\dagger + W_i^\dagger U_i - U_i^\dagger W_i - W_iU_i^\dagger = 2i\left(\mathcal{J}^2 - \mathcal{S}^2\right) + i\,(d-2)^2 \tag{7.6}$$

leading to following expression for the quadratic Casimir of $\mathfrak{so}(d,2)$

$$\mathcal{C}_2\left[\mathfrak{so}(d,2)\right] = \frac{(d+2)}{(d-2)}\,\mathcal{S}^2 - \frac{1}{2}\left(d^2-4\right) \tag{7.7}$$

which agrees with the quadratic Casimir of the undeformed minrep of $\mathfrak{so}(d,2)$ when $\mathcal{S}^2 = 0$. The minrep and its deformations are unitary lowest weight (positive energy) representations of $SO(d,2)$ that are uniquely determined by the lowest energy irrep of $SO(d)$ subgroup. Therefore we shall label them as

$$|E_0, (n_1, \cdots n_r)\rangle$$

where E_0 is the eigenvalue of H and $(n_1, \cdots n_r)$ are the Dynkin labels of lowest energy irrep of $SO(d)$.

7.1. *Deformations in odd dimensions*

It is well-known that the AdS_4/CFT_3 Lie algebra $SO(3,2)$ is isomorphic to the symplectic Lie algebra $Sp(4,\mathbb{R})$ whose minimal unitary representation that describes a conformal scalar admits a single deformation which describes a conformal spinor in 3 dimensions. They are simply the scalar and spinor singletons of Dirac[70] which describe the only conformally massless fields in $d = 3$. Similarly, the minrep of AdS_6/CFT_5 group $SO(5,2)$ admits a unique deformation.[63] The minrep of $SO(5,2)$ and its deformation are the analogs of Dirac singletons and decribe conformally massless scalar and spinor fields in $d = 5$. This phenomenon extends to all conformal algebras $\mathfrak{so}(d,2)$ in odd dimensions as a consequence of the fact that the constraint (7.4) for spin terms S_{ij} of $SO(d-2)$ that determine the deformations of the minrep has a unique non-trivial solution for odd d. It is simply the $2^{\left(\frac{d-3}{2}\right)}$ dimensional spinor irrep of $SO(d-2)$ generated by

$$S_{ij} = \frac{1}{4}[\gamma_i, \gamma_j] \tag{7.8}$$

where γ_i are the $2^n \times 2^n$ $(n = (d-3)/2)$ Euclidean Dirac gamma matrices in $(d-2)$ dimensions. The minrep and its unique deformations describe the conformally massless scalar and spinor fields. They exhaust the list of conformally masless fields for odd d.[69] The lowest energy irrep of the minrep has the labels

$$|\frac{(d-2)}{2}, (0, \cdots, 0)_D\rangle$$

while the unique spinorial deformation of the minrep has the lowest energy irrep

$$|\frac{(d-1)}{2}, (0, \cdots, 0, 1)_D\rangle\,.$$

7.2. *Deformations in even dimensions*

The minrep of the conformal group in four dimensions admits a one parameter family of deformations labelled by helicity.[21] For integer and half integer values of helicity the resulting unitary representations are isomorphic to the doubleton representations that describe massless conformal fields of arbitrary spin in four dimensions that were constructed and studied in Refs. 24–26. Similarly one finds that the minrep of $SO(6,2)$ admits a discrete infinite family of deformations[22,23] that are isomorphic to doubleton representations that describe higher spin massless conformal fields in six dimensions studied earlier in Refs. 27, 58, 71.

In contrast to odd dimensions, the constraints on the spin terms that describe the deformations of the minrep of $SO(d,2)$ for even d admit an infinite set of solutions that describe conformally massless fields. They exhaust the list of conformally massless fields as a consequence of the fact that Eq. (7.4) is a necessary and sufficient condition for a massless representation of the Poincare group to be extendable to a unitary representation of the conformal group.[68,69] These deformations are all unitary lowest weight representations of $SO(d,2)$ whose lowest energy irreps have the $SO(2) \times SO(d)$ labels[64]

$$|E_0 = \frac{(d+s-2)}{2}; (0, \cdots, 0, 0, s)_D\rangle$$

or

$$|E_0 = \frac{(d+s-2)}{2}; (0, \cdots, 0, s, 0)_D\rangle$$

where s is a non-negative integer. The spinorial generators S_{ij} $(i, j, \cdots = 1, \ldots, (d-2))$ can always be realized in terms of fermionic oscillators.[64]

8. Quasiconformal Construction of Higher Spin Algebras

8.1. *Bosonic $AdS_{(d+1)}/CFT_d$ higher spin algebras*

To construct the $AdS(d+1)/CFT_d$ higher spin algebra , let us first express the $\mathfrak{so}(d,2)$ generators in the "canonical basis" M_{AB} $(A,B=0,1,\ldots,d+1)$ in terms of the generators of the compact 3-grading

$$M_{0M}=\frac{1}{2}\left(C_M^+ + C_M^-\right) \qquad M_{0,d+1}=H$$
$$M_{MN}=C_{MN} \qquad M_{M,d+1}=\frac{i}{2}\left(C_M^+ - C_M^-\right). \tag{8.1}$$

They satisfy the commutation relations:

$$[M_{AB}\,,\,M_{CD}]=i\left(\eta_{BC}M_{AD}-\eta_{AC}M_{BD}-\eta_{BD}M_{AC}+\eta_{AD}M_{BC}\right) \tag{8.2}$$

where $\eta_{AB}=\text{diag}\,(-,+,\ldots,+,-)$ is the $SO(d,2)$ invariant metric.

In terms of the Lorentz covariant generators $\mathcal{D}$, $\mathcal{M}_{\mu\nu}$, $\mathcal{P}_\mu$, $\mathcal{K}_\mu$ $(\mu,\nu=0,\ldots,d-1)$, the generators M_{AB} take the form:

$$M_{\mu\nu}=\mathcal{M}_{\mu\nu} \qquad M_{\mu,d+1}=\frac{1}{2}\left(\mathcal{P}_\mu+\mathcal{K}_\mu\right)$$
$$M_{\mu,d}=\frac{1}{2}\left(\mathcal{P}_\mu-\mathcal{K}_\mu\right) \qquad M_{d,d+1}=-\mathcal{D}. \tag{8.3}$$

The $AdS_{(d+1)}/CFT_d$ higher spin algebra is simply the quotient of the universal enveloping algebra $\mathscr{U}(d,2)$ of $SO(d,2)$ by its Joseph ideal $\mathscr{J}(d,2)$.[46,52,60–63,72] We recall that the Joseph ideal of the universal enveloping algebra of a Lie algebra is a two-sided ideal that annihilates its minimal unitary representation. In fact , in the mathematics literature minrep is defined by this property. We shall denote the corresponding $AdS_{(d+1)}/CFT_d$ higher spin algebra as $hs(d,2)$:

$$hs(d,2)=\frac{\mathscr{U}(d,2)}{\mathscr{J}(d,2)}. \tag{8.4}$$

In terms of M_{AB} the explicit expression for the generators J_{ABCD} of the Joseph ideal of $SO(d,2)$ was given in:[72]

$$\begin{aligned}J_{ABCD}&=M_{AB}M_{CD}-M_{AB}\circledcirc M_{CD}-\frac{1}{2}\left[M_{AB}\,,\,M_{CD}\right]+\frac{(d-2)}{4d(d+1)}\langle M_{AB},M_{CD}\rangle\,\mathbf{1}\\&=\frac{1}{2}M_{AB}\cdot M_{CD}-M_{AB}\circledcirc M_{CD}+\frac{(d-2)}{4d(d+1)}\langle M_{AB},M_{CD}\rangle\,\mathbf{1}\end{aligned} \tag{8.5}$$

where the symbol $\cdot$ denotes the symmetric product

$$M_{AB}\cdot M_{CD}=M_{AB}M_{CD}+M_{CD}M_{AB}\,, \tag{8.6}$$

and the symbol $\circledcirc$ denotes the Cartan product of two generators of $\mathfrak{so}(d,2)$:[73]

$$\begin{aligned} M_{AB} \circledcirc M_{CD} = &\frac{1}{3} M_{AB} M_{CD} + \frac{1}{3} M_{DC} M_{BA} \\ &+ \frac{1}{6} M_{AC} M_{BD} - \frac{1}{6} M_{AD} M_{BC} + \frac{1}{6} M_{DB} M_{CA} - \frac{1}{6} M_{CB} M_{DA} \\ &- \frac{1}{2d} \left(\eta_{BD} M_{AE} M_C{}^E - \eta_{AD} M_{BE} M_C{}^E + \eta_{AC} M_{BE} M_D{}^E - \eta_{BC} M_{AE} M_D{}^E \right) \\ &- \frac{1}{2d} \left(\eta_{BD} M_{CE} M_A{}^E - \eta_{AD} M_{CE} M_B{}^E + \eta_{AC} M_{DE} M_B{}^E - \eta_{BC} M_{DE} M_A{}^E \right) \\ &+ \frac{1}{d(d+1)} M_{EF} M^{EF} \left(\eta_{AC} \eta_{BD} - \eta_{BC} \eta_{AD} \right) , \end{aligned} \tag{8.7}$$

The bilinear product $\langle M_{AB}, M_{CD} \rangle$ is defined by the Killing form of $SO(d,2)$:

$$\langle M_{AB}, M_{CD} \rangle = -\frac{2d}{(d^2-4)} M_{EF} M_{GH} \left(\eta^{EG} \eta^{FH} - \eta^{EH} \eta^{FG} \right) \left(\eta_{AC} \eta_{BD} - \eta_{AD} \eta_{BC} \right). \tag{8.8}$$

Under the adjoint action of $\mathfrak{so}(d,2)$ the decomposition of the enveloping algebra $\mathscr{U}(d,2)$ is given by the symmetrized products of the adjoint representation with Young tableau

$M_{AB} \sim$ 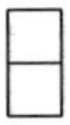 . (8.9)

The symmetrized tensor product of two copies of the adjoint representation decomposes as:

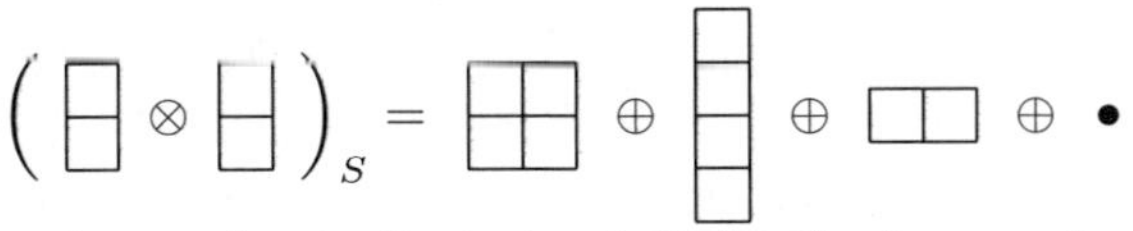

(8.10)

where $\bullet$ represents the quadratic Casimir of $SO(d,2)$. As was shown by Vasiliev,[46] the higher spin gauge fields in $AdS_{(d+1)}$ transform in the traceless two-row Young tableaux representations of $SO(d,2)$:

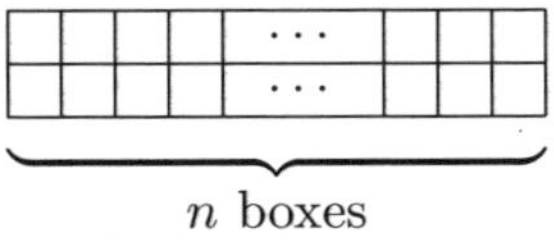

(8.11)

Therefore one has to mod out all the representations on the right hand side of equation (8.10) except for the window diagram. Quotienting the enveloping algebra by the Joseph ideal does precisely this and the generators of the resulting infinite higher spin algebra transform under $\mathfrak{so}(d,2)$ as follows:

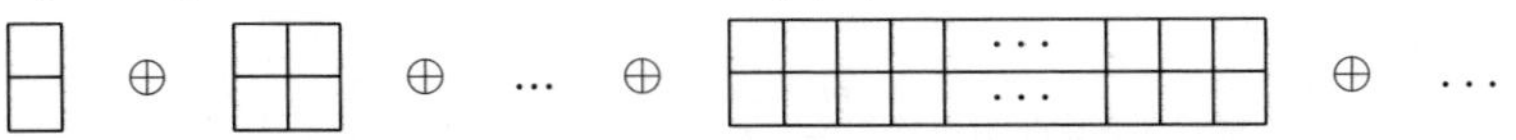

(8.12)

For the minimal unitary realization of $SO(d,2)$ obtained via the quasiconformal method the generators J_{ABCD} of the Joseph ideal vanish identically as operators. For $d = 3, 4, 6$ dimensions this was established in Refs. 61, 62. These results were extended to $d = 5$ in Ref. 63 and to general $d > 6$ in Ref. 64. Hence the enveloping algebra of the minrep of $SO(d,2)$ that results from quantization of its geometric quasiconformal action yields directly the bosonic higher spin $AdS_{(d+1)}/\mathrm{Conf}_d$ algebra in all dimensions.

8.2. *Deformations and supersymmetric extensions of $AdS_{(d+1)}/CFT_d$ higher spin algebras*

The minimal unitary representation of $SO(d,2)$ and its deformations as obtained via the quasiconformal approach are in one-to-one correspondence with conformally massless fields in d-dimensional Minkowskian spacetimes. Therefore one defines the *deformations* of the bosonic $AdS_{(d+1)}/CFT_d$ higher spin algebras as enveloping algebras of the deformed minimal unitary representations.[61–64] These deformed higher spin algebras can be interpreted as quotients of the universal enveloping algebra $\mathscr{U}(d,2)$ of $SO(d,2)$ by a deformed Joseph ideal $\mathscr{J}_{Def}(d,2)$:

$$hs(d,2)_{Def} = \frac{\mathscr{U}(d,2)}{\mathscr{J}_{Def}(d,2)}. \tag{8.13}$$

The deformations of the Joseph ideal are best studied by decomposing the generators J_{ABCD} of the Joseph ideal with respect to the Lorentz group $SO(d-1,1)$ as was done in Refs. 61–64. Vanishing of the generators J_{ABCD} as operators within quasiconformal realization of the minrep are equivalent to the following Lorentz covariant conditions:

$$\begin{aligned}
d\,\mathcal{D}\cdot\mathcal{D} + \mathcal{M}^{\mu\nu}\cdot M_{\mu\nu} + \frac{(d-2)}{2}\,\mathcal{P}^{\mu}\cdot\mathcal{K}_{\mu} &= 0\\
\mathcal{P}^{\mu}\cdot(\mathcal{M}_{\mu\nu} + \eta_{\mu\nu}\,\mathcal{D}) &= 0\\
\mathcal{K}^{\mu}\cdot(\mathcal{M}_{\mu\nu} - \eta_{\mu\nu}\,\mathcal{D}) &= 0\\
\eta^{\mu\nu}\,\mathcal{M}_{\mu\rho}\cdot\mathcal{M}_{\nu\sigma} - \mathcal{P}_{(\rho}\cdot\mathcal{K}_{\sigma)} + (d-2)\,\eta_{\rho\sigma} &= 0\\
\mathcal{M}_{\mu\nu}\cdot\mathcal{M}_{\rho\sigma} + \mathcal{M}_{\mu\sigma}\cdot\mathcal{M}_{\nu\rho} + \mathcal{M}_{\mu\rho}\cdot\mathcal{M}_{\sigma\nu} &= 0\\
\mathcal{D}\cdot\mathcal{M}_{\mu\nu} + \mathcal{P}_{[\mu}\cdot\mathcal{K}_{\nu]} &= 0\\
\mathcal{M}_{[\mu\nu}\cdot\mathcal{P}_{\rho]} &= 0\\
\mathcal{M}_{[\mu\nu}\cdot\mathcal{K}_{\rho]} &= 0
\end{aligned} \tag{8.14}$$

where symmetrizations (round brackets) and anti-symmetrizations (square brackets) are of weight one.

Under deformation some of the generators of $SO(d,2)$ get modified by spin dependent terms. More specifically only the first three identities above remain unchanged and the momentum and special conformal generators remain lightlike under deformations:

$$P^2 = \mathcal{P}^{\mu}\mathcal{P}_{\mu} = 0 \quad , \quad K^2 = \mathcal{K}^{\mu}\mathcal{K}_{\mu} = 0. \tag{8.15}$$

The fourth identity above takes the form[64]

$$\eta^{\mu\nu}\mathcal{M}_{\mu\rho}\cdot\mathcal{M}_{\nu\sigma} - \mathcal{P}_{(\rho}\cdot\mathcal{K}_{\sigma)} - \left[\frac{2}{(d-2)}\mathcal{S}^2 - (d-2)\right]\eta_{\rho\sigma} = 0 \tag{8.16}$$

where $\mathcal{S}^2$ is the Casimir operator of the little group $SO(d-2)_S$ generated by the spin terms S_{ij}.

Physical interpretation of some of the deformed identities depends on the space-time dimension. In four dimensions one finds that the identities that depend on the deformation parameter ζ , which is twice the helicity, take the form[61]

$$W^{\mu} = \frac{1}{2}\epsilon^{\mu\nu\rho\sigma} M_{\nu\rho} \cdot P_{\sigma} = \zeta P^{\mu} \tag{8.17}$$

$$V^{\mu} = \frac{1}{2}\epsilon^{\mu\nu\rho\sigma} M_{\nu\rho} \cdot K_{\sigma} = -\zeta K^{\mu} \tag{8.18}$$

$$\eta^{\mu\nu} M_{\mu\rho} \cdot M_{\nu\sigma} - P_{(\rho} \cdot K_{\sigma)} + 2\eta_{\rho\sigma} = \frac{\zeta^2}{2}\eta_{\rho\sigma} \tag{8.19}$$

$$M_{\mu\nu} \cdot M_{\rho\sigma} + M_{\mu\sigma} \cdot M_{\nu\rho} + M_{\mu\rho} \cdot M_{\sigma\nu} = \zeta \epsilon_{\mu\nu\rho\sigma} \Delta \tag{8.20}$$

$$\Delta \cdot M_{\mu\nu} + P_{[\mu} \cdot K_{\nu]} = -\frac{\zeta}{2}\epsilon_{\mu\nu\rho\sigma} M^{\rho\sigma} \tag{8.21}$$

and W^{μ} and V^{μ} are the Pauli-Lubansky vector and its special conformal analog, respectively. The eigenvalues of the Casimir operators of the deformed minreps of $SO(4,2)$ depend only on ζ:[61]

$$C_2 = M_B{}^A M_A{}^B = 6 - \frac{3\zeta^2}{2} \tag{8.22}$$

$$C_3 = \epsilon^{ABCDEF} M_{AB} M_{CD} M_{EF} = 6\zeta\left(\zeta^2 - 4\right) = -8C_2\sqrt{1 - \frac{C_2}{6}} \tag{8.23}$$

$$C_4 = M_B{}^A M_C{}^B M_D{}^C M_A{}^D = \frac{3}{8}\left(\zeta^4 + 8\zeta^2 - 48\right) = \frac{C_2^2}{6} - 4C_2. \tag{8.24}$$

Deformed Joseph ideal generators of $6d$ conformal group $SO(6,2)$ were given in Ref. 62. One finds that the analog of Pauli-Lubanski vector in $6d$ is a tensor of rank 3 and its conformal analogue defined as

$$A_{\mu\nu\rho} = \frac{1}{3!}\epsilon_{\mu\nu\rho\sigma\delta\tau} M^{[\sigma\delta} \cdot P^{\tau]} \qquad B_{\mu\nu\rho} = \frac{1}{3!}\epsilon_{\mu\nu\rho\sigma\delta\tau} M^{[\sigma\delta} \cdot K^{\tau]}. \tag{8.25}$$

They vanish identically for the minrep

$$A_{\mu\nu\rho} = 0 \qquad B_{\mu\nu\rho} = 0. \tag{8.26}$$

For the deformed minreps they do not vanish, but satisfy self-duality and anti-self-duality conditions:

$$\begin{aligned} A_{\mu\nu\rho} &= \widetilde{A}_{\mu\nu\rho} \\ B_{\mu\nu\rho} &= -\widetilde{B}_{\mu\nu\rho} \end{aligned} \tag{8.27}$$

where the dual rank three tensors are defined as follows:

$$\widetilde{A}_{\mu\nu\rho} = \frac{1}{3!}\epsilon_{\mu\nu\rho\sigma\delta\tau} A^{\sigma\delta\tau}, \qquad \widetilde{B}_{\mu\nu\rho} = \frac{1}{3!}\epsilon_{\mu\nu\rho\sigma\delta\tau} B^{\sigma\delta\tau}. \tag{8.28}$$

Similarly one finds the following identity for the generators of the deformed minreps:

$$M_{\mu\nu} \cdot M_{\rho\sigma} + M_{\mu\sigma} \cdot M_{\nu\rho} + M_{\mu\rho} \cdot M_{\sigma\nu} = \epsilon_{\mu\nu\rho\sigma}{}^{\delta\tau}(P_{[\delta} \cdot K_{\tau]} + M_{\delta\tau} \cdot \Delta) \tag{8.29}$$

while each side of this equation vanishes separately for the true minrep.

The quadratic Casimir operator of the deformed minrep of $SO^*(8)$is given by:[22]

$$C_2\left[\mathfrak{so}^*(8)\right]_{\text{deformed}} = -8\left(2-\mathcal{T}^2\right) \tag{8.30}$$

where $\mathcal{T}^2$ is the quadratic Casimir operator of an $SU(2)$ subgroup of the little group $SO(4)$ of massless particles whose eigenvalues label the deformations.

In general the representation

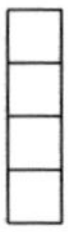

that vanished in the symmetric tensor product of the adjoint representation for the true minrep of $SO(d,2)$ no longer vanishes for the deformed minrep. On the other hand the representation 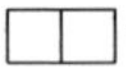 that occurs in the tensor product still vanishes for the deformed minreps:

$$\eta^{CD} M_{AC} \cdot M_{DB} = 0. \tag{8.31}$$

Hence the gauge fields of deformed higher spin theories defined by $hs(d,2)_{Def}$ will not consist only of fields corresponding to traceless two-row Young tableaux. Their Young tableaux will be those that occur in the symmetrized products of the adjoint tableau 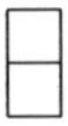 of $SO(d,2)$ subject to the constraint:

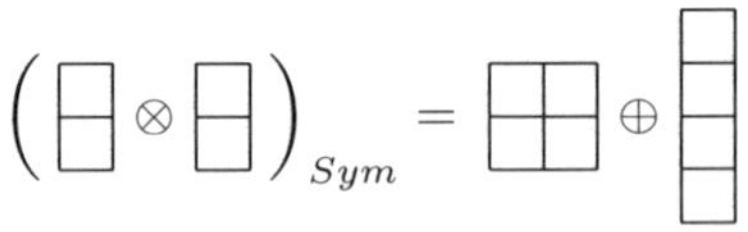

8.3. *Supersymmetric deformations of higher spin algebras*

Maximal dimension for the existence of simple superconformal algebras over the field of real or complex numbers is six.[74] In $d=3$ the conformal group $SO(3,2)$ admits supersymmetric extensions to supergroups $OSp(N|4,\mathbb{R})$ with even subgroups $SO(N)\times Sp(4,\mathbb{R})$. The minimal unitary supermultiplet of $OSp(N|4,\mathbb{R})$ for even N consists of a scalar singleton transforming in a chiral spinor representation of $SO(N)$ and a spinor singleton transforming in the conjugate spinor representation of $SO(N)$. For odd N the minimal unitary supermultiplet contains a scalar and a spinor singleton both transforming in the unique spinor representation of $SO(N)$.

That the higher spin algebra of Fradkin and Vasiliev in AdS_4 as well as its supersymmetric extensions are given by the enveloping algebras of the singletonic realization of $SO(3,2)$ and its supersymmetric extensions was first pointed out in Ref. 52. In higher dimensions $AdS_{(d+1)}/CFT_d$ higher spin superalgebras are defined as enveloping algebras of the quasiconformal realization of the minimal unitary representation of the correponding superconformal algebras.[61–63] The minimal unitary supermultiplet contains the minrep of $SO(d,2)$ and certain deformations of the minrep. As such it corresponds to a massless conformal supermultiplet of the underlying conformal superalgebra.

In $d = 4$ the conformal algebra $SU(2,2)$ can be extended to an infinite family of superalgebras $SU(2,2\,|\,N)$ with the even subalgebra $SU(2,2)\oplus U(N)$. The minimal unitary representation of $SU(2,2|N)$ and its deformations were constructed via the quasiconformal method in.[21] These representations were later reformulated in terms of helicity deformed twistors that transform non-linearly under the Lorentz group and the corresponding AdS_5/CFT_4 higher spin superalgebras were studied in Ref. 61.

The conformal algebra $SO(6,2) = SO^*(8)$ in $d = 6$ also admits an infinite family of superconformal extensions $OSp(8^*\,|\,2N)$ with even subalgebras $SO^*(8)\oplus USp(2N)$. The quasiconformal construction of the minimal unitary supermultiplets of $OSp(8^*\,|\,2N)$ and their deformations were given in Refs. 22, 23. These representaions were reformulated in terms of deformed twistors and the corresponding AdS_7/CFT_6 higher spin superalgebras were studied in Ref. 62.

In five dimensions there exists a unique superconformal algebra , namely the exceptional superalgebra with the even subalgebra $SO(5,2)\oplus SU(2)$. The minimal unitary supermultiplet of $F(4)$ was constructed via the quasiconformal method and the unique higher spin AdS_6/CFT_5 superalgbera was studied in Ref. 64.

Acknowledgments

I would like to thank Sudarshan Fernando, Karan Govil and Oleksandr Pavlyk for most enjoyable and stimulating scientific collaborations on topics covered in this review. I would also like to thank Evgeny Skvortsov, Massimo Taronna and Misha Vasiliev for many helpful discussions on higher spin theories. Thanks are also due to Lars Brink and other organizers for the invitation to the Conference on Higher Spin Gauge Theories at the Institute of Advanced Studies in Nanyang Technological University in Singapore. The research reviewed here was supported in part by the US Department of Energy under DOE Grant No: DE-SC0010534 and US National Science Foundation under grants PHY-1213183 and PHY-08-55356.

References

1. A. Joseph, Minimal realizations and spectrum generating algebras, *Comm. Math. Phys.* **36**, 325–338, (1974). ISSN 0010-3616.
2. D. A. Vogan, Jr. Singular unitary representations. In *Noncommutative harmonic analysis and Lie groups (Marseille, 1980)*, vol. 880, *Lecture Notes in Math.*, pp. 506–535. Springer, Berlin, (1981).
3. B. Kostant. The vanishing of scalar curvature and the minimal representation of $SO(4,4)$. In *Operator algebras, unitary representations, enveloping algebras, and invariant theory (Paris, 1989)*, vol. 92, *Progr. Math.*, pp. 85–124. Birkhäuser Boston, Boston, MA, (1990).
4. B. Binegar and R. Zierau, Unitarization of a singular representation of $so(p,q)$, *Communications in Mathematical Physics.* **138**(2), 245–258, (1991).
5. D. Kazhdan and G. Savin. The smallest representation of simply laced groups. In *Festschrift in honor of I. I. Piatetski-Shapiro on the occasion of his sixtieth birthday,*

Part I (Ramat Aviv, 1989), vol. 2, *Israel Math. Conf. Proc.*, pp. 209–223. Weizmann, Jerusalem, (1990).
6. R. Brylinski and B. Kostant. Lagrangian models of minimal representations of E_6, E_7 and E_8. In *Functional analysis on the eve of the 21st century, Vol. 1 (New Brunswick, NJ, 1993)*, vol. 131, *Progr. Math.*, pp. 13–63. Birkhäuser Boston, Boston, MA, (1995).
7. R. Brylinski and B. Kostant, Minimal representations, geometric quantization, and unitarity, *Proc. Nat. Acad. Sci. U.S.A.* **91**(13), 6026–6029, (1994). ISSN 0027-8424.
8. B. H. Gross and N. R. Wallach. A distinguished family of unitary representations for the exceptional groups of real rank = 4. In *Lie theory and geometry*, vol. 123, *Progr. Math.*, pp. 289–304. Birkhäuser Boston, Boston, MA, (1994).
9. J.-S. Li. Minimal representations & reductive dual pairs. In *Representation theory of Lie groups (Park City, UT, 1998)*, vol. 8, *IAS/Park City Math. Ser.*, pp. 293–340. Amer. Math. Soc., Providence, RI, (2000).
10. T. Kobayashi and B. Ørsted, Analysis on the minimal representation of $O(p,q)$. I. Realization via conformal geometry, *Adv. Math.* **180**(2), 486–512, (2003). ISSN 0001-8708. doi: 10.1016/S0001-8708(03)00012-4.
11. T. Kobayashi and B. Ørsted, Analysis on the minimal representation of $O(p,q)$. II. Branching laws, *Adv. Math.* **180**(2), 513–550, (2003). ISSN 0001-8708. doi: 10.1016/S0001-8708(03)00013-6.
12. T. Kobayashi and B. Ørsted, Analysis on the minimal representation of $O(p,q)$. III. Ultrahyperbolic equations on $\mathbf{R}^{p-1,q-1}$, *Adv. Math.* **180**(2), 551–595, (2003). ISSN 0001-8708. doi: 10.1016/S0001-8708(03)00014-8.
13. A. R. Gover and A. Waldron, The so(d+2,2) Minimal Representation and Ambient Tractors: the Conformal Geometry of Momentum Space. (2009).
14. D. Kazhdan, B. Pioline, and A. Waldron, Minimal representations, spherical vectors, and exceptional theta series. I, *Commun. Math. Phys.* **226**, 1–40, (2002).
15. M. Günaydin, K. Koepsell, and H. Nicolai, Conformal and quasiconformal realizations of exceptional Lie groups, *Commun. Math. Phys.* **221**, 57–76, (2001).
16. M. Günaydin and O. Pavlyk, Generalized spacetimes defined by cubic forms and the minimal unitary realizations of their quasiconformal groups, *JHEP.* **08**, 101, (2005).
17. M. Günaydin, K. Koepsell, and H. Nicolai, The minimal unitary representation of $E_{8(8)}$, *Adv. Theor. Math. Phys.* **5**, 923–946, (2002).
18. M. Günaydin, G. Sierra, and P. K. Townsend, Exceptional supergravity theories and the magic square, *Phys. Lett.* **B133**, 72, (1983).
19. M. Günaydin and O. Pavlyk, Minimal unitary realizations of exceptional U-duality groups and their subgroups as quasiconformal groups, *JHEP.* **01**, 019, (2005).
20. M. Günaydin and O. Pavlyk, A unified approach to the minimal unitary realizations of noncompact groups and supergroups, *JHEP.* **09**, 050, (2006).
21. S. Fernando and M. Gunaydin, Minimal unitary representation of SU(2,2) and its deformations as massless conformal fields and their supersymmetric extensions, *J.Math.Phys.* **51**, 082301, (2010). doi: 10.1063/1.3447773.
22. S. Fernando and M. Gunaydin, Minimal unitary representation of SO*(8) = SO(6,2) and its SU(2) deformations as massless 6D conformal fields and their supersymmetric extensions, *Nucl.Phys.* **B841**, 339–387, (2010). doi: 10.1016/j.nuclphysb.2010.07.001.
23. S. Fernando and M. Gunaydin, SU(2) deformations of the minimal unitary representation of $OSp(8^*|2N)$ as massless 6D conformal supermultiplets, *Nucl.Phys.* **B843**, 784–815, (2011). doi: 10.1016/j.nuclphysb.2010.10.019.
24. M. Gunaydin and N. Marcus, The Spectrum of the S^5 Compactification of the Chiral $N = 2, D = 10$ Supergravity and the Unitary Supermultiplets of $U(2,2/4)$, *Class. Quant. Grav.* **2**, L11, (1985).

25. M. Gunaydin, D. Minic, and M. Zagermann, 4D doubleton conformal theories, CPT and II B string on AdS(5) x S(5), *Nucl. Phys.* **B534**, 96–120, (1998). doi: 10.1016/S0550-3213(98)00543-4.
26. M. Gunaydin, D. Minic, and M. Zagermann, Novel supermultiplets of $SU(2,2|4)$ and the AdS_5/CFT_4 duality, *Nucl. Phys.* **B544**, 737–758, (1999). doi: 10.1016/S0550-3213(99)00007-3.
27. M. Gunaydin, P. van Nieuwenhuizen, and N. P. Warner, General Construction of the Unitary Representations of Anti-De Sitter Superalgebras and the Spectrum of the S^4 Compactification of Eleven-Dimensional Supergravity, *Nucl. Phys.* **B255**, 63, (1985). doi: 10.1016/0550-3213(85)90129-4.
28. P. A. M. Dirac, A remarkable representation of the 3+ 2 de sitter group, *Journal of Mathematical Physics.* **4**, 901, (1963).
29. M. Gunaydin, Oscillator-Like Unitary Representations Of Non-Compact Groups And Supergroups And Extended Supergravity Theories, *Lect. Notes Phys.* **180**, 192–213, (1983).
30. M. Günaydin, Oscillator-like unitary representations of noncompact groups and supergroups and extended supergravity theories. Extended version presented at 11th Int. Colloq. on Group Theoretical Methods in Physics, Intanbul, Turkey, Aug 23- 28, 1982.
31. M. Günaydin and C. Saclioglu, Oscillator-like unitary representations of noncompact groups with a Jordan structure and the noncompact groups of supergravity, *Commun. Math. Phys.* **87**, 159, (1982). doi: 10.1007/BF01218560.
32. I. Bars and M. Gunaydin, Unitary Representations of Noncompact Supergroups, *Commun. Math. Phys.* **91**, 31, (1983). doi: 10.1007/BF01206048.
33. M. Gunaydin and S. Hyun, Unitary lowest weight representations of the noncompact supergroup $OSp(2n|2m, R)$, *J.Math.Phys.* **29**, 2367, (1988). doi: 10.1063/1.528120.
34. M. Flato and C. Fronsdal, One Massless Particle Equals Two Dirac Singletons: Elementary Particles in a Curved Space. 6., *Lett. Math. Phys.* **2**, 421–426, (1978). doi: 10.1007/BF00400170.
35. M. Flato and C. Fronsdal, Quantum field theory of singletons. the rac, *Journal of Mathematical Physics.* **22**, 1100, (1981).
36. C. Fronsdal, The Dirac Supermultiplet, *Phys. Rev.* **D26**, 1988, (1982). doi: 10.1103/PhysRevD.26.1988.
37. M. Flato and C. Fronsdal, Quantum Field Theory of Singletons: The Rac, *J. Math. Phys.* **22**, 1100, (1981). doi: 10.1063/1.524993.
38. E. Angelopoulos, M. Flato, C. Fronsdal, and D. Sternheimer, Massless Particles, Conformal Group and De Sitter Universe, *Phys. Rev.* **D23**, 1278, (1981). doi: 10.1103/PhysRevD.23.1278.
39. E. Fradkin and M. A. Vasiliev, Cubic Interaction in Extended Theories of Massless Higher Spin Fields, *Nucl.Phys.* **B291**, 141, (1987). doi: 10.1016/0550-3213(87)90469-X.
40. S. Konshtein and M. A. Vasiliev, Massless representations and admissibility condition for higher spin superalgebras, *Nucl.Phys.* **B312**, 402, (1989). doi: 10.1016/0550-3213(89)90301-5.
41. I. Klebanov and A. Polyakov, AdS dual of the critical O(N) vector model, *Phys.Lett.* **B550**, 213–219, (2002). doi: 10.1016/S0370-2693(02)02980-5.
42. M. A. Vasiliev, More on equations of motion for interacting massless fields of all spins in (3+1)-dimensions, *Phys. Lett.* **B285**, 225–234, (1992). doi: 10.1016/0370-2693(92)91457-K.
43. S. Giombi and X. Yin, Higher Spin Gauge Theory and Holography: The Three-Point

Functions, *JHEP.* **1009**, 115, (2010). doi: 10.1007/JHEP09(2010)115.
44. S. Giombi and X. Yin, Higher Spins in AdS and Twistorial Holography, *JHEP.* **1104**, 086, (2011). doi: 10.1007/JHEP04(2011)086.
45. M. A. Vasiliev, Higher spin gauge theories in four-dimensions, three-dimensions, and two-dimensions, *Int.J.Mod.Phys.* **D5**, 763–797, (1996). doi: 10.1142/S0218271896000473.
46. M. A. Vasiliev, Higher spin gauge theories: Star product and AdS space. (1999).
47. X. Bekaert, S. Cnockaert, C. Iazeolla, and M. Vasiliev, Nonlinear higher spin theories in various dimensions. (2005).
48. C. Iazeolla, On the Algebraic Structure of Higher-Spin Field Equations and New Exact Solutions. (2008).
49. A. Sagnotti, Notes on Strings and Higher Spins, *J.Phys.* **A46**, 214006, (2013). doi: 10.1088/1751-8113/46/21/214006.
50. S. Giombi and X. Yin, The Higher Spin/Vector Model Duality, *J.Phys.* **A46**, 214003, (2013). doi: 10.1088/1751-8113/46/21/214003.
51. V. Didenko and E. Skvortsov, Elements of Vasiliev theory. (2014).
52. M. Gunaydin, Singleton and doubleton supermultiplets of space-time supergroups and infinite spin superalgebras. Invited talk given at Trieste Conf. on Supermembranes and Physics in (2+1)-Dimensions, Trieste, Italy, Jul 17-21, 1989 ed. by M. J. Duff, et. al., World Scientific (1990), p. 442-456.
53. E. S. Fradkin and V. Ya. Linetsky, Conformal superalgebras of higher spins, *Ann. Phys.* **198**, 252–292, (1990). doi: 10.1016/0003-4916(90)90252-J.
54. E. Sezgin and P. Sundell, Doubletons and 5-D higher spin gauge theory, *JHEP.* **09**, 036, (2001). doi: 10.1088/1126-6708/2001/09/036.
55. E. Sezgin and P. Sundell, 7-D bosonic higher spin theory: Symmetry algebra and linearized constraints, *Nucl. Phys.* **B634**, 120–140, (2002). doi: 10.1016/S0550-3213(02)00299-7.
56. E. Sezgin and P. Sundell, Towards massless higher spin extension of D=5, N=8 gauged supergravity, *JHEP.* **09**, 025, (2001). doi: 10.1088/1126-6708/2001/09/025.
57. E. Sezgin and P. Sundell, Massless higher spins and holography, *Nucl. Phys.* **B644**, 303–370, (2002). doi: 10.1016/S0550-3213(02)00739-3. [Erratum: Nucl. Phys.B660,403(2003)].
58. S. Fernando, M. Gunaydin, and S. Takemae, Supercoherent states of $OSp(8^*|2N)$, conformal superfields and the AdS(7) / CFT(6) duality, *Nucl.Phys.* **B628**, 79–111, (2002). doi: 10.1016/S0550-3213(02)00076-7.
59. M. Vasiliev, Higher spin superalgebras in any dimension and their representations, *JHEP.* **0412**, 046, (2004). doi: 10.1088/1126-6708/2004/12/046.
60. M. G. Eastwood, Higher symmetries of the Laplacian, *Annals Math.* **161**, 1645–1665, (2005). doi: 10.4007/annals.2005.161.1645.
61. K. Govil and M. Gunaydin, Deformed Twistors and Higher Spin Conformal (Super-)Algebras in Four Dimensions. (2013).
62. K. Govil and M. Gunaydin, Deformed Twistors and Higher Spin Conformal (Super-)Algebras in Six Dimensions, *JHEP.* **1407**, 004, (2014). doi: 10.1007/JHEP07(2014)004.
63. S. Fernando and M. Gunaydin, Minimal unitary representation of 5d superconformal algebra $F(4)$ and AdS_6/CFT_5 higher spin (super)-algebras, *Nucl. Phys.* **B890**, 570–605, (2014). doi: 10.1016/j.nuclphysb.2014.11.015.
64. S. Fernando and M. Gunaydin, Massless conformal fields, AdS_{d+1}/CFT_d higher spin algebras and their deformations, *Nucl. Phys.* **B904**, 494–526, (2016). doi: 10.1016/j.nuclphysb.2016.01.024.

65. V. de Alfaro, S. Fubini, and G. Furlan, Conformal invariance in quantum mechanics, *Nuovo Cim.* **A34**, 569, (1976).
66. F. Calogero and C. Marchioro, Lower bounds to the ground-state energy of systems containing identical particles, *J.Math.Phys.* **10**, 562–569, (1969). doi: 10.1063/1.1664877.
67. F. Calogero, Solution of the one-dimensional N body problems with quadratic and/or inversely quadratic pair potentials, *J.Math.Phys.* **12**, 419–436, (1971). doi: 10.1063/1.1665604.
68. E. Angelopoulos and M. Laoues, Masslessness in n-dimensions, *Rev.Math.Phys.* **10**, 271–300, (1998). doi: 10.1142/S0129055X98000082.
69. M. Laoues, Massless particles in arbitrary dimensions, *Rev. Math. Phys.* **10**, 1079–1109, (1998). doi: 10.1142/S0129055X98000355.
70. P. A. M. Dirac, A Remarkable respresentation of the 3+2 de Sitter group, *J. Math. Phys.* **4**, 901–909, (1963).
71. M. Gunaydin and S. Takemae, Unitary supermultiplets of $OSp(8*|4)$ and the AdS_7/CFT_6 duality, *Nucl.Phys.* **B578**, 405–448, (2000). doi: 10.1016/j.nuclphysb.2004.07.022.
72. M. Eastwood, P. Somberg, and V. Soucek, The uniqueness of the Joseph ideal for the classical groups. Arxiv:0512296 [math] (2005).
73. M. Eastwood, The Cartan product, *Bulletin of the Belgian Mathematical Society-Simon Stevin.* **11**(5), 641–651, (2005).
74. W. Nahm, Supersymmetries and their representations, *Nucl. Phys.* **B135**, 149, (1978). doi: 10.1016/0550-3213(78)90218-3.

BMS Modules in Three Dimensions

A. Campoleoni[a,*], H.A. González[a], B. Oblak[a,b,†] and M. Riegler[c]

[a] *Université Libre de Bruxelles and International Solvay Institutes ULB-Campus Plaine CP231, 1050 Brussels, Belgium*

[b] *DAMTP, Centre for Mathematical Sciences, University of Cambridge Wilberforce Road, Cambridge CB3 0WA, United Kingdom*

[c] *Institute for Theoretical Physics, Vienna University of Technology Wiedner Hauptstrasse 8-10, 1040 Vienna, Austria*

We build unitary representations of the BMS algebra and its higher-spin extensions in three dimensions, using induced representations as a guide. Our prescription naturally emerges from an ultrarelativistic limit of highest-weight representations of Virasoro and $\mathcal{W}$ algebras, which is to be contrasted with non relativistic limits that typically give non-unitary representations. To support this dichotomy, we also point out that the ultrarelativistic and non-relativistic limits of generic $\mathcal{W}$ algebras differ in the structure of their non-linear terms.

1. Introduction

It has long been known that asymptotically flat gravitational theories in three and four space-time dimensions enjoy powerful symmetries at null infinity, given by an infinite-dimensional extension of the Poincaré group known as the Bondi-Metzner-Sachs (BMS) group [1–4]. The latter and its local generalisation [5] have been the focus of renewed interest in the last few years due to their relation e.g. to holography [6], soft graviton theorems [7] and black holes [8].

In three space-time dimensions this extension of Minkowski isometries is closely related to the infinite-dimensional symmetry enhancement of Anti-de Sitter space at spatial infinity [9]. In the latter case asymptotic symmetries are generated by two copies of the Virasoro algebra and admit an İnönü-Wigner contraction that reproduces the $\mathfrak{bms}_3$ algebra.

In three dimensions one can also broaden the purely gravitational setup to in-

*Postdoctoral Researcher of the Fund for Scientific Research-FNRS Belgium.
†Research Fellow of the Fund for Scientific Research-FNRS Belgium.

clude "higher-spin" gauge fields[a] on both flat and AdS backgrounds. The equations that, for $D > 3$, describe the propagation of a massless field of spin s indeed imply the absence of local degrees of freedom in $D = 3$ when $s \geq 2$. This feature is manifest in the Chern-Simons formulation of both Einstein gravity [10, 11] and higher-spin theories [12]. In this approach one can also accommodate flat backgrounds [13–15], thus bypassing the no-go results severely constraining higher-spin interactions in $D > 3$ Minkowski space (see e.g.[16]). The presence of higher-spin fields further enhances the asymptotic symmetries. Around AdS these generically consist of two copies of a non-linear $\mathcal{W}$ algebra [17, 18]; around flat space they consist of a non-linear higher-spin extension of the $\mathfrak{bms}_3$ algebra [14, 15], which can be obtained as a contraction of the direct sum of two $\mathcal{W}$ algebras.

Identifying a symmetry does not suffice to control its implementation at the quantum level: to this end one also needs to select the representations that are physically relevant in a given context. With this motivation in mind, in this note we study a class of representations of the $\mathfrak{bms}_3$ algebra and of its higher-spin extensions. In the gravitational case, the representations we are going to describe are the algebraic counterpart of the unitary representations of the BMS_3 group built and classified in [19, 20] along the lines originally used by Wigner for the Poincaré group [21]. The Hilbert space of each such representation consists of wavefunctionals in supermomentum space, in direct analogy to standard quantum mechanics. In the higher-spin case, analogous induced representations were built in [22] and their Lie-algebraic version was briefly investigated. The main purpose of the present paper is to delve deeper in the details of that algebraic construction.

We stress that the inclusion of higher-spin fields is a highly non-trivial extension of the basic setup, due to the non-linearities that appear in the symmetry algebra on both AdS and flat backgrounds. As a result, standard group-theoretic methods fail to apply. One faces a similar situation when analyzing extended BMS symmetry in four dimensions: representations of the globally well defined BMS group have been classified [23, 24], but a corresponding result for its local counterpart is still missing. Aside from their applications in three-dimensional higher-spin theories, we thus hope that our techniques will also prove useful in this challenging domain.

We will first build representations in a basis suggested by the theory of induced representations of Lie algebras, before showing how one can move to a basis of supermomentum eigenstates analogous to one-particle states with definite momentum. For that reason, in Section 2 we briefly review the construction of representations of the three-dimensional Poincaré algebra in a way that simplifies generalisations to the $\mathfrak{bms}_3$ algebra and its higher-spin extensions. We also recall how standard unitary representations of the Poincaré algebra emerge from an ultrarelativistic limit of highest-weight representations of $\mathfrak{so}(2,2)$. We then apply a similar construction to

[a]The little group of massless particles does not admit arbitrary discrete helicities in $D = 3$, but in this context it is customary to use the word "spin" to label the representations of the Lorentz group under which fields transform.

the $\mathfrak{bms}_3$ algebra in Section 3, and to its higher-spin extensions in Section 4. In both cases we also comment on the emergence of unitary representations from an ultra-relativistic limit of highest-weight representations of the (extended) local conformal algebra, while stressing that non-relativistic limits naturally lead to non-unitary representations as those considered in [25, 26].

2. Poincaré Modules in Three Dimensions

The unitary representations of the Poincaré group in three dimensions were classified in [27] and recently reviewed e.g. in [19] due to their relation with BMS$_3$ representations. Here we discuss how Wigner's standard method for the construction of irreducible, unitary representations of the Poincaré group (as presented e.g. in [28]) can be recovered from induced representations of the Poincaré *algebra*, also known as Poincaré modules. The advantage of this approach is that such modules can also be built for the $\mathfrak{bms}_3$ algebra and its non-linear higher-spin generalisations.

2.1. *The Poincaré algebra*

In three dimensions, the Lie algebra of the Poincaré group is spanned by three Lorentz generators J_m and three translation generators P_m $(m = -1, 0, 1)$ whose Lie brackets read

$$[J_m, J_n] = (m - n)\, J_{m+n}\,, \tag{1a}$$

$$[J_m, P_n] = (m - n)\, P_{m+n}\,, \tag{1b}$$

$$[P_m, P_n] = 0\,. \tag{1c}$$

Our conventions are such that these basis elements generate the complexification of the Poincaré algebra. Real translations, for example, are generated by linear combinations $\alpha_m P_m$ with complex coefficients satisfying $(\alpha_m)^* = \alpha_{-m}$; similarly real boosts are generated by combinations $zJ_1 + z^* J_{-1}$ while rotations are generated by θJ_0, with θ real. Accordingly, in any unitary representation, the operators representing Poincaré generators must satisfy the hermiticity conditions

$$(P_m)^\dagger = P_{-m}\,, \quad (J_m)^\dagger = J_{-m}\,. \tag{2}$$

Note that the P_m's correspond to the standard translation generators $\mathcal{P}^\mu$ (with $\mu = 0, 1, 2$ a Lorentz index) as $P_0 = \mathcal{P}^0$, $P_1 = \mathcal{P}^1 + i\mathcal{P}^2$ and $P_{-1} = \mathcal{P}^1 - i\mathcal{P}^2$.

The three-dimensional Poincaré algebra is thus the semi-direct sum

$$\mathfrak{iso}(2,1) = \mathfrak{sl}(2,\mathbb{R}) \in_{\text{ad}} (\mathfrak{sl}(2,\mathbb{R}))_{\text{Ab}} \tag{3}$$

where $\mathfrak{sl}(2,\mathbb{R}) \cong \mathfrak{so}(2,1)$ is the Lorentz algebra (generated by J_m's) and $(\mathfrak{sl}(2,\mathbb{R}))_{\text{Ab}}$ is an Abelian Lie algebra of translations (generated by P_m's) isomorphic to the

Lorentz algebra as a vector space, and acted upon by Lorentz transformations according to the adjoint representation. The Poincaré algebra admits two quadratic Casimir operators: the mass squared

$$\mathcal{M}^2 = P_0^2 - P_1 P_{-1} \tag{4}$$

and the three-dimensional analogue of the square of the Pauli-Lubanski vector,

$$\mathcal{S} = P_0 J_0 - \frac{1}{4}\left(J_1 P_{-1} + J_{-1} P_1 + P_1 J_{-1} + P_{-1} J_1\right). \tag{5}$$

The eigenvalues of these operators can be used to classify irreducible representations.

2.2. *Poincaré modules*

Irreducible unitary representations of the Poincaré group are obtained by considering the orbit of a given momentum under Lorentz transformations — i.e. all momenta $p^\mu = (p^0, p^1, p^2)$ that satisfy $p^2 = -M^2$ for some mass M — and building a Hilbert space of wavefunctions on that orbit. The eigenvalue of P_0 gives the energy of the corresponding particle and inspection of (1) shows that the operators that commute with P_0 are P_1, P_{-1} and J_0. It is therefore natural to build a basis of eigenstates of momentum for the Hilbert space of wavefunctions on the orbit; we will denote such eigenstates by $|p^\mu, s\rangle$. These correspond to plane waves of definite momentum p^μ, while $s \in \mathbb{R}$ is a spin label related to the eigenvalue of J_0 in a particular frame (see Eq. (7)). Different values of s yield inequivalent irreducible representations [19, 27]. Under a Lorentz transformation parametrised by $\Lambda^\mu{}_\nu$ these wavefunctions transform as

$$U(\Lambda)|p^\mu, s\rangle = e^{is\theta}|\Lambda^\mu{}_\nu p^\nu, s\rangle\,, \tag{6}$$

where $U(\Lambda)$ is a unitary operator and θ is a Λ-dependent Wigner angle. The components p^μ with $\mu = 0, 1, 2$ are related to the eigenvalues p_m of the generators P_m by $p^0 = p_0$, $p^1 = (p_1 + p_{-1})/2$ and $p^2 = (p_1 - p_{-1})/2i$.

Lorentz transformations act transitively on the momentum orbit, so for each fixed value of the mass squared one can choose a "standard" momentum k^μ and obtain all plane waves by acting with Lorentz boosts on the corresponding wavefunction $|k^\mu, s\rangle$. For *massive* representations — on which we focus for the sake of comparison with $\mathfrak{bms}_3$ and its higher-spin extensions — one can choose as a representative the momentum $k^\mu = (M, 0, 0)$ of the particle at rest. We denote by $|M, s\rangle$ the corresponding wavefunction, which satisfies

$$P_0|M, s\rangle = M|M, s\rangle\,, \quad P_{-1}|M, s\rangle = P_1|M, s\rangle = 0\,, \quad J_0|M, s\rangle = s|M, s\rangle\,, \tag{7}$$

and call it the *rest-frame state* of the representation. To obtain a plane wave $|p^\mu, s\rangle$ with boosted momentum, one can act with a Lorentz transformation implemented by the unitary operator

$$U(\omega) = \exp\left[\, i\left(\omega J_1 + \omega^* J_{-1}\right)\right], \tag{8}$$

where ω is the complex rapidity

$$\omega = \frac{i}{2} \operatorname{arcsinh}\left(\frac{\sqrt{p_1 p_{-1}}}{M}\right) \frac{p_{-1}}{\sqrt{p_1 p_{-1}}} \tag{9}$$

that one can obtain by inverting the relation $p^\mu = \Lambda^\mu{}_\nu k^\nu$ taking into account (2).

The previous discussion is standard, but note that (7) defines a one-dimensional representation of the subalgebra generated by $\{P_m, J_0\}$. Given a representation of a subalgebra $\mathfrak{h}$ of the Lie algebra $\mathfrak{g}$ on a vector space V, one can always build a representation of $\mathfrak{g}$ on a suitable quotient of the space $\mathcal{U}(\mathfrak{g}) \otimes V$, where $\mathcal{U}(\mathfrak{g})$ is the universal enveloping algebra of $\mathfrak{g}$ (see e.g. Section 10.7 of [29]). Representations of this kind are called *induced modules.* With this method one can construct an irreducible representation of the Poincaré *algebra* on the vector space $\mathcal{H}_M$ with basis vectors[b]

$$|k, l\,\rangle = (J_{-1})^k (J_1)^l |M, s\rangle\,. \tag{10}$$

Upon acting from the left on the states (10) one obtains indeed linear operators on $\mathcal{H}_M$ whose commutators coincide with (1). Moreover, the Casimir operators (4) and (5) have the same eigenvalue on each state (10), since they commute by construction with all elements of the algebra. This readily implies that the representation thus obtained is irreducible.

Unitarity, on the other hand, is far less obvious: it is not clear how to define a scalar product on the space $\mathcal{H}_M$ spanned by the states (10), even after enforcing the standard hermiticity conditions (2). Fortunately, experience with the Poincaré group suggests a way to circumvent the problem.[c] Upon acting on the rest frame state $|M, s\rangle$ with a Lorentz boost (8) one obtains (possibly up to an irrelevant phase) a plane wave

$$|p^\mu, s\rangle = U(\Lambda)|M, s\rangle\,. \tag{11}$$

Here p^μ can be any momentum belonging to the orbit with mass M, provided one chooses properly the Lorentz parameter Λ as in (9). Such plane wave states can be normalised so that

$$\langle\, p^\mu, s\,|\, q^\mu, s\,\rangle = \delta_\mu(p, q)\,, \tag{12}$$

where δ_μ is the Dirac distribution associated with the Lorentz-invariant measure

$$d\mu(q) = \frac{dq_1 dq_{-1}}{2i\sqrt{M^2 + q_1 q_{-1}}}\,. \tag{13}$$

---

[b]The states (10) form a basis of the induced $\mathfrak{iso}(2,1)$-module

$$\mathrm{Ind}_{\mathfrak{h}}^{\mathfrak{iso}(2,1)}(\rho) \equiv (\mathcal{U}(\mathfrak{p}_3) \otimes \mathbb{C}) \,/\, \{X \otimes \lambda - 1 \otimes \rho[X]\lambda \,|\, X \in \mathfrak{h},\, \lambda \in \mathbb{C}\}\,,$$

where $\mathfrak{h} = \mathrm{Span}\{P_n, J_0\}$ and ρ is the one-dimensional $\mathbb{C}$-valued representation

$$\rho[P_0] = M\,, \quad \rho[P_{-1}] = \rho[P_1] = 0\,, \quad \rho[J_0] = s$$

defined by (7). The quotient amounts to the rule that when one acts by the left with any element in $\mathfrak{iso}(2,1)$ one moves P_m and J_0 to the right by computing commutators and then lets them act on $|M, s\rangle$, as is implicit in the ket notation (10).

[c]In Section 2.3 we will also see an alternative way to define a scalar product on Poincaré modules from limits of representations of the $\mathfrak{so}(2,2)$ algebra.

In mapping the rest-frame state onto the states $|p^\mu, s\rangle$ we applied *finite* Lorentz transformations, so that we secretly brought the discussion back to the group-theoretic level. Nevertheless, to perform the "change of basis" from states of the form (10) to eigenstates of momentum, one does not need to control the full group structure; rather, it suffices to ensure that the boost (8) is well defined and that one can define a measure on the momentum orbit such that (12) is satisfied (see e.g.[19] for more details). The states obtained by acting with boosts on $|M, s\rangle$ can then be seen as infinite linear combinations of states (10). Unitarity finally follows from the fact that plane waves form an orthonormal basis of the Hilbert space (cf. Eq. (12)).

2.3. *Ultrarelativistic limit of* $\mathfrak{sl}(2,\mathbb{R})$ *modules*

In addition to being convenient for generalisations to infinite-dimensional extensions of the Poincaré algebra, Poincaré modules can be seen to arise as a limit of unitary representations of the AdS$_3$ isometry algebra, namely $\mathfrak{so}(2,2)$. Owing to the isomorphism $\mathfrak{so}(2,2) \cong \mathfrak{sl}(2,\mathbb{R}) \oplus \mathfrak{sl}(2,\mathbb{R})$, the generators of this algebra can be divided in two groups, $\mathcal{L}_m$ and $\bar{\mathcal{L}}_m$ with $m = -1, 0, 1$, and their Lie brackets read

$$[\mathcal{L}_m, \mathcal{L}_n] = (m-n)\,\mathcal{L}_{m+n}\,, \qquad [\bar{\mathcal{L}}_m, \bar{\mathcal{L}}_n] = (m-n)\,\bar{\mathcal{L}}_{m+n}\,. \tag{14}$$

As in (1) our conventions are such that this is a basis of the complexification of $\mathfrak{sl}(2,\mathbb{R})$, so that real $\mathfrak{sl}(2,\mathbb{R})$ matrices are linear combinations $i\,x_m \mathcal{L}_m$ with $(x_m)^* = x_{-m}$. In particular, in any unitary representation the operators representing the generators $\mathcal{L}_m$ and $\bar{\mathcal{L}}_m$ must satisfy the hermiticity conditions

$$(\mathcal{L}_m)^\dagger = \mathcal{L}_{-m}\,, \qquad (\bar{\mathcal{L}}_m)^\dagger = \bar{\mathcal{L}}_{-m}\,. \tag{15}$$

In terms of these basis elements the quadratic Casimir of each copy of $\mathfrak{sl}(2,\mathbb{R})$ reads

$$\mathcal{C} = \mathcal{L}_0^2 - \frac{1}{2}\left(\mathcal{L}_1\mathcal{L}_{-1} + \mathcal{L}_{-1}\mathcal{L}_1\right). \tag{16}$$

The Poincaré algebra (1) can be recovered from an İnönü-Wigner contraction of (14) by introducing a lenght scale ℓ (to be identified with the AdS radius) and by defining the new generators

$$P_m \equiv \frac{1}{\ell}\left(\mathcal{L}_m + \bar{\mathcal{L}}_{-m}\right), \qquad J_m \equiv \mathcal{L}_m - \bar{\mathcal{L}}_{-m}\,. \tag{17}$$

The Lie brackets of $\mathfrak{sl}(2,\mathbb{R}) \oplus \mathfrak{sl}(2,\mathbb{R})$ are turned into

$$[J_m, J_n] = (m-n)\,J_{m+n}\,, \tag{18a}$$

$$[J_m, P_n] = (m-n)\,P_{m+n}\,, \tag{18b}$$

$$[P_m, P_n] = \ell^{-2}(m-n)\,J_{m+n}\,, \tag{18c}$$

and in the limit $\ell \to \infty$ one recovers the Poincaré algebra. In addition the quadratic Casimir (16) can be combined with its counterpart $\bar{\mathcal{C}}$ in the second copy of $\mathfrak{sl}(2,\mathbb{R})$, producing

$$\frac{2}{\ell^2}\left(\mathcal{C} + \bar{\mathcal{C}}\right) = \mathcal{M}^2 + \mathcal{O}(\ell^{-2})\,, \qquad \frac{1}{\ell}\left(\mathcal{C} - \bar{\mathcal{C}}\right) = \mathcal{S}\,, \tag{19}$$

where $\mathcal{M}^2$ and $\mathcal{S}$ are the Poincaré Casimirs (4) and (5).

Aside from comparing Casimir operators, one can track how Poincaré modules (defined by (7) and (10)) emerge from the corresponding limit of highest-weight representations of $\mathfrak{so}(2,2)$. These are built out of highest-weight representations of $\mathfrak{sl}(2,\mathbb{R})$, which are defined starting from a state $|h\rangle$ that satisfies the conditions

$$\mathcal{L}_0|h\rangle = h\,|h\rangle\,, \qquad \mathcal{L}_1|h\rangle = 0\,. \tag{20}$$

The carrier space of the representation is then spanned by all descendant states $(\mathcal{L}_{-1})^m|h\rangle$,[d] and the Casimir (16) takes the value $h(h-1)$. If one builds a similar representation with weight $\bar{h}$ for a second copy of $\mathfrak{sl}(2,\mathbb{R})$, one can produce a representation of $\mathfrak{sl}(2,\mathbb{R})\oplus\mathfrak{sl}(2,\mathbb{R})$ from the tensor product.

To relate this tensor product — spanned by the states $(\mathcal{L}_{-1})^m(\bar{\mathcal{L}}_{-1})^n|h,\bar{h}\rangle$ — to a Poincaré module, we rewrite it in the new basis given by (10), where M and s are related to the $\mathfrak{so}(2,2)$ weights as

$$M \equiv \frac{h+\bar{h}}{\ell}\,, \qquad s \equiv h-\bar{h}\,, \tag{21}$$

since in terms of the operators (17) one has

$$P_0|h,\bar{h}\rangle = \frac{h+\bar{h}}{\ell}|h,\bar{h}\rangle\,, \qquad J_0|h,\bar{h}\rangle = (h-\bar{h})|h,\bar{h}\rangle\,. \tag{22}$$

This change of basis is invertible because no J_n annihilate the vacuum. Each $\mathfrak{so}(2,2)$ representation now takes the form

$$P_n|k,l\rangle = \sum_{k',l'} \mathsf{P}^{(n)}_{k',l';\,k,l}(M,s,\ell)|k',l'\rangle\,, \tag{23a}$$

$$J_n|k,l\rangle = \sum_{k',l'} \mathsf{J}^{(n)}_{k',l';\,k,l}(M,s)|k',l'\rangle \tag{23b}$$

where $\mathsf{P}^{(n)}$ and $\mathsf{J}^{(n)}$ are infinite matrices and where only negative powers of ℓ appear in (23a). These only arise because of the highest-weight conditions, which can be rewritten as

$$\left(P_{\pm1} \pm \frac{1}{\ell}J_{\pm1}\right)|h,\bar{h}\rangle = 0\,, \tag{24}$$

allowing to express the action of P_n in terms of the states $|k,l\rangle$. As a result, the matrix elements $\mathsf{P}^{(n)}_{k',l';\,k,l}$ and $\mathsf{J}^{(n)}_{k',l';\,k,l}$ have a well defined limit for $\ell\to\infty$. By computing the action of the generators P_n and J_n on the Poincaré module spanned

[d]Note that $\mathfrak{sl}(2,\mathbb{R})$ highest-weight representations can also be interpreted as induced modules. Eq. (20) defines indeed a one-dimensional representation of the subalgebra spanned by $\{\mathcal{L}_0,\mathcal{L}_1\}$, while the vector space of descendant states can be identified with a quotient of $\mathcal{U}(\mathfrak{sl}(2,\mathbb{R}))\otimes\mathbb{C}$ as discussed in footnote b. The main difference with respect to the Poincaré case is the splitting of the algebra as $\mathfrak{n}^-\oplus\mathfrak{h}\oplus\mathfrak{n}^+$, where $\mathfrak{n}^\pm$ are nilpotent subalgebras. This decomposition allows one to define a scalar product by enforcing the hermiticity condition (15). One can then verify that $\langle h|(\mathcal{L}_1)^m(\mathcal{L}_{-1})^m|h\rangle$ is positive for $h>0$.

by (10), one obtains the same outcome provided that the conformal weights scale as

$$h = \frac{M\ell + s}{2} + \lambda + \mathcal{O}(\ell^{-1}), \quad \bar{h} = \frac{M\ell - s}{2} + \lambda + \mathcal{O}(\ell^{-1}), \tag{25}$$

where λ is an arbitrary parameter independent of ℓ. In particular, the latter condition implies that in the limit the $\mathfrak{sl}(2,\mathbb{R})$ highest-weight conditions (24) turn into the rest-frame conditions (7). Note also that the Poincaré Casimirs $\mathcal{M}^2$ and $\mathcal{S}$ take the values dictated by the rest frame conditions. Moreover, in principle one could also define a scalar product on Poincaré modules starting from the limit of the scalar product $\langle h|(\mathcal{L}_1)^m(\mathcal{L}_{-1})^n|h\rangle$. This procedure will lead in general to a complicated non-diagonal quadratic form. We already know, however, that the plane-wave basis (12) diagonalises it, thus appearing as a natural alternative also from this vantage point.

Relation (25) shows that the flat limit defined via (17) can be interpreted as an ultrarelativistic/high-energy limit from the viewpoint of AdS$_3$. Poincaré modules are thus remnants of $\mathfrak{so}(2,2)$ representations whose energy becomes large in the limit $\ell \to \infty$. In Section 3.4 we shall also discuss a different contraction from $\mathfrak{so}(2,2)$ to $\mathfrak{iso}(2,1)$, to be interpreted as a non-relativistic limit giving rise to representations of the type discussed in [25, 26].

3. Induced Modules for the $\mathfrak{bms}_3$ Algebra

In this section, we remark that one can readily obtain representations of the $\mathfrak{bms}_3$ algebra by exploiting the induced module construction introduced in Section 2.2. We then show how one can move to a basis of supermomentum eigenstates by following analogous steps to those that we reviewed for the Poincaré case. This basis then allows one to discuss the irreducibility and unitarity of the induced representations. We finally display how the previous representations can be obtained from an ultrarelativistic limit of Virasoro Verma modules, while recalling why Galilean limits typically lead to non-unitary representations of a different kind.

3.1. $\mathfrak{bms}_3$ *algebra*

The $\mathfrak{bms}_3$ algebra is an infinite-dimensional algebra spanned by superrotation generators J_m and supermomentum generators P_m ($m \in \mathbb{Z}$) whose Lie brackets read

$$[J_m, J_n] = (m-n)J_{m+n} + \frac{c_1}{12}\, m(m^2-1)\,\delta_{m+n,0}\,, \tag{26a}$$

$$[J_m, P_n] = (m-n)P_{m+n} + \frac{c_2}{12}\, m(m^2-1)\,\delta_{m+n,0}\,, \tag{26b}$$

$$[P_m, P_n] = 0\,, \tag{26c}$$

where c_1 and c_2 are central charges. The central charge c_2 plays a key role for representation theory and it is e.g. non-vanishing in three-dimensional gravity [4], where it takes the value $c_2 = \frac{3}{G}$ with G being Newton's constant. The Poincaré algebra (1) is a subalgebra of $\mathfrak{bms}_3$. Similarly to (3), the $\mathfrak{bms}_3$ algebra is the semi-direct sum

$$\mathfrak{bms}_3 = \mathfrak{vir} \inplus_{\text{ad}} (\mathfrak{vir})_{\text{Ab}} \tag{27}$$

where $\mathfrak{vir}$ denotes the Virasoro algebra. In contrast with Poincaré, the operators (4) and (5) no longer commute with all generators of the algebra.

To the best of our knowledge, the classification of $\mathfrak{bms}_3$ Casimir operators is unknown. However, it was shown in [30] that the only Casimirs of the Virasoro algebra are functions of its central charges. If one assumes that all $\mathfrak{bms}_3$ Casimirs can be obtained as flat limits of Virasoro Casimirs (in the same way that the Poincaré Casimirs can be seen as limits, cf. (19)), then there are no $\mathfrak{bms}_3$ Casimirs other than its central charges.

3.2. $\mathfrak{bms}_3$ *modules*

Irreducible unitary representations of the BMS3 group are classified by orbits of supermomenta under the action of finite superrotations, that is, by coadjoint orbits of the Virasoro group [19]. In analogy with the Poincaré example, each orbit consists of supermomenta obtained by acting with superrotations on a given supermomentum p. The latter is a function on the circle,

$$p(\varphi) = \sum_{n\in\mathbb{Z}} p_n e^{in\varphi}\,, \tag{28}$$

and can be interpreted, from the gravitational viewpoint, as the Bondi mass aspect associated with an asymptotically flat metric in three dimensions — i.e. the energy density carried by the gravitational field at null infinity. It transforms as a quadratic density (or equivalently as a CFT stress tensor) under superrotations. The corresponding representation is then obtained by assuming the existence of a (quasi-)invariant measure on the orbit and by building a Hilbert space of square-integrable wavefunctionals on that orbit [19]. This Hilbert space admits a basis of eigenstates of the operators P_m, which generalise plane waves of definite momentum, and that we will denote as $|p(\varphi), s\rangle$. Here $s \in \mathbb{R}$ is a spin label directly analogous to its Poincaré counterpart; BMS3 representations with identical supermomentum orbits but different spins are mutually inequivalent.

Massive modules

An important class of representations is provided by supermomentum orbits that contain a constant $p(\varphi) = M - c_2/24$, where $M > 0$ is a mass parameter and c_2 is the central charge entering (26). This class contains e.g. the vacuum representation

$M = 0$, that accounts for all perturbative boundary excitations around the vacuum [20]. The Hilbert space of any such representation contains a wavefunction $|M, s\rangle$ that satisfies

$$P_0|M, s\rangle = M|M, s\rangle\,, \quad P_m|M, s\rangle = 0 \text{ for } m \neq 0\,, \quad J_0|M, s\rangle = s|M, s\rangle\,, \tag{29}$$

i.e. which is a supermomentum eigenstate for the constant eigenvalue $p(\varphi) = M - c_2/24$. In analogy with (7), we will call $|M, s\rangle$ the *rest-frame state* of the representation.

As in the Poincaré case, (29) defines a one-dimensional representation of the subalgebra of (26) spanned by $\{P_n, J_0, c_1, c_2\}$. This representation can be used to define an induced $\mathfrak{bms}_3$ module $\mathcal{H}_M$ with basis vectors

$$J_{n_1} J_{n_2} \cdots J_{n_N} |M, s\rangle\,, \tag{30}$$

where the n_i's are non-zero integers such that $n_1 \geq n_2 \geq \ldots \geq n_N$. With this ordering, states (30) with different combinations of n_i's are linearly independent within the universal enveloping algebra of $\mathfrak{bms}_3$, and acting on them from the left with the generators of the algebra provides linear operators on $\mathcal{H}_M$ whose commutators coincide with (26).

It is again unclear, however, how to define from scratch a scalar product on the space spanned by (30). Without scalar product one cannot look for null states to identify reducible modules, and the operators (4) and (5) can no longer be used to check irreducibility. Analogously to Poincaré, one can nevertheless consider the scalar product inherited from the limit of representations of the conformal algebra or, more naturally, reach a basis of supermomentum eigenstates. These states can be built (up to an irrelevant phase) from the rest-frame wavefunctional as

$$|p(\varphi), s\rangle = U(\omega)|M, s\rangle\,, \tag{31}$$

where[e]

$$U(\omega) = \exp\left(i \sum_{n\in\mathbb{Z}^*} \omega_n J_n\right), \quad \text{with } \omega_n^* = \omega_{-n} \tag{32}$$

is a unitary operator implementing a finite superrotation. The complex coefficients ω_n are the Fourier modes of a vector field on the circle $\omega(\varphi)\partial_\varphi$. One readily verifies that the semi-direct structure (27) implies that the states (31) are eigenstates of supermomentum. Indeed, the Baker-Campbell-Hausdorff formula yields

$$P_m|p(\varphi), s\rangle = U\left(U^{-1} P_m U\right)|M, s\rangle = U \cdot \exp\left(i\, \mathrm{ad}_\omega\right)[P_m]|M, s\rangle\,, \tag{33}$$

[e] The exponential map from the Virasoro algebra to the Virasoro group is *not* locally surjective (see e.g.[31]), so exponential operators such as (32) cannot map the rest-frame state $|M, s\rangle$ on *all* other plane waves in the supermomentum orbit. Nevertheless, superrotations do act transitively on the orbit: (31) is always correct for some unitary operator U, although the latter cannot always be written as an exponential (32). For the sake of simplicity, we assume that this subtlety does not affect our arguments; some additional comments can be found in appendix A.

where $\sum_n \omega_n J_n$ acts on P_m according to the adjoint representation,

$$\mathrm{ad}_\omega[P_m] = \sum_{n\in\mathbb{Z}^*} [\,\omega_n J_n, P_m\,]\,. \tag{34}$$

The bracket (26b) then implies that $U^{-1}P_mU$ is a certain combination of products of P_m's and central charges c_2. This combination acts multiplicatively on the rest-frame state thanks to conditions (29), from which we conclude as announced that $U|M,s\rangle$ is an eigenstate of supermomentum.

In fact we can even be more precise and say something about the eigenvalue of $U|M,s\rangle$ under P_m. (The complete argument is presented in appendix A, while here we only discuss its salient aspects.) By construction, supermomenta transform under superrotations according to the coadjoint representation of the Virasoro group, which coincides with the standard transformation law of CFT stress tensors under conformal transformations. Explicitly, a superrotation $f(\varphi)$ (satisfying $f(\varphi+2\pi)=f(\varphi)+2\pi$) maps a supermomentum $p(\varphi)$ on a new supermomentum $(f\cdot p)(\varphi)$ given by

$$(f\cdot p)(f(\varphi)) = \frac{1}{(f'(\varphi))^2}\left[p(\varphi)+\frac{c_2}{12}\{f;\varphi\}\right], \tag{35}$$

where $\{f;\varphi\}$ is the Schwarzian derivative of f at φ. Accordingly we know that the supermomentum of the state $U(\omega)|M,s\rangle$ takes the form (35) with $p(\varphi)$ replaced by $M-c_2/24$, in which case the eigenvalue of $U|M,s\rangle$ under P_m is the mth Fourier mode of $f\cdot p$. The diffeomorphism f is the exponential of the vector field $\omega(\varphi)\partial_\varphi$; in other words f is given by the flow of $\omega(\varphi)\partial_\varphi$ (see Eq. (A.12) for the exact correspondence). This is, in principle at least, the relation between the Fourier modes ω_n and the corresponding finite diffeomorphism. It is the BMS_3 analogue of the relation (9) that we displayed in the Poincaré case.

The basis of plane waves $|p(\varphi),s\rangle$ has the virtue of making BMS_3 representations manifestly unitary [19], since their scalar product takes the form (12). The only difference is that now the measure μ is a path integral measure on a supermomentum orbit; such measures were shown to exist in [32]. In addition, this setup provides a simple argument for showing that the representation is irreducible. Indeed, the supermomenta $p(\varphi)$ of these states span a superrotation orbit; the latter is a homogeneous space for the Virasoro group, so any plane wave can be mapped on any other one thanks to a suitably chosen superrotation. If we think of plane waves as a basis of the space of the representation, then this property of transitivity implies that the space of the representation admits no non-trivial invariant subspace, which is to say that the representation is irreducible. We will also expose further arguments for irreducibility in Section 3.3.

Vacuum module

The vacuum $\mathfrak{bms}_3$ module can be characterised in a similar way, starting from a vacuum state $|0\rangle$ such that

$$P_m|0\rangle = 0 \text{ for all } m \in \mathbb{Z} \quad \text{and} \quad J_n|0\rangle = 0 \text{ for } n = -1, 0, 1. \tag{36}$$

Here the condition $P_0|0\rangle = 0$ says that the vacuum has zero mass, while the additional conditions $J_{\pm 1}|0\rangle = 0$ reflect Lorentz-invariance. They imply that the stability group of the constant momentum $p(\varphi) = M - c_2/24$ is enhanced for $M = 0$. In the language of induced representations, this means that the little group of the vacuum is the Lorentz group instead of the group of rotations.

If we were dealing with the Poincaré algebra, such conditions would produce a trivial representation. Here, by contrast, there exist non-trivial "boosted vacua" of the form (30), where now $n_1, ..., n_N$ are integers different from $-1, 0, 1$. These vacua are analogous to the boundary gravitons of AdS_3 [33]. The fact that the vacuum is not invariant under the full BMS symmetry, but only under its Poincaré subgroup, implies that the boosted states (30) (with all n_i's $\neq -1, 0, 1$) can be interpreted as Goldstone-like states associated with broken symmetry generators. Equivalently each state (30) can be interpreted as a vacuum dressed with the three-dimensional analogue of soft graviton degrees of freedom created by superrotations.[f] This also provides a natural interpretation for non-vacuum BMS_3 representations as particles dressed with soft gravitons. It should be noted that, in contrast to the realistic four-dimensional case, supertranslations here do *not* create new states.

3.3. *Ultrarelativistic limit of Virasoro modules*

In analogy with the discussion in Section 2.3, $\mathfrak{bms}_3$ modules emerge as limits of irreducible unitary representations of the local conformal algebra, which are built as tensor products of irreducible Verma modules of the Virasoro algebra. We still denote the generators of the local conformal algebra by two sets of commuting $\mathcal{L}_m$ and $\bar{\mathcal{L}}_m$ as in (14), but now $m \in \mathbb{Z}$ and the generators obey the centrally extended algebra

$$[\mathcal{L}_m, \mathcal{L}_n] = (m-n)\,\mathcal{L}_{m+n} + \frac{c}{12}\,m(m^2-1)\delta_{m+n,0}\,, \tag{37a}$$

$$[\bar{\mathcal{L}}_m, \bar{\mathcal{L}}_n] = (m-n)\,\bar{\mathcal{L}}_{m+n} + \frac{\bar{c}}{12}\,m(m^2-1)\delta_{m+n,0}\,. \tag{37b}$$

Highest weight representations of this algebra are built upon an eigenstate $|h, \bar{h}\rangle$ of $\mathcal{L}_0$ and $\bar{\mathcal{L}}_0$ that satisfies

$$\mathcal{L}_n|h, \bar{h}\rangle = 0\,, \qquad \bar{\mathcal{L}}_n|h, \bar{h}\rangle = 0 \quad \text{when } n > 0\,. \tag{38}$$

[f]Since three-dimensional gravity has no local degrees of freedom, the notion of "soft graviton" is ambiguous. (There are no genuine gravitons whose zero-frequency limit would be soft gravitons.) Our viewpoint here is that "soft" degrees of freedom coincide with the boundary degrees of freedom defined by non-trivial asymptotic symmetries, in accordance with their relation to soft theorems [7].

The carrier space of the representation is then spanned by the states

$$\mathcal{L}_{-n_1} \cdots \mathcal{L}_{-n_k} \bar{\mathcal{L}}_{-\bar{n}_1} \cdots \bar{\mathcal{L}}_{-\bar{n}_l} |h, \bar{h}\rangle \tag{39}$$

with $n_1 \geq n_2 \geq \cdots \geq n_k > 0$ and a similar ordering for the $\bar{n}_i > 0$. As suggested by the analysis in Section 2.3, we will be interested in large values of h and $\bar{h}$, for which these representations are irreducible. In addition the standard hermiticity condition

$$(\mathcal{L}_m)^\dagger = \mathcal{L}_{-m} \tag{40}$$

yields a scalar product on this space, allowing one to discuss unitarity.

As for the Poincaré case, one can define the new generators (17) and rewrite this vector space in the basis (30), where M and s are the eigenvalues of P_0 and J_0 related to h and $\bar{h}$ by (21). The change of basis is again invertible because no J_n annihilates the vacuum. Each representation of the conformal algebra is still specified by an analogue of (23), where now each state is labelled by the quantum numbers n_i, $\bar{n}_j$ and the matrices $\mathsf{P}^{(n)}$ and $\mathsf{J}^{(n)}$ also depend on the central charges c_1 and c_2 defined by

$$c_1 = c - \bar{c}, \qquad c_2 = \frac{c + \bar{c}}{\ell}. \tag{41}$$

As before, only negative powers of ℓ enter $\mathsf{P}^{(n)}$ via the highest-weight conditions (38) written in the new basis:

$$\left(P_{\pm n} \pm \frac{1}{\ell} J_{\pm n} \right) |h, \bar{h}\rangle = 0. \tag{42}$$

A limit $\ell \to \infty$ performed at fixed M, s and c_1, c_2 (rather than e.g. at fixed h, $\bar{h}$) then gives the $\mathfrak{bms}_3$ module that we built from scratch in Section 3.2. Note, in particular, that the highest-weight state (38) is mapped to the rest-frame state (29) in this limit. In this sense a $\mathfrak{bms}_3$ module is just a high-energy limit of the tensor product of two Virasoro modules. Since Virasoro representations are irreducible for large h, it is reasonable to expect that the same is true of $\mathfrak{bms}_3$ modules.

3.4. *Galilean limit of Virasoro modules*

In this section we consider another possible group contraction, to be interpreted as a non-relativistic limit of the conformal symmetry. This limiting procedure yields an infinite-dimensional extension of the Galilean algebra known as Galilean conformal algebra or $\mathfrak{gca}_2$ (see e.g.[25]), which is isomorphic to $\mathfrak{bms}_3$. In spite of the algebras being isomorphic, the representations that one obtains in the ultrarelativistic or Galilean limits are significantly different. In particular, the Galilean contraction we are going to review generically leads to non-unitary representations [25].

As in the previous section, the generators $\mathcal{L}_n$ and $\bar{\mathcal{L}}_n$ satisfy the algebra (37), and we consider Virasoro highest-weight representations as in (38). In order to perform the nonrelativistic limit we introduce a dimensionless contraction parameter ϵ and the new generators

$$M_n \equiv \epsilon \left(\bar{\mathcal{L}}_n - \mathcal{L}_n\right), \qquad L_n \equiv \bar{\mathcal{L}}_n + \mathcal{L}_n . \tag{43}$$

We stress that the combinations of $\mathcal{L}_m$'s appearing in this definition are different from those of the ultrarelativistic contraction (17). In this basis the conformal algebra reads

$$[L_m, L_n] = (m-n)\, L_{m+n} + \frac{c_L}{12}\, m(m^2-1)\, \delta_{m+n,0} , \tag{44a}$$

$$[L_m, M_n] = (m-n)\, M_{m+n} + \frac{c_M}{12}\, m(m^2-1)\, \delta_{m+n,0} , \tag{44b}$$

$$[M_m, M_n] = \epsilon^2 \left((m-n)\, L_{m+n} + \frac{c_L}{12}\, m(m^2-1)\, \delta_{m+n,0}\right) , \tag{44c}$$

where the central charges are given by

$$c_L = \bar{c} + c , \qquad c_M = \epsilon\, (\bar{c} - c) . \tag{45}$$

In the limit $\epsilon \to 0$ one obtains an algebra isomorphic to (26).

We denote the eigenvalues of M_0 and L_0 on a highest-weight state $|h, \bar{h}\rangle$ by

$$\Delta = \bar{h} + h , \qquad \xi = \epsilon \left(\bar{h} - h\right) , \tag{46}$$

and we use them to label the state as $|\Delta, \xi\rangle$. In terms of the operators (43) the highest-weight conditions (38) become

$$L_n|\Delta, \xi\rangle = 0 , \qquad M_n|\Delta, \xi\rangle = 0 , \quad n > 0 . \tag{47}$$

Note that these constraints hold for any value of ϵ, including the limit $\epsilon \to 0$. One can then consider the descendant states

$$|\{\mathfrak{l}_i\}, \{\mathfrak{m}_j\}\rangle = L_{-\mathfrak{l}_1} \dots L_{-\mathfrak{l}_i} M_{-\mathfrak{m}_1} \dots M_{-\mathfrak{m}_j} |\Delta, \xi\rangle , \tag{48}$$

with $\mathfrak{l}_1 \geq \ldots \geq \mathfrak{l}_i > 0$ and $\mathfrak{m}_1 \geq \ldots \geq \mathfrak{m}_j > 0$, and compute the matrix elements of the operators M_n and L_n in this basis, by using the commutators (44). Only positive powers of ϵ^2 appear due to (44c), while in contrast to the ultrarelativistic case the highest-weight conditions (47) do not bring any power of ϵ. The matrix elements also depend on the central charges c_L and c_M of (45). In the limit $\epsilon \to 0$ at Δ, ξ, c_L and c_M fixed one finds the same matrix elements that one would obtain by working directly with the $\mathfrak{gca}_2$ algebra.

We stress that the highest-weight conditions (47) significantly differ from the rest-frame conditions (38) that we obtained in the ultrarelativistic limit. Consequently, the corresponding representations have very different features. In the Galilean case one can readily define a scalar product by imposing the hermiticity conditions $(M_m)^\dagger = M_{-m}$ and $(L_m)^\dagger = L_{-m}$. This allows one to compute $\langle\{\mathfrak{l}_i\}, \{\mathfrak{m}_j\}|\{\mathfrak{l}_k\}, \{\mathfrak{m}_l\}\rangle$ by taking advantage of (47). One realises in this way that, in contrast with $\mathfrak{bms}_3$ modules, these representations are typically reducible and non-unitary [25].

4. Higher-Spin Modules in Flat Space

We now turn to the higher-spin analogue of the algebraic constructions described above. For concreteness and simplicity we focus on the spin-3 extension of $\mathfrak{bms}_3$ but our considerations apply, mutatis mutandis, to other higher-spin extensions as well. We will start by defining the quantum flat $\mathcal{W}_3$ algebra as an ultrarelativistic limit of $\mathcal{W}_3 \oplus \mathcal{W}_3$, which will produce a specific ordering of operators in the non-linear terms of the commutators. Section 4.2 will then be devoted to the construction of induced modules along the lines described above for Poincaré and $\mathfrak{bms}_3$, and we will see there that the ordering that emerges in the ultrarelativistic limit may be seen as a normal ordering with respect to rest-frame conditions. Along the way we will compare our results to those of the non-relativistic limit described in [26], and we will see that the two limits lead to different quantum algebras.

4.1. *Spin-3 extension of the* $\mathfrak{bms}_3$ *algebra*

One can add to the $\mathfrak{bms}_3$ algebra two sets of generators W_n and Q_n that transform under superrotations as the modes of primary fields of conformal weight 3. This gives a non-linear algebra that can be obtained as an İnönü-Wigner contraction of the direct sum of two $\mathcal{W}_3$ algebras. This contraction has been discussed at the *semiclassical* level in [14, 15] and a Galilean limit of the quantum algebra has been considered in [26]. Here we are interested instead in an ultrarelativistic limit of $\mathcal{W}_3 \oplus \mathcal{W}_3$. The key difference between the Galilean and ultrarelativistic contractions is that the latter mixes generators with positive and negative mode numbers, whereas the former does not. For linear algebras, such as the Virasoro algebra for example, this does not pose a problem and thus the two contractions yield isomorphic algebras. As soon as non-linear algebras are involved in the contraction, however, Galilean and ultrarelativistic limits do not necessarily yield isomorphic (quantum) algebras anymore.

Ultrarelativistic contraction

The quantum $\mathcal{W}_3$ algebra is spanned by two sets of generators $\mathcal{L}_m$ and $\mathcal{W}_m$ ($m \in \mathbb{Z}$) whose commutation relations read

$$[\mathcal{L}_m, \mathcal{L}_n] = (m-n)\,\mathcal{L}_{m+n} + \frac{c}{12}\,(m^3-m)\,\delta_{m+n,0}\,, \tag{49a}$$

$$[\mathcal{L}_m, \mathcal{W}_n] = (2m-n)\,\mathcal{W}_{m+n}\,, \tag{49b}$$

$$\begin{aligned}[\mathcal{W}_m, \mathcal{W}_n] = {}&(m-n)(2m^2+2n^2-mn-8)\,\mathcal{L}_{m+n} + \frac{96}{c+\frac{22}{5}}\,(m-n)\,{:}\mathcal{L}\mathcal{L}{:}_{m+n}\\ &+ \frac{c}{12}\,(m^2-4)(m^3-m)\,\delta_{m+n,0}\,,\end{aligned} \tag{49c}$$

with the usual normal ordering prescription[g]

$$:\mathcal{L}\mathcal{L}:_m = \sum_{p\geq -1} \mathcal{L}_{m-p}\mathcal{L}_p + \sum_{p<-1} \mathcal{L}_p\mathcal{L}_{m-p} - \frac{3}{10}(m+3)(m+2)\mathcal{L}_m\,. \tag{50}$$

The standard hermiticity conditions on the generators of this algebra are

$$(\mathcal{W}_m)^\dagger = \mathcal{W}_{-m} \tag{51}$$

together with (40); these conditions must hold in any unitary representation of the $\mathcal{W}_3$ algebra.

We consider a direct sum $\mathcal{W}_3 \oplus \mathcal{W}_3$ where the generators and the central charge of the other copy of $\mathcal{W}_3$ will be denoted with a bar on top ($\bar{\mathcal{L}}_m$, $\bar{\mathcal{W}}_m$ and $\bar{c}$). Introducing a length scale ℓ (to be interpreted as the AdS_3 radius), we define new generators P_m and J_m as in (17) together with

$$W_m \equiv \mathcal{W}_m - \bar{\mathcal{W}}_{-m}\,, \qquad Q_m \equiv \frac{1}{\ell}\left(\mathcal{W}_m + \bar{\mathcal{W}}_{-m}\right). \tag{52}$$

We also define central charges c_1 and c_2 as in (41). In the limit $\ell \to \infty$, and provided the central charges scale in such a way that both c_1 and c_2 be finite, one finds that J_m and P_m satisfy the brackets (26) and

$$[J_m, W_n] = (2m-n)W_{m+n}\,, \qquad [J_m, Q_n] = (2m-n)Q_{m+n}\,, \tag{53a}$$

$$[P_m, W_n] = (2m-n)Q_{m+n}\,, \qquad [P_m, Q_n] = 0\,. \tag{53b}$$

The remaining brackets involving higher-spin generators are

$$\begin{aligned}[W_m, W_n] &= (m-n)(2m^2+2n^2-mn-8)J_{m+n} + \frac{96}{c_2}(m-n)\Lambda_{m+n} \\ &\quad - \frac{96\,c_1}{c_2^2}(m-n)\Theta_{m+n} + \frac{c_1}{12}(m^2-4)(m^3-m)\,\delta_{m+n,0}\,,\end{aligned} \tag{53c}$$

$$\begin{aligned}[W_m, Q_n] &= (m-n)(2m^2+2n^2-mn-8)P_{m+n} + \frac{96}{c_2}(m-n)\Theta_{m+n} \\ &\quad + \frac{c_2}{12}(m^2-4)(m^3-m)\,\delta_{m+n,0}\,,\end{aligned} \tag{53d}$$

$$[Q_m, Q_n] = 0\,, \tag{53e}$$

where we have introduced the following notation for non-linear terms:

$$\Theta_m \equiv \sum_{p=-\infty}^{\infty} P_{m-p}P_p\,, \qquad \Lambda_m \equiv \sum_{p=-\infty}^{\infty} \left(P_{m-p}J_p + J_{m-p}P_p\right). \tag{54}$$

One can check that with this definition the algebra (26), (53) satisfies Jacobi identities. We will call this algebra the (quantum) *flat* $\mathcal{W}_3$ *algebra*. In any unitary representation, its generators satisfy the hermiticity conditions

$$(Q_m)^\dagger = Q_{-m}\,, \quad (W_m)^\dagger = W_{-m}\,. \tag{55}$$

[g]The term linear in $\mathcal{L}_m$ ensures that the resulting normal-ordered operator $:\mathcal{L}\mathcal{L}:_m$ is quasi-primary with respect to the action of $\mathcal{L}_m$'s.

supplemented with (2) for $m \in \mathbb{Z}$.

The expressions (54) for the quadratic terms follow from the identities

$$:\mathcal{L}\mathcal{L}:_m + :\bar{\mathcal{L}}\bar{\mathcal{L}}:_{-m} = \frac{\ell^2}{2}\,\Theta_m + \mathcal{O}(\ell)\,, \tag{56a}$$

$$:\mathcal{L}\mathcal{L}:_m - :\bar{\mathcal{L}}\bar{\mathcal{L}}:_{-m} = \frac{\ell}{2}\,\Lambda_m + \mathcal{O}(1)\,. \tag{56b}$$

Note, in particular, that both the linear term in (50) and the mixing between positive and negative modes in (17)–(52) are necessary to reorganize the sum of quadratic terms with the precise order of (54). We shall see in Section 4.2 that (54) can be considered as a normal-ordered polynomial with respect to our definition of the vacuum.

Galilean contraction

By contrast, a Galilean contraction of $\mathcal{W}_3 \oplus \mathcal{W}_3$ can be obtained by defining central charges c_L and c_M as in (45), introducing new generators M_m and L_m as in (43) and writing

$$Q_m \equiv \epsilon\left(\bar{\mathcal{W}}_m - \mathcal{W}_m\right), \qquad W_m \equiv \bar{\mathcal{W}}_m + \mathcal{W}_m\,. \tag{57}$$

In the limit $\epsilon \to 0$ one obtains brackets of the same form as in (53) after the substitutions $J_m \to L_m$, $P_m \to M_m$ and $c_1 \to c_L$, $c_2 \to c_M$. However, there are two important differences: the coefficient in front of Θ_{m+n} in (53c) contains a shifted central charge $c_L + 44/5$ and the quadratic terms read

$$\Theta_m = \sum_{p=-\infty}^{\infty} M_p M_{m-p}\,, \tag{58a}$$

$$\begin{aligned}\Lambda_m &= \sum_{p\geq -1} \left(M_{m-p}L_p + L_{m-p}M_p\right) + \sum_{p<-1} \left(M_p L_{m-p} + L_p M_{m-p}\right) \\ &\quad - \frac{3}{5}\,(m+3)(m+2)M_m \end{aligned} \tag{58b}$$

(see Eq. (2.18) of [26]). They can be interpreted as normal-ordered operators with respect to a highest-weight vacuum defined by conditions of the type (47) with $\Delta = \xi = 0$. These differences show that the two contractions lead to different *quantum* algebras, despite the fact that the corresponding classical algebras coincide.[h] Thus, in the presence of higher-spin fields, the difference between ultrarelativistic and Galilean limits manifests itself directly in the symmetry algebras and not only at the level of the representations surviving in the limit.

In the following we restrict our attention to irreducible unitary representations of the ultrarelativistic quantum algebra (53), built once again according to the induced

[h] An interesting problem is to understand if these algebras are merely different because of an unfortunate choice of basis, or if they are genuinely distinct in the sense that they are not isomorphic. We will not address this issue here.

module prescription. On the other hand, highest-weight representations of Galilean contractions of two copies of non-linear $\mathcal{W}$ algebras were discussed in [26], where it was shown that unitary representations with higher-spin states do not exist.

4.2. *Flat $\mathcal{W}_3$ modules*

The non-linearities in (53) make standard group-theoretic techniques inapplicable to the construction of representations of the flat $\mathcal{W}_3$ algebra. In spite of this limitation, it was suggested in [22] that unitary irreducible representations can be built as Hilbert spaces of wavefunctionals defined on orbits of extended supermomenta

$$p(\varphi) = \sum_{n\in\mathbb{Z}} p_n e^{in\varphi}\,, \qquad q(\varphi) = \sum_{n\in\mathbb{Z}} q_n e^{in\varphi}\,, \tag{59}$$

where $p(\varphi)$ is the Bondi mass aspect/supermomentum already introduced in (28), while $q(\varphi)$ is its spin-3 analogue. The latter is a function on the circle transforming as a cubic density (or equivalently as a current of spin 3) under superrotations. In analogy with the $\mathfrak{bms}_3$ case, we will further assume that the Hilbert space of wavefunctionals on a given orbit admits a basis of eigenstates of P_m, Q_m, denoted by $|p(\varphi), q(\varphi)\rangle$, where for brevity we omitted the spin labels (such as s in (29)) that are fixed in any given irreducible representation. This proposal was tested in [22], for arbitrary spin, by matching suitable products of one-loop higher-spin partition functions with group characters derived using the Frobenius formula that follows from the orbit construction. In the remainder of this section we investigate the algebraic counterpart of that proposal.

Massive modules

We focus on orbits that contain a constant extended supermomentum $p(\varphi) = M - c_2/24$, $q(\varphi) = q_0$. The Hilbert space of the corresponding representation then contains a plane wave state $|M, q_0\rangle$ that satisfies

$$P_m|M, q_0\rangle = 0\,, \qquad Q_m|M, q_0\rangle = 0 \quad \text{for } m \neq 0\,, \tag{60a}$$

and is an eigenstate of zero-mode charges:

$$P_0|M, q_0\rangle = M|M, q_0\rangle\,, \qquad J_0|M, q_0\rangle = s|M, q_0\rangle\,, \tag{60b}$$

$$Q_0|M, q_0\rangle = q_0|M, q_0\rangle\,, \qquad W_0|M, q_0\rangle = w|M, q_0\rangle\,. \tag{60c}$$

Here M and s are the mass and spin labels encountered earlier, while q_0 and w are their spin-3 counterparts. As before we will call $|M, q_0\rangle$ the *rest-frame state* of the representation.

The conditions (60) define a one-dimensional representation of the subalgebra spanned by $\{P_m, Q_m, J_0, W_0\}$. They can be used to define an induced module $\mathcal{H}_{M,q_0}$ with basis elements

$$W_{k_1} \cdots W_{k_m} J_{l_1} \cdots J_{l_n} |M, q_0\rangle\,, \tag{61}$$

where $k_1 \geq \cdots \geq k_m$ and $l_1 \geq \cdots \geq l_n$ are non-zero integers. This provides an explicit representation of the quantum flat $\mathcal{W}_3$ algebra. Note that the presence of non-linearities in the commutators (53) does not affect the construction of the induced module, which involves the universal enveloping algebra anyway.

As in the previous examples, the basis (61) is very useful to prove the existence of a given representation, but not very illuminating if one wants to understand its properties. Following our mantra, we thus move to a basis of eigenstates of supermomentum by acting on the rest-frame state as

$$|p(\varphi), q(\varphi)\rangle = U(\omega, \Omega)|M, q_0\rangle\,, \tag{62}$$

where

$$U(\omega, \Omega) = \exp\left(i \sum_{n\in\mathbb{Z}^*} (\omega_n J_n + \Omega_n W_n)\right) \quad \text{with } \omega_n^* = \omega_{-n},\ \Omega_n^* = \Omega_{-n} \tag{63}$$

is a unitary operator implementing a finite higher-spin superrotation. The complex coefficients ω_n and Ω_n can be interpreted as the Fourier modes of the tensor fields $\omega(\varphi)\partial_\varphi$ and $\Omega(\varphi)\partial_\varphi\partial_\varphi$ on the circle.

To prove that the states (62) diagonalise all P_m's and Q_m's we can use the Baker-Campbell-Hausdorff formula as in (33) to obtain

$$P_m|p(\varphi), q(\varphi)\rangle = U \cdot \exp\left(i\, \mathrm{ad}_{\omega,\Omega}\right)[P_m]|M, q_0\rangle\,, \tag{64a}$$

$$Q_n|p(\varphi), q(\varphi)\rangle = U \cdot \exp\left(i\, \mathrm{ad}_{\omega,\Omega}\right)[Q_n]|M, q_0\rangle\,, \tag{64b}$$

where

$$\mathrm{ad}_{\omega,\Omega}[\,\cdot\,] = \sum_{n\in\mathbb{Z}^*} [\,\omega_n J_n + \Omega_n W_n,\,\cdot\,]\,. \tag{65}$$

Inspection of (53) then shows that the nested commutators that appear in (64) always produce either P_n or Q_n modes. As a result the right-hand sides of (64) turn into series of products of P_n and Q_n operators, which act diagonally on the rest-frame state by virtue of the rest-frame conditions (60). Computing the eigenvalue associated with a given set of modes ω_n and Ω_n requires instead a control of the finite action of $\mathcal{W}_3$ superrotations, that was discussed e.g. in [34].

As before the plane waves (62) provide an orthonormal basis of the Hilbert space of the representation, thus making unitarity manifest (provided a suitable measure exists on the orbit). Irreducibility can be inferred from the same argument we used for $\mathfrak{bms}_3$: by construction, a supermomentum orbit is a homogeneous space for the action of superrotations, and this carries over to the higher-spin setting. This implies that $\mathcal{W}_3$ superrotations can map any plane wave state on any other one, which in turn implies that the space of the representation has no non-trivial invariant subspace.

Vacuum module

The vacuum module of the flat $\mathcal{W}_3$ algebra can be built in direct analogy to its $\mathfrak{bms}_3$ counterpart discussed around (36). The only subtlety is the enhancement of the little group, which leads to additional conditions on superrotations. Indeed the vacuum state $|0\rangle$ is now an eigenstate of all modes P_m and Q_m with zero eigenvalue, and satisfies in addition

$$J_n|0\rangle = 0 \quad \text{for } n = -1, 0, 1, \quad W_m|0\rangle = 0 \quad \text{for } m = -2, -1, 0, 1, 2. \tag{66}$$

These conditions ensure that the vacuum is invariant under the $\mathfrak{sl}(3, \mathbb{R})$ wedge algebra of the $\mathcal{W}_3$ subalgebra (which includes in particular the Lorentz algebra). The corresponding module can then be built as usual by acting with higher-spin superrotation generators on the vacuum state and producing states of the form (61), where now all l_i's must be different from $-1, 0, 1$ and all k_i's must be different from $-2, -1, 0, 1, 2$.

The definition of the flat $\mathcal{W}_3$ vacuum allows us to interpret the quadratic terms in (54) as being normal-ordered. Indeed, the expectation value of the operators Θ_n and Λ_n vanishes on the vacuum $|0\rangle$:

$$\langle 0|\Theta_n|0\rangle = \langle 0|\Lambda_n|0\rangle = 0. \tag{67}$$

By contrast, for a highest-weight vacuum of the type (47) one obtains non-vanishing expectation values. Thus the additional non-linear structure introduced by higher spins stresses once more that the natural representations to be considered in the ultrarelativistic limit are the induced ones discussed above, rather than the highest-weight ones of [25, 26].

These considerations appear to be a robust feature of "flat $\mathcal{W}$ algebras". Ultrarelativistic contractions of $\mathcal{W}_N \oplus \mathcal{W}_N$ algebras always take the form

$$\text{"flat } \mathcal{W}_N\text{"} = \mathcal{W}_N \in_{\text{ad}} (\mathcal{W}_N)_{\text{Ab}} \tag{68}$$

and therefore contain an Abelian ideal. In addition the structure constants of the non-linear terms are always proportional to inverse powers of the central charge. Indeed, for a non-linear operator of nth order the structure constants for large c are proportional to $\frac{1}{c^{n-1}}$. When expanding them in powers of the contraction parameter ℓ, this implies that the leading term is proportional to ℓ^{1-n} thanks to (41). In order to obtain a finite result, it is thus necessary that the resulting non-linear operator consists of at least $n-1$ Abelian generators. Terms of this kind always have a vanishing expectation value on our rest-frame vacuum, although the precise ordering in the polynomial should be fixed by other means, e.g. by defining the algebra via a contraction of the quantum algebra or by imposing Jacobi identities.

5. Conclusion

In this work we have seen how induced representations of groups can be recovered from induced modules of Lie algebras. When applied to $\mathfrak{bms}_3$ and its higher-spin extensions, this approach confirms and expands the prescription previously described in [22]. Each such module consists of a rest-frame state and of its "descendants" obtained by acting with all superrotation generators. It provides an explicit irreducible unitary representation of the corresponding symmetry algebra. It can also be seen as a high-energy, high central charge limit of highest weight representations of direct sums of Virasoro or $\mathcal{W}$ algebras.

This approach has also allowed us to discuss the differences between ultrarelativistic and Galilean limits of Virasoro/$\mathcal{W}$ representations. As argued above, the ultrarelativistic limit always produces unitary induced modules, while the Galilean limit generically produces non-unitary highest-weight representations. In fact, higher-spin considerations allow us to see this dichotomy already at the level of Lie algebras, regardless of representation theory: the ultrarelativistic limit of the quantum $\mathcal{W}_3 \oplus \mathcal{W}_3$ algebra produces non-linear terms of the type (54), which are normal-ordered with respect to the rest-frame conditions adapted to induced modules. By contrast, its Galilean limit leads to terms such as (58), which are *not* normal-ordered with respect to rest-frame conditions, but are indeed normal-ordered with respect to the highest-weight conditions of [26]. We stress that this difference is a genuine quantum higher-spin effect: it is not apparent at the classical level, and it does not appear in pure gravity either.

The difference between ultrarelativistic and Galilean limits also emphasizes their physical distinction: the ultrarelativistic limit is adapted to gravity, and more generally to models of fundamental interactions, where unitarity is a crucial requirement. In particular, flat space holography (at least in the framework of Einstein gravity) should rely on the unitary construction of [19, 20], as confirmed in [22, 35] by the matching between group characters and one-loop partition functions. By contrast, the Galilean viewpoint is suited to condensed matter applications and, more generally, to situations where unitarity need not hold — as was indeed argued in [25].

Acknowledgments

The authors are grateful to G. Barnich for illuminating discussions on BMS representations. A.C. would like to thank the Nanyang Technological University in Singapore for hospitality, and the participants in the NTU Workshop on Higher-Spin Gauge Theories for feedback on a presentation of some of the results contained in this paper. The work of A.C. and H.A.G. is partially supported by the ERC Advanced Grant "SyDuGraM", by FNRS-Belgium (convention FRFC PDR T.1025.14 and convention IISN 4.4503.15) and by the "Communauté Française de Belgique" through the ARC program. The work of B.O. is supported by a doctoral fellowship of the Wiener-Anspach Foundation and by the Fund for Scientific Research-FNRS under grant number FC-95570. This research of M.R. is supported by the FWF projects P27182-N27 and the START project Y435-N16. M.R. is also supported by a *DOC* fellowship of the Austrian Academy of Sciences, the *Doktoratskolleg Particles and Interactions* (FWF project DKW1252-N27) and the FWF project I 1030-N27.

Appendix A. Superrotations and supermomenta

In this appendix we revisit the arguments surrounding (35) according to which supermomenta transform under superrotations as Virasoro coadjoint vectors. We also display the exact relation between a vector field $\omega(\varphi)\partial_\varphi$ representing an infinitesimal superrotation, and its exponential.

A.1. Action of superrotations on supermomenta

We consider a state $|p(\varphi), s\rangle$ and act on it with a superrotation f using a unitary operator $U[f]$. (The discussion below (31) is recovered upon taking $|p(\varphi), s\rangle = |M, s\rangle$ and $f = \exp[\omega]$, but our argument will be independent of these details.) We wish to show that the state $U[f]|p(\varphi), s\rangle$ is an eigenstate of supermomentum whose eigenvalues are the Fourier modes of the supermomentum $f \cdot p$, where the dot denotes the coadjoint representation (35).

We start by combining the generators P_m in a "quantum supermomentum"

$$P(\varphi) \equiv \sum_{m\in\mathbb{Z}} P_m e^{im\varphi} \tag{A.1}$$

and act with it on the state $U[f]|p(\varphi), s\rangle$. We find

$$P(\varphi)U[f]|p(\varphi), s\rangle = U[f]\left(U[f]^{-1}P(\varphi)U[f]\right)|p(\varphi), s\rangle\,. \tag{A.2}$$

Now recall how the supermomentum generators P_m are defined: a supertranslation α is represented by a unitary operator

$$U[\alpha] = \exp\left[\frac{i}{2\pi}\int_0^{2\pi} d\varphi\,\alpha(\varphi)P(\varphi)\right] \tag{A.3}$$

so we have

$$P(\varphi) = -2\pi i\,\frac{\delta}{\delta\alpha(\varphi)}\,(U[\alpha])|_{\alpha=0} \tag{A.4}$$

and the definition of P_m follows after taking the mth Fourier mode of this expression. Now, plugging this definition in (A.2) we get

$$\begin{aligned}P(\varphi)U[f]|p(\varphi), s\rangle &= -2\pi i\,\frac{\delta}{\delta\alpha(\varphi)}\,U[f]\left(U[f]^{-1}U[\alpha]U[f]\right)\big|_{\alpha=0}|p(\varphi), s\rangle\\ &= -2\pi i\,\frac{\delta}{\delta\alpha(\varphi)}\,U[f]\left(U[\sigma_{f^{-1}}\alpha]\right)\big|_{\alpha=0}|p(\varphi), s\rangle\end{aligned} \tag{A.5}$$

where σ denotes the adjoint representation of the Virasoro group. In writing this we have used the group operation of BMS_3,

$$(f, \alpha) \cdot (g, \beta) = (f \circ g, \alpha + \sigma_f \beta) . \tag{A.6}$$

Now using (A.3) we find

$$U[\sigma_{f^{-1}}\alpha] = \exp\left[\frac{i}{2\pi}\int_0^{2\pi} d\varphi\,(\sigma_{f^{-1}}\alpha)(\varphi)P(\varphi)\right] = \exp\left[\frac{i}{2\pi}\int_0^{2\pi} d\varphi\,\alpha(\varphi)(\sigma_f^* P)(\varphi)\right] \tag{A.7}$$

where σ^* is the dual representation corresponding to σ and thus coincides with the coadjoint representation of the Virasoro group. Plugging this in (A.5) and taking the functional derivative we finally find

$$P(\varphi)U[f]|p(\varphi), s\rangle = U[f](\sigma_f^* P)(\varphi)|p(\varphi), s\rangle . \tag{A.8}$$

Since $|p(\varphi), s\rangle$ is an eigenstate of supermomentum, it follows that

$$P(\varphi)U[f]|p(\varphi), s\rangle = (\sigma_f^* p)(\varphi)U[f]|p(\varphi), s\rangle \tag{A.9}$$

which was to be proven.

A.2. Exponential superrotations

As mentioned in footnote e, the exponential map of the Virasoro group is not locally surjective. Nevertheless one can consider superrotations given by exponentials of vector fields as in (32). It is then natural to ask what is the explicit form of the diffeomorphism $f = \exp[\omega]$ generated by a vector field $\omega = \omega(\varphi)\partial_\varphi$.

As mentioned below (35) the relation between f and ω is given by the flow of the latter. Explicitly, ω defines a family of integral curves, which are paths $\varphi(t)$ on the circle that satisfy the differential equation

$$\frac{d\varphi(t)}{dt} = \omega(\varphi(t)). \tag{A.10}$$

Given an initial condition $\varphi(0) = \varphi_0$, the solution of this equation is unique; this defines a one-parameter family of diffeomorphisms f_t such that

$$f_t(\varphi_0) = \varphi(t) \tag{A.11}$$

and known as the flow of ω. In particular, by definition of the exponential map, the diffeomorphism $f = \exp[\omega]$ generated by ω is just $f = f_1$, that is, the flow evaluated at time $t = 1$. Using the evolution equation (A.10) this can also be restated as the condition

$$\int_{\varphi_0}^{f(\varphi_0)} \frac{d\varphi}{\omega(\varphi)} = 1, \tag{A.12}$$

required to hold for any φ_0 on the circle. This is the exact correspondence between a superrotation generator ω and its exponential $f = \exp[\omega]$. Note that (A.12) automatically ensures $f(\varphi_0 + 2\pi) = f(\varphi_0) + 2\pi$, in accordance with the fact that f is a diffeomorphism of the circle.

References

[1] H. Bondi, M. G. J. van der Burg and A. W. K. Metzner, Gravitational waves in general relativity. 7. Waves from axisymmetric isolated systems, *Proc. Roy. Soc. Lond.* **A269**, pp. 21–52 (1962), doi:10.1098/rspa.1962.0161.

[2] R. K. Sachs, Gravitational waves in general relativity. 8. Waves in asymptotically flat space-times, *Proc. Roy. Soc. Lond.* **A270**, pp. 103–126 (1962), doi:10.1098/rspa.1962.0206.

[3] A. Ashtekar, J. Bicak and B. G. Schmidt, Asymptotic structure of symmetry reduced general relativity, *Phys. Rev.* **D55**, pp. 669–686 (1997), doi:10.1103/PhysRevD.55.669, `arXiv:gr-qc/9608042 [gr-qc]`.

[4] G. Barnich and G. Compere, Classical central extension for asymptotic symmetries at null infinity in three spacetime dimensions, *Class. Quant. Grav.* **24**, pp. F15–F23 (2007), doi:10.1088/0264-9381/24/5/F01,10.1088/0264-9381/24/11/C01, `arXiv:gr-qc/0610130 [gr-qc]`.

[5] G. Barnich and C. Troessaert, Symmetries of asymptotically flat 4 dimensional spacetimes at null infinity revisited, *Phys. Rev. Lett.* **105**, p. 111103 (2010a), doi:10.1103/PhysRevLett.105.111103, `arXiv:0909.2617 [gr-qc]`.

[6] G. Barnich and C. Troessaert, Aspects of the BMS/CFT correspondence, *JHEP* **05**, p. 062 (2010b), doi:10.1007/JHEP05(2010)062, `arXiv:1001.1541 [hep-th]`.

[7] A. Strominger, On BMS Invariance of Gravitational Scattering, *JHEP* **07**, p. 152 (2014), doi:10.1007/JHEP07(2014)152, `arXiv:1312.2229 [hep-th]`.

[8] S. W. Hawking, M. J. Perry and A. Strominger, Soft Hair on Black Holes, (2016), `arXiv:1601.00921 [hep-th]`.

[9] J. D. Brown and M. Henneaux, Central Charges in the Canonical Realization of Asymptotic Symmetries: An Example from Three-Dimensional Gravity, *Commun. Math. Phys.* **104**, pp. 207–226 (1986), doi:10.1007/BF01211590.

[10] A. Achucarro and P. K. Townsend, A Chern-Simons Action for Three-Dimensional anti-De Sitter Supergravity Theories, *Phys. Lett.* **B180**, p. 89 (1986), doi:10.1016/0370-2693(86)90140-1.

[11] E. Witten, (2+1)-Dimensional Gravity as an Exactly Soluble System, *Nucl. Phys.* **B311**, p. 46 (1988), doi:10.1016/0550-3213(88)90143-5.

[12] M. P. Blencowe, A Consistent Interacting Massless Higher Spin Field Theory in $D = (2+1)$, *Class. Quant. Grav.* **6**, p. 443 (1989), doi:10.1088/0264-9381/6/4/005.

[13] A. Campoleoni, Higher Spins in D = 2 + 1, *Subnucl. Ser.* **49**, pp. 385–396 (2013), doi:10.1142/9789814522519_0020, `arXiv:1110.5841 [hep-th]`.

[14] H. Afshar, A. Bagchi, R. Fareghbal, D. Grumiller and J. Rosseel, Spin-3 Gravity in Three-Dimensional Flat Space, *Phys. Rev. Lett.* **111**, 12, p. 121603 (2013), doi:10.1103/PhysRevLett.111.121603, `arXiv:1307.4768 [hep-th]`.

[15] H. A. Gonzalez, J. Matulich, M. Pino and R. Troncoso, Asymptotically flat spacetimes in three-dimensional higher spin gravity, *JHEP* **09**, p. 016 (2013), doi:10.1007/JHEP09(2013)016, `arXiv:1307.5651 [hep-th]`.

[16] X. Bekaert, N. Boulanger and P. Sundell, How higher-spin gravity surpasses the spin two barrier: no-go theorems versus yes-go examples, *Rev. Mod. Phys.* **84**, pp. 987–1009 (2012), doi:10.1103/RevModPhys.84.987, `arXiv:1007.0435 [hep-th]`.

[17] M. Henneaux and S.-J. Rey, Nonlinear $W_{infinity}$ as Asymptotic Symmetry of Three-Dimensional Higher Spin Anti-de Sitter Gravity, *JHEP* **12**, p. 007 (2010), doi:10.1007/JHEP12(2010)007, `arXiv:1008.4579 [hep-th]`.

[18] A. Campoleoni, S. Fredenhagen, S. Pfenninger and S. Theisen, Asymptotic symmetries of three-dimensional gravity coupled to higher-spin fields, *JHEP* **11**, p. 007

(2010), doi:10.1007/JHEP11(2010)007, `arXiv:1008.4744 [hep-th]`.

[19] G. Barnich and B. Oblak, Notes on the BMS group in three dimensions: I. Induced representations, *JHEP* **06**, p. 129 (2014), doi:10.1007/JHEP06(2014)129, `arXiv:1403.5803 [hep-th]`.

[20] G. Barnich and B. Oblak, Notes on the BMS group in three dimensions: II. Coadjoint representation, *JHEP* **03**, p. 033 (2015), doi:10.1007/JHEP03(2015)033, `arXiv:1502.00010 [hep-th]`.

[21] E. Wigner, On unitary representations of the inhomogeneous lorentz group, *Annals of Mathematics* **40**, 1, pp. 149–204 (1939), `http://www.jstor.org/stable/1968551`.

[22] A. Campoleoni, H. A. Gonzalez, B. Oblak and M. Riegler, Rotating Higher Spin Partition Functions and Extended BMS Symmetries, (2015), `arXiv:1512.03353 [hep-th]`.

[23] P. J. McCarthy, Representations of the Bondi-Metzner-Sachs group. I. Determination of the representations, *Proc. R. Soc. Lond. A* **330**, 1583, pp. 517–535 (1972), doi: 10.1098/rspa.1972.0157.

[24] P. J. McCarthy, Representations of the Bondi-Metzner-Sachs group. II. Properties and classification of the representations, *Proc. R. Soc. Lond. A* **333**, 1594, pp. 317–336 (1973), doi:10.1098/rspa.1973.0065.

[25] A. Bagchi, R. Gopakumar, I. Mandal and A. Miwa, GCA in 2d, *JHEP* **08**, p. 004 (2010), doi:10.1007/JHEP08(2010)004, `arXiv:0912.1090 [hep-th]`.

[26] D. Grumiller, M. Riegler and J. Rosseel, Unitarity in three-dimensional flat space higher spin theories, *JHEP* **07**, p. 015 (2014), doi:10.1007/JHEP07(2014)015, `arXiv:1403.5297 [hep-th]`.

[27] B. Binegar, Relativistic Field Theories in Three-dimensions, *J. Math. Phys.* **23**, pp. 1511–1517 (1982), doi:10.1063/1.525524.

[28] S. Weinberg, *The Quantum Theory of Fields*, v. 1. Cambridge University Press (1995), ISBN 9780521550017.

[29] M. Schottenloher, *A Mathematical Introduction to Conformal Field Theory*, Lecture Notes in Physics. Springer Berlin Heidelberg (2008), ISBN 9783540686255.

[30] B. Feigin and D. Fuks, Casimir-operators in modules over virasoro algebra, *DOKLADY AKADEMII NAUK SSSR* **269**, 5, pp. 1057–1060 (1983).

[31] L. Guieu and C. Roger, *L'algèbre et le groupe de Virasoro*. Publications du CRM, Université de Montréal (2007).

[32] E. Shavgulidze, An example of a measure quasi-invariant under the action of the diffeomorphism group of the circle, *Funkts. Anal. Prilozh.* **12**, 3, pp. 55–60 (1978).

[33] A. Maloney and E. Witten, Quantum Gravity Partition Functions in Three Dimensions, *JHEP* **02**, p. 029 (2010), doi:10.1007/JHEP02(2010)029, `arXiv:0712.0155 [hep-th]`.

[34] J. Gomis, J. Herrero, K. Kamimura and J. Roca, Finite W(3) transformations in a multitime approach, *Phys. Lett.* **B339**, pp. 59–64 (1994), doi:10.1016/0370-2693(94)91132-0, `arXiv:hep-th/9409024 [hep-th]`.

[35] G. Barnich, H. A. Gonzalez, A. Maloney and B. Oblak, One-loop partition function of three-dimensional flat gravity, *JHEP* **04**, p. 178 (2015), doi:10.1007/JHEP04(2015)178, `arXiv:1502.06185 [hep-th]`.

Action Principles for Higher and Fractional Spin Gravities

Cesar Arias[(1)], Roberto Bonezzi[(2)], Nicolas Boulanger[(2)*], Ergin Sezgin[(3)], Per Sundell[(1)], Alexander Torres-Gomez[(1)] and Mauricio Valenzuela[(4)]

[(1)] *Departamento de Ciencias Físicas, Universidad Andres Bello, Republica 220, Santiago, Chile*
ce.arias@uandresbello.edu, per.anders.sundell@gmail.com, alexander.torres.gomez@gmail.com

[(2)] *Group of Mechanics and Gravitation, Theoretical and Mathematical Physics, University of Mons – UMONS, 20 place du Parc, 7000 Mons, Belgium*
roberto.bonezzi@umons.ac.be, nicolas.boulanger@umons.ac.be

[(3)] *George and Cynthia Woods Mitchell Institute for Fundamental Physics and Astronomy*
Texas A& M University, College Station, TX 77843, USA
sezgin@tamu.edu

[(4)] *Facultad de Ingeniería y Tecnología, Universidad San Sebastián, General Lagos 1163, Valdivia 5 110693, Chile*
valenzuela.u@gmail.com

We review various off-shell formulations for interacting higher-spin systems in dimensions 3 and 4. Associated with higher-spin systems in spacetime dimension 4 is a Chern–Simons action for a superconnection taking its values in a direct product of an infinite-dimensional algebra of oscillators and a Frobenius algebra. A crucial ingredient of the model is that it elevates the rigid closed and central two-form of Vasiliev's theory to a dynamical 2-form and doubles the higher-spin algebra, thereby considerably reducing the number of possible higher spin invariants and giving a nonzero effective functional on-shell. The two action principles we give for higher-spin systems in 3D are based on Chern–Simons and BF models. In the first case, the theory we give unifies higher-spin gauge fields with fractional-spin fields and an internal sector. In particular, Newton's constant is related to the coupling constant of the internal sector. In the second case, the BF action we review gives the fully nonlinear Prokushkin–Vasiliev, bosonic equations for matter-coupled higher spins in 3D. We present the truncation to a single, real matter field relevant in the Gaberdiel–Gopakumar holographic duality. The link between the various actions we present is the fact that they all borrow ingredients from Topological Field Theory. It has bee conjectured that there is an underlying

*Associate Researcher of the F.R.S-FNRS (Belgium).

and unifying 2-dimensional first-quantised description of the previous higher-spin models in 3D and 4D, in the form of a Cattaneo–Felder-like topological action containing fermionic fields.

1. Introduction

For our contribution to the proceedings of the International Workshop on Higher Spin Gauge Theories that took place in Singapore on 5–7 November 2015, we review a number of results presented in Refs. 1–4 where off-shell formulations of nonlinear higher-spin systems have been found that describe interacting higher-spin fields in spacetime dimensions 3 and 4.

In Section 2 we first review the work[3] giving a Frobenius–Chern–Simons model for nonlinear higher-spin theory in 4D. This model makes use of ingredients provided in the geometrical formulation of higher spin gravity[5] and the action principle proposed in Ref. 6. The important new properties are the introduction of a dynamical 2-form and the attendant phenomenon of the higher spin algebra doubling. This leads to a more predictive power, as it restricts the possible higher spin invariant functionals.

In Section 3 based on Ref. 2, we review the construction of 3D higher-spin (HS) models coupled to an internal $U(\infty) \otimes U(\infty)$ sector and fractional-spin fields. The latter fields generalise the gravitini and our model can be seen as an extension of the Achucaro–Townsend Chern–Simons supergravity.[7] In particular, the couping constant in the $U(\infty) \otimes U(\infty)$ internal sector is proportional to the gravitational Newton constant.

In Section 4 based on Ref. 4, we review the construction of an action that reproduces the fully nonlinear and bosonic Prokushkin–Vasiliev equations.[8] The action is shown to restrict to the Blencowe action[9,10] thereby reproducing the standard kinetic terms in the higher-spin sector of the action.[a] We also review from Ref. 4 the unfolded nonlinear equations for 3D HS fields coupled to a single real scalar field in the matter sector, as this is relevant in the context of the Gaberdiel–Gopakumar duality.[12,13]

Finally, in Section 5 we review the works[1,14] where a topological open-string model of the Cattaneo–Felder type[15] was proposed as an underlying first-quantised model for higher-spin gauge theory.

2. Frobenius-Chern-Simons Action for 4D Higher Spin Gravity

An outstanding problem in higher spin gravity in four (and higher) dimensions[16–18] is to find an action principle with desirable properties. Treating this problem as nonlinear completion of Fronsdal kinetic terms in a Noether procedure approach runs into considerable technical difficulties. Indeed, in the metric-like[19] and the related frame-like[20–22] approaches, long term efforts — see for example Refs. 23, 24

[a]For asymptotic symmetries of three-dimensional gravity coupled to higher-spin field, on top of Ref. 10, see also Ref. 11.

and Refs. 25–33 in the case of *AdS* background — has so far led to primarily cubic interactions. Beyond the cubic order, the fact that higher spin gravity has a mass scale set by the bare cosmological constant while nonabelian higher spin symmetries require higher derivative vertices, lead to intractable abelian vertices built from curvatures and their higher derivatives; see Ref. 29 and the review Ref. 34.

The Noether procedure approach does not exploit the fact that Vasiliev's equations[16–18] provide a fully non-linear description of higher spin gravity on-shell. Furthermore, it is background dependent procedure since it is based on perturbation around AdS_4. Both drawbacks can be avoided by considering covariant Hamiltonian actions from which the background independent full Vasiliev equations follow. These equations are Cartan integrable systems of differential forms on special noncommutative manifolds taking their values in associative higher spin algebras. Treating these forms as the fundamental fields following the AKSZ approach,[35] one is led to a path integral formulation[36] based on covariant Hamiltonian actions[6] on noncommutative manifolds with boundaries. The importance of the boundaries, which are absent in the related proposals,[37,38] is that they facilitate the deformation of the bulk action by boundary terms[5] that contribute to the action but not its variation on-shell. The resulting higher spin amplitudes reproduce desired holographic correlation functions,[39–41] which are suggestive of an underlying topological open string.[1,42] However, the presence of a large number of free parameters impede the predictive power of the model.

A more predictive model has been proposed in Ref. 3 in which the closed and central holomorphic two form in Vasiliev equations is elevated to a dynamical two-form master field, and the higher spin algebra is necessarily extended to includ new master one-form field. The model also employs an eight dimensional Frobenius algebra and provides an action which takes the form of a Chern–Simons term for a superconnection that accommodates all the master fields. For this reason the model is referred to as the Frobenius–Chern–Simons (FCS) gauge theory. The construction of the model has been described in detail in Ref. 3. Here we shall summarize its salient features.

2.1. *Base manifold*

The model is formulated in terms of differential forms on the direct product space $\mathcal{M}_9 = \mathcal{X}_5 \times \mathcal{Z}_4$, where $\mathcal{X}_5$ is a five-dimensional commutative manifold with boundary $\mathcal{X}_4 = \partial\mathcal{X}_5$, containing the original spacetime manifold $\mathcal{M}_4$ as a possibly open subset, and $\mathcal{Z}_4$ is a four-dimensional noncommutative space without boundary. Thus, $\partial\mathcal{M}_9 = \mathcal{X}_4 \times \mathcal{Z}_4$, where $\mathcal{X}_5 = \mathcal{X}_4 \times [0, \infty[$.

The topology of $\mathcal{Z}_4$ may be chosen in a variety ways with nontrivial and interesting consequences to be investigated. In Ref. 3, $\mathcal{Z}_4$ is obtained from the standard noncommutative $\mathbb{C}^4$ by choosing a real form and a compatible convolution formula for the star product and then adding points at infinity to create a compact noncommutative space that can be used to define a (graded cyclic) trace operation. Moreover, $\mathcal{Z}_4$ is taken to be closed to avoid boundary terms, and that its structure admits a certain closed two-form and a global $SL(2;\mathbb{C})$ symmetry, in order to make contact with Vasiliev's theory. To this end, one introduces canonical coordinates $(z^\alpha, \bar{z}^{\dot\alpha})$ $(\alpha, \dot\alpha = 1,2)$ and anti-commuting differentials $(dz^\alpha, d\bar{z}^{\dot\alpha})$ on $\mathbb{C}^4$. We then consider a formally defined associative star product algebra given by the space $\Omega(\mathbb{C}^4)$ of differential forms equipped with two associative composition rules, namely the standard graded commutative wedge product rule, denoted by juxtaposition, and the graded noncommutative rule

$$f \star g = f \exp\left(-i(\overleftarrow{\partial}^\alpha \overrightarrow{\partial}_\alpha + \overleftarrow{\bar\partial}^{\dot\alpha} \overrightarrow{\bar\partial}_{\dot\alpha})\right) g \,, \tag{1}$$

The star product is thus the representation using Weyl ordering symbols of the associative algebra of composite operators built from anti-commuting line elements and noncommutative coordinates with canonical commutation rules

$$[z^\alpha, z^\beta]_\star = -2i\epsilon^{\alpha\beta} \,, \qquad [z^\alpha, \bar{z}^{\dot\alpha}]_\star = 0 \,, \qquad [\bar{z}^{\dot\alpha}, \bar{z}^{\dot\beta}]_\star = -2i\epsilon^{\dot\alpha\dot\beta} \,. \tag{2}$$

In models with four-dimensional Lorentz symmetry, it is natural to select real forms on the real slice

$$\mathcal{R}_4^{\mathbb{C}} = \{(z^\alpha, \bar{z}^{\dot\alpha}) : (z^\alpha)^\dagger = -\bar{z}^{\dot\alpha} \,, \quad (\bar{z}^{\dot\alpha})^\dagger = -z^\alpha\} \cong \mathbb{C}^2 \times \overline{\mathbb{C}^2} \subset \mathbb{C}^4 \,, \tag{3}$$

on which z^α is thus a complex doublet.

In order to include Gaussian elements and distributions, it is useful to first introduce auxiliary integral representations of the star product (1) as follows

$$f \star g = \int_{\mathcal{R}_4^{\mathbb{R}}} \frac{d^2\xi d^2\tilde\xi}{(2\pi)^2} \int_{\mathcal{R}_4^{\mathbb{R}}} \frac{d^2\eta d^2\tilde\eta}{(2\pi)^2} e^{i(\eta^\alpha \xi_\alpha + \tilde\eta^{\dot\alpha} \tilde\xi_{\dot\alpha})} f(z+\xi, \bar{z}+\tilde\xi; dz, d\bar{z}) g(z-\eta, \bar{z}-\tilde\eta; dz, d\bar{z}), \tag{4}$$

where the integration domain is chosen conveniently as the real

$$\mathcal{R}_4^{\mathbb{R}} = \left\{(\xi^\alpha, \tilde\xi^{\dot\alpha}) : \xi^\alpha, \tilde\xi^{\dot\alpha} \in \mathbb{R}^2\right\} \cong \mathbb{R}^2 \times \mathbb{R}^2 \tag{5}$$

The graded cyclic trace operation on $\Omega(\mathcal{Z}_4)$ is defined as

$$\mathrm{STr}_{\Omega(\mathcal{Z}_4)} f = \int_{\mathcal{R}_4^{\mathbb{R}}} f \,, \tag{6}$$

which projects onto the top form in f. Thus, if f is a top form in $\Omega(\mathcal{Z}_4)$ then its representative in $\Omega(\mathcal{R}_4^{\mathbb{R}})$ must fall off sufficiently fast at infinity for the integral to be convergent. Therefore, $\mathcal{Z}_4$ must be a compact manifold obtained by adding points to $\mathcal{R}_4^{\mathbb{R}}$ at infinity to extend its differential Poisson algebra structure[43–46] (see also[1,14]). This can be achieved by assuming that $\mathcal{Z}_4$ admits a Poisson structure and a compatible pre-connection. The latter is assumed to be trivial for simplicity. In addition, $\mathcal{Z}_4$ is required to be closed, such that $\mathrm{STr}_{\Omega(\mathcal{Z}_4)}\, df = 0$, in order to avoid boundary terms from $\mathcal{Z}_4$ in varying the FCS action. Assuming that f and g are two smooth symbols that fall off sufficiently fast, it follows from (4) that

$$\mathrm{STr}_{\Omega(\mathcal{Z}_4)} f \star g = \int_{\mathcal{R}_4^{\mathbb{R}}} f \star g = \int_{\mathcal{R}_4^{\mathbb{R}}} fg \,, \tag{7}$$

which is graded cyclic. As this property will be useful in analyzing boundary conditions in the FCS model arising from the variational principle, we shall assume that (7) holds for all elements in $\Omega(\mathcal{Z}_4)$ including distributions. Finally, in order to obtain Vasiliev's equations from the FCS action, it is assumed that $\Omega(\mathcal{Z}_4)$ admits global $SL(2;\mathbb{C})$ symmetry and contains the (globally defined) closed two-forms

$$j_z = -\frac{i}{4} dz^\alpha dz_\alpha \kappa_z \ , \qquad \bar{j}_{\bar{z}} = (j_z)^\dagger \ , \tag{8}$$

where the inner Klein operator

$$\kappa_z = 2\pi \delta^2(z^\alpha) \ . \tag{9}$$

A choice of topology that satisfies all of the requirements as stated above iis given by[b]

$$\Omega(\mathcal{Z}_4) = \bigoplus_{m,\bar{m}=0,1} (\Omega(S^2) \star (j_z)^{\star m}) \otimes (\Omega(S^2) \star (\bar{j}_{\bar{z}})^{\star \bar{m}}) \ , \tag{10}$$

where $\Omega(S^2)$ consists of globally defined forms on S^2 with Poisson structure obtained by extending the Poisson structure of (1) to the point at ∞. At this point, the resulting Poisson bivector and all its derivatives vanish. Hence, provided it is possible to exchange the order of differentiation and summation in (1) and using the fact that increasing number of derivatives of a form that falls off yields forms that fall off even faster, it follows that if $f, g \in \Omega(S^2)$ then $(f \star g)|_\infty = f|_\infty g|_\infty$ i.e. the point at infinity is a commuting point of $\Omega(S^2)$. In other words, one is working with a topological two-sphere equipped with a differential Poisson algebra with trivial pre-connection. Moreover, in order for the elements in $\Omega(\mathcal{Z}_4)$ to have finite traces, it is assumed that the top forms on each two-sphere fall off sufficiently fast at infinity working in the original $\mathbb{R}^2 \times \mathbb{R}^2$ coordinate chart. For this fall-off condition to be embeddable in a differential star product algebra, also the forms in lower degrees must fall off appropriately at infinity. In particular, the only forms that can have finite values at infinity are the zero-forms. Thus, in effect, one has

$$\mathcal{Z}_4 = S^2 \times S^2 \ , \tag{11}$$

by assuming boundary conditions at the commuting points at infinity and allowing for delta function distributions at the origins so as to create a space of forms that is closed under exterior differentiation and star products, and has a space of top forms with finite traces that vanish for exact elements and obey (7). For this to hold true it is important that the delta function κ_z always appears together with line elements in the combination j_z given in (8), which obeys $j_z \star j_z = 0$, whereas the inclusion of κ_z into the algebra would require the inclusion of $j_z \star \kappa_z$ as well which is not integrable.

[b]The manifolds S^4 and $S^3 \times S^1$ do not admit j, while $T^2 \times T^2$ breaks global $SL(2;\mathbb{C})$ symmetry.

2.2. *Superconnection and Frobenius-Chern-Simons action*

The construction of the FCS action employs an eight-dimensional, 3-graded Frobenius algebra

$$\mathcal{F} = \mathcal{F}^{(-1)} \oplus \mathcal{F}^{(0)} \oplus \mathcal{F}^{(+1)} , \qquad (e_{ij}, h\, e_{ij}) \in \mathcal{F}^{(i-j)} , \tag{12}$$

where e_{ij} is the 2×2 matrix whose only non-vanishing entry is a 1 at the ith row and jth column, and h is a Klein element satisfying $[h, e_{11}] = 0 = [h, e_{22}]$, $\{h, e_{12}\} = 0 = \{h, e_{12}\}$. The coordinate-like master fields are assembled into

$$X = \sum_{i,j} X^{ij} e_{ij} = \begin{pmatrix} A & B \\ \widetilde{B} & \widetilde{A} \end{pmatrix} , \tag{13}$$

and the momentum-like master fields, which will play the role of Lagrange multipliers, into

$$P = \sum_{i,j} P^{ij} e_{ij} = \begin{pmatrix} V & U \\ \widetilde{U} & \widetilde{V} \end{pmatrix} . \tag{14}$$

The above master fields are duality extended in the sense that they are formal sums of forms with degrees

$$\begin{aligned} \deg(B, A, \widetilde{A}, \widetilde{B}) &= \{(2n, 1+2n, 1+2n, 2+2n)\}_{n=0,1,2,3} , \\ \deg(\widetilde{U}, V, \widetilde{V}, U) &= \{(8-2n, 7-2n, 7-2n, 6-2n)\}_{n=0,1,2,3} . \end{aligned} \tag{15}$$

One then proceeds by defining a superconnection and superdifferential[47] as

$$Z = hX + P , \tag{16}$$

respectively, which are thus objects with odd total degree given by form degree plus Frobenius 3-degree. The space $\Omega(\mathcal{M}_9)$ of differential forms is equipped with two associative composition rules, namely the standard graded commutative wedge product rule, denoted by juxtaposition, and Weyl ordered star products of functions. The need for Weyl ordering is due to the property (7) that is crucial for the boundary conditions to make sense in a noncommutative set up. Once one has put any star-product expression in its factorized form in Y and Z, say $F(Y) \star G(Z)$ equal to $F(Y)G(Z)$ in Weyl order, one makes assumptions about the functional classes to which $F(Y)$ and $G(Z)$ belong. See Section 3.7 of[48] for details.

For reasons explained in detail in,[3] the action function is expressed in terms of globally defined configurations in

$$\mathcal{E} = \Omega(\mathcal{X}_5) \otimes \frac{1}{2}(1+\pi\bar{\pi}) \left[\Omega(\mathcal{Z}_4) \otimes \mathcal{A} \otimes \tfrac{1}{2}(1 + k \star \bar{k})\right] , \tag{17}$$

where

$$\mathcal{A} = \mathcal{F} \otimes \mathcal{W}_0 \otimes \mathcal{K} . \tag{18}$$

Furthermore, $\mathcal{W}_0$ is an extended Weyl algebra

$$\mathcal{W}_0 = \bigoplus_{r,\bar{r}=0,1} \mathrm{Aq}(2) \star (\kappa_y)^{\star r} \star (\bar{\kappa}_{\bar{y}})^{\star \bar{r}} \tag{19}$$

where $Aq(2)$ consists of star polynomials in two complex doublets $(y^\alpha, \bar{y}^{\dot\alpha})$, $\alpha, \dot\alpha = 1, 2$, obeying the oscillator algebra

$$[y^\alpha, y^\beta]_\star = 2i\epsilon^{\alpha\beta} \ , \quad [y^\alpha, \bar{y}^{\dot\beta}]_\star = 0 \ , \quad [\bar{y}^{\dot\alpha}, \bar{y}^{\dot\beta}]_\star = 2i\epsilon^{\dot\alpha\dot\beta} \ . \tag{20}$$

The inner Klein operators in Weyl order are defined as

$$\kappa_y := 2\pi\delta^2(y) \ , \qquad \bar\kappa_{\bar y} := 2\pi\delta^2(\bar y) \ , \tag{21}$$

so that $P \in \mathcal{W}_0$ obey

$$\kappa_y \star P \star \kappa_y = \pi_y(P) \ , \qquad \bar\kappa_{\bar y} \star P \star \bar\kappa_{\bar y} = \bar\pi_{\bar y}(P) \ , \tag{22}$$

where π_y and $\bar\pi_{\bar y}$ are inner automorphisms whose action in Weyl order is given by

$$\pi_y(y) = -y \ , \qquad \bar\pi_{\bar y}(\bar y) = -\bar y \ , \tag{23}$$

leaving intact all the Klein operators. Thus, the generic elements of $\mathcal{W}_0$ is of the form

$$P = \sum_{r,\bar r=0,1} P^{r,\bar r}_{\alpha(n),\dot\alpha(\bar n)} (\kappa_y)^{\star r} \star (\bar\kappa_{\bar y})^{\star \bar r} y^{(\alpha_1} \star \cdots \star y^{\alpha_n)} \star \bar y^{(\dot\alpha_1} \star \cdots \star \bar y^{\dot\alpha_{\bar n})} \ . \tag{24}$$

Turning to $\mathcal{K}$, it is the collection of outer Klein operators

$$\mathcal{K} = \{1, k, \bar k, k \star \bar k\} \ . \tag{25}$$

where for $f \in \mathcal{E}$, the adjoint action

$$k \star f \star k = \pi(f) \ , \qquad \bar k \star f \star \bar k = \bar\pi(f) \ , \tag{26}$$

and the outer automorphisms $\bar\pi$ and $\bar\pi$ with the only nontrivial actions

$$\pi(y, z) = (-y, -z) \ , \qquad \bar\pi(\bar y, \bar z) = (-\bar y, -\bar z) \ . \tag{27}$$

Employing the ingredients summarized above, the following action has been proposed[3]

$$S = \int_{\mathcal{M}_9} \mathrm{Tr}_{\mathcal{A}} \left(\tfrac{1}{2}\, Z \star qZ + \tfrac{1}{3}\, Z \star Z \star Z \right) - \frac{1}{4} \int_{\partial\mathcal{M}_9} \mathrm{Tr}_{\mathcal{A}} \left[h\pi_h(Z) \star Z \right] \ , \tag{28}$$

where π_h is the automorphism sending h to $-h$, and

$$q := hd \ . \tag{29}$$

Keeping in mind that $f \in \mathcal{E}$, the operation $\mathrm{Tr}_{\mathcal{A}}$ is defined as

$$S = \int_{\mathcal{M}_9} \mathrm{Tr}_{\mathcal{A}}\, f := \int_{\mathcal{X}_5} \mathrm{STr}_{\Omega(\mathcal{Z}_4)}\, \mathrm{Tr}_{\mathcal{F}}\, \mathrm{Tr}_{\mathcal{W}_0}\, \mathrm{Tr}_{\mathcal{K}} f \ , \tag{30}$$

where $\mathrm{STr}_{\Omega(\mathcal{Z}_4)}$ is defined in (6), and the remaining trace operations are defined as

$$\mathrm{Tr}_{\mathcal{F}} \sum_{i,j} e_{ij} M^{ij}(h) = M^{11}(0) + M^{22}(0) \ , \tag{31}$$

$$\mathrm{Tr}_{\mathcal{W}_0} P = P^{1,1}(0,0) \ , \tag{32}$$

$$\mathrm{Tr}_{\mathcal{K}} f = f|_{k=0=\bar k} \ , \tag{33}$$

with $P \in \mathcal{W}_0$ from (24) which furnishes the definition of $P^{1,1}(0,0)$. Defining $\mathrm{Tr}_{\mathcal{E}} f := \int_{\mathcal{M}_9} \mathrm{Tr}_{\mathcal{A}}\, f$, it can be shown that $\mathrm{Tr}_{\mathcal{E}}\, f \star g = \mathrm{Tr}_{\mathcal{E}}\, g \star f$.[3]

The total variation of the FCS action gives

$$\delta S = \int_{\mathcal{M}_9} \mathrm{Tr}_{\mathcal{A}}\, \delta Z \star R + \frac{1}{2} \oint_{\mathcal{M}_9} \mathrm{Tr}_{\mathcal{A}}\, h\, \delta Z \star (Z + \pi_h(Z)) \ , \tag{34}$$

where the Cartan curvature

$$R := qZ + Z \star Z \ . \tag{35}$$

Thus, imposing the boundary condition

$$(Z + \pi_h(Z))\, |_{\partial\mathcal{M}_9} = 0 \ , \tag{36}$$

one has the equation of motion $R = 0$. This equation is Cartan integrable, hence gauge invariant, with transformations

$$\delta Z = q\theta + [Z, \theta]_\star \ , \qquad \delta R = [R, \theta]_\star \ . \tag{37}$$

One the other hand, the requirement of gauge invariance of the action gives the following boundary conditions

$$(\theta - \pi_h(\theta))\, |_{\partial\mathcal{M}_9} = 0 \ . \tag{38}$$

In obtaining this result, the property (7) plays an important role. Setting aside nontrivial flat connections due to the noncommutativity of the base manifold, Z can be given on-shell in terms of a gauge function L (which contains forms in different degrees) and a zero-form integration constant C, viz.

$$Z = L^{\star(-1)} \star (q + C) \star L \ , \qquad qC = C \star C = 0 \ , \tag{39}$$

where the algebraic condition on C is a consequence of the fact that the form content is as given in (15).

The superconnection Z is assumed to be globally defined. However, if it is rather given by a set of representatives defined locally on charts that cover $\mathcal{M}_9$, the appropriate global definition of the action is described in Ref. 3. In doing so, it proves convenient to write the action (28) in terms of master fields (X, P) defined via $Z = hX + P$. Thanks to the boundary term in (28), one finds[3]

$$S = \int_{\mathcal{M}_9} \mathrm{Tr}_{\mathcal{A}} \left(P \star F^X + \tfrac{1}{3}\, P \star P \star P \right) \ , \tag{40}$$

where $F^X := dX + hXh \star X$. The general variation of this action (28) reads

$$\delta S = \int_{\mathcal{M}_9} \mathrm{Tr}_{\mathcal{A}} \left(\delta X \star R^P h + \delta P \star R^X + d(\delta X \star P) \right) \ , \tag{41}$$

where the total derivatives cancel between neighboring patches (in the interior of $\mathcal{M}_9$) since δX and P belong to sections. Writing $R = R^X + R^P$ where

$$R^X := F^X + P \star P \ , \qquad R^P := qP + hX \star P + P \star hX \ , \tag{42}$$

then on shell we have $R^X = 0$ and $R^P = 0$, and we are left with

$$\delta S = \int_{\mathcal{M}_9} \mathrm{Tr}_{\mathcal{A}}\, d(\delta X \star P) = \oint_{\partial\mathcal{M}_9} \mathrm{Tr}_{\mathcal{A}}\, \delta X \star P\ . \tag{43}$$

Using crucially the property (7) to replace the star product in $\Omega(\mathcal{Z}_4)$ by the classical product (keeping in mind that $j_z \star j_z = 0$), the variation becomes

$$\delta S = \oint_{\partial\mathcal{M}_9} \mathrm{Tr}_{\mathcal{A}}\, \delta X \star_{\mathcal{A}} P\ . \tag{44}$$

Hence, if X is free to fluctuate at $\partial\mathcal{M}_9$, it follows from the variational principle that

$$P|_{\partial\mathcal{M}_9} = 0\ . \tag{45}$$

Finally, while action is invariant under the gauge transformations with parameters ϵ^X, it transforms into a total derivative under transformations with parameters ϵ^P, viz.

$$\delta_{\epsilon^P} S = \int_{\mathcal{M}_9} \mathrm{Tr}_{\mathcal{A}}\, d\left(\epsilon^P \star R^X\right)\ , \tag{46}$$

that vanishes provided that ϵ^P belongs to the same section as P, and

$$\epsilon^P|_{\partial\mathcal{M}_9} = 0 \tag{47}$$

that is indeed equivalent to (38).

Using $Z = hX + P$, the gauge transformations (37) read

$$\begin{aligned} \delta X =& d\epsilon^X + X \star \epsilon^X - h\epsilon^X h \star X + hPh \star \epsilon^P - \epsilon^P \star P\ , \\ \delta P =& d\epsilon^P + hXh \star \epsilon^P - h\epsilon^P h \star X + [P, \epsilon^X]_\star\ . \end{aligned} \tag{48}$$

2.3. *Component formulation*

The action (28), upon using the definitions (13), (14) and (16), takes the form

$$\begin{aligned} S = \int_{\mathcal{M}_9} \mathrm{Tr}_{\mathcal{W}\otimes\mathcal{K}} \Big[& \widetilde{U} \star DB + V \star \left(F - B \star \widetilde{B} + \tfrac{1}{3} V^{\star 2} + U \star \widetilde{U}\right) \\ & + U \star \widetilde{D}\widetilde{B} + \widetilde{V} \star \left(\widetilde{F} - \widetilde{B} \star B + \tfrac{1}{3} \widetilde{V}^{\star 2} + \widetilde{U} \star U\right) \Big]\ , \end{aligned} \tag{49}$$

where

$$\begin{aligned} F &:= dA + A \star A\ , & \widetilde{F} &:= d\widetilde{A} + \widetilde{A} \star \widetilde{A}\ , \\ DB &:= dB + A \star B - B \star \widetilde{A}\ , & \widetilde{D}\widetilde{B} &:= d\widetilde{B} + \widetilde{A} \star \widetilde{B} - \widetilde{B} \star A\ , \\ DU &:= dU + A \star U - U \star \widetilde{A}\ , & \widetilde{D}\widetilde{U} &:= d\widetilde{U} + \widetilde{A} \star \widetilde{U} - \widetilde{U} \star A\ , \\ DV &:= dV + A \star V + V \star A\ , & \widetilde{D}\widetilde{V} &:= d\widetilde{V} + \widetilde{A} \star \widetilde{V} + \widetilde{V} \star \widetilde{A}\ . \end{aligned} \tag{50}$$

The bulk equations of motion, which amount to vanishing Cartan curvatures, read

$$\begin{aligned} & F - B \star \widetilde{B} + V \star V + U \star \widetilde{U} = 0\ , && DB + V \star U + U \star \widetilde{V} = 0\ , \\ & \widetilde{F} - \widetilde{B} \star B + \widetilde{V} \star \widetilde{V} + \widetilde{U} \star U = 0\ , && \widetilde{D}\widetilde{B} + \widetilde{V} \star \widetilde{U} + \widetilde{U} \star V = 0\ , \\ & DU + B \star \widetilde{V} + V \star B = 0\ , && DV + B \star \widetilde{U} - U \star \widetilde{B} = 0\ , \\ & \widetilde{D}\widetilde{U} + \widetilde{B} \star V + \widetilde{V} \star \widetilde{B} = 0\ , && \widetilde{D}\widetilde{V} + \widetilde{B} \star U - \widetilde{U} \star B = 0\ . \end{aligned} \tag{51}$$

The gauge parameter can be written as $\theta = \epsilon^X + h\epsilon^P$, where

$$\epsilon^X = \begin{pmatrix} \epsilon & \eta \\ \widetilde{\eta} & \widetilde{\epsilon} \end{pmatrix} , \qquad \epsilon^P = \begin{pmatrix} \epsilon^V & \eta^U \\ \eta^{\widetilde{U}} & \epsilon^{\widetilde{V}} \end{pmatrix} . \tag{52}$$

The transformation rules for the component fields can be readily obtained from (48) by using the definitions (13) and (14). Thus, on $\partial\mathcal{M}_9$, where $(U, \widetilde{U}; V, \widetilde{V})$ vanish, one finds

$$\begin{aligned} F - B \star \widetilde{B} &= 0 , \qquad & DB &= 0 , \\ \widetilde{F} - \widetilde{B} \star B &= 0 , \qquad & \widetilde{D}\widetilde{B} &= 0 , \end{aligned} \tag{53}$$

which is the desired modification of Vasiliev's original system.[16,17] Alternatively, going to the basis

$$\widetilde{A} = W + K , \qquad A = W - K , \tag{54}$$

the equations of motion on $\partial\mathcal{M}_9$ read

$$\begin{aligned} &F_W + K \star K - \frac{1}{2}\{B, \widetilde{B}\}_\star = 0 , \qquad D_W K - \frac{1}{2}[\widetilde{B}, B]_\star = 0 , \\ &D_W B - \{K, B\}_\star = 0 , \qquad D_W \widetilde{B} + \{K, \widetilde{B}\}_\star = 0 , \end{aligned} \tag{55}$$

where we have defined $D_W f = df + W \star f - (-1)^{\deg(f)} f \star W$ and $F_W = dW + W^2$. Since $\epsilon^P|_{\partial\mathcal{M}_9} = 0$, recalling the notation (52), and splitting the gauge parameters $(\epsilon, \widetilde{\epsilon})$ as

$$\epsilon = \alpha - \beta , \qquad \widetilde{\epsilon} = \alpha + \beta , \tag{56}$$

the gauge transformations under which the field equations (55) are invariant can be written as

$$\begin{aligned} \delta W &= D_W \alpha + [K, \beta]_\star + \tfrac{1}{2}\{\widetilde{\eta}, B\}_\star + \tfrac{1}{2}\{\widetilde{B}, \eta\}_\star , \\ \delta K &= D_W \beta + [K, \alpha]_\star + \tfrac{1}{2}[\widetilde{\eta}, B]_\star + \tfrac{1}{2}[\widetilde{B}, \eta]_\star , \\ \delta B &= D_W \eta + [B, \alpha]_\star - [K, \eta]_\star + \{B, \beta\}_\star , \\ \delta \widetilde{B} &= D_W \widetilde{\eta} + [\widetilde{B}, \alpha]_\star + [K, \widetilde{\eta}]_\star - \{\widetilde{B}, \beta\}_\star . \end{aligned} \tag{57}$$

2.4. *Comparison with the duality extended Vasiliev system*

In Ref. 3 it has been shown that bay taking $K = 0$ and choosing $\widetilde{B}$ appropriately, the equations of motion on $\partial\mathcal{M}_9$ given in (55) take the form

$$\begin{aligned} &F_W - \mathcal{V} \star J + \overline{\mathcal{V}} \star \overline{J} + \mathcal{U}_0 \star J \star \overline{J} + \mathcal{U}_1 \star J_{[2]} + \mathcal{U}_2 \star J_{[4]} = 0 , \\ &D_W B = 0 , \end{aligned} \tag{58}$$

where

$$J := -\frac{i}{8} dz^\alpha dz_\alpha \, \kappa_z \star \kappa_y \star k \star (1 + k \star \bar{k}) , \qquad \overline{J} := J^\dagger , \tag{59}$$

with j_z defined in (8). Furthermore, $J_{[2]}$ and $J_{[4]}$ are forms that belong to the de Rham cohomology on $\mathcal{X}_4$, and $(\mathcal{V}, \bar{\mathcal{V}}, U_0, U_1, U_2)$ are star polynomial function of the form $f(B) = \sum_{n=0}^{\infty} f_n B^{\star(n+1)}$.

It is interesting to compare this system with Vasiliev's recently proposed extended system,[49] adapted to our notation, given by

$$\begin{aligned} &F_W - \mathcal{V} \star J + \overline{\mathcal{V}} \star \overline{J} + \mathcal{U}_0 \star J \star \overline{J} + gJ \star \overline{J} + \mathcal{L}_{[2]} + \mathcal{L}_{[4]} = 0 \,, \\ &D_W B = 0 \,, \end{aligned} \tag{60}$$

where $\mathcal{L}_{[2]}$ and $\mathcal{L}_{[4]}$ are two new dynamical fields, referred to as Lagrangian forms, given by globally defined central and closed elements of degrees two and four, respectively. As far as the local dynamics is concerned, the two systems are equivalent in form degrees zero and one, since one can always choose a representative for $\mathcal{L}_{[2]}$ that vanishes in a given coordinate chart. In higher form degrees, the duality extended Vasiliev system contains the term $gJ \star \overline{J}$ and the Lagrangian forms, which are not present in the FCS system[c]. In Ref. 49, the integral $\oint \mathcal{L}_{[2]}$ has been interpreted as a black hole charge, as has been substantiated by its evaluation[49] on the Didenko–Vasiliev black hole solution.[50] As for the integral of $\mathcal{L}_{[4]}$ over spacetime, it has been proposed[49] as the generating functional of correlators within the context of holography [d]. An important open problem in this framework is how to account for loop corrections. It has been suggested that the quantum mechanical effects may emerge from classical dynamics in an infinite dimensional space that has enough room to describe all multiparticle states in the system.[51] If true, this would be a drastically new way of looking at quantum gravity. The tests of these proposals remain to be seen.

In the approach of Ref. 3 a path integral formulation of the FCS model is proposed along the same lines as the AKSZ construction of Ref. 36 within the geometric framework of Ref. 5. In this approach, the terms proportional to the closed and central elements in (58), which are similar to the Lagrangian form terms in (60) but play a different role, as the computation of the effective action proceeds in this case by means of path integral quantization rules which necessarily involves the FCS action itself. The advantage of this approach is the availability of path integral formulation for quantization. The computation of quantum effects are left to future work but an outline of the the role of certain topological invariants in the construction of the on-shell effective action is given in Ref. 3, which we summarize below.

2.5. *On-shell actions from topological invariants*

Starting from an AKSZ path integral on $\mathcal{M}_9 = [0, \infty[\times \mathcal{X}_4 \times \mathcal{Z}_4$, where all fields vanish at $\{\infty\} \times \mathcal{X}_4 \times \mathcal{Z}_4$ and in addition $P|_{\{0\}\times\mathcal{X}_4\times\mathcal{Z}_4} = 0$, as required by the

[c] Whether such coupling can be obtained either by expanding B around a constant background value or allowing the dependence of $\widetilde{B}$ on B to contain a simple pole, remains to be seen.

[d] Another proposal for the black hole entropy and generating functional of correlators in higher spin gravity has been made in Ref. 5.

Batalin–Vilkovisky master equation, one finds that $S_{\rm H}$ vanishes on-shell. Following,[5] one may generate an on-shell action by adding to $S_{\rm H}$ a globally defined boundary term $S_{\rm top} = \oint_{\partial\mathcal{M}_9} \mathcal{V}(X, dX)$, whose total variation vanishes off-shell, i.e. $S_{\rm top}$ is a topological invariant. Assuming that $S_{\rm top}$ does not affect the boundary condition on P nor the equations of motion, one may argue that the on-shell action is given by $S_{\rm top}$.

Aspects of topological invariants for a general structure group are discussed in Ref. 3, where it is also shown that taking it to be generated by α-transformation displayed in (57), one has the invariants

$$S_{\rm top}[W, K] = \sum_{p=0}^{2}\sum_{n=1}^{p+2} \oint_{\mathcal{X}_{2p}\times\mathcal{Z}_4} \beta_{n,p}\left(\tfrac{d}{dt}\right) {\rm Tr}_{\mathcal{W}_0\otimes\mathcal{K}} \left(F_{W_t}\right)^{\star n}\Big|_{t=0} . \tag{61}$$

where $\mathcal{X}_{2p} \subset \mathcal{X}_4$ are closed subsets of dimension $2p$ for $p = 0, 1, 2$;

$$W_t = W + tK , \qquad F_{W_t} = F_W + tD_W K + t^2 K \star K , \tag{62}$$

and $\beta_{n,p}$ are linear differential operators of order $(2n-1)$ in d/dt with constant coefficients.Thus, there are $2, 3, 4$ invariants for $p = 0, 1, 2$, respectively. The on-shell value of $S_{\rm top}[W, K]$ is built out of integrals of traces of $B\star\widetilde{B}$, $\widetilde{B}\star B$ and $K\star K$ forming a finite set of invariants. The observables are invariant off shell under gauge transformations with parameter α, and on shell using parameters $(\beta, \eta, \widetilde{\eta})$. In the semi-classical limit, one has the partition function[3]

$$Z_{\rm FCS} = \sum_{\rm saddles} \mathcal{N} e^{iS_{\rm top}} . \tag{63}$$

2.6. *Linearized fluctuations*

The theory on the boundary of $\mathcal{M}_9$ admits the vacuum solutions

$$\widetilde{B}^{(0)} = I , \qquad W^{(0)} = L^{-1}\star dL , \qquad K^{(0)} = 0 , \qquad B^{(0)} = 0 , \tag{64}$$

where L is a gauge function (consisting of forms) and I is a closed and central element on $\partial\mathcal{M}_9$. In particular, to describe Vasiliev's phase of the theory, it is assumeed that

$$I = J_{\mathcal{X}} + e^{i\theta_0} J - e^{-i\theta_0}\overline{J} , \tag{65}$$

where $J_{\mathcal{X}}$ is a closed a central element on $\mathcal{X}_4$, J and $\overline{J}$ are the closed and central elements on $\mathcal{Z}_4$ defined in (59), and θ_0 is an arbitrary real constant. The fluctuations in the boundary fields can be expanded as

$$(W - W^{(0)}, B, K, \widetilde{B} - \widetilde{B}^{(0)}) = \sum_{n\geqslant 1}(W^{(n)}, B^{(n)}, K^{(n)}, \widetilde{B}^{(n)}) . \tag{66}$$

At the first order, the equations of motion (55) read

$$\begin{aligned} &D^{(0)}W^{(1)} - \tfrac{1}{2}\{I, B^{(1)}\}_\star = 0 , \qquad D^{(0)}K^{(1)} = 0 , \\ &D^{(0)}B^{(1)} = 0 , \qquad D^{(0)}\widetilde{B}^{(1)} + \{I, K^{(1)}\}_\star = 0 , \end{aligned} \tag{67}$$

and the abelian gauge transformations following from (57) are given by

$$\delta W^{(1)} = D^{(0)}\alpha^{(1)} + \tfrac{1}{2}\{I,\eta^{(1)}\}_\star\ , \qquad \delta B^{(1)} = D^{(0)}\eta^{(1)}\ , \tag{68}$$

$$\delta K^{(1)} = D^{(0)}\beta^{(1)}\ , \qquad \delta \widetilde{B}^{(1)} = D^{(0)}\widetilde{\eta}^{(1)} - \{I,\beta^{(1)}\}_\star\ . \tag{69}$$

Expressing the first order fluctuations as

$$(W^{(1)}, B^{(1)}, K^{(1)}, \widetilde{B}^{(1)}) = L^{-1}\star(W^{(1)\prime}, B^{(1)\prime}, K^{(1)\prime}, \widetilde{B}^{(1)\prime})\star L\ , \tag{70}$$

where L is a gauge function and the primed fields are independent of x, the linearized equations (67) are solved by[3]

$$B^{(1)\prime} = B^{(1)\prime}_{[0]} + \mathrm{d}\rho_v B^{(1)\prime}\ , \tag{71}$$

$$W^{(1)\prime} = d\rho_v W^{(1)\prime} - \tfrac{1}{2}(d\rho_v - 1)\left(\{\rho_v I, B^{(1)\prime}_{[0]}\}_\star + \{I,\rho_v B^{(1)\prime}\}_\star\right)\ , \tag{72}$$

$$K^{(1)\prime} = d\rho_v K^{(1)\prime}\ , \tag{73}$$

$$\widetilde{B}^{(1)\prime} = d\rho_v \widetilde{B}^{(1)\prime} + (d\rho_v - 1)\{I,\rho_v K^{(1)\prime}\}_\star\ , \tag{74}$$

where $B^{(1)\prime}_{[0]}$ the zero-form integration constant which harbours the local degrees of freedom of the system and the homotopy contractor ρ_v, with the convenient choice of $v = z^\alpha\partial_\alpha$ is defined by

$$\rho_v f(Z,Y,dZ) = Z^{\underline{\alpha}}\frac{\partial}{\partial dZ^{\underline{\alpha}}}\int_0^1 dt\frac{1}{t}f(tZ,Y,tdZ)\ . \tag{75}$$

The connection $W^{(1)}_{[1]}$ consists of a pure gauge solution, as its gauge function and gauge parameter belong to the same spaces, plus a a set particular solutions that carrying the aforementioned local massless degrees of freedom.

The fields $K^{(1)}$ and $\widetilde{B}^{(1)\prime}$, on the other hand, may introduce new topological degrees of freedom arising in cohomological spaces given by spaces of gauge functions over the spaces of gauge parameters.

In particular, $\widetilde{B}^{(1)\prime}_{[2]}$ contains moduli associated to the gauge function $\rho_v\widetilde{B}^{(1)\prime}_{[2]} = \rho_v(e^{i\theta_0}J - e^{-i\theta_0}\bar{J})$, as $d\rho_v(e^{i\theta_0}J - e^{-i\theta_0}\bar{J}) = e^{i\theta_0}J - e^{-i\theta_0}\bar{J}$ belongs to an admissible section for $\widetilde{B}^{(1)\prime}_{[2]}$ while $\rho_v(e^{i\theta_0}J - e^{-i\theta_0}\bar{J})$ does not belong to an admissible section for $\widetilde{\eta}^{(1)}$. In more detail it is shown in Ref. 3 that the moduli of $\widetilde{B}^{(1)\prime}_{[2]}$ can be associated to modes that blow up at infinity, i.e. at the commutative point of $\mathcal{Z}_4$.

Going to more general backgrounds for $\mathcal{M}_9$, it follows from the fact that the fields K, B and $\widetilde{B}$ belong to sections of the structure group that they can contain topological degrees of freedom provided that there are matching elements in the de Rham cohomology, whose rôle remains to be investigated further. Likewise, going to higher order in perturbation theory, the moduli of $\widetilde{B}$ will generate interaction terms which are expected to have important consequences in the perturbative expansion of the theory and the computation of the correlation functions.

The above linearization suffices to show that the perturbative degrees of freedom of the system are contained in the initial data for the Weyl zero-form. However, in order to obtain the field equations for the Fronsdal fields one has to perform one last step, namely going from Weyl to normal order and change gauge in order to make direct contact with Vasiliev's perturbative expansion. In the latter, one obtains the so-called Central On Mass Shell Theorem (COMST) which states that the first-order equations that arise in the perturbative expansion of Vasiliev equations about anti-de Sitter space are the Fronsdal equations for massless higher spin fields, and it makes crucial use of normal ordering scheme in the definition of the star product. On the other hand, the FCS model employs Weyl ordered product for reasons explained above. Since the physical spectrum content of the FCS model is the same as that of Vasiliev theory, its perturbative expansion should yield the COMST as well. In doing so, the following observations are pertinent. A naive transformation that maps the perturbative solution obtained in the normal ordering scheme to Weyl order is known to be singular.[52e] However, in the FCS model, the master fields belong to an extended class of symbols defined *a priori* in Weyl order and include the inner Klein operators as well so as to produce a well-defined perturbation theory in a specific gauge. Thus, the passage back to the normal ordering scheme involves a further subtlety as it also involves a change of gauge. At the linearized level, and working with definite boundary conditions (corresponding to generalized Type D solutions[48,54]), it possible to map back to the Vasiliev's normal ordering scheme and verify the COMST in a specific gauge.

3. Fractional Spin Gravity Theory

In 2 + 1 dimensions Vasiliev's higher-spin gravity, or more specifically the Prokushkin–Vasiliev model,[8] admits truncation to Chern–Simons higher spin gravity.[9,55] In Ref. 55 the gauge connection is valued in a higher spin algebra that consists of monomials of Wigner-Heisenberg deformed oscillator operators. Monomials of the same order transform in spinor-tensorial representations of the AdS_3 algebra, with arbitrary half-integer spins (which includes integer and half-an-integer values), and the correspondent fields have standard boson or fermion statistics. However, in $2+1$ dimensional spacetimes the representations of the $so(2,1)$ algebra admit spin interpolating half-integer numbers[56,57] and are referred to as fractional. The physical realisations of fractional spins are known as anyons, and their statistics interpolates between bosons and fermions.[58–60]

As higher spin gravity aims at describing fields with arbitrary spin, for completeness, in three dimensions it should be extended to incorporate fundamental fractional-spin fields. The first step to achieve this goal was given in Ref. 61 using operator formalism. Later this was done in Ref. 2 by means of deformation quantization methods — i.e. using star-products.[62]

The model constructed in Ref. 2 is a Chern–Simons theory for a gauge field that

[e] For a general discussion of ordering schemes and the map from one to another, see e.g. Ref. 53.

can be expressed in the form

$$\mathbb{A} = \begin{bmatrix} W & \psi \\ \overline{\psi} & U \end{bmatrix} \cong \begin{bmatrix} \text{HS gravity} & \text{Fractional spin} \\ \text{Fractional spin} & \text{Internal interactions} \end{bmatrix}, \tag{76}$$

where the blocks correspond to four different sectors of the gauge algebra of the theory, $\mathcal{A}(2;\nu|w)$, dubbed fractional-spin algebra, which we shall introduce below. To these sectors we associate a higher-spin gravity connection W, an internal connection 1-form U and the "fractional-spin gravitinos" $(\psi, \overline{\psi})$. Indeed, the Chern–Simons action obtained for (76) resembles the Achucarro–Townsend theory[7] of supergravity in three dimensions. However, any attempt to extend standard supergravity with fractional-spin fields would lead to higher spin gravity. Since fractional-spin Lorentz representations are infinite dimensional, there appear infinitely many additional symmetries that can be gauged: The higher spin symmetries. The naive matrix (super)trace gives rise, in this context, to divergences that cannot readily be regularised consistently without changing the definition of super-traces of matrices. One achievement of[2] can be regarded as a solution to this problem; see[63] and[64] for a related discussion and an extension of the naive matrix trace.[f]

In what follows we present the main points that lead to the formulation of fractional-spin gravity as presented in Ref. 2. Here we give a somewhat simplified presentation compared the more technical work[2] to which we refer for a complete and precise treatment.

3.1. *The fractional spins algebra*

The building block of the fractional spin algebra is the algebra $Aq(2;\nu)$ introduced by Vasiliev[55] and identified with the universal enveloping algebra of the deformed oscillator algebra[65,66] (see also the Refs. 67,68), in turn presented by

$$[q_\alpha, q_\beta]_\star = 2i(1+\nu k)\epsilon_{\alpha\beta}\ , \qquad \{k, q_\alpha\}_\star = 0\ , \qquad k \star k = 1\ , \tag{77}$$

$$(q_\alpha)^\dagger = q_\alpha\ , \qquad k^\dagger = k\ , \quad \nu \in \mathbb{R}\ . \tag{78}$$

The associative algebra $Aq(2;\nu)$ consists of arbitrary star polynomials (of finite degree) in (q_α, k), which in Weyl order read

$$T_{\alpha(n)} := q_{\alpha_1} \cdots q_{\alpha_n} \equiv q_{(\alpha_1} \star \cdots \star q_{\alpha_n)}\ , \qquad T_{\alpha(n)} \star k\ , \tag{79}$$

where the symmetrisation has unit strength. We split the algebra in four sectors, using the projected elements

$$T^{\sigma,\sigma'}_{\alpha(n)} := [q_{\alpha_1} \cdots q_{\alpha_n}]^{\sigma,\sigma'} := \Pi^\sigma \star q_{(\alpha_1} \star \cdots \star q_{\alpha_n)} \star \Pi^{\sigma'}\ , \qquad \Pi^\pm = \frac{1}{2}(1 \pm k)\,, \tag{80}$$

which are non-vanishing iff $\sigma\sigma' = (-1)^n$. These projections belong to the subalgebras $Aq(2;\nu)^{\sigma,\sigma'} = \Pi^\sigma \star Aq(2;\nu) \star \Pi^{\sigma'} \in Aq(2;\nu)$. Formally, the space $Aq(2;\nu)$

[f] We are grateful to A. Campoleoni and T. Procházka for discussions on this issue.

does not contain distribution (non-polynomial) class of functions. It is necessary, in order to support fractional spins, to extend $Aq(2;\nu)$ with certain "w-class distributions", where w refers to the operator which is diagonalised by them. We will refer to the algebra of these elements as $Aw(2;\nu)$. The operator w appears in the definition of the spin operator

$$J_0 = \frac{1}{2}\, w \star \Pi^+ \ , \qquad w := \frac{1}{4}\,(\tau_0)^{\alpha\beta} q_\alpha \star q_\beta \ , \tag{81}$$

which is here chosen in a non-standard form (cf.[55]), as it involves a projector Π^+. J_0 generates the rotations in the spatial plane, and the projection Π^+ plays an essential role in order to create fractional spins in the connection. More generally, the spin part of the Lorentz transformation is generated by

$$J_a = \frac{1}{4}\,(\tau_a)^{\alpha\beta} J_{\alpha\beta} \ , \qquad J_{\alpha\beta} = \frac{1}{2}\, q_{(\alpha} \star q_{\beta)} \ \star \Pi^+ \quad \in \quad Aq(2;\nu)^{+,+} \ , \quad a = 0,1,2 \ , \tag{82}$$

where in terms of Pauli matrices,

$$(\tau_a)^{\alpha\beta} = (\tau_a)^{\beta\alpha} = (\mathbb{1}, \sigma^1, \sigma^3) \ , \qquad (\tau^a)_\alpha{}^\beta = (\tau^a)^{\alpha'\beta}\epsilon_{\alpha'\alpha} = (-i\sigma^2, -\sigma^3, \sigma^1) \ . \tag{83}$$

Note that here $(\tau^a)_\alpha{}^\beta$ generates a real Clifford algebra and that the conjugation matrix ϵ rises and lower spinor indices. For more detail about these conventions the reader may consult the Ref. 2.

The distribution needed to describe fractional spin are solutions of the "stargenvalue" problem,

$$2J_0 \star T_E = (w \star \Pi^+) \star T_E = E\, T_E \ . \tag{84}$$

The operator w can also be expressed in terms of ladder operators defined as

$$w = a^+ a^- = \tfrac{1}{2}\,\{a^-, a^+\}_\star \ , \qquad [w, a^\pm]_\star = \pm a^\pm \ , \tag{85}$$

where

$$a^\pm = u^{\pm\alpha} q_\alpha \ , \quad u^{+\alpha} u^-_\alpha \ = \ -\frac{i}{2} \ , \quad (u^\pm_\alpha)^\dagger \ = \ u^\mp_\alpha \ . \tag{86}$$

Since powers of w form a closed subalgebra, as we shall see below, in order to solve (84) we expand their $\star$-genfunctions as[g]

$$T_E = \sum_{m=0}^{\infty} f_m w^m \star \Pi^+ \ , \qquad w^m = (a^+)^m (a^-)^m \ , \tag{87}$$

where f_m are constants. We can verify that

$$a^\pm \star [w^m]^{\sigma,\sigma} = \left[a^\pm \left(w^m \mp \frac{m(2m+1-\nu\sigma)}{2(2m+1)}\, w^{m-1}\right)\right]^{-\sigma,\sigma} \ , \tag{88}$$

and hence

$$w \star [w^m]^{\sigma,\sigma} = \left[w^{m+1} + \lambda^\sigma_m w^{m-1}\right]^{\sigma,\sigma} \ , \tag{89}$$

[g]Notice that this amount to taking $\sigma = +1$ in the conventions of Ref. 2.

where we have defined

$$\lambda_m^\sigma = -\frac{m^2}{4}\frac{(2m+1-\nu\sigma)(2m-1+\nu\sigma)}{(2m+1)(2m-1)} \,. \tag{90}$$

Using (89) in (84) yields a recursive formulas. To restrict the space of solutions, and using our intuition on the harmonic oscillator in Fock space, we can solve the lowest weight condition to identify the ground state[h]

$$a^- \star T_{E_0} = 0\,, \tag{91}$$

for which we find a unique solution with $f_0 = 1$, given by

$$f_m = \frac{(-2)^m}{m!}\frac{\left(\frac{3}{2}\right)_m}{\left(\frac{3-\nu}{2}\right)_m}\,, \tag{92}$$

where the Pochhammer symbol $(a)_n$ is given by 1 if $n = 0$ and by $a(a+1)\cdots(a+n-1)$ if $n = 1, 2, \ldots$ Hence, using the definition of the confluent hypergeometric function, viz.

$${}_1F_1(a;b;z) = \sum_{n\geqslant 0}\frac{(a)_n}{(b)_n}\frac{z^n}{n!}\,, \tag{93}$$

we have

$$T_{E_0} = {}_1F_1\left(\frac{3}{2};\frac{3-\nu}{2};-2w\right)\star\Pi^\sigma\,, \tag{94}$$

which obeys

$$(w - E_0)\star T_{E_0} = 0\,, \qquad E_0 = \frac{1+\nu}{2}\,, \tag{95}$$

by virtue of (89). T_{E_0} can be identified with the ground state non-normalised projector $|0)(0|$ of the *deformed harmonic oscillator* introduced by Wigner.[65,66] Higher states can be generated by the left and right $\star$-multiplication of elements of $Aq(2;\nu)$, in terms of $\star$-powers of ladder operators,

$$|m)(n| = (a^+)^{\star n}\star|0)(0|\star(a^-)^{\star n}\,, \qquad |0)(0| := T_{E_0}\,, \tag{96}$$

such that

$$(w - E_m)\star|m)(n| \;=\; 0 \;=\; |m)(n|\star(w - E_n)\,, \qquad E_m = \left(m + \frac{1+\nu}{2}\right). \tag{97}$$

The projector Π^+ in the definition of J_a makes its action trivial on odd parity labels since $\Pi^+\star|2n+1)(m| = |m)(2n+1|\star\Pi^+ = 0\,.$

Introducing the projections of the algebra $Aw(2;\nu)$

$$Aw(2;\nu)^{\sigma,\sigma'} = \Pi^\sigma\star Aw(2;\nu)\star\Pi^{\sigma'}\,, \tag{98}$$

the fractional spin algebra is defined by specification of four sectors

$$\mathcal{A}(2;\nu|w) := \begin{bmatrix} Aq(2;\nu)^{+,+} & Aw(2;\nu)^{+,-} \\ Aw(2;\nu)^{-,+} & Aw(2;\nu)^{-,-}\end{bmatrix} \qquad \ni \qquad \mathbb{A} = \begin{bmatrix} W & \psi \\ \overline{\psi} & U\end{bmatrix}, \tag{99}$$

[h]Here we are taking the choice $\epsilon = +1$ in the notation of.[2]

to which the gauge connection (76) belongs to.

Thus, in each sector the gauge connection must be expanded as follows,

$$\mathbb{A} = \begin{bmatrix} W & \psi \\ \overline{\psi} & U \end{bmatrix} = \begin{bmatrix} W = \sum_n W^{\alpha(n)} \, q_{\alpha(n)} \star \Pi^+ & \psi = \sum_{m,n\geq 0} \psi^{mn} |2m)(2n+1| \\ \overline{\psi} = \sum_{n,m\geq 0} \overline{\psi}^{mn} |2m+1)(2n| & U = \sum_{n,m\geq 0} U^{mn} |2m+1)(2n+1| \end{bmatrix} . \tag{100}$$

These sectors satisfy the "fusion rules"

$$\mathcal{A}(2;\nu|w)^{\sigma,\sigma'} \star \mathcal{A}(2;\nu|w)^{\tau,\tau'} \cong \delta^{\sigma',\tau} \mathcal{A}(2;\nu|w)^{\sigma,\tau'} , \tag{101}$$

where the projections

$$\mathcal{A}(2;\nu|w)^{\sigma,\sigma'} = \Pi^{\sigma} \star \mathcal{A}(2;\nu|w) \star \Pi^{\sigma'} , \tag{102}$$

fall in the correspondent blocks of (99).

Let us perform now a (global) Lorentz transformation

$$\mathbb{A}' = g_e \star \mathbb{A} \star g_{\epsilon}^{-1} , \qquad g_{\epsilon} = \exp_{\star}(i\epsilon^a J_a) . \tag{103}$$

Because the projector Π^+ in its definition, J_a acts non-trivially only in the upper diagonal block (of higher spin gravity) and on the off-diagonal blocks ("fractional spin gravitinos"), while the lower diagonal block U (of internal interactions) does not transform. Internal transformations, whose basis of generators is given by $|2m+1)(2n+1|$ (up to normalisations and reality conditions), act non-trivialy on fractional spin fields and on itself.

As we are interested to show how fractional spins make their appearance, let us perform a rotation by 2π, focusing in the sector $\mathcal{A}(2;\nu|w)^{\pm,\mp}$. The parameter of transformation is given by $\epsilon = 2\pi J_0$, hence single elements of the basis of the sectors $\mathcal{A}(2;\nu|w)^{\pm,\mp}$ transform as

$$g_{2\pi} \star |2m)(2n+1| \star g_{-2\pi} = e^{i\pi(m+\frac{1+\nu}{4})} \, |2m)(2n+1| , \tag{104}$$

$$g_{2\pi} \star |2n+1)(2m| \star g_{-2\pi} = e^{-i\pi(m+\frac{1+\nu}{4})} \, |2n+1)(2m| , \tag{105}$$

hence it follows that the projector basis $|\text{odd})(\text{even}| \oplus |\text{even})(\text{odd}|$ have non-fermionic/bosonic statistical phases. Note that for odd values of ν the phases become fermionic or bosonic. The cases ν negative-odd are critical, in the sense that the representations of the Lorentz algebra become non-unitary and decouple in two sectors, of finite dimension and non-unitary, and of infinite dimension and unitary. For positive-odd ν the representations of half-integer spin of the Lorentz algebra are unitary and infinite dimensional. Thus, for half-integer values of the spin, there appear different representations of the Lorentz group. Indeed, the second (infinite dimensional case) is more exotic, since the field theory of these type of representation is less known, while in the finite dimensional case the field theories are standard, including e.g. Dirac, Rarita–Schwinger, or Bargmann–Wigner, and well known gauge gravity models including $SL(N)$-like Chern–Simons higher spin gravities in three dimensions. For field theories of infinite dimensional representations with half-integer spins the reader can consult the Refs. 69–73. More details and complete analysis on the fractional spin algebra and its critical limits can be found in Refs. 2,61.

3.2. *Gravitational and gauge couplings from Chern-Simons fractional spin gravity*

For a polynomial class of functions, with elements $f(y;k)$ Vasiliev's super-trace is given by,

$$\mathrm{STr}_{Aq(2;\nu)} f(y;k) = f(0;-\nu) \tag{106}$$

It is not straightforward that Vasiliev's super-trace (106) will be consistent when operating on elements of the fractional spin algebra, because the presence of non-polynomial functions. It suffices to establish two consistency conditions: (i) Finite star products, ii) Finite Vasiliev supertraces (which together with (i) implies cyclicity).

The condition (i) implies that

$$|0)(0| \star |0)(0| = \mathcal{N}^{-1}\,|0)(0|\;, \qquad \mathcal{N} \in\,]0,\infty[\;, \tag{107}$$

where $\mathcal{N}$ is a normalisation constant. This calculation was verified up to first order in w-power series and at all order in ν. Assuming that this remains true for all order in w implies the existence of a normalised projector

$$|0\rangle\langle 0| := \mathcal{N}\,|0)(0|\;, \qquad |0\rangle\langle 0| \star |0\rangle\langle 0| = |0\rangle\langle 0|\;, \tag{108}$$

and the normalisation can be obtained trusting on the supertrace applied to the latter product, for which we need to compute just the zero order term in the w-expansion. Doing so we obtain

$$\mathcal{N}^{-1} = \frac{1-\nu}{2}\;. \tag{109}$$

Before writing down the Chern–Simon action, we should introduce a Clifford algebra $\{\mathbb{1}, \Gamma\}$, $\Gamma^2 = \mathbb{1}$, which will allow us to double[i] the fractional higher spin algebra and embed in it the AdS_3 isometry algebra $so(2,2) \cong so(2,1) \oplus so(2,1)$. The supertrace of functions in oscillator variables and Γ is now defined as

$$\mathrm{STr} f(y;k;\Gamma) = f(0;-\nu;0)\,. \tag{110}$$

It can be verified that

$$\mathrm{STr}\,(J_a \star J_b) = \frac{1}{32}\,(1-\nu^2)(1-\frac{\nu}{3})\,\eta_{ab}\;, \tag{111}$$

and

$$\mathrm{STr}\left(P_m{}^n \star P_{m'}{}^{n'}\right) = -\delta_m^{n'}\delta_{m'}^n\;, \qquad P_m{}^n := |m\rangle\langle n| \;\in\; u(\infty)\,. \tag{112}$$

[i]The generator Γ was denoted γ in Ref. 2.

Comparing with normalised trace operations operations, such that

$$\mathrm{Tr}_{\mathrm{grav}}(J_a \star J_b) = \tfrac{1}{2}\,\eta_{ab}\ , \qquad \mathrm{Tr}_{\mathrm{int}}(P_m{}^{n} \star P_{m'}{}^{n'}) = \tfrac{1}{2}\,\delta^{n'}_{m}\delta^{n}_{m'}\ , \tag{113}$$

it follows that

$$\mathrm{STr}|_{\mathrm{grav}} = \tfrac{1}{16}\,(1-\nu^2)(1-\frac{\nu}{3})\,\mathrm{Tr}_{\mathrm{grav}}\ , \tag{114}$$

$$\mathrm{STr}|_{\mathrm{int}} = -\,2\,\mathrm{Tr}_{\mathrm{int}}\ , \tag{115}$$

where STr|. means restriction of the supertrace of the fractional spin gravity (110) either to the gravity sector or the internal sector respectively. Hence the Chern–Simon action for the fractional spin theory reads

$$S[\mathbb{A}] = \frac{\varkappa}{2\pi}\int_{M_3}\mathrm{Tr}_{\mathcal{A}}\left(\tfrac{1}{2}\,\mathbb{A}\star d\mathbb{A} + \tfrac{1}{3}\,\mathbb{A}\star\mathbb{A}\star\mathbb{A}\right)\ , \tag{116}$$

where the trace $\mathrm{Tr}_{\mathcal{A}}$ of $\mathbb{F}\in\mathcal{A}(2;\nu|w)$ is defined by[j]

$$\mathrm{Tr}_{\mathcal{A}}[\mathbb{F}] := \mathrm{Str}[\Gamma\star(\mathbb{F}^{+,+}+\mathbb{F}^{-,-})]\ . \tag{117}$$

With the decomposition

$$W(y,k,\Gamma) = \tfrac{1+\Gamma}{2}\star W_L(y,k) + \tfrac{1-\Gamma}{2}\star W_R(y,k)\ , \quad \textit{idem}\quad U\ , \tag{118}$$

the action (116) produces

$$S[\mathbb{A}] = \tfrac{1}{2}\,S[\mathbb{A}_L] - \tfrac{1}{2}\,S[\mathbb{A}_R]\ , \tag{119}$$

where $(c = L, R)$

$$S[\mathbb{A}_c] = \frac{\varkappa}{2\pi}\int\Big[\mathcal{L}_{\mathrm{CS}}(W_c) + \mathcal{L}_{\mathrm{CS}}(U_c) + \tfrac{1}{2}\,\mathrm{Str}\Big(\psi_c\star D\overline{\psi}_c + \overline{\psi}_c\star D\psi_c\Big)\Big]\ , \tag{120}$$

in terms of the Chern–Simons Lagrangian

$$\mathcal{L}_{\mathrm{CS}}(W) = \mathrm{STr}\left[\tfrac{1}{2}\,W\star dW + \tfrac{1}{3}\,W\star W\star W\right]\ , \tag{121}$$

idem $\mathcal{L}_{\mathrm{CS}}(U)$, and the covariant derivatives

$$D\psi = d\psi + W\star\psi + \psi\star U\ , \qquad D\overline{\psi} = d\overline{\psi} + U\star\overline{\psi} + \overline{\psi}\star W\ . \tag{122}$$

By comparison with the sum of the standard gravity action and gauge interactions in absence of fractional spin gravitinos and higher spin interactions,

$$\begin{aligned} S_{\mathrm{grav}}[W] + S_{\mathrm{int}}[U] = {} & \frac{k_{\mathrm{grav}}}{2\pi}\int_{M_3}\mathrm{Tr}_{\mathrm{grav}}\left[\tfrac{1}{2}\,W_{\mathrm{grav}}\star dW_{\mathrm{grav}} + \tfrac{1}{3}\,W^{\star 3}_{\mathrm{grav}}\right] \\ & + \frac{k_{\mathrm{int}}}{2\pi}\int_{M_3}\mathrm{Tr}_{\mathrm{int}}\left[\tfrac{1}{2}\,U\star dU + \tfrac{1}{3}\,U^{\star 3}\right]\ , \end{aligned} \tag{123}$$

that gives, up to boundary terms in the gravitational sector, the sum of the Einstein–Hilbert action and an internal Chern–Simons theory for the group $U(\infty)\otimes U(\infty)$,

$$S_{\mathrm{grav}}[W] + S_{\mathrm{int}}[U] = \frac{k_{\mathrm{hs}}}{4\pi\ell}\int d^3x\,\sqrt{-g}\left(R + \frac{2}{\ell^2}\right) + \frac{k_{\mathrm{int}}}{4\pi}\int_{M_3}\mathrm{Tr}_{\mathrm{int}}\left(U\star dU + \tfrac{2}{3}\,U^{\star 3}\right), \tag{124}$$

[j] Out of the two possibilities for the fractional-spin algebra denoted by $\mathcal{A}_{\pm}$ in Ref.,[2] here we choose $\mathcal{A}_{+}$ that we call $\mathcal{A}$ for short. This choice entails the introduction of a fermionic Klein operator that was denoted ξ in Ref. 2.

we find that the higher spin gravity and the internal couplings are given by[k]

$$k_{\rm hs} = \frac{\varkappa}{32}\,(1-\nu^2)(1-\frac{\nu}{3})\ , \qquad k_{\rm int} = -\varkappa\ . \tag{125}$$

The Newton constant is given by $G_N = \ell/(4k_{\rm hs})$. The relation between the coupling constants of the (fractional-spin) gravity sector and the internal interaction sector is therefore given by

$$k_{\rm hs} = -\tfrac{1}{32}\,(1-\nu^2)(1-\frac{\nu}{3})\,k_{\rm int}\ . \tag{126}$$

The interactions predicted by the model (116) can be read from the resulting equations of motion:

$$d\mathbb{A} + \mathbb{A}\star\mathbb{A} = 0\ , \tag{127}$$

which in components are given by

$$dW + W\star W + \psi\star\overline{\psi} = 0\ , \qquad dU + U\star U + \overline{\psi}\star\psi = 0\ , \tag{128}$$

$$d\psi + W\star\psi + \psi\star U = 0\ , \qquad d\overline{\psi} + U\star\overline{\psi} + \overline{\psi}\star W = 0\ . \tag{129}$$

Here we observe how the fractional-spin fields source the field strength of higher-spin gravity and the internal interactions, while they couple minimally to the latter interactions, either from the left or right actions. We can visualise this result as saying that fractional-spin charges carry higher-spin gravity fluxes.

To conclude this section, we would like to mention that we have omitted many details for the sake of simplicity. The complete treatment can be found in Ref. 2. In terms of the notation of Ref. 2, here we took the choice $\sigma = +1$, $\epsilon = +1$ and considered the fractional-spin algebra $\mathcal{A}_+$ where the component fields in the "gravitino" sector are fermionic, which requires the introduction of a fermionic oscillator denoted ξ.

4. Matter-Coupled 3D Higher-Spin Gravity

In this section, we review the results of Ref. 4 where an action was provided for matter-coupled 3D higher-spin gravity, that reproduces upon variation the full non-linear Prokushkin–Vasiliev (PV) equations. We take the opportunity of this review to give a review of the PV equations. We also spell out the truncation of the PV spectrum of matter fields to a single *real* scalar field, something that was not made explicit elsewhere. We think that this truncation fills a gap in the literature and can be useful in the context of the Gaberdiel–Gopakumar conjecture.[12] See Ref. 13 for a review.

[k]Here we take the opportunity to correct a typo appearing in the expression for $k_{\rm hs}$ given in the first equation (4.30) of Ref. 2.

4.1. *Geometric formulation of Prokushkin–Vasiliev's system*

In this subsection we are going to present a geometric formulation of Prokushkin–Vasiliev systems,[8] describing matter coupled to gauge fields in three dimensional spacetime. Master fields consist of a one form A and a zero form B, defined on a so-called correspondence space $\mathcal{M}_5 = \mathcal{M}_3 \times \mathcal{Z}_2$, where $\mathcal{M}_3$ is the three dimensional spacetime manifold with local coordinates x^μ, and $\mathcal{Z}_2$ is a non-commutative manifold with coordinates z^α, $\alpha = 1, 2$. The fields take value in the higher spin algebra that extends $sp(2,\mathbb{R})$, generated by twistor variables y^α, and are tensored with elements Γ_i that generate a Clifford algebra: $\{\Gamma_i, \Gamma_j\} = 2\delta_{ij}$, for $i, j = 1, .., N$.

$$A = dx^\mu U_\mu(x, z|y; \Gamma_i) + dz^\alpha V_\alpha(x, z|y; \Gamma_i) \ , \quad B = B(x, z|y; \Gamma_i) \ . \tag{130}$$

The dependence of the master fields on (y^α, z^α) is treated using symbol calculus, whereby they belong to classes of functions (or distributions) on $\mathcal{Y}_2 \times \mathcal{Z}_2$ that can be composed using two associative products: the standard commutative product rule, denoted by juxtaposition, and an additional noncommutative product rule, denoted by a $\star$. In what follows, we shall use the normal ordered basis in which the star product rule is defined formally by

$$(f \star g)(y, z) := \int_{\mathbb{R}^4} \frac{d^2u d^2v}{(2\pi)^2} e^{iv^\alpha u_\alpha} f(y + u, z + u)\, g(y + v, z - v) \ , \tag{131}$$

whereas a more rigorous definition requires a set of fusion rules. In particular, the above composition rule rigorously defines the associative Weyl algebra Aq(4). This algebra consists of arbitrary polynomials in y^α and z^α, modulo

$$y_\alpha \star y_\beta = y_\alpha y_\beta + i\epsilon_{\alpha\beta} \ , \qquad y_\alpha \star z_\beta = y_\alpha z_\beta - i\epsilon_{\alpha\beta} \ , \tag{132}$$

$$z_\alpha \star y_\beta = z_\alpha y_\beta + i\epsilon_{\alpha\beta} \ , \qquad z_\alpha \star z_\beta = z_\alpha z_\beta - i\epsilon_{\alpha\beta} \ , \tag{133}$$

whose symmetric and anti-symmetric parts, respectively, define the normal order and the (ordering independent) commutation rules, viz.[1]

$$[y^\alpha, y^\beta]_\star = -[z^\alpha, z^\beta]_\star = 2i\epsilon^{\alpha\beta} \ , \quad [y^\alpha, z^\beta]_\star = 0 \ . \tag{134}$$

The basis one-forms (dx^μ, dz^α) obey

$$[dx^\mu, f]_\star = 0 = [dz^\alpha, f]_\star \ , \tag{135}$$

where the graded star commutator of differential forms is given by

$$[f, g]_\star = f \star g - (-1)^{\deg(f)\deg(g)} g \star f \ , \tag{136}$$

with deg denoting the total form degree on $\mathcal{M}_3 \times \mathcal{Z}_2$. To describe bosonic models, we impose

$$\pi(A) = A \ , \quad \pi(B) = B \tag{137}$$

[1]The doublet variables y^α and z^α form Majorana spinors once the equations are cast into a manifestly Lorentz covariant form.

where π is the automorphism of the differential star product algebra defined by

$$\pi(x^\mu, dx^\mu, z^\alpha, dz^\alpha, y^\alpha, \Gamma_i) = (x^\mu, dx^\mu, -z^\alpha, -dz^\alpha, -y^\alpha, \Gamma_i) \ . \tag{138}$$

The hermitian conjugation is defined by

$$(f \star g)^\dagger = (-1)^{\text{deg(f)deg(g)}} g^\dagger \star f^\dagger \ , \qquad (z_\alpha, dz^\alpha; y_\alpha, \Gamma_i)^\dagger = (-z_\alpha, -dz^\alpha; y_\alpha, \Gamma_i) \ . \tag{139}$$

and the reality conditions on the master fields read

$$A^\dagger = -A \ , \qquad B^\dagger = B \ . \tag{140}$$

Defining

$$F = dA + A \star A \ , \quad DB = dB + A \star B - B \star A \ , \quad d = dx^\mu \partial_\mu + dz^\alpha \frac{\partial}{\partial z^\alpha} \ , \tag{141}$$

where the differential obeys

$$d(f \star g) \ = \ (df) \star g + (-1)^{\text{deg(f)}} f \star dg \ , \quad (df)^\dagger = d(f^\dagger) \ , \tag{142}$$

the PV field equations can be written as

$$F + B \star J = 0 \ , \qquad DB = 0 \ , \tag{143}$$

where

$$J := -\tfrac{i}{4}\, dz^\alpha dz_\alpha \, \kappa \qquad \kappa := e^{iy^\alpha z_\alpha} \ . \tag{144}$$

The element J is closed and central in the space of π-invariant forms, viz.

$$dJ = 0 \ , \qquad J \star f = \pi(f) \star J \ , \tag{145}$$

as can be seen from the fact that κ, which is referred to as the inner Klein operator, obeys

$$\kappa \star f(x, dx, z, dz, y, \Gamma_i) \star \kappa = f(x, dx, -z, dz, -y, \Gamma_i) \ . \tag{146}$$

It follows that (143) defines a universally Cartan integrable system (i.e. a set of generalized curvature constraints compatible with $d^2 \equiv 0$ in any dimension). The Cartan gauge transformations take the form

$$\delta_\epsilon A = d\epsilon + [A, \epsilon]_\star \ , \quad \delta_\epsilon B = [B, \epsilon]_\star \ . \tag{147}$$

In order to see the equivalence with Prokushkin–Vasiliev systems, let us introduce the oscillator-like fields $S_\alpha := z_\alpha - 2i\, V_\alpha$ and split the field equations in dx and dz directions, thus obtaining

$$\begin{aligned} &d_X U + U \star U = 0 \ , \quad d_X B + [U, B]_\star = 0 \ , \quad d_X S_\alpha + [U, S_\alpha]_\star = 0 \ , \\ &[S_\alpha, B]_\star = 0 \ , \quad [S_\alpha, S_\beta]_\star = -2i\epsilon_{\alpha\beta} \left(1 - B \star \kappa\right) \ , \end{aligned} \tag{148}$$

where $d_X := dx^\mu \partial_\mu$ is the spacetime differential. We stress that, due to the bosonic projection, one has $\{S_\alpha, \kappa\}_\star = 0$, that will be crucial for the discussion of massive vacua.

Massless vacua. Let us analyse the above system around the vacuum solution $B_0 = 0$. From the vacuum equation $[S_{0\alpha}, S_{0\beta}]_\star = -2i\epsilon_{\alpha\beta}$ one can take $S_{0\alpha} = z_\alpha$ and hence

$$0 = [z_\alpha, U_0]_\star = -2i\,\frac{\partial U_0}{\partial z_\alpha} \quad \to \quad U_0 = \Omega(x|y;\Gamma_i)\,, \tag{149}$$

and the remaining equation is the flatness condition $d\Omega + \Omega \star \Omega = 0$. Bilinears in y variables generate $sp(2,\mathbb{R})$ under star-commutators:

$$T_{\alpha\beta} := \frac{1}{2i} y_\alpha y_\beta\,, \quad [T_{\alpha\beta}, T_{\gamma\delta}]_\star = 4\epsilon_{(\alpha(\gamma}T_{\beta)\delta)}\,, \tag{150}$$

but some outer element is needed in order to double $sp(2,\mathbb{R})$ and thus represent the AdS_3 isometry algebra $sp(2,\mathbb{R}) \oplus sp(2,\mathbb{R})$. In order to describe massless vacua, it turns out that the minimal dimension of the Clifford algebra is $N = 2$, and one can write the vacuum spacetime connection as

$$\Omega(x|y;\Gamma_i) = \frac{1}{4i}\Big(\omega^{\alpha\beta}(x)\, y_\alpha y_\beta + \Gamma_1\, e^{\alpha\beta}(x)\, y_\alpha y_\beta\Big) \equiv \omega + \Gamma_1\, e\,, \tag{151}$$

where $\omega^{\alpha\beta}$ and $e^{\alpha\beta}$ are the background Lorentz connection and dreibein, respectively. The flatness condition amounts then to

$$de_{\alpha\beta} + 2\omega_{(\alpha}{}^{\gamma} \wedge e_{\beta)\gamma} = 0\,, \quad d\omega_{\alpha\beta} + \omega_\alpha{}^\gamma \wedge \omega_{\beta\gamma} = -e_\alpha{}^\gamma \wedge e_{\beta\gamma}\,, \tag{152}$$

that indeed describes AdS_3 spacetime with unit radius. In order to study fluctuations around this vacuum we expand the master fields as $U = \Omega + w_1 + ..$, $B = 0 + B_1 + ..$ and $S_\alpha = z_\alpha + S_{1\alpha} + ..$, and the field equations for first order fluctuations read

$$\begin{aligned} &D_0 w_1 = 0\,, \quad D_0 B_1 = 0\,, \quad D_0 S_\alpha + 2i\,\frac{\partial w_1}{\partial z^\alpha} = 0\,, \\ &\frac{\partial B_1}{\partial z^\alpha} = 0\,, \quad \frac{\partial S_1^\alpha}{\partial z^\alpha} = B_1 \star \kappa\,, \end{aligned} \tag{153}$$

from which we see that $B_1 = C(x|y;\Gamma_i)$ is z-independent, and the background covariant derivative is defined by $D_0 f := df + [\Omega, f]_\star$, with Ω given by (151). If we make explicit the dependence on Γ_i in C as

$$C(x|y;\Gamma_i) = C_{\rm aux}(x|y;\Gamma_1) + C_{\rm dyn}(x|y;\Gamma_1)\,\Gamma_2\,, \tag{154}$$

we can see that $D_0 C = 0$ splits into

$$D_L C_{\rm aux} + \Gamma_1 [e, C_{\rm aux}]_\star = 0\,, \quad D_L C_{\rm dyn} + \Gamma_1 \{e, C_{\rm dyn}\}_\star = 0\,, \tag{155}$$

where the Lorentz covariant derivative is $D_L := d + [\omega, \bullet]_\star$. It is known[74] that the equation for $C_{\rm aux}$ describes non-propagating degrees of freedom, and $C_{\rm aux}$ is indeed referred to as auxiliary. On the other hand, by expanding the equation for $C_{\rm dyn}$ in power series in y^α, one can see that $C_{\rm dyn}$ contains two real AdS$_3$ massless scalar fields together with all their on-shell nontrivial derivatives. At this stage, the next steps would be

(1) Use $\frac{\partial S_1^\alpha}{\partial z^\alpha} = C \star \kappa$ to solve $S_{1\alpha}$ in terms of C, up to some pure-gauge contribution;

(2) Substitute for S_1^α into $\frac{\partial w_1}{\partial z^\alpha} = -D_0 S_{1\alpha}$ to solve the z-dependence of w_1 in terms of C. One gets $w_1 = \omega_1(x|y;\Gamma_i) + f_1(x,z|y;\Gamma_i)$, where f_1 is a known function linear in C;

(3) Use $D_0\omega_1 = -\Big(D_0 f_1\Big)|_{z=0}$ as higher spin equation for ω_1.

It is known,[75] however, that the linearized curvature $D_0\omega_1$ does not receive non-trivial sources linear in C, and indeed the higher spin fields do not propagate. For a more complete and thorough analysis of the 3D equations, including the issue of field redefinition, locality and elimination of the auxiliary zero-form $C_{\rm aux}$ up to second order in the weak fields, see Ref. 76. See also Ref. 77 for another review and exact solutions of the PV equations.

Massive vacua. Let us study the model around a different vacuum, namely $B_0 = \nu\,\Gamma$, where ν is a constant and Γ is some element generated by the Γ_i, obeying $\Gamma^\dagger = \Gamma$ and $\Gamma^2 = 1$. The vacuum equations now read

$$\begin{aligned} &dU_0 + U_0 \star U_0 = 0\,, \quad [\Gamma, U_0] = 0\,, \quad [\tilde z_\alpha, \Gamma]_\star = 0 \\ &[\tilde z_\alpha, U_0]_\star = 0\,, \quad [\tilde z_\alpha, \tilde z_\beta]_\star = -2i\,\epsilon_{\alpha\beta}(1 - \nu\,\Gamma\kappa)\,, \end{aligned} \tag{156}$$

where we defined $\tilde z_\alpha := S_{0\alpha}$, that obey[m] $\{\tilde z_\alpha, \Gamma\kappa\}_\star = 0$ and the deformed oscillator algebra. The strategy is first to find a solution for $\tilde z_\alpha$, and then solve $[\tilde z_\alpha, U_0]_\star = 0$ by finding other deformed variables $\tilde y_\alpha$ that star commute with $\tilde z_\alpha$. To this end, let us introduce, as done in Ref. 8,

$$\sigma_\alpha := \nu \int_0^1 dt\, t\, e^{ityz}(y_\alpha + z_\alpha)\,, \quad \tau_\alpha := \nu \int_0^1 dt\,(t-1)\, e^{ityz}(y_\alpha + z_\alpha)\,, \tag{157}$$

with $yz := y^\alpha z_\alpha$. One can check that they obey the following relations:

$$\begin{aligned} &[z_{[\alpha}, \sigma_{\beta]}]_\star = -i\nu\,\epsilon_{\alpha\beta}\,\kappa\,, \quad [\sigma_\alpha, \sigma_\beta]_\star = 0\,, \quad \{\sigma_\alpha, \tau_\beta\}_\star = 0\,, \\ &[z_\alpha, \tau_\beta]_\star = \{\sigma_\alpha, y_\beta\}_\star\,, \quad \{y_{[\alpha}, \tau_{\beta]}\}_\star = i\nu\,\epsilon_{\alpha\beta}\,, \quad [\tau_\alpha, \tau_\beta]_\star = 0\,. \end{aligned} \tag{158}$$

Let us make the following Ansatz for the deformed oscillators:

$$\tilde z_\alpha = X\, z_\alpha + Y\, \sigma_\alpha\,, \quad \tilde y_\alpha = A\, y_\alpha + B\, \tau_\alpha\,, \tag{159}$$

where X, Y, A and B are built out of gamma matrices Γ_i. By demanding

$$[\tilde z_\alpha, \tilde z_\beta]_\star = -2i\epsilon_{\alpha\beta}(1 - \nu\Gamma\kappa)\,, \quad [\tilde y_\alpha, \tilde z_\beta]_\star = 0\,, \tag{160}$$

one has the following constraints,

$$X^2 = 1\,, \quad XY = YX = -\Gamma\,, \quad [A, X] = \{A, Y\} = [B, X] = \{B, Y\} = 0\,. \tag{161}$$

[m]Since $\tilde z_\alpha$ commutes with Γ and anticommutes with κ from the necessary bosonic projection.

As in the original Prokushkin–Vasiliev model, we demand that $\tilde{y}_\alpha$ obey a deformed oscillator algebra:

$$[\tilde{y}_\alpha, \tilde{y}_\beta]_\star = 2i\epsilon_{\alpha\beta}(1-\nu\Gamma)\ , \quad \{\tilde{y}_\alpha, \Gamma\} = 0\ . \tag{162}$$

This imposes

$$\{A,\Gamma\} = 0\ , \quad A^2 = 1\ , \quad AB = -BA = -\Gamma\ . \tag{163}$$

A convenient solution, that admits propagating degrees of freedom, is given by $\Gamma = \Gamma_{1234}$

$$X = 1\ , \quad Y = -\Gamma\ , \quad A = \Gamma_1\ , \quad B = -\Gamma_{234} = -\Gamma_1\Gamma\ ,$$

where we have chosen the number of generators of the Clifford algebra to be $N = 4$ and $\Gamma_{i_1..i_k} := \Gamma_{[i_1}..\Gamma_{i_k]}$.

With this solution, fields obeying $[f, \tilde{z}_\alpha]_\star = 0$ and $[f,\Gamma] = 0$ are given by $f(x|\tilde{y};\Gamma_{ij})$, i.e. star functions of the deformed $\tilde{y}$'s and of all bilinears Γ_{ij}, together with Γ itself, that is generated by bilinears. One can check that undeformed $sp(2,\mathbb{R})$ is still generated by $\tilde{y}$'s as $T_{\alpha\beta} := \frac{1}{4i}\{\tilde{y}_\alpha, \tilde{y}_\beta\}_\star$, and one can take the background connection $U_0 = \Omega$ to be

$$\Omega(x|\tilde{y};\Gamma_{ij}) = \frac{1}{4i}\Big(\omega^{\alpha\beta}(x)\,\tilde{y}_\alpha \star \tilde{y}_\beta + i\Gamma_{23}\, e^{\alpha\beta}(x)\,\tilde{y}_\alpha \star \tilde{y}_\beta\Big) = \omega + i\Gamma_{23}\, e\ , \tag{164}$$

such that $d\Omega + \Omega^{\star 2} = 0$ is solved by the AdS_3 background. Fluctuations B_1 defined by $B = \nu\Gamma + B_1 + ..$ obey the linearized equations

$$[\tilde{z}_\alpha, B_1]_\star = 0 \quad \Rightarrow \quad B_1 = C(x|\tilde{y};\Gamma_{ij})\ , \quad D_0 C = 0\ , \tag{165}$$

where, as before, $D_0 f := df + [\Omega, f]_\star$, but now with Ω given by (164). In order to find the propagating degrees of freedom, let us explicitate the Γ_{ij} dependence of C:

$$C(x|\tilde{y};\Gamma_{ij}) = C_{\text{aux}}(x|\tilde{y};\Gamma, i\Gamma_{23}) + C_{\text{dyn}}(x|\tilde{y};\Gamma, i\Gamma_{23}) i\Gamma_{24}\ , \tag{166}$$

and use it in the equation $D_0 C = 0$, obtaining

$$D_L C_{\text{aux}} + i\Gamma_{23}\,[e, C_{\text{aux}}]_\star = 0\ , \quad D_L C_{\text{dyn}} + i\Gamma_{23}\,\{e, C_{\text{dyn}}\}_\star = 0\ , \tag{167}$$

from which one can find that C_{dyn} contains four real massive propagating scalars, while C_{aux} yields four topological deformations. The analysis is now equivalent to the original Prokushkin–Vasiliev model, since at this level one can identify

$$(k)_{\text{PV}} = \Gamma\ , \quad (\nu)_{\text{PV}} = -\nu\ , \quad (\rho)_{\text{PV}} = \Gamma_1\ , \quad (y_\alpha)_{\text{PV}} = \Gamma_1\, y_\alpha\ , \quad (z_\alpha)_{\text{PV}} = \Gamma_1\, z_\alpha\ , \tag{168}$$

$$(\psi_1)_{\text{PV}} = i\Gamma_{23}\ , \qquad (\psi_2)_{\text{PV}} = i\Gamma_{24}\ . \tag{169}$$

One can truncate the master fields as

$$\Pi^+_\Gamma A = A\ , \quad \Pi^+_\Gamma B = B\ ,$$

with the projector $\Pi^+_\Gamma := \frac{1\pm\Gamma}{2}$, yielding a model with two propagating scalars contained in $C_{\text{dyn}}(x|\tilde{y};\Gamma, i\Gamma_{23}) = \frac{1+\Gamma}{2}\, C_+(x|\tilde{y};i\Gamma_{23})$.

Minimal truncation. We now want to further truncate the matter sector contained in C_+ to a single, real scalar field. By using the anti-automorphism defined by

$$\tau(f \star g) = (-1)^{\mathrm{deg(f)deg(g)}} \tau(g) \star \tau(f) \ , \tag{170}$$

$$\tau(z^\alpha, dz^\alpha; y^\alpha, \Gamma_i) = (-iz^\alpha, -idz^\alpha; iy^\alpha, \epsilon_{(i)}\Gamma_i) \ , \qquad \epsilon_{(i)} = (+,+,-,-) \ , \tag{171}$$

one can truncate further, by requiring

$$\tau(A) = -A \ , \quad \tau(B) = B \ . \tag{172}$$

This last truncation leaves a single real propagating scalar in the spectrum, since now the two chains of fields multiplying 1 and $i\Gamma_{23}$ only contain even and odd numbers of derivatives of the physical field:

$$C_+(x|\tilde{y}; i\Gamma_{23}) = \sum_{n=0}^{\infty} C^0_{\alpha_1...\alpha_{4n}}(x)\, \tilde{y}^{\alpha_1}...\tilde{y}^{\alpha_{4n}} + i\Gamma_{23} \sum_{n=0}^{\infty} C^1_{\alpha_1...\alpha_{4n+2}}(x)\, \tilde{y}^{\alpha_1}...\tilde{y}^{\alpha_{4n+2}} \ . \tag{173}$$

Finally, it is possible to find solution for the deformed oscillator algebra such that the deformed oscillators obey the same reality conditions and transformation properties under the tau map as the undeformed y^α and z^α, and therefore it is understood above that the master fields are expanded in terms of these deformed oscillators.

4.2. *Covariant Hamiltonian action*

In this section we begin by discussing some generalities on covariant Hamiltonian actions on $\mathcal{X}_4 \times \mathcal{Z}_2$. We then determine the constraints on the Hamiltonian such that it leads to a master action in which the master field content, including the Lagrange multipliers, are extended to consist of sum of even and odd forms of appropriate degree, and central elements. This action yields a generalized version of the PV field equations.

Generalities. In order to formulate the theory within the AKSZ framework[35] using its adaptation to noncommutative higher spin geometries proposed in,[36] we assume a formulation of the PV system that treats $\mathcal{Z}_2$ as being closed and introduce an open six-manifold $\mathcal{M}_6$ with boundary

$$\partial \mathcal{M}_6 = \mathcal{X}_3 \times \mathcal{Z}_2 \ , \tag{174}$$

where $\mathcal{X}_3$ is a closed manifold containing $\mathcal{M}_3$ as an open submanifold. On $\mathcal{M}_6$, we introduce a two-fold duality extended[6,49,78 n] set of differential forms given by

$$A = A_{[1]} + A_{[3]} + A_{[5]} \ , \qquad B = B_{[0]} + B_{[2]} + B_{[4]} \ , \tag{175}$$

$$T = T_{[4]} + T_{[2]} + T_{[0]} \ , \qquad S = S_{[5]} + S_{[3]} + S_{[1]} \ , \tag{176}$$

[n] Starting from a universally Cartan integrable system and replacing each p-form by a sum of forms of degrees p, $p+2$, ..., $p+2N$, and each structure constant by a function of off-shell closed and central terms, i.e. elements in the de Rham cohomology valued in the center of the fiber algebra, with a decomposition into degrees 0, 2, ..., $2N$, yields a new universally Cartan integrable system, referred to as the N-fold duality extension of the original system. More generally, one may consider on-shell duality extensions by including on-shell closed complex-valued functionals into the extension of the structure constants.[49,79]

valued in $\mathcal{A}\otimes\mathcal{C}_4^+$, $\mathcal{C}_4^+$ being the even subspace of the four dimensional Clifford algebra, and where the subscript denotes the form degree. The restriction to the even Clifford subalgebra, i.e. to fields obeying $[f(\Gamma_i),\Gamma]=0$, is required by demanding integrability of the field equations coming from the action that we will present in the following. We let $\{J^I\}$ denote the generators of the ring of off-shell closed and central terms, i.e. elements in the de Rham cohomology of $\mathcal{M}_6$ valued in the center of $\mathcal{A}\otimes\mathcal{C}_4^+$, which hence obey

$$dJ^I \;=\; 0\;, \quad \left[J^I,f\right]_\star \;=\; 0\;, \tag{177}$$

(off-shell) for any differential form f on $\mathcal{M}_6$ valued in $\mathcal{A}\otimes\mathcal{C}_4^+$. Following the approach of,[6] we consider actions of the form

$$S_{\rm H} \;=\; \int_{\mathcal{M}_6}\mathrm{Tr}_{\mathcal{A}\otimes\mathcal{C}_4^+}\left[S\star DB+T\star F+\mathcal{V}(S,T;B;J^I)\right] \tag{178}$$

$$=\int_{\mathcal{M}_6}\mathrm{Tr}_{\mathcal{A}\otimes\mathcal{C}_4^+}\left[S\star dB+T\star dA-\mathcal{H}(S,T;A,B;J^I)\right] \tag{179}$$

where $\mathrm{Tr}_{\mathcal{A}\otimes\mathcal{C}_4^+}$ denotes a cyclic trace operation on $\mathcal{A}\otimes\mathcal{C}_4^+$. We assume a structure group gauged by A and that S, T and B belong to sections, and (179) makes explicit the covariant Hamiltonian form, with

$$\mathcal{H}(S,T;A,B;J^I)=-S\star[A,B]_\star-T\star A\star A-\mathcal{V}(S,T;B;J^I)\;. \tag{180}$$

Thus, the coordinate and momentum master fields, defined by

$$(X^\alpha;P_\alpha):=(A,B;T,S)\;, \tag{181}$$

lie in subspaces of $\mathcal{A}$ that are dually paired using $\mathrm{Tr}_{\mathcal{A}}$, which leads to distinct models depending on whether these subspaces are isomorphic or not. In the reductions that follow, we shall consider the first type of models, moreover, for definiteness, we shall assume that

$$\mathcal{M}_6=\mathcal{X}_4\times\mathcal{Z}_2\;, \tag{182}$$

and the associative bundle is chosen such that

$$\check{\mathcal{L}}=\oint_{\mathcal{Z}_2}\mathrm{Tr}_{\mathcal{A}\otimes\mathcal{C}_4^+}\left[S\star DB+T\star F+\mathcal{V}(S,T;B;J^I)\right]\;, \tag{183}$$

is finite and globally defined on $\mathcal{X}_4$. The action can then be written as

$$S_{\rm H}=\int_{\mathcal{X}_4}\check{\mathcal{L}}\;. \tag{184}$$

The master action. The Hamiltonian is constrained by gauge invariance, or equivalently, by universal on-shell Cartan integrability[o]. In addition, it is constrained by the requirement that the equations of motion on $\mathcal{M}_6$ reduce to a desired set of equations of motion on $\partial\mathcal{M}_6$ upon assuming natural boundary conditions. In

[o]Covariant Hamiltonian actions are gauge invariant iff their equations of motion form universally Cartan integrable systems.

order to obtain a model that admits consistent truncations to three-dimensional CS higher spin gravities, we need to assume that $\mathcal{V}$ contains a term that is quadratic in T. The simplest possible such action is given by

$$S_{\rm H} = \int_{\mathcal{M}_6} {\rm Tr}_{\mathcal{A}\otimes\mathcal{C}_4^+} \left[S \star DB + T \star \left[F + g + h \star (B - \tfrac{1}{2}\mu \star T)\right] + \mu \star B \star S \star S\right] \tag{185}$$

where

$$g = g(J^I) \ , \qquad h = h(J^I) \ , \qquad \mu = \mu(J^I) \tag{186}$$

are even closed and central elements on $\mathcal{M}_6$ in degrees

$$\deg(g, h, \mu) = (2 \bmod 2, 2 \bmod 2, 0 \bmod 2) \ . \tag{187}$$

The reality conditions are given by

$$(A, B; T, S; g, h, \mu)^\dagger = (-A, B; -T, S; -g, -h, -\mu) \ , \tag{188}$$

The total variation yields

$$\delta S_{\rm H} = \int_{\mathcal{M}_6} {\rm Tr}_{\mathcal{A}\otimes\mathcal{C}_4^+} \left(\delta T \star \mathcal{R}^A + \delta S \star \mathcal{R}^B + \delta A \star \mathcal{R}^T + \delta B \star \mathcal{R}^S\right) + \oint_{\partial\mathcal{M}_6} {\rm Tr}_{\mathcal{A}\otimes\mathcal{C}_4^+} (T \star \delta A - S \star \delta B) \ , \tag{189}$$

where the Cartan curvatures read

$$\begin{aligned} \mathcal{R}^A &= F + g + h \star (B - \mu \star T) \approx 0 \\ \mathcal{R}^B &= DB + \mu \star [S, B]_\star \approx 0 \\ \mathcal{R}^T &= DT + [S, B]_\star \approx 0 \\ \mathcal{R}^S &= DS + h \star T + \mu \star S \star S \approx 0 \end{aligned} \tag{190}$$

The generalized Bianchi identities are

$$D\mathcal{R}^A \equiv h \star (\mathcal{R}^B - \mu \star \mathcal{R}^T) \ , \tag{191}$$

$$D\mathcal{R}^B \equiv [(\mathcal{R}^A + \mu \star \mathcal{R}^S), B]_\star - \mu \star \{\mathcal{R}^B, S\}_\star \ , \tag{192}$$

$$D\mathcal{R}^T \equiv [\mathcal{R}^A, T]_\star + [\mathcal{R}^S, B]_\star - \{\mathcal{R}^B, S\}_\star \ , \tag{193}$$

$$D\mathcal{R}^S \equiv [\mathcal{R}^A, S]_\star + \mu \star [\mathcal{R}^S, S]_\star + h \star \mathcal{R}^T \ . \tag{194}$$

The gauge transformations

$$\delta_{\epsilon,\eta} A = D\epsilon^A - h \star (\epsilon^B - \mu \star \eta^T) \ , \tag{195}$$

$$\delta_{\epsilon,\eta} B = D\epsilon^B - [\epsilon^A, B]_\star - \mu \star [\eta^S, B]_\star + \mu \star \{S, \epsilon^B\}_\star \ , \tag{196}$$

$$\delta_{\epsilon,\eta} T = D\eta^T - [\epsilon^A, T]_\star - [\eta^S, B]_\star + \{S, \epsilon^B\}_\star \ , \tag{197}$$

$$\delta_{\epsilon,\eta} S = D\eta^S - [\epsilon^A, S]_\star - \mu \star [\eta^S, S]_\star - h \star \eta^T \ , \tag{198}$$

which transform the Cartan curvatures into each other, induce

$$\delta_{\epsilon,\eta} S_{\rm H} = \oint_{\partial\mathcal{M}_6} {\rm Tr}_{\mathcal{A}\otimes\mathcal{C}_4^+} \left(\eta^T \star [F + g + h \star B] + \eta^S \star DB\right) . \tag{199}$$

We take $(\epsilon^B; \eta^T, \eta^S)$ to belong to sections of the structure group and impose[p]

$$(\eta^T, \eta^S)|_{\partial\mathcal{M}_6} = 0 . \tag{200}$$

We have also assumed that (A, B) fluctuate on $\partial\mathcal{M}_6$, which implies

$$T|_{\partial\mathcal{M}_6} \approx 0 \approx S|_{\partial\mathcal{M}_6} . \tag{201}$$

The resulting boundary equations of motion

$$F + g + h \star B \approx 0 , \quad DB \approx 0 \tag{202}$$

thus provide a duality extended version of the Prokushkin–Vasiliev equations, that is free from any interaction ambiguity, following a variational principle. Let us notice that in the action (185), the relative coefficient of the BSS and TT terms is fixed uniquely by Cartan integrability.

4.3. *Consistent truncations*

In this section we review[4] the consistent truncation of the above covariant Hamiltonian master action in six dimensions down to a BF-like model on $\mathcal{X}_4$ that reproduces Blencowe's action. The truncation consists in integrating out the fluctuations in B around its vacuum expectation value followed by reductions on $\mathcal{X}_4$. Starting from the equations of motion (190) and setting $B = 0$ yields

$$F + g - h \star \mu \star T = 0 , \qquad DT = 0 , \tag{203}$$

and

$$DS + h \star T + \mu \star S \star S = 0 , \tag{204}$$

which together form a Cartan integrable system containing (203) as a subsystem, i.e. the free differential algebra generated by (A, T, S) contains a subalgebra generated by (A, T). Assuming $\partial\mathcal{M}_6$ to consist of a single component, it follows from $S|_{\partial\mathcal{M}_6} = 0$ that S can be reconstructed from (A, T) on-shell [q] from (204). Therefore, the system (203) is a consistent truncation of the original system (190) on-shell.

Rewriting the full action (185) by integrating by parts in its SDB-term yields

$$S_{\rm H} = \int_{\mathcal{M}_6} {\rm Tr}_{\mathcal{A}\otimes\mathcal{C}_4^+} \left[T \star (F + g - \tfrac{1}{2} h \star \mu \star T) + B \star (DS + h \star T + \mu \star S \star S)\right] . \tag{205}$$

[p] Following the AKSZ approach, the Batalin–Vilkovisky classical master equation requires that the ghosts corresponding to (η^T, η^S) vanish at $\partial\mathcal{M}_6$ off-shell.

[q] Since $T|_{\partial\mathcal{M}_6} = 0$ on-shell as well it follows that both S and T can be taken to vanish on $\mathcal{M}_6$ on-shell.

It follows that $B = 0$ is a saddle point of the path integral at which B and S can be integrated out in a perturbative expansion. Schematically, modulo gauge fixing, one has

$$\int_{\langle B\rangle=0} [DB][DS] e^{\frac{i}{\hbar}S_{\rm H}} \sim e^{\frac{i}{\hbar}S_{\rm eff}[A,T]} , \tag{206}$$

where the effective action

$$S_{\rm eff}[A,T] = S_{\rm red}[A,T] + O(\hbar) , \tag{207}$$

consists of loop corrections (comprising attendant functional determinants on non-commutative manifolds) and

$$S_{\rm red} = \int_{\mathcal{M}_6} {\rm Tr}_{\mathcal{A}\otimes\mathcal{C}_4^+} T \star (F + g - \tfrac{1}{2} h \star \mu \star T) . \tag{208}$$

The latter is a consistently reduced classical action in the sense that it reproduces the subsystem (203). The reduced system, which thus consists of the gauge sector of the original system, is a topological theory with local symmetries

$$\delta A = D\epsilon + \mu \star h \star \eta , \quad \delta T = D\eta - [\epsilon, T]_\star , \tag{209}$$

and equations of motion and boundary conditions given by

$$F + g - \mu \star h \star T = 0 , \quad DT = 0 , \tag{210}$$

$$T|_{\partial\mathcal{M}_6} = 0 . \tag{211}$$

The boundary equations are thus given by

$$(F + g)|_{\partial\mathcal{M}_6} = 0 . \tag{212}$$

To address Blencowe's theory, we truncate once more by reducing (203) under the assumptions that

$$g = 0 , \quad \mu = \mu_{[0]} \equiv \mu_0 , \quad h = J , \tag{213}$$

where μ_0 is an imaginary constant, and

$$A = \check{W}_{[1]} - \check{K}_{[1]} - \mu_0 J \star \check{K}_{[1]} , \quad T = \check{T}_{[2]} + \check{K}_{[1]} \star \check{K}_{[1]} - \mu_0 J \star \check{T}_{[2]} , \tag{214}$$

where we define checked fields to be z-independent: $\check{f} \in \Omega(\mathcal{X}_4) \otimes \mathcal{A} \otimes \mathcal{C}_4^+$. By defining

$$\check{F} = d_X \check{W} + \check{W} \star \check{W} , \quad \check{D}\check{K} = d_X \check{K} + [\check{W}, \check{K}]_\star , \quad \check{D}\check{T} = d_X \check{T} + [\check{W}, \check{T}]_\star , \tag{215}$$

and suppressing the subscripts indicating form degrees, the reduction of (203) yields

$$\check{F} + \check{T} = 0 , \qquad \check{D}\check{T} = 0 , \tag{216}$$

$$D\check{K} - \check{K} \star \check{K} + \check{T} = 0 , \tag{217}$$

which is a Cartan integrable system containing (216) as a subsystem. From (211) and (214), we deduce the boundary conditions

$$\check{T}|_{\partial\mathcal{X}_4} = 0 = (\check{K} \star \check{K})|_{\partial\mathcal{X}_4} , \tag{218}$$

Substituting (214) into (208) and using (218) we obtain

$$\check{S}_{\rm red}[\check{W},\check{T}] = -\mu_0 \int_{\mathcal{X}_4}\int_{\mathcal{Z}_2} {\rm Tr}_{\mathcal{A}\otimes\mathcal{C}_4^+} J\star\check{T}\left(\check{F}+\tfrac{1}{2}\check{T}\right) . \tag{219}$$

At this stage, we truncate the models further as follows:

$$\check{W} = \Pi_\Gamma^+ W_+ + \Pi_\Gamma^- W_- \ , \quad \check{T} = \Pi_\Gamma^+ T_+ + \Pi_\Gamma^- T_- \ , \tag{220}$$

where $W_\pm$ and $T_\pm$ are independent of Γ_i. Inserting (220) into (219) and using

$$\int_{\mathcal{Z}_2} {\rm Tr}_{\mathcal{A}\otimes\mathcal{C}_4^+} J\star\Pi_\Gamma^\pm \check{f} =: \pm\tfrac{i\pi}{2}\,{\rm STr}_{{\rm Aq}(2)}\,\check{f} \ , \tag{221}$$

for $\check{f}$ independent of Γ_i and belonging to Aq(2), the associative algebra given by the universal enveloping algebra of the undeformed oscillators y^α. We thereby obtain the following four-dimensional Hamiltonian extension of Blencowe's action:

$$S_{\rm Bl} = -\tfrac{i\pi}{2}\,\mu_0 \int_{\mathcal{X}_4} {\rm STr}_{{\rm Aq}(2)}\left[T_+(F_+ + \tfrac{1}{2}T_+) - T_-(F_- + \tfrac{1}{2}T_-)\right] . \tag{222}$$

Assuming that $\mathcal{X}_4 = \mathcal{X}_3\times[0,\infty[$ and that all fields fall off at $\mathcal{X}_3\times\infty$, and assuming furthermore that $\mathcal{X}_3$ has a simple topology, the elimination of the Lagrange multipliers yields

$$S_{\rm Bl} = \tfrac{i\pi}{2}\,\mu_0\left(S_{\rm CS}[W_+] - S_{\rm CS}[W_-]\right) \ , \tag{223}$$

with

$$S_{\rm CS}[W] = \oint_{\mathcal{X}_3} {\rm STr}_{{\rm Aq}(2)}\left[\tfrac{1}{2}W\star dW + \tfrac{1}{3}W\star W\star W\right] , \tag{224}$$

where now d denotes the exterior derivative on $\mathcal{X}_3$. Equivalently,

$$S_{\rm Bl} = i\mu_0\pi \oint_{\mathcal{X}_3} {\rm STr}_{{\rm Aq}(2)}\left[E\star(d\Omega+\Omega\star\Omega)+\tfrac{1}{3}\,E\star E\star E\right] , \qquad W_\pm = \Omega\pm E \ , \tag{225}$$

from which we identify

$$\mu_0 = -\frac{4i}{\pi^2}\frac{\ell_{\rm AdS}}{G_{\rm N}} \tag{226}$$

using the conventions of.[2]

Truncation to the Prokushkin–Segal–Vasiliev action. In the paper[4] that we are reviewing here, another consistent truncation of the action (185) is shown to reproduce the action principle[80] for the PV equation that is formulated in a 2-dimensional base space for the z^α oscillators. The interested reader can find all the details in.[4]

5. Higher Spins and Topological Strings

In this section we shall argue, based on the formal structure of higher spin equations and actions, that higher spin dynamics can be described in terms of first-quantized topological open strings.

Vasiliev's equations exhibit two basic properties akin to open string field theory: First of all, we recall that in Cartan's formulation of field theories, as free differential algebras, the spaces of forms in degrees one and zero correspond to gauge Lie algebras and spectra of local degrees of freedom, respectively. In this respect, a remarkable feature of Vasiliev's higher spin gravities is that their one- and zero-form modules are unified into associative algebras[r]. Second, they can be embedded into a Frobenius–Chern–Simons theory with a cubic action principle that only refers to the trace and star product operations of the associative algebra and a nilpotent differential containing the de Rham differential. These algebraic structures can be naturally encoded into a class of two-dimensional Poisson sigma models with gauged supersymmetry, corresponding to the de Rham operator, which are our candidate topological open string models.

In order to explain the rational behind these models in more detail, let us start from a particle on a symplectic manifold N with symplectic potential one-form $\vartheta = d\phi^\alpha \vartheta_\alpha$ and Hamiltonian H, consisting of constraints and Lagrange multipliers, as described by an action of the form

$$S_\vartheta = \int_C (\vartheta - H) \ ,$$

integrated along a worldline C. Assigning physical states of the system to Hilbert spaces it is natural to consider open worldlines whose boundaries are allowed to fluctuate in Lagrangian submanifolds of N. Alternatively, and more generally, one may assign the physical states to density matrices, in which case it is natural to take C to be closed, and choose boundary conditions such that the path integral provides a trace operation. Considering initially the case of a trivial Hamiltonian, and letting $\Pi^{\alpha\beta}$ denote the Poisson bi-vector on N, *i.e.* the inverse of the symplectic two-form $\omega_{\alpha\beta} = \partial_\alpha \vartheta_\beta - \partial_\beta \vartheta_\alpha$, the resulting path integral over worldlines C can be extended into a path integral over open worldsheets Σ with boundary C, weighted by $\exp \frac{i}{\hbar} S_\Pi$ where S_Π is the action of the Ikeda–Schaller–Strobl Poisson sigma model

$$S_\Pi = \int_\Sigma (\eta_\alpha \wedge d\phi^\alpha + \tfrac{1}{2}\, \Pi^{\alpha\beta} \eta_\alpha \wedge \eta_\beta) \ ,$$

subject to the boundary condition $\eta_\alpha|_C = 0$. (which can be derived on-shell using the variational principle but that actually must be imposed off-shell as well in order for the classical Batalin–Vilkovisky master equation to hold). The resulting path integral can be performed in two steps: First over discs with a point of C attached

[r] An interesting consequence is that the spectra of massless particles and generalized Type D solutions in four-dimensional higher spin gravity are related by a $\mathbb{Z}_2$-symmetry,[48] unlike in ordinary gauge theories, in which the Type D solution spaces are finite dimensional.

to a fixed point $p_0 \in N$ (which one may think of as a $(D-1)$-brane), and then by integrating over p_0 using the symplectic measure. Inserting vertex operators along points $p_i \in C$, given by (the pull-back of) functions f_i on N, the resulting path integral can be viewed as a formal definition of $\mathrm{Tr}_N \prod_i^\star f_i$, where $f \star g$ is an associative noncommutative product, referred to as the star product. Provided that f and g belong to sufficiently smooth classes of functions (*e.g.* polynomials or formal power series with coefficients given by power series in $\hbar$), then the star product has an expansion in terms of pointwise derivatives, *viz.*

$$f \star g = fg + i\hbar\{f,g\} + \cdots \ , \qquad \{f,g\} := \Pi^{\alpha\beta}\partial_\alpha f \partial_\beta g \ ,$$

where the higher order terms in the $\hbar$ expansion are given in terms of multiple derivatives of the functions and the Poisson structure (but not its inverse), while in order to compose more general elements, including (nonperturbative) distributions on N and (which arise as density matrices related to unitary representations), nonlocal versions of the star product, based on auxiliary integrations, are required. Unlike the particle action, the Poisson sigma model remains well-defined on Poisson manifolds, where $\Pi^{\alpha\beta}$ is assumed to be a bi-vector obeying

$$\Pi^{\delta[\alpha}\partial_\delta \Pi^{\beta\gamma]} = 0 \ ,$$

as this ensures the invariance of the action under η_α gauge transformations, while there is no requirement on its invertibility.

Historically, the existence of a star product on general symplectic manifolds was first established formally by De Wilde and Lecomte.[81] Its pointwise form was given explicitly and on a manifestly covariant form by Fedosov.[82] His construction resembles that of Vasiliev, though it does not provide any dynamics, which instead will require a suitable gauging leading to topological open strings, as we shall propose below. Later, Kontsevich used his formality theorem to show the existence and uniqueness (up to similarity transformations and changes of basis corresponding to different ordering schemes) of the star product on general Poisson manifolds. He also provided an explicit formula in the case of $N \cong \mathbb{R}^n$, derived soon after by means of perturbative AKSZ quantization of the Poisson sigma model by Cattaneo and Felder,[83] who also provided a globally defined star product for any (N, Π) in.[84]

In order to spell out our proposal, we need one more basic ingredient of Vasiliev's models and their FCS generalizations. In addition to the aforementioned fusion of their Cartan modules into associative fiber algebras, their simplicity relies on yet one more unification, namely of spacetime and noncommutative symplectic manifolds into extended base manifolds of Poisson type so as to facilitate the construction of closed and central two-forms whose star products with the zero-form provide (nontrivial) co-cycles for the curvatures. The resulting framework is that of differential Poisson manifolds, which are natural generalizations of Poisson manifolds on which classes of differential forms (and distributions) in different degrees, and not just functions, can be equipped by star products by deforming the classical wedge product along differential Poisson brackets. In the case of sufficiently smooth objects, the resulting pointwise star product reads

$$f \star g = f \wedge g + i\hbar\{f, g\} + \cdots , \tag{227}$$

where the differential Poisson bracket, that now involves the bi-vector as well as a compatible connection, is of the form[s]

$$\{\omega, \eta\} = \Pi^{\alpha\beta}\, \nabla_\alpha \omega \wedge \nabla_\beta \eta + (-1)^{\deg(\omega)}\, \widetilde{R}^{\alpha\beta} \wedge i_\alpha \omega \wedge i_\beta \eta , \tag{228}$$

where $\widetilde{R}^{\alpha\beta} = \Pi^{\alpha\gamma}\widetilde{R}^\alpha{}_\gamma$ and $\widetilde{R}^\alpha{}_\beta$ is the curvature of the connection shifted by the torsion. When written on the above form, the bracket is graded skew-symmetric, of vanishing intrinsic degree, compatible with the de Rham differential and obeying Leibniz rule, while the graded Jacobi identity requires the additional conditions [t]

$$\Pi^{\delta[\alpha} T^{\beta}_{\delta\epsilon} \Pi^{\gamma]\epsilon} = 0 , \qquad \Pi^{\alpha\rho}\Pi^{\sigma\beta} R_{\rho\sigma}{}^{\gamma}{}_{\delta} = 0 , \tag{229}$$

$$\Pi^{\alpha\lambda}\nabla_\lambda \widetilde{R}_{\beta\gamma}{}^{\rho\sigma} = 0 , \qquad \widetilde{R}_{\epsilon[\rho}{}^{(\alpha\beta}\widetilde{R}_{\sigma\lambda]}{}^{\gamma)\epsilon} = 0 . \tag{230}$$

By extending the formality theorem to the graded supermanifold with coordinates $(\phi^\alpha, \theta^\alpha)$ (see[46] for a related discussion), the $\hbar$-expansion of the star product can be determined together with a nilpotent operator, given by an $\hbar$-deformation of the de Rham differential, by requiring associativity and compatibility. In particular, in degree zero, the star product must yield Cattaneo and Felder's manifestly covariant form of Kontsevich product (which in its turn reproduces Fedosov's product in the symplectic case).

Motivated by the above considerations, it was proposed in[1] that the differential star product algebra can be obtained by perturbative quantization of the supersymmetric two-dimensional topological sigma model based on the (classical) action

$$S = \int_\Sigma \left(\eta_\alpha \wedge d\phi^\alpha + \chi_\alpha \wedge \nabla\theta^\alpha + \tfrac{1}{2}\,\Pi^{\alpha\beta}\eta_\alpha \wedge \eta_\beta + \tfrac{1}{4}\theta^\alpha\theta^\beta \widetilde{R}_{\alpha\beta}{}^{\gamma\delta}\, \chi_\gamma \wedge \chi_\delta \right) . \tag{231}$$

Indeed, the conditions on the background fields required by the Jacobi identity, as listed above, can equivalently be derived by requiring the appropriate two-dimensional gauge symmetries, which serve to gauge away all local degrees of freedom except the constant modes in ϕ^α and θ^α. Moreover, the coefficient of the four-Fermi coupling is fixed by the requirement that the action has a global nilpotent supersymmetry given by

$$\delta_{\rm f}\phi^\alpha = \theta^\alpha , \qquad \delta_{\rm f}\theta^\alpha = 0, \tag{232}$$

$$\delta_{\rm f}\eta_\alpha = \tfrac{1}{2}\widetilde{R}_{\beta\gamma}{}^{\delta}{}_{\alpha}\, \chi_\delta\, \theta^\beta\theta^\gamma - \Gamma^\gamma_{\alpha\beta}\, \eta_\gamma\, \theta^\beta , \qquad \delta_{\rm f}\chi_\alpha = -\eta_\alpha + \Gamma^\gamma_{\alpha\beta}\, \chi_\gamma\, \theta^\beta . \tag{233}$$

This transformations can be identified as an avatar for the de Rham differential upon representing the forms on N as functions on the parity shifted bundle $T[1]N$

[s]Strictly speaking, there remains one possible tensorial deformation; see[1] for a more detailed form of the Poisson bracket.

[t]The quadratic constraint on the curvature tensor is a Yang–Baxter equation and the two-dimensional differential Poisson sigma model can thus be used as a framework for associative algebras including Hopf algebras.

coordinatized by $(\phi^\alpha, \theta^\alpha)$. It can be shown that the model can be reformulated on more general $(n|n)$ supermanifolds equipped with super Poisson bi-vectors. The global symmetries of this model, which include the original Hamiltonian vector fields as well as more general super Killing vectors, such as δ_{f}, can be gauged.

The special case of gauging of δ_{f} was studied in.[14] It remains to be examined whether it is consistent at the quantum level, which may require extra conditions on differential Poisson geometry (*e.g.* its Ricci tensor), though the absence of obstructions at the first sub-leading order in $\hbar$[46] suggests their absence to all orders in perturbation theory. Assuming quantum consistency, it is natural to propose that finite deformations of the background can be modelled by a topological open string field Ψ obeying

$$\{Q, \Psi\}_\star + \Psi \star \Psi = 0 \ , \tag{234}$$

where the BRST operator Q contains a sector that gauges the de Rham differential acting on the zero-modes of ϕ^α. We claim that the FCS formulation of four-dimensional higher spin gravity in twistor space can be obtained by a reduction of the above system down to a finite set of modes describing stretched strings, whereas in order to obtain the minimal bosonic models[u] based on vector oscillators one has to gauge the additional $sp(2)$ Killing vectors generated by the moment functions of the conformal particle.

Finally, let us remark on the connection to tensionless strings in anti-de Sitter spacetime. To this end, it is important that the associative algebras in higher spin gravity are of a special type related to singletons. These algebras can be realized either via the group algebras (or enveloping algebras) of the underlying finite-dimensional isometry groups or oscillator algebras over ideals given by singleton annihilators [v], which can be thought of as the equations of motions for conformal particles on the embedding space of anti-de Sitter spacetime. Thus one may view the topological open string as the germ of a tensionless string, consisting of two string partons orbiting (as conformal particles) around a center-of-mass. Indeed, similar excitations, carrying the quantum numbers of singletons, are known to arise in the form of cusps on tensionful closed boconic strings in D-dimensional anti-de Sitter space time in the semi-classical limit (described by soliton solutions of the of the Nambu–Goto action). Thus, as proposed in,[42] one may think of the Hagedorn transition in flat spacetime as switching on a small negative cosmological constant whereby the cyclically ordered one-string states of tensionful strings break up in the tensionless limit into totally symmetric multi-singleton states. In particular, on physical grounds, long (folded) string states, which thus connect two cusps at the opposite side of the center-of-mass (with closed worldlines in periodic anti-de Sitter

[u]These models are known as the D-dimensional Type A models. The D-dimensional Type B models should arise in a similar fashion from gauging an $osp(1|2)$ algebra.

[v]The topological open string approach may be useful in further elucidating the whether there exist two dual underlying first-quantized formulations, one on the group manifold, thought of as a Poisson manifold, and another one directly on the singleton phase space, thought of as a symplectic manifold.

spacetime), should remain self-interacting in this limit, at least in the classical limit of the string field theory, and admit a first-quantized description as an $sp(2)$-gauged topological open bosonic string with en effective field theory description in terms of minimal bosonic higher spin gravity.

In summary, one further piece of motivation for studying topological open strings of the type proposed above is thus that they may provide a link between higher spin gravity and tensionless strings in anti-de Sitter spacetime, and possibly also a glimpse of what one might expect from a background independent formulation of closed string field theory.

Acknowledgments

We thank Th. Basile, A. Campoleoni, V. Didenko, C. Iazeolla, E. Skvortsov, Ph. Spindel, T. Procházka and M. Vasiliev for discussions. N.B., Associate F.R.S.-FNRS Researcher, is very grateful to the Institute of Advanced Studies from Nanyang Technological University in Singapore for hospitality, and to the organisers of the workshop "Higher Spin Gauge Theories" for providing the opportunity to present this work and contribute to the proceedings. The work of E.S. is supported in part by NSF grant PHY-1214344. P. S. would like to acknowledge the support of CONICYT grant DPI 20140115 and FONDECYT grant Regular 1140296.

References

1. C. Arias, N. Boulanger, P. Sundell, and A. Torres-Gomez, 2D sigma models and differential Poisson algebras, *JHEP.* **08**, 095, (2015). doi: 10.1007/JHEP08(2015)095.
2. N. Boulanger, P. Sundell, and M. Valenzuela, Gravitational and gauge couplings in Chern-Simons fractional spin gravity, *JHEP.* **01**, 173, (2016). doi: 10.1007/JHEP01(2016)173.
3. N. Boulanger, E. Sezgin, and P. Sundell, 4D Higher Spin Gravity with Dynamical Two-Form as a Frobenius–Chern–Simons Gauge Theory, (2015) [arXiv:1505.04957 [hep-th]].
4. R. Bonezzi, N. Boulanger, E. Sezgin, and P. Sundell, An Action for Matter Coupled Higher Spin Gravity in Three Dimensions, *JHEP* **1605**, 003 (2016). doi: 10.1007/JHEP05(2016)003.
5. E. Sezgin and P. Sundell, Geometry and Observables in Vasiliev's Higher Spin Gravity, *JHEP.* **07**, 121, (2012). doi: 10.1007/JHEP07(2012)121.
6. N. Boulanger and P. Sundell, An action principle for Vasiliev's four-dimensional higher-spin gravity, *J.Phys.* **A44**, 495402, (2011). doi: 10.1088/1751-8113/44/49/495402.
7. A. Achucarro and P. Townsend, A Chern-Simons Action for Three-Dimensional anti-De Sitter Supergravity Theories, *Phys.Lett.* **B180**, 89, (1986). doi: 10.1016/0370-2693(86)90140-1.
8. S. F. Prokushkin and M. A. Vasiliev, Higher-spin gauge interactions for massive matter fields in 3D AdS space-time, *Nucl. Phys.* **B545**, 385, (1999). doi: 10.1016/S0550-3213(98)00839-6.
9. M. P. Blencowe, A Consistent Interacting Massless Higher Spin Field Theory in $D = (2+1)$, *Class. Quant. Grav.* **6**, 443, (1989). doi: 10.1088/0264-9381/6/4/005.

10. A. Campoleoni, S. Fredenhagen, S. Pfenninger, and S. Theisen, Asymptotic symmetries of three-dimensional gravity coupled to higher-spin fields, *JHEP.* **1011**, 007, (2010). doi: 10.1007/JHEP11(2010)007.
11. M. Henneaux and S. J. Rey, Nonlinear $W_{infinity}$ as asymptotic symmetry of three-dimensional higher spin anti-de Sitter gravity, *JHEP* **1012**, 007, (2010). doi:10.1007/JHEP12(2010)007 [arXiv:1008.4579 [hep-th]].
12. M. R. Gaberdiel and R. Gopakumar, An AdS_3 Dual for Minimal Model CFTs, *Phys.Rev.* **D83**, 066007, (2011). doi: 10.1103/PhysRevD.83.066007.
13. M. R. Gaberdiel and R. Gopakumar, Minimal Model Holography, *J. Phys.* **A46**, 214002, (2013). doi: 10.1088/1751-8113/46/21/214002.
14. R. Bonezzi, P. Sundell, and A. Torres-Gomez, 2D Poisson Sigma Models with Gauged Vectorial Supersymmetry, *JHEP.* **08**, 047, (2015). doi: 10.1007/JHEP08(2015)047.
15. A. S. Cattaneo and G. Felder, Poisson sigma models and deformation quantization, *Mod. Phys. Lett.* **A16**, 179–190, (2001). ISSN 0217-7323. doi: 10.1142/S0217732301003255. URL `http://dx.doi.org/10.1142/S0217732301003255`.
16. M. A. Vasiliev, Consistent equation for interacting gauge fields of all spins in (3+1)-dimensions, *Phys. Lett.* **B243**, 378–382, (1990). ISSN 0370-2693. doi: 10.1016/0370-2693(90)91400-6.
17. M. A. Vasiliev, More on equations of motion for interacting massless fields of all spins in (3+1)-dimensions, *Phys. Lett.* **B285**, 225–234, (1992). doi: 10.1016/0370-2693(92)91457-K.
18. M. A. Vasiliev, Nonlinear equations for symmetric massless higher spin fields in (A)dS(d), *Phys. Lett.* **B567**, 139–151, (2003). doi: 10.1016/S0370-2693(03)00872-4.
19. C. Fronsdal, Massless Fields with Integer Spin, *Phys. Rev.* **D18**, 3624, (1978). doi: 10.1103/PhysRevD.18.3624.
20. M. A. Vasiliev, Free massless fields of arbitrary spin in the de Sitter space and initial data for a higher spin superalgebra, *Fortsch.Phys.* **35**, 741–770, (1987).
21. V. Lopatin and M. A. Vasiliev, Free massless bosonic fields of arbitrary spin in d-dimensional de sitter space, *Mod. Phys. Lett.* **A3**, 257, (1988). doi: 10.1142/S0217732388000313.
22. M. A. Vasiliev, Cubic interactions of bosonic higher spin gauge fields in AdS(5), *Nucl. Phys.* **B616**, 106–162, (2001). doi: 10.1016/S0550-3213(01)00433-3.
23. E. S. Fradkin and M. A. Vasiliev, Cubic Interaction in Extended Theories of Massless Higher Spin Fields, *Nucl. Phys.* **B291**, 141, (1987). doi: 10.1016/0550-3213(87)90469-X.
24. M. Vasiliev, Cubic Vertices for Symmetric Higher-Spin Gauge Fields in $(A)dS_d$, *Nucl.Phys.* **B862**, 341–408, (2012). doi: 10.1016/j.nuclphysb.2012.04.012.
25. I. L. Buchbinder, A. Fotopoulos, A. C. Petkou and M. Tsulaia, Constructing the cubic interaction vertex of higher spin gauge fields, *Phys. Rev. D* **74**, 105018, (2006). doi:10.1103/PhysRevD.74.105018 [hep-th/0609082].
26. R. R. Metsaev, Gravitational and higher-derivative interactions of massive spin 5/2 field in (A)ds space, *Phys. Rev. D* **77**, 025032, (2008). doi:10.1103/PhysRevD.77.025032 [hep-th/0612279].
27. A. Fotopoulos, N. Irges, A. C. Petkou and M. Tsulaia, Higher-spin gauge fields interacting with scalars: The Lagrangian cubic vertex, *JHEP* **0710**, 021, (2007). doi:10.1088/1126-6708/2007/10/021 [arXiv:0708.1399 [hep-th]].
28. Y. M. Zinoviev, On spin 3 interacting with gravity, *Class. Quant. Grav.* **26**, 035022, (2009). doi:10.1088/0264-9381/26/3/035022 [arXiv:0805.2226 [hep-th]].
29. N. Boulanger, S. Leclercq and P. Sundell, On the uniqueness of minimal coupling in higher-spin gauge theory, *JHEP* **0808**, 056, (2008). doi: 10.1088/1126-6708/2008/08/056 [arXiv:0805.2764 [hep-th]].

30. N. Boulanger, E. D. Skvortsov and Y. M. Zinoviev, Gravitational cubic interactions for a simple mixed-symmetry gauge fields in AdS and flat backgrounds, *J. Phys. A* **44**, 415403, (2011). doi:10.1088/1751-8113/44/41/415403 [arXiv:1107.1872 [hep-th]].
31. E. Joung, L. Lopez, and M. Taronna, Generating functions of (partially-)massless higher-spin cubic interactions, *JHEP.* **1301**, 168, (2013). doi: 10.1007/JHEP01(2013)168.
32. N. Boulanger, D. Ponomarev, and E. Skvortsov, Non-abelian cubic vertices for higher-spin fields in anti-de Sitter space, *JHEP.* **1305**, 008, (2013). doi: 10.1007/JHEP05(2013)008.
33. E. Joung and M. Taronna, Cubic-interaction-induced deformations of higher-spin symmetries, *JHEP.* **1403**, 103, (2014). doi: 10.1007/JHEP03(2014)103.
34. X. Bekaert, N. Boulanger, and P. A. Sundell, How higher-spin gravity surpasses the spin-two barrier, *Rev. Mod. Phys.* **84**, 987–1009 (Jul, 2012). doi: 10.1103/RevModPhys.84.987.
35. M. Alexandrov, M. Kontsevich, A. Schwartz, and O. Zaboronsky, The Geometry of the master equation and topological quantum field theory, *Int. J. Mod. Phys.* **A12**, 1405–1430, (1997). ISSN 0217-751X. doi: 10.1142/S0217751X97001031.
36. N. Boulanger, N. Colombo, and P. Sundell, A minimal BV action for Vasiliev's four-dimensional higher spin gravity, *JHEP.* **1210**, 043, (2012). doi: 10.1007/JHEP10(2012)043.
37. M. A. Vasiliev, Consistent equations for interacting massless fields of all spins in the first order in curvatures, *Annals Phys.* **190**, 59–106, (1989). doi: 10.1016/0003-4916(89)90261-3.
38. N. Doroud and L. Smolin, An Action for higher spin gauge theory in four dimensions. (2011).
39. N. Colombo and P. Sundell, Twistor space observables and quasi-amplitudes in 4D higher spin gravity, *JHEP.* **1111**, 042, (2011). doi: 10.1007/JHEP11(2011)042.
40. N. Colombo and P. Sundell, Higher Spin Gravity Amplitudes From Zero-form Charges. (2012).
41. V. Didenko and E. Skvortsov, Exact higher-spin symmetry in CFT: all correlators in unbroken Vasiliev theory, *JHEP.* **1304**, 158, (2013). doi: 10.1007/JHEP04(2013)158.
42. J. Engquist and P. Sundell, Brane partons and singleton strings, *Nucl. Phys.* **B752**, 206–279, (2006). doi: 10.1016/j.nuclphysb.2006.06.040.
43. C.-S. Chu and P.-M. Ho, Poisson algebra of differential forms, *Int. J. Mod. Phys.* **A12**, 5573–5587, (1997). doi: 10.1142/S0217751X97002929.
44. E. Beggs and S. Majid, Semiclassical differential structures. (2003).
45. A. Tagliaferro, The Star Product for Differential Forms on Symplectic Manifolds. (2008).
46. S. McCurdy and B. Zumino, Covariant Star Product for Exterior Differential Forms on Symplectic Manifolds, *AIP Conf.Proc.* **1200**, 204–214, (2010). doi: 10.1063/1.3327559.
47. D. Quillen, Superconnections and the chern character, *Topology.* **24**(1), 89 – 95, (1985). ISSN 0040-9383. doi: http://dx.doi.org/10.1016/0040-9383(85)90047-3. URL `http://www.sciencedirect.com/science/article/pii/0040938385900473`.
48. C. Iazeolla and P. Sundell, Families of exact solutions to Vasiliev's 4D equations with spherical, cylindrical and biaxial symmetry, *JHEP.* **1112**, 084, (2011). doi: 10.1007/JHEP12(2011)084.
49. M. Vasiliev, Invariant Functionals in Higher-Spin Theory. (2015).
50. V. E. Didenko, A. S. Matveev, and M. A. Vasiliev, Unfolded Description of AdS_4 Kerr Black Hole, *Phys. Lett.* **B665**, 284–293, (2008). doi: 10.1016/j.physletb.2008.05.067.

51. M. A. Vasiliev, Holography, Unfolding and Higher-Spin Theory, *J. Phys.* **A46**, 214013, (2013). doi: 10.1088/1751-8113/46/21/214013.
52. M. A. Vasiliev, Star-Product Functions in Higher-Spin Theory and Locality, *JHEP.* **06**, 031, (2015). doi: 10.1007/JHEP06(2015)031.
53. C. Zachos, D. Fairlie, and T. Curtright, *Quantum mechanics in phase space: an overview with selected papers.* vol. 34, (World Scientific Publishing Company Incorporated, 2005).
54. C. Iazeolla, E. Sezgin and P. Sundell, Real forms of complex higher spin field equations and new exact solutions, *Nucl. Phys.* **B791**, 231, (2008). doi: 10.1061/j.nuclphysb.2007.08.002 [arXiv:0706.2983 [hep-th]].
55. M. A. Vasiliev, Higher spin algebras and quantization on the sphere and hyperboloid, *Int. J. Mod. Phys.* **A6**, 1115–1135, (1991). doi: 10.1142/S0217751X91000605.
56. V. Bargmann, Irreducible unitary representations of the Lorentz group, *Annals Math.* **48**, 568–640, (1947). doi: 10.2307/1969129.
57. A. O. Barut and C. Fronsdal, On non-compact groups, ii. representations of the 2+1 lorentz group, *Proc. Roy. Soc. London.* **A287**, 532–548, (1965).
58. J. Leinaas and J. Myrheim, On the theory of identical particles, *Nuovo Cim.* **B37**, 1–23, (1977). doi: 10.1007/BF02727953.
59. F. Wilczek, Quantum Mechanics of Fractional Spin Particles, *Phys.Rev.Lett.* **49**, 957, (1982). doi: 10.1103/PhysRevLett.49.957.
60. S. Forte, Quantum mechanics and field theory with fractional spin and statistics, *Rev.Mod.Phys.* **64**, 193–236, (1992). doi: 10.1103/RevModPhys.64.193.
61. N. Boulanger, P. Sundell, and M. Valenzuela, Three-dimensional fractional-spin gravity, *JHEP.* **1402**, 052, (2014). doi: 10.1007/JHEP02(2014)052.
62. M. A. Vasiliev, Higher spin gauge theories: Star-product and AdS space, in *The Many Faces of Superworld. Yuri Golfand Memorial Volume*, ed. M. A. Shifman (World Scientific, 2000), pp. 533–610.
63. A. Campoleoni, T. Prochazka, and J. Raeymaekers, A note on conical solutions in 3D Vasiliev theory, *JHEP.* **05**, 052, (2013). doi: 10.1007/JHEP05(2013)052.
64. B. Khesin and F. Malikov, Universal Drinfeld-Sokolov reduction and matrices of complex size, *Commun. Math. Phys.* **175**, 113–134, (1996). doi: 10.1007/BF02101626.
65. E. P. Wigner, Do the Equations of Motion Determine the Quantum Mechanical Commutation Relations?, *Phys. Rev.* **77**, 711–712, (1950). doi: 10.1103/PhysRev.77.711.
66. L. M. Yang, A Note on the Quantum Rule of the Harmonic Oscillator, *Phys. Rev.* **84**, 788–790, (1951). doi: 10.1103/PhysRev.84.788.
67. M. Plyushchay, Deformed Heisenberg algebra, fractional spin fields and supersymmetry without fermions, *Annals Phys.* **245**, 339–360, (1996). doi: 10.1006/aphy.1996.0012.
68. M. S. Plyushchay, Deformed Heisenberg algebra with reflection, *Nucl.Phys.* **B491**, 619–634, (1997). doi: 10.1016/S0550-3213(97)00065-5.
69. E. Majorana, Relativistic theory of particles with arbitrary intrinsic momentum, *Nuovo Cim.* **9**, 335–344, (1932). doi: 10.1007/BF02959557.
70. E. C. G. Sudarshan and N. Mukunda, Quantum theory of the infinite-component majorana field and the relation of spin and statistics, *Phys. Rev.* **D1**, 571–583, (1970). doi: 10.1103/PhysRevD.1.571.
71. P. A. M. Dirac, A positive-energy relativistic wave equation, *Proc. Roy. Soc. Lond.* **A322**, 435–445, (1971). doi: 10.1098/rspa.1971.0077.
72. P. A. M. Dirac, A positive-energy relativistic wave equation. 2., *Proc. Roy. Soc. Lond.* **A328**, 1–7, (1972). doi: 10.1098/rspa.1972.0064.
73. L. P. Staunton, A Spin 1/2 Positive Energy Relativistic Wave Equation, *Phys. Rev.* **D10**, 1760, (1974). doi: 10.1103/PhysRevD.10.1760.

74. M. A. Vasiliev, Unfolded representation for relativistic equations in (2+1) anti-De Sitter space, *Class. Quant. Grav.* **11**, 649–664, (1994).
75. M. A. Vasiliev, Equations of motion for d = 3 massless fields interacting through Chern-Simons higher spin gauge fields, *Mod. Phys. Lett.* **A7**, 3689–3702, (1992). doi: 10.1142/S0217732392003116.
76. P. Kessel, G. L. Gomez, E. Skvortsov, and M. Taronna, Higher Spins and Matter Interacting in Dimension Three. (2015).
77. C. Iazeolla and J. Raeymaekers, On big crunch solutions in Prokushkin-Vasiliev theory, *JHEP.* **01**, 177, (2016). doi: 10.1007/JHEP01(2016)177.
78. M. Vasiliev, On Conformal, SL(4,R) and Sp(8,R) Symmetries of 4d Massless Fields, *Nucl.Phys.* **B793**, 469–526, (2008). doi: 10.1016/j.nuclphysb.2007.10.017.
79. E. Sezgin and P. Sundell, Supersymmetric Higher Spin Theories, *J.Phys.* **A46**, 214022, (2013). doi: 10.1088/1751-8113/46/21/214022.
80. S. F. Prokushkin, A. Y. Segal, and M. A. Vasiliev, Coordinate free action for AdS(3) higher spin matter systems, *Phys.Lett.* **B478**, 333–342, (2000). doi: 10.1016/S0370-2693(00)00258-6.
81. M. De Wilde and P. B. Lecomte, Existence of star-products and of formal deformations of the poisson lie algebra of arbitrary symplectic manifolds, *Letters in Mathematical Physics.* **7**(6), 487–496, (1983).
82. B. v. Fedosov, A Simple geometrical construction of deformation quantization, *J. Diff. Geom.* **40**(2), 213–238, (1994).
83. A. S. Cattaneo and G. Felder, A path integral approach to the Kontsevich quantization formula, *Commun. Math. Phys.* **212**, 591–611, (2000). ISSN 0010-3616. doi: 10.1007/s002200000229.
84. A. S. Cattaneo and G. Felder, On the globalization of Kontsevich's star product and the perturbative Poisson sigma model, *Prog. Theor. Phys. Suppl.* **144**, 38–53, (2001). doi: 10.1143/PTPS.144.38.

Deriving Field Theories for Particles of Arbitrary Spin With and Without Supersymmetry

Sudarshan Ananth
Indian Institute of Science Education and Research
Pune 411008, India

We review the derivation of light-cone interaction vertices for fermionic and bosonic fields of arbitrary spin. The resulting amplitudes and their factorization properties are discussed. We then show how this symmetry-based approach works for theories with extended supersymmetry like $\mathcal{N} = 4$ Yang-Mills theory and $\mathcal{N} = 8$ supergravity.

1. Introduction

At the level of the equations of motion, there has been considerable progress in our understanding of higher spin theories [1]. However, a description of higher spin fields ($\lambda > 2$) in a Lagrangian formalism, essential to quantization, remains elusive. There are various no-go theorems relating to higher spin fields and one of the things they tell us is that we cannot write down a consistent interacting theory of massless higher spin fields in flat spacetime. In contradiction to this statement is the result of [2] where a consistent cubic interaction vertex was constructed in the light-cone formalism. Given that the cubic interaction vertex is in some sense trivial and may be written down based on helicity and dimensional considerations [3] the question then becomes whether a quartic interaction vertex can be arrived at within the same formalism. The discussions in [3] outline why a BCFW [4] type construction is bound to fail but this does not rise to the level of a proof. It is worth noting, at this stage, that most of the no-go literature pertaining to higher spin theories [5] assumed theories with manifest locality and manifest Lorentz invariance. Neither of these properties is manifest in light-cone gauge and this is one motivation to study higher spin fields in this gauge. Another motivation is the exclusive focus on the physical degrees of freedom. Since the first paper on light-cone cubic interaction vertices, there has been considerable work on the subject [6–9].

The procedure we review here, is that initiated in [2]. We revisited their study [9] in the light of recent advances in field theory [10]. These new methods, when introduced to the older analysis of [2] yielded among other results, a Lagrangian origin [11] for the KLT relations [12] and factorization properties [8]. The formalism in [2] starts with just the Poincaré algbera, for four dimensional flat spacetime, in light-cone gauge. Among the generators of this algebra is + component of momentum which is the light-cone Hamiltonian. An ansatz for this Hamiltonian is made based on dimensional analysis and helicity and then refined by requiring closure of the Poincaré algebra.[a] This leads to a class of higher spin cubic vertices.

This framework is particularly interesting in the context of aribtrary spin theories in non-flat spacetime backgrounds. The formalism described here may be extended to AdS_4 (for the spin=2 case, see [13]). The cubic interaction vertices are straightforward to derive in this background but the quartic interaction vertices are challenging. This approach is likely to yield key ingredients necessary to establish a Lagrangian origin to the Vasiliev program [1]. In other words, a derivation of consistent quartic interaction vertices involving higher spin fields in AdS_4 should necessitate the inclusion of a tower of higher spin fields thus providing a Lagrangian origin to the Vasiliev equations of motion.

The approach we review is particularly powerful when applied to supersymmetric theories where we have more than just the Poincaré algebra to work with. Supergravity in eleven dimensions [14], when reduced to $d = 4$, becomes the maximally supersymmetric $\mathcal{N} = 8$ theory. Its light-cone formulation in superspace is incomplete because its higher-point interactions are not known in terms of a light-cone superfield. A simpler, yet in many ways similar, theory in $d = 4$ is the maximally supersymmetric $\mathcal{N} = 4$ Yang-Mills theory. The $\mathcal{N} = 8$ and $\mathcal{N} = 4$ theories are each described by a single light-cone superfield which captures their physical degrees of freedom. Both theories may be oxidized [15] to their higher-dimensional parent theories, yielding superspace descriptions without auxiliary fields. However, while there exist two distinct algebraic methods to derive the entire classical Hamiltonian of $\mathcal{N} = 4$ Yang-Mills there is considerable difficulty in deriving the $\mathcal{N} = 8$ theory even to quartic order.

2. Poincaré Generators

We define light-cone co-ordinates in $(-,+,+,+)$ Minkowski space-time by

$$x^{\pm} = \frac{x^0 \pm x^3}{\sqrt{2}}, \quad x = \frac{x^1 + ix^2}{\sqrt{2}}, \quad \bar{x} = \frac{x^1 - ix^2}{\sqrt{2}}. \tag{1}$$

The corresponding derivatives are $\partial_{\pm}$, $\bar{\partial}$ and ∂. One of the reasons $d = 4$ is so special is that *all* massless fields have exactly two physical degrees of freedom, which we call ϕ and $\bar{\phi}$. The field ϕ has helicity λ while the field $\bar{\phi}$ has helicity $-\lambda$.

[a]In light-cone gauge, Poincaré invariance needs to be checked as the formalism is not manifestly covariant.

The light-cone generators of the Poincaré algebra are

$$p^- = i\frac{\partial\bar{\partial}}{\partial_-} = -p_+ \quad p^+ = -i\partial^+ = -p_- \quad \bar{p} = -i\bar{\partial}, \tag{2}$$

$$\begin{aligned} j &= i(x\bar{\partial} - \bar{x}\partial - \lambda), \qquad j^+ = (x^+\partial - x\partial^+), \\ j^{+-} &= (x^+\frac{\partial\bar{\partial}}{\partial^+} - x^-\partial^+), \quad j^- = (x^-\partial - x\frac{\partial\bar{\partial}}{\partial^+} + \lambda\frac{\partial}{\partial^+}), \end{aligned} \tag{3}$$

and their complex conjugates with $\frac{1}{\partial_-}$ defined following [16]. For the free theory, $\partial_+ = \frac{\partial\bar{\partial}}{\partial_-}$ and the Hamiltonian reads

$$H \equiv \int d^3x\,\mathcal{H} = -\int d^3x\,\bar{\phi}\,\partial\bar{\partial}\,\phi, \tag{4}$$

also written as

$$H \equiv \int d^3x\,\mathcal{H} = \int d^3x\,\partial_-\bar{\phi}\,\delta_{\mathcal{H}}\phi, \tag{5}$$

where the time translation operator is introduced through the Poisson bracket

$$\delta_{\mathcal{H}}\phi \equiv \partial_+\phi = \{\phi, \mathcal{H}\}. \tag{6}$$

On the light-cone, spacetime symmetries split into two types. Kinematical symmetries are unaltered by interactions while dynamical symmetries pick up corrections and are *non-linearly* realized on the fields. In supersymmetric theories, the supersymmetries also separate into dynamical and kinematical supersymmetries. For the interacting theory, $\delta_{\mathcal{H}}$ picks up corrections, order by order in the coupling constant α, as do

$$\delta_{j^{+-}}\phi; \quad \delta_{j^-}\phi; \quad \delta_{\bar{j}^-}\phi. \tag{7}$$

These corrections, non-linear in nature, need to be constructed.

3. Bosonic Fields

We review here, the procedure to derive cubic interaction vertices for three bosonic fields of arbitrary spin. For additional details, we refer the reader to [8]. At cubic order, the following structures (at order α) appear in the Hamiltonian

$$\delta^{\alpha}_{\mathcal{H}}\phi_1 \sim \phi_2\phi_3; \quad \delta^{\alpha}_{\mathcal{H}}\phi_2 \sim \phi_1\phi_3; \quad \delta^{\alpha}_{\mathcal{H}}\phi_3 \sim \phi_1\phi_2. \tag{8}$$

The fields ϕ_1, ϕ_2 and ϕ_3 have integer spins λ_1, λ_2 and λ_3 respectively. In terms of the action, the first structure would correspond to

$$S \sim \int d^4x\,\bar{\phi}_1\,\phi_2\,\phi_3 + c.c. \tag{9}$$

We sprinkle in derivatives to arrive at the ansatz

$$\delta^{\alpha}_{\mathcal{H}}\phi_1 = \alpha\,A\,\partial^{+\mu}\left[\bar{\partial}^a\partial^{+\rho}\phi_2\bar{\partial}^b\partial^{+\sigma}\phi_3\right] + c.c. \tag{10}$$

μ, ρ, σ, a, b are integers and A is a number. The commutators

$$\begin{aligned} [\delta_j, \delta^\alpha_{\mathcal{H}}]\,\phi_1 &= 0, \\ \left[\delta_{j^{+-}}, \delta^\alpha_{\mathcal{H}}\right]\phi_1 &= -\delta_{\mathcal{H}}\phi_1, \end{aligned} \tag{11}$$

imply that

$$\begin{aligned} a+b &= \lambda_2+\lambda_3-\lambda_1 \\ \mu+\rho+\sigma &= -1. \end{aligned} \tag{12}$$

As $a, b > 0$, the first condition implies that the vertex exists only if $\lambda_2+\lambda_3 > \lambda_1$. Let $\lambda \equiv \lambda_2+\lambda_3-\lambda_1$ so $a+b=\lambda$. The ansatz is now a sum of the $(\lambda+1)$ terms

$$\delta^\alpha_H\phi_1 = \alpha\sum_{n=0}^{\lambda} A_n\,\partial^{+\mu_n}\left[\bar\partial^n\partial^{+\rho_n}\phi_2\,\bar\partial^{(\lambda-n)}\partial^{+\sigma_n}\phi_3\right] + c.c. \tag{13}$$

The commutators

$$\left[\delta_{\bar{j}^-}, \delta_H\right]^\alpha\phi_1 = 0 \quad \left[\delta_{j^+}, \delta_H\right]^\alpha\phi_1 = 0, \tag{14}$$

yield the conditions

$$\begin{aligned} \sum_{n=0}^{\lambda} A_n\,&\{(\mu_n+1-\lambda_1)\partial^{+(\mu_n-1)}\bar\partial(\bar\partial^n\partial^{+\rho_n}\phi_2\bar\partial^{(\lambda-n)}\partial^{+\sigma_n}\phi_3) \\ &+(\rho_n+\lambda_2)\partial^{+\mu_n}(\bar\partial^{(n+1)}\partial^{+(\rho_n-1)}\phi_2\bar\partial^{(\lambda-n)}\partial^{+\sigma_n}\phi_3) \\ &+(\sigma_n+\lambda_3)\partial^{+\mu_n}(\bar\partial^n\partial^{+\rho_n}\phi_2\bar\partial^{(\lambda-n+1)}\partial^{+(\sigma_n-1)}\phi_3)\} = 0, \end{aligned} \tag{15}$$

$$\begin{aligned} \sum_{n=0}^{\lambda} A_n\{&n\,\partial^{+\mu_n}(\bar\partial^{(n-1)}\partial^{+(\rho_n+1)}\phi_2\bar\partial^{(\lambda-n)}\partial^{+\sigma_n}\phi_3) \\ &+(\lambda-n)\partial^{+\mu_n}(\bar\partial^n\partial^{+\rho_n}\phi_2\bar\partial^{(\lambda-n-1)}\partial^{+(\sigma_n+1)}\phi_3)\} = 0, \end{aligned} \tag{16}$$

which are satisfied when the coefficients obey

$$\begin{aligned} A_{n+1} &= -\frac{(\lambda-n)}{(n+1)}A_n = (-1)^{(n+1)}\binom{\lambda}{n+1}A_0, \\ \rho_{n+1} &= \rho_n-1; \quad \sigma_{n+1}=\sigma_n+1; \quad \mu_{n+1}=\mu_n, \end{aligned} \tag{17}$$

with the conditions

$$\rho_{\,n=\lambda} = -\lambda_2 \quad \sigma_{\,n=0} = -\lambda_3. \tag{18}$$

We find

$$\rho_n = \lambda-\lambda_2-n; \quad \sigma_n = n-\lambda_3; \quad \mu_n = \lambda_1-1. \tag{19}$$

Thus (13) becomes

$$\delta^\alpha_H\phi_1 = \alpha\sum_{n=0}^{\lambda}(-1)^n\binom{\lambda}{n}\partial^{+(\lambda_1-1)}\left[\bar\partial^n\partial^{+(\lambda-\lambda_2-n)}\phi_2\,\bar\partial^{(\lambda-n)}\partial^{+(n-\lambda_3)}\phi_3\right] + c.c. \tag{20}$$

Using

$$H = \int d^3x \; \partial_- \bar{\phi}_1 \, \delta_{\mathcal{H}} \phi_1, \tag{21}$$

the interacting Hamiltonian is

$$H^\alpha = \alpha \int d^3x \sum_{n=0}^{\lambda} (-1)^n \binom{\lambda}{n} \bar{\phi}_1 \, \partial^{+\lambda_1} \left[\bar{\partial}^n \partial^{+(\lambda-\lambda_2-n)} \phi_2 \, \bar{\partial}^{(\lambda-n)} \partial^{+(n-\lambda_3)} \phi_3 \right] + c.c. \tag{22}$$

The power of this approach is clear. One example: H^α vanishes for $\lambda_1 = \lambda_2 = \lambda_3 = \lambda'$ in (22) for odd λ' making the introduction of a gauge group mandatory for odd integer spins.

3.1. *Factorization and perturbative ties*

The above results may be written in the language of spinor helicity products [10]

$$\langle kl \rangle = \sqrt{2} \, \frac{(kl_- - lk_-)}{\sqrt{k_- l_-}} \qquad [kl] = \sqrt{2} \, \frac{(\bar{k}l_- - \bar{l}k_-)}{\sqrt{k_- l_-}}. \tag{23}$$

Equation (22) contains two kinds of terms: $\bar{\phi}\phi\phi$ and $\phi\bar{\phi}\bar{\phi}$ and the Fourier coefficient of the second variety $\phi_1(p)\bar{\phi}_2(k)\bar{\phi}_3(l) \, \delta^4(p+k+l)$ is

$$\frac{p_-^{\lambda_1}}{k_-^{\lambda_2} l_-^{\lambda_3}} (lk_- - l_-k)^{\lambda_2+\lambda_3-\lambda_1}, \tag{24}$$

which simplifies to

$$\frac{1}{\sqrt{2^\lambda}} \, \langle pk \rangle^{(-\lambda_1+\lambda_2-\lambda_3)} \, \langle kl \rangle^{(\lambda_1+\lambda_2+\lambda_3)} \, \langle lp \rangle^{(-\lambda_1-\lambda_2+\lambda_3)}. \tag{25}$$

This means that given vertices for $(\lambda_1, \lambda_2, \lambda_3)$ and $(\lambda_1', \lambda_2', \lambda_3')$, their product yields the vertex for $(\lambda_1+\lambda_1', \lambda_2+\lambda_2', \lambda_3+\lambda_3')$. Further, the coefficient for $(n\lambda_1, n\lambda_2, n\lambda_3)$ is that for $(\lambda_1, \lambda_2, \lambda_3)$ raised to the power n [8]. In particular, $\lambda_1 = \lambda_2 = \lambda_3 = 1$ and $\lambda_1 = \lambda_2 = \lambda_3 = 2$ offers a field-theoretic realization [11] of the KLT relations [12].

4. Fermionic Fields

We now introduce fermions into this formalism so we can construct interactions involving matter. The 4×4 Gamma matrices are

$$\gamma^0 = \begin{pmatrix} 0 & 1 \\ 1 & 0 \end{pmatrix} \qquad \gamma^i = \begin{pmatrix} 0 & \sigma^i \\ -\sigma^i & 0 \end{pmatrix} \qquad C = \begin{pmatrix} i\sigma^2 & 0 \\ 0 & -i\sigma^2 \end{pmatrix}, \tag{26}$$

and we define

$$P_+ \equiv \frac{1}{2} \gamma_+ \gamma_- \tag{27}$$

$$P_- \equiv \frac{1}{2} \gamma_- \gamma_+, \tag{28}$$

where

$$\gamma_+ = \frac{1}{\sqrt{2}}(\gamma_0 + \gamma_3)$$
$$\gamma_- = \frac{1}{\sqrt{2}}(\gamma_0 - \gamma_3). \tag{29}$$

To a Dirac spinor with Grassmann valued components

$$\Psi = \begin{pmatrix} \chi_1 \\ \chi_2 \\ \chi_3 \\ \chi_4 \end{pmatrix}. \tag{30}$$

we apply the Majorana condition $\Psi = C\bar{\Psi}^T$ to find

$$\chi_1 = \bar{\chi}_4; \quad \chi_2 = -\bar{\chi}_3. \tag{31}$$

We introduce

$$\Psi_+ = P_+\Psi; \quad \Psi_- = P_-\Psi, \tag{32}$$

which satisfy

$$\partial_-\Psi_- = \frac{1}{2}\gamma_-\gamma_i\partial_i\Psi_+ \quad i = 1,2, \tag{33}$$

$$\partial_+\Psi_+ = \frac{1}{2}\gamma_+\gamma_i\partial_i\Psi_- \quad i = 1,2. \tag{34}$$

The first (kinematic) relation yields

$$\chi_1 = \frac{\bar{\partial}}{\partial_-}\chi_2. \tag{35}$$

The entire spinor, in terms of $\chi_3 \equiv \chi$, is

$$\Psi = \begin{pmatrix} -\frac{\bar{\partial}}{\partial_-}\bar{\chi} \\ -\bar{\chi} \\ \chi \\ -\frac{\partial}{\partial_-}\chi \end{pmatrix}. \tag{36}$$

The free light-cone action for a fermionic field of half-integer spin λ is

$$S = \int d^4x\, i\bar{\psi}\,\frac{\Box}{\partial_-}\,\psi. \tag{37}$$

The Poincaré generators are

$$\delta_j\psi = i(x\bar{\partial} - \bar{x}\partial + \lambda)\psi, \qquad \delta_{j^+}\psi = (x^+\partial - x\partial^+)\psi,$$
$$\delta_{j^{+-}}\psi = \left(x^+\frac{\partial\bar{\partial}}{\partial^+} - x^-\partial^+ - \frac{1}{2}\right)\psi, \quad \delta_{j^-}\psi = \left(x^-\partial - x\frac{\partial\bar{\partial}}{\partial^+} + (\lambda + \frac{1}{2}\right)\frac{\partial}{\partial^+})\psi, \tag{38}$$

and their conjugates. Unlike the bosonic case, $\bar{\psi}$ has helicity λ, a positive half integer.

4.1. *Interacting arbitrary spin fields*

The free Hamiltonian involving a massless spin λ boson and a massless spin λ' fermion is

$$H = \int d^3x \left(-\bar{\phi}\,\partial\bar{\partial}\,\phi \; + i\bar{\psi}\,\frac{\partial\bar{\partial}}{\partial_-}\,\psi \right) = \int d^3x \left(\partial_-\bar{\phi}\,\delta_{\mathcal{H}}\phi + i\bar{\psi}\,\delta_{\mathcal{H}}\,\psi \right). \tag{39}$$

We start with an ansatz for the interactions from $\delta^{\alpha}_{\mathcal{H}}\phi_1$ having the structure $\bar{\phi}_1\psi_2\psi_3$ where the fields ϕ_1, ψ_2 and ψ_3 carry spins λ_1 (integer), λ_2 and λ_3 (both half integers) respectively. For a detailed description of the following, we refer the reader to [17]. The ansatz reads

$$\delta^{\alpha}_{\mathcal{H}}\phi_1 = \alpha\, A\, \partial^{+\mu}\left[\partial^{a}\partial^{+\sigma}\psi_2\partial^{b}\partial^{+\rho}\psi_3\right]. \tag{40}$$

Again, μ, ρ, σ, a, b are integers and A is a number. To ensure Lorentz invariance, we demand closure of the Poincaré algebra to this order in α. The kinematical constraints

$$[\delta_j, \delta^{\alpha}_{\mathcal{H}}]\,\phi_1 = 0 \quad \left[\delta_{j^{+-}}, \delta_{\mathcal{H}}\right]^{\alpha}\phi_1 = -\delta^{\alpha}_{\mathcal{H}}\phi_1,$$

impose the constraints

$$a + b = \lambda_1 + \lambda_2 + \lambda_3 \equiv \lambda$$

$$\mu + \rho + \sigma = -2. \tag{41}$$

While

$$\left[\delta^{-}_{j}, \delta_{H}\right]^{\alpha}\phi_1 = 0 \quad \left[\delta_{\bar{j}^{+}}, \delta_{H}\right]^{\alpha}\phi_1 = 0, \tag{42}$$

require that

$$\begin{aligned}\sum_{n=0}^{\lambda} A_n \Bigg\{ & -(\mu_n + 1 + \lambda_1)\partial^{+(\mu_n - 1)}\partial(\partial^{n}\partial^{+\rho_n}\psi_2\partial^{(\lambda-n)}\partial^{+\sigma_n}\psi_3) \\ & - \left(\sigma_n + \lambda_2 + \frac{1}{2}\right)\partial^{+\mu_n}(\partial^{(n+1)}\partial^{+(\sigma_n-1)}\psi_2\partial^{(\lambda-n)}\partial^{+\rho_n}\psi_3) \\ & - \left(\rho_n + \lambda_3 + \frac{1}{2}\right)\partial^{+\mu_n}(\partial^{n}\partial^{+\sigma_n}\psi_2\partial^{(\lambda-n+1)}\partial^{+(\rho_n-1)}\psi_3)\Bigg\} = 0,\end{aligned} \tag{43}$$

$$\begin{aligned}\sum_{n=0}^{\lambda} A_n \{ & n\,\partial^{+\mu_n}(\partial^{(n-1)}\partial^{+(\sigma_n+1)}\psi_2\partial^{(\lambda-n)}\partial^{+\rho_n}\psi_3) \\ & + (\lambda - n)\partial^{+\mu_n}(\partial^{n}\partial^{+\sigma_n}\psi_2\bar{\partial}^{(\lambda-n-1)}\partial^{+(\rho_n+1)}\psi_3)\} = 0.\end{aligned} \tag{44}$$

We find

$$A_n = (-1)^{(n)}\binom{\lambda}{n}; \quad \sigma_n = -\lambda - \lambda_2 + n - \frac{1}{2}; \quad \rho_n = n - (\lambda_3 + \frac{1}{2}); \quad \mu_n = -(\lambda_1 + 1),$$

so (40) becomes

$$\delta_H^\alpha \phi_1 = \alpha \sum_{n=0}^{\lambda} (-1)^n \binom{\lambda}{n} \frac{1}{\partial^{+(\lambda_1+1)}} \left[\partial^n \frac{\partial^{+(\lambda-n)}}{\partial^{+(\lambda_2+\frac{1}{2})}} \psi_2 \, \partial^{(\lambda-n)} \frac{\partial^{+n}}{\partial^{+(\lambda_3+\frac{1}{2})}} \psi_3 \right]. \tag{45}$$

From which we obtain the Hamiltonian and thus the action

$$S = \int d^4x \Bigg[\frac{1}{2} \bar{\phi}_1 \Box \phi_2 + i\bar{\psi}_2 \frac{\Box}{\partial_-} \psi_2 + i\bar{\psi}_3 \frac{\Box}{\partial_-} \psi_3 + \alpha \bar{\phi}_1 \sum_{n=0}^{\lambda} (-1)^n \binom{\lambda}{n} \frac{1}{\partial^{+\lambda_1}} \left[\partial^n \frac{\partial^{+(\lambda-n)}}{\partial^{+(\lambda_2+\frac{1}{2})}} \psi_2 \, \partial^{(\lambda-n)} \frac{\partial^{+n}}{\partial^{+(\lambda_3+\frac{1}{2})}} \psi_3 \right] \Bigg]. \tag{46}$$

Once again, the algebra teaches us that for odd helicity ϕ_1, self-interactions require an internal symmetry group with an antisymmetric structure constant. This permits two varieties of vertices: $t_a \bar{\phi}_1^a \psi_2^b \psi_{3b}$ and $\bar{\phi}_1^a \psi_2^b \psi_3^c f_{abc}$. As an example, consider $(\lambda_1, \lambda_2, \lambda_3) = (1, \frac{1}{2}, -\frac{1}{2})$. With an internal SU(3) symmetry, the first variety represents the coupling of gluons to quarks in QCD while the second variety corresponds to the cubic coupling in $\mathcal{N} = 4$ Yang-Mills [18].

4.2. *Scattering amplitudes*

Rewrite (46) in momentum space using spinor helicity

$$\alpha \int \frac{d^4p}{(2\pi)^4} \frac{d^4k}{(2\pi)^4} \frac{d^4l}{(2\pi)^4} (2\pi)^4 \delta^4(p+k+l) \frac{(k_- l - l_- k)^\lambda}{p_-^{\lambda_1} k_-^{\lambda_2+\frac{1}{2}} l_-^{\lambda_3+\frac{1}{2}}} \tilde{\bar{\phi}}_1(p) \tilde{\psi}_2(k) \tilde{\psi}_3(l), \tag{47}$$

The corresponding amplitude is

$$\langle pk \rangle^{(-\lambda_1+\lambda_2-\lambda_3)} \; \langle kl \rangle^{(\lambda_1+\lambda_2+\lambda_3)} \; \langle lp \rangle^{(-\lambda_1-\lambda_2+\lambda_3)}, \tag{48}$$

of the same form as the three-boson case [8] and consistent with [3, 6, 7, 10].

5. Theories with Supersymmetry: $\mathcal{N} = 4$ Yang-Mills Theory

In this section we discuss this formalism in the context of maximally supersymmetric field theories [2]. The ten-dimensional $\mathcal{N} = 1$ supermultiplet has eight vectors and eight spinors of the little group $SO(8)$. Reduction to four dimensions involves

$$SO(8) \supset SO(2) \times SO(6), \tag{49}$$

yielding

$$\mathbf{8}_v = \mathbf{6}_0 + \mathbf{1}_1 + \mathbf{1}_{-1}, \quad \mathbf{8}_s = \mathbf{4}_{1/2} + \bar{\mathbf{4}}_{-1/2}, \tag{50}$$

with $SO(2)$ subscripts. The ($\mathcal{N} = 4, d = 4$) theory contains six scalar fields, one vector field and four spinor fields (and conjugates). Introduce Grassmann variables θ^m and $\bar{\theta}_m$ ($m, n, \ldots = 1, \ldots 4$) which transform as the $\mathbf{4}$ and $\bar{\mathbf{4}}$ of $SU(4)$. Their derivatives are

$$\bar{\partial}_m \equiv \frac{\partial}{\partial \theta^m}; \quad \partial^m \equiv \frac{\partial}{\partial \bar{\theta}_m}. \tag{51}$$

The physical degrees of freedom of $\mathcal{N}=4$ Yang-Mills are captured in one superfield [18]

$$\phi\,(y)=\frac{1}{\partial^+}\,A\,(y)+\frac{i}{\sqrt{2}}\,\theta^m\,\theta^n\,\overline{C}_{mn}\,(y)+\frac{1}{12}\,\theta^m\,\theta^n\,\theta^p\,\theta^q\,\epsilon_{mnpq}\,\partial^+\,\bar{A}\,(y)$$
$$+\frac{i}{\partial^+}\,\theta^m\,\bar{\chi}_m(y)+\frac{\sqrt{2}}{6}\theta^m\,\theta^n\,\theta^p\,\epsilon_{mnpq}\,\chi^q(y), \tag{52}$$

where the superfield ϕ is not to be confused with the bosonic field, also denoted ϕ, used in the sections $2-4$. The original eight gauge fields become

$$A=\frac{1}{\sqrt{2}}\,(A_1+i\,A_2),\quad \bar{A}=\frac{1}{\sqrt{2}}\,(A_1-i\,A_2), \tag{53}$$

and the scalars, written as $SU(4)$ bi-spinors

$$C^{m\,4}\;=\frac{1}{\sqrt{2}}\,(A_{m+3}+i\,A_{m+6}),\quad \overline{C}_{m\,4}=\frac{1}{\sqrt{2}}\,(A_{m+3}-i\,A_{m+6}), \tag{54}$$

for $m\neq 4$. These satisfy

$$\overline{C}_{mn}\;=\frac{1}{2}\,\epsilon_{mnpq}\,C^{pq}. \tag{55}$$

The fermion fields are χ^m and $\bar{\chi}_m$. All fields carry gauge indices and are local in

$$y=\left(x,\,\bar{x},\,x^+,\,y^-\equiv x^- - \frac{i}{\sqrt{2}}\,\theta^m\,\bar{\theta}_m\right). \tag{56}$$

The chiral derivatives are

$$d^{\,m}\;=-\partial^m-\frac{i}{\sqrt{2}}\,\theta^m\,\partial^+;\quad \bar{d}_{\,n}=\bar{\partial}_n+\frac{i}{\sqrt{2}}\,\bar{\theta}_n\,\partial^+. \tag{57}$$

The superfield satisfies both the chiral constraint

$$d^{\,m}\,\phi=0, \tag{58}$$

and the inside-out relation

$$\bar{d}_m\,\bar{d}_n\,\phi=\frac{1}{2}\,\epsilon_{mnpq}\,d^p\,d^q\,\bar{\phi}, \tag{59}$$

The action for $\mathcal{N}=4$ Yang-Mills is

$$\int d^4x\int d^4\theta\,d^4\bar{\theta}\,\mathcal{L}, \tag{60}$$

where

$$\mathcal{L}=-\bar{\phi}\,\frac{\Box}{\partial^{+2}}\,\phi+\frac{4g}{3}\,f^{abc}\left(\frac{1}{\partial^+}\,\bar{\phi}^a\,\phi^b\,\bar{\partial}\,\phi^c+\frac{1}{\partial^+}\,\phi^a\,\bar{\phi}^b\,\partial\,\bar{\phi}^c\right)$$
$$-\,g^2 f^{abc}\,f^{ade}\left(\frac{1}{\partial^+}(\phi^b\,\partial^+\phi^c)\frac{1}{\partial^+}\,(\bar{\phi}^d\,\partial^+\,\bar{\phi}^e)+\frac{1}{2}\,\phi^b\bar{\phi}^c\,\phi^d\,\bar{\phi}^e\right). \tag{61}$$

5.1. *Generators*

The Lorentz generators, introduced earlier, now accommodate the superspace variables

$$j = x\,\bar\partial - \bar x\,\partial + \frac{1}{2}\,(\theta^p\,\bar\partial_p - \bar\theta_p\,\partial^p) - \lambda, \tag{62}$$

with

$$\lambda = \frac{i}{4\sqrt{2}\,\partial^+}\,(d^p\,\bar d_p - \bar d_p\,d^p). \tag{63}$$

$\lambda = +1$ for a chiral superfield. On a chiral superfield, we have

$$\delta\,\phi = i\,\omega\,j\,\phi, \quad \delta\,\bar\phi = -i\,\omega\,j\,\bar\phi. \tag{64}$$

The other kinematical generators read

$$j^+ = i\,x\,\partial^+, \quad \bar j^+ = i\,\bar x\,\partial^+, \quad j^{+-} = i\,x^-\,\partial^+ - \frac{i}{2}\,(\theta^p\bar\partial_p + \bar\theta_p\,\partial^p) + i. \tag{65}$$

The boosts now read

$$\begin{aligned} j^- &= i\,x\,\frac{\partial\bar\partial}{\partial^+} - i\,x^-\,\partial + i\left(\theta^p\bar\partial_p - \lambda - 1\right)\frac{\partial}{\partial^+}\,, \\ \bar j^- &= i\,\bar x\,\frac{\partial\bar\partial}{\partial^+} - i\,x^-\,\bar\partial + i\left(\bar\theta_p\partial^p + \lambda - 1\right)\frac{\bar\partial}{\partial^+}\,. \end{aligned} \tag{66}$$

For further details, we refer the reader to [19]. Half of the supersymmetry generators

$$q^m_+ = -\partial^m + \frac{i}{\sqrt{2}}\,\theta^m\,\partial^+; \quad \bar q_{+n} = \bar\partial_n - \frac{i}{\sqrt{2}}\,\bar\theta_n\,\partial^+, \tag{67}$$

are kinematical while the others are dynamical

$$q^m_- \equiv i\,[\,\bar j^-\,,\,q^m_+\,] = \frac{\bar\partial}{\partial^+}\,q^m_+, \quad \bar q_{-m} \equiv i\,[\,j^-\,,\,\bar q_{+m}\,] = \frac{\partial}{\partial^+}\,\bar q_{+m}. \tag{68}$$

These are "square-roots" of the Hamiltonian

$$\{q^m_-\,,\,\bar q_{-n}\} = i\,\sqrt{2}\,\delta^m{}_n\,\frac{\partial\bar\partial}{\partial^+}. \tag{69}$$

5.2. *Superconformal algebra*

The $\mathcal{N} = 4$ Yang-Mills theory has a much larger symmstery group than just the usual Poincaré symmetry. To build this group: $PSU(2,2|\,4)$, we start with the "plus" conformal generator [19]

$$K^+ = 2i\,x\,\bar x\,\partial^+, \tag{70}$$

which along with j^{+-} yields

$$[K^+, p^-] = -2i\,D + 2i\,j^{+-}, \tag{71}$$

the dilatation generator

$$D = i\left(x^-\partial^+ - x\bar{\partial} - \bar{x}\partial - \frac{1}{2}\theta\frac{\partial}{\partial\theta} - \frac{1}{2}\bar{\theta}\frac{\partial}{\partial\bar{\theta}}\right). \tag{72}$$

Boosting K^+ results in

$$K = i\,[\,j^-, K^+] = 2ix\left(x^-\partial^+ - x\bar{\partial} - \theta\frac{\partial}{\partial\,\theta} + \lambda\right), \tag{73}$$

$$\bar{K} = i\,[\,\bar{j}^-, K^+] = 2i\bar{x}\left(x^-\partial^+ - \bar{x}\partial - \bar{\theta}\frac{\partial}{\partial\bar{\theta}} - \lambda\right). \tag{74}$$

The supersymmetry generators now include conformal supersymmetries obtained from

$$[\,K^+\,,\,q_-^m\,] = -\sqrt{2}\,(\,i\,\sqrt{2}\,\bar{x}\,q_+^m) = -\sqrt{2}\,s_+^m, \tag{75}$$

and their conjugates, both kinematical. The dynamical conformal supersymmetries read

$$s_-^m = i\,[\,j^-\,,\,s_+^m\,] = i\,\sqrt{2}\left(x^-\,\partial^+ - x\,\bar{\partial} - \theta\,\frac{\partial}{\partial\,\theta} + \lambda + 1\right)\frac{1}{\partial^+}\,q_+^m, \tag{76}$$

and their conjugates. The dynamical conformal generator K^- is

$$K^- = i\,[\,\bar{j}^-,\,K]. \tag{77}$$

5.3. *Deriving the theory*

As before, we begin with an ansatze for the order g Hamiltonian (we use g instead of α for the coupling constant in supersymmetric Yang-Mills theory). For additional details, we refer the reader to [2, 19].

$$\delta^g_{p^-}\,\phi = -\,i\,g\,\partial^{+\mu}\,[\,\bar{\partial}^a\,\partial^{+\rho}\,\phi\,\bar{\partial}^b\,\partial^{+\sigma}\,\phi]. \tag{78}$$

Exactly like in the non-supersymmetric case, closure of the algebra to order g yields

$$a + b = 1, \quad \mu + \rho + \sigma = 0. \tag{79}$$

The need for a gauge structure function follows as before. The variation that satisfies Poincaré invariance to order g is then

$$\delta^g_{p^-}\,\phi^a = -\,i\,g\,f^{abc}\,\frac{1}{\partial^+}\,(\,\bar{\partial}\,\phi^b\,\partial^+\,\phi^c). \tag{80}$$

In a similar manner, the non-linear boosts are [19]

$$\delta^g_{j^-}\,\phi^a = -\,x\,\delta^g_{p^-}\,\phi^a + i\,g\,f^{abc}\,\frac{1}{\partial^+}\,\left\{\left(\theta\,\frac{\partial}{\partial\,\theta} - 1\right)\phi^b\,\partial^+\,\phi^c)\right\}. \tag{81}$$

One approach to deriving the Hamiltonian for this theory is to use chirality, dimensional analysis, helicity and elementary commutators to argue that the first order dynamical supersymmetry has the form

$$\delta^g_{\bar{q}_-}\,\phi^a = -g\,f^{abc}\,\frac{1}{\partial^{+\,(2\nu+1)}}\{\,\bar{d}\,\partial^{+\,\nu}\,\phi^b\,\partial^{+\,(\nu+1)}\,\phi^c\}, \tag{82}$$

where $(d)^4 \equiv \epsilon_{mnpq}\,d^m d^n d^p d^q$ and ν is a free parameter. Its conjugate with (59) yields

$$\delta^g_{q_-}\,\phi^a = -g\,f^{abc}\,\frac{(d)^4}{48\,\partial^{+\,(2\nu+3)}}\{d\partial^{+\,\nu}\,\bar{\phi}^b\,\partial^{+\,(\nu+1)}\,\bar{\phi}^c\}, \tag{83}$$

Evaluate

$$\{\,\delta_{q^m_-}\,,\,\delta_{\bar{q}_{-\,n}}\,\}^g\,\phi^a = -\sqrt{2}\,\delta^m_n\,\delta^g_{p^-}\,\phi^a, \tag{84}$$

to first order in g to obtain the Hamiltonian

$$\delta_{p^-}\,\phi^a = -i\,\frac{\partial\bar{\partial}}{\partial^+}\,\phi^a - i\,g\,f^{abc}\left\{\frac{1}{\partial^{+\,(2\nu+1)}}\,(\bar{\partial}\,\partial^{+\,(\nu)}\phi^b\,\partial^{+\,(\nu+1)}\phi^c)\right.$$
$$\left.+\,\frac{(d)^4}{48\,\partial^{+(2\nu+3)}}\,(\partial\,\partial^{+\,(\nu)}\,\bar{\phi}^b\,\partial^{+\,(\nu+1)}\,\bar{\phi}^c)\right\} + \mathcal{O}(g^2). \tag{85}$$

The dynamical supersymmetry does not extend to g^2 [19] and as a consequence, the classical Hamiltonian terminates at order g^2. Rather than extend the procedure here to order g^2, which is lengthy, we adopt a different approach to arrive at the quartic interaction vertex.

We simply ask that the supersymmetry variations leave the Hamiltonian invariant [19]

$$\delta_{\bar{q}_-}\,H = 0. \tag{86}$$

This yields the three conditions

$$\delta^0_{\bar{q}_-}\,H^0 = 0, \tag{87}$$

$$\delta^g_{\bar{q}_-}\,H^0 + \delta^0_{\bar{q}_-}\,H^g = 0, \tag{88}$$

$$\delta^g_{\bar{q}_-}\,H^g + \delta^0_{\bar{q}_-}\,H^{g^2} = 0, \tag{89}$$

and hence a systematic link to H^g and H^{g^2} from $\delta_{\bar{q}_-}$ and H^0. The second condition gives

$$\delta^g_{\bar{q}_-}\,H^0 = \delta^g_{q^-}\left\{\int\bar{\phi}^a\,\frac{2\,\partial\bar{\partial}}{\partial^{+2}}\,\phi^a\right\}, \tag{90}$$

implying that

$$\delta^g_{q^-}\left\{\int\bar{\phi}^a\,\frac{2\,\partial\bar{\partial}}{\partial^{+2}}\,\phi^a\right\} = 2\,g\,f^{abc}\int\bar{\phi}^b\,d\bar{\phi}^c\,\frac{\partial\,\bar{\partial}}{\partial^{+2}}\,\phi^a. \tag{91}$$

so

$$\delta^0_{\bar{q}_-}\,H^g = -2\,g\int f^{abc}\,\bar{\phi}^b\,d\bar{\phi}^c\,\frac{\partial\,\bar{\partial}}{\partial^{+2}}\,\phi^a. \tag{92}$$

Now consider

$$\delta^0_{q\,-}\,\left\{g\,f^{abc}\int\frac{1}{\partial^+}\,\phi^a\,\overline{\phi}^b\,\partial\,\overline{\phi}^c\right\},\tag{93}$$

which yields two terms [19] which after some manipulations yield

$$g\,f^{abc}\int\frac{1}{\partial^+}\,\phi^a\,\frac{\bar\partial}{\partial^+}\,d\,\overline{\phi}^b\,\partial\,\overline{\phi}^c=\frac{1}{2}\left\{g\,f^{abc}\int\frac{1}{\partial^+}\,\phi^a\,\overline{\phi}^b\,\frac{\partial\,\bar\partial}{\partial^+}\,d\,\overline{\phi}^c\right\}.\tag{94}$$

Hence the variation

$$\delta^0_{q\,-}\,\left\{g\,f^{abc}\int\frac{1}{\partial^+}\,\phi^a\,\overline{\phi}^b\,\partial\,\overline{\phi}^c\right\}=\frac{3}{2}\,g\,f^{abc}\int\frac{1}{\partial^+}\,\phi^a\,\overline{\phi}^b\,\frac{\partial\,\bar\partial}{\partial^+}\,d\,\overline{\phi}^c,\tag{95}$$

leads to the previously derived cubic vertex in (85). Next, vary the cubic vertex to obtain the quartic vertex using (89). The cubic vertex involves a part with the transverse derivative ∂ and a part with $\bar\partial$. But H^{g^2} does not contain transverse derivatives since it stems from supersymmetries (at order g) which do not carry transverse derivatives. Thus

$$\delta^g_{q-}\;H^g_{\partial}=0,\tag{96}$$

since δ^0_{q-} contains no ∂. For supersymmetries to commute with the Hamiltonian then

$$\left[\delta^0_{q-}\,,\,\delta^{g^2}_{p-}\right]+\left[\delta^g_{q-}\,,\,\delta^g_{p-}\right]=0.\tag{97}$$

The Hamiltonian contains both ∂ and $\bar\partial$ while δ^0_{q-} has only $\bar\partial$. Thus one requirement is

$$\left[\delta^g_{q-}\,,\,\delta^g_{p-}\right]\Big|_{\partial}=0,\tag{98}$$

which holds as long as the structure functions are antisymmetric and obey the Jacobi identity [19]. The other requirement

$$\left[\delta^0_{q-}\,,\,\delta^{g^2}_{p-}\right]+\left[\delta^g_{q-}\,,\,\delta^g_{p-}\right]\Big|_{\bar\partial}=0,\tag{99}$$

leads to H^{g^2}

$$H^{g^2}=i\int\frac{1}{\partial^+}\bar\phi^a\,\delta^{g^2}_{p-}\,\phi^a=-\frac{i}{4\sqrt{2}}\int\frac{1}{\partial^+}\bar\phi^a\,\left\{\delta^g_{q-}\,,\,\delta^g_{\bar q-}\right\}\,\phi^a.\tag{100}$$

The detailed form of the Hamiltonian is presented in [19] and is not relevant here since our aim is to describe the method. Interestingly, this Hamiltonian has the structure of a quadratic form [19].

6. Theories with Supersymmetry: $\mathcal{N}=8$ Supergravity

Having described $\mathcal{N}=4$ Yang-Mills, we now move to the other maximally supersymmetric theory in four dimensions, $\mathcal{N}=8$ supergravity [20]. All the fields in $\mathcal{N}=8$ supergravity are similarly captured by a single superfield [18]. θ^m now transforms as the 8 of $SU(8)$.

$$\begin{aligned}\phi\,(\,y\,) = &\ \frac{1}{\partial^{+2}}\,h\,(y) + i\,\theta^m\,\frac{1}{\partial^{+2}}\,\bar{\psi}_m\,(y) + \frac{i}{2}\,\theta^m\,\theta^n\,\frac{1}{\partial^+}\,\bar{A}_{mn}\,(y),\\ &-\frac{1}{3!}\,\theta^m\,\theta^n\,\theta^p\,\frac{1}{\partial^+}\,\bar{\chi}_{mnp}\,(y) - \frac{1}{4!}\,\theta^m\,\theta^n\,\theta^p\,\theta^q\,\bar{C}_{mnpq}\,(y),\\ &+\frac{i}{5!}\,\theta^m\,\theta^n\,\theta^p\,\theta^q\,\theta^r\,\epsilon_{mnpqrstu}\,\chi^{stu}\,(y),\\ &+\frac{i}{6!}\,\theta^m\,\theta^n\,\theta^p\,\theta^q\,\theta^r\,\theta^s\,\epsilon_{mnpqrstu}\,\partial^+\,A^{tu}\,(y),\\ &+\frac{1}{7!}\,\theta^m\,\theta^n\,\theta^p\,\theta^q\,\theta^r\,\theta^s\,\theta^t\,\epsilon_{mnpqrstu}\,\partial^+\,\psi^u\,(y),\\ &+\frac{4}{8!}\,\theta^m\,\theta^n\,\theta^p\,\theta^q\,\theta^r\,\theta^s\,\theta^t\,\theta^u\,\epsilon_{mnpqrstu}\,\partial^{+2}\,\bar{h}\,(y),\end{aligned} \tag{101}$$

where the two-component graviton is

$$h = \frac{1}{\sqrt{2}}\,(h_{11} + i\,h_{12});\quad \bar{h} = \frac{1}{\sqrt{2}}\,(h_{11} - i\,h_{12}). \tag{102}$$

$\bar{\psi}_m$ are the spin-$\frac{3}{2}$ gravitinos, $\bar{A}_{mn}$ the 28 gauge fields and $\bar{\chi}_{mnp}$, the gauginos. $\bar{C}_{mnpq}$ represents the 70 scalar fields. Apart from the chiral condition, we have

$$\phi = \frac{1}{4}\,\frac{(d\,)^8}{\partial^{+4}}\,\bar{\phi}, \tag{103}$$

the inside-out relation, with $(d\,)^8 = d^1\,d^2\,\ldots\,d^8$. The action to first order in the gravitational coupling constant κ reads

$$\beta\int d^4x\int d^8\theta\,d^8\bar{\theta}\,\mathcal{L}, \tag{104}$$

where $\beta = -\frac{1}{64}$ and

$$\mathcal{L} = -\bar{\phi}\,\frac{\Box}{\partial^{+4}}\,\phi - 2\,\kappa\left(\frac{1}{\partial^{+2}}\,\bar{\phi}\;\bar{\partial}\phi\,\bar{\partial}\phi + \frac{1}{\partial^{+2}}\,\phi\,\partial\bar{\phi}\,\partial\bar{\phi}\right), \tag{105}$$

first derived in [2] and subsequently simplified in [15].

It is technically very challenging to extend this derivation to order κ^2 for supergravity. For this reason, we are forced to consider other approaches. As mentioned earlier, the Hamiltonian of =4 Yang-Mills is a quadratic form [19] and this turns out to be a feature shared by all maximally supersymmetric theories and hence valid for $\mathcal{N}=8$ supergravity [21].

Using $\mathcal{N} = 4$ Yang-Mills as a guide, the light-cone Hamiltonian for $\mathcal{N} = 8$ supergravity is of the form

$$\mathcal{H} = \frac{1}{4\sqrt{2}}\,(\mathcal{W}_m,\,\mathcal{W}_m), \tag{106}$$

where

$$\mathcal{W}_m = \bar{q}_{-\,m}\,\phi, \tag{107}$$

and the product is

$$(\phi,\,\xi) \equiv 2i\int d^4x\,d^8\theta\,d^8\,\bar{\theta}\,\bar{\phi}\,\frac{1}{\partial^{+3}}\xi. \tag{108}$$

At lowest order

$$\begin{aligned}\mathcal{H}^0 &= \frac{1}{4\sqrt{2}}\,(\mathcal{W}^0_m,\,\mathcal{W}^0_m),\\ &= \frac{2i}{4\sqrt{2}}\int d^4x\,d^8\theta\,d^8\,\bar{\theta}\,q_-^{\,m}\,\bar{\phi}\,\frac{1}{\partial^{+3}}\,\bar{q}_{-\,m}\,\phi,\end{aligned} \tag{109}$$

which can be simplified using (103) to

$$\mathcal{H}^0 = \frac{i}{4\sqrt{2}}\int d^4x\,d^8\theta\,d^8\,\bar{\theta}\left(q_-^{\,m}\,\bar{\phi}\,\frac{1}{\partial^{+3}}\,\bar{q}_{-\,m}\,\phi + \frac{1}{\partial^{+4}}\,q_-^{\,m}\,\phi\,\partial^+\,\bar{q}_{-\,m}\,\bar{\phi}\right). \tag{110}$$

Putting in the expressions for the supersymmetries we obtain [21]

$$\mathcal{H}^0 = \int d^4x\,d^8\theta\,d^8\bar{\theta}\,\bar{\phi}\,\frac{2\,\partial\bar{\partial}}{\partial^{+4}}\,\phi, \tag{111}$$

the appropriate kinetic term in the superspace Hamiltonian of $\mathcal{N} = 8$ supergravity [2]. Moving to order κ, where the dynamical supersymmetry generators are known [2] we have

$$\mathcal{W}_m = \frac{\partial}{\partial^+}\bar{q}_{+\,m}\,\phi + \kappa\,\frac{1}{\partial^+}\left(\bar{\partial}\,\bar{d}_m\,\phi\,\partial^{+2}\,\phi - \partial^+\,\bar{d}_m\,\phi\,\partial^+\,\bar{\partial}\,\phi\right) + \mathcal{O}(\kappa^2), \tag{112}$$

$$\overline{\mathcal{W}}^m = \frac{\bar{\partial}}{\partial^+}\,q_+^{\,m}\,\bar{\phi} + \kappa\,\frac{1}{\partial^+}\left(\partial\,d^m\,\bar{\phi}\,\partial^{+2}\,\bar{\phi} - \partial^+\,d^m\,\bar{\phi}\,\partial^+\,\partial\,\bar{\phi}\right) + \mathcal{O}(\kappa^2). \tag{113}$$

which when put into

$$\frac{1}{4\sqrt{2}}\,(\mathcal{W},\,\mathcal{W}) = \frac{2\,i}{4\sqrt{2}}\int d^4x\,d^8\theta\,d^8\,\bar{\theta}\,\overline{\mathcal{W}}\,\frac{1}{\partial^{+3}}\,\mathcal{W}. \tag{114}$$

yield [21]

$$\int d^4x\,d^8\theta\,d^8\bar{\theta}\,2\,\kappa\left(\frac{1}{\partial^{+2}}\,\overline{\phi}\;\bar{\partial}\phi\,\bar{\partial}\phi + \frac{1}{\partial^{+2}}\,\phi\,\partial\overline{\phi}\,\partial\overline{\phi}\right), \tag{115}$$

which is the cubic interaction vertex — and matches that derived by gauge fixing the covariant theory in (105).

This quadratic form structure leads us naturally to a quartic interaction vertex [21] but the result is too cumbersome to be useful in calculations. There is a more recent coherent state strategy that seems to make more sense but again its use in explicit computations is unclear at this point. The way forward here appears to be through the exceptional symmetry in the theory. In particular, using the $E_{7(7)}$ symmetry in $\mathcal{N}=8$ supergravity in conjunction with the quadratic form approach yields a much simpler quartic interaction vertex [22].

As far as $\mathcal{N}=8$ supergravity is concerned, it seems fairly clear that there are two distinct approaches to building the theory. The first is to use the super-Poincaré algebra exclusively while the other is to follow the exceptional symmetry in the theory [23]. Clearly a field redefinition should relate the results of these two approaches.

It is also interesting to note that this valuable exceptional symmetry grows under dimensional reduction [24]. In $d=3$ it becomes an $E_{8(8)}$ symmetry, an E_9 symmetry in $d=2$, an E_{10} symmetry in $d=1$ and an E_{11} symmetry in $d=0$. It seems possible that some or all of these higher symmetry groups could be present in the eleven-dimensional theory itself. Their link to clarifying the finiteness issue [25] relating to the $\mathcal{N}=8$ model is certainly worth examining further.

* * *

There are a number of open questions worth exploring. First, can this light-cone symmetry-based approach provide a Lagrangian origin to the Vasiliev equations of motion in AdS_4? Second, can we derive consistent quartic interaction vertices in flat spacetime? If not, where exactly does the procedure fail and how does this fit in with the exisiting no-go literature pertaining to higher spins in flat spacetime. Third, can this approach offer hints regarding the missing ingredients for the $(\mathcal{N}=2,0)$ theory in $d=6$? In particular, since we know the six-dimensional supercofrmal algebra in light-cone gauge can we close commutators and zero in on what structures are missing [26]?

Acknowledgments

I thank the organizers of the Higher Spin Gauge Theories workshop at NTU and the Institute of Advanced Studies, NTU for the hospitality. I am grateful to Y. S. Akshay, Lars Brink, Mahendra Mali, Sucheta Majumdar, Pierre Ramond and Hidehiko Shimada for discussions.

References

[1] E. S. Fradkin and M. A. Vasiliev *Phys. Lett.* **B189**, 89 (1987).
E. S. Fradkin and M. A. Vasiliev, *Nucl. Phys.* **B291**, 141 (1987).
M. A. Vasiliev, *Phys. Lett.* **B243**, 378 (1990).

M. A. Vasiliev, *Phys. Lett.* **B285**, 225 (1992).
M. A. Vasiliev, arXiv:hep-th/0104246.

[2] A. K. H. Bengtsson, I. Bengtsson and L. Brink, *Nucl. Phys.* **B227**, 31 (1983).
A. K. H. Bengtsson, I. Bengtsson and L. Brink, *Nucl. Phys.* **B227**, 41 (1983).
A. K. H. Bengtsson, I. Bengtsson and N. Linden, *Class. Quant. Grav.* **4**, 1333 (1987).

[3] P. Benincasa and F. Cachazo, arXiv:0705.4305.
M. Porrati, *Phys. Rev.* **D78**, 065016 (2008).

[4] R. Britto, F. Cachazo, B. Feng and E. Witten, *Phys. Rev. Lett.* **94**, 181602 (2005).

[5] S. Weinberg, *Phys. Rev* **138**, B988–B1002 (1965).
S.R. Coleman and J. Mandula, *Phys. Rev* **159**, 1251–1256 (1967).
C. Aragone S. Deser, *Phys. Lett.* **B86**, 161 (1979).
X. Bekaert, N. Boulanger and P. Sundell, *Rev. Mod. Phys.* **84**, 987 (2012), arXiv:1007.0435.
E. Witten and S. Weinberg, *Phys. Lett. B* **96**, 59 (1980).
Loyal Durand III, *Phys. Rev.* **128**, 434 (1962).
K. Case and S. Gasiorowicz, *Phys. Rev.* **125**, 1055 (1962).
C. Aragone and S. Deser, *Nuovo Cim. B*, **57**, 33 (1980).
A. K. H. Bengtsson and I. Bengtsson, *Class. Quant. Grav.* **3**, 927 (1986).

[6] R. R. Metsaev, *Nucl. Phys.* **B759**, 147 (2006), arXiv:hep-th/0512342.
R. R. Metsaev, *Nucl. Phys.* **B859**, 13 (2012), arXiv:hep-th/0712.3526.
R. R. Metsaev, *Phys. Lett.* **B720**, 237–243 (2013).
R. R. Metsaev, *Nucl. Phys.* **B859**, 13 (2012), arXiv:hep-th/0712.3526.
R. R. Metsaev and E. S. Fradkin, *Phys. Rev.* **D52**, 4660 (1995).

[7] A. K. H. Bengtsson, arXiv:1205.6117.
A. K. H. Bengtsson, *J. Math. Phys.* **48**, 072302 (2007).
A. K. H. Bengtsson, *JHEP* **1409**, 105 (2014).

[8] Y. S. Akshay and S. Ananth, *J. Phys.* **A47**, 045401 (2014).
Y. S. Akshay and S. Ananth, *Nucl. Phys.* **B887**, 168 (2014).

[9] S. Ananth, *JHEP* **1211** (2012).

[10] L. Dixon, arXiv:9601359.
H. Elvang and Y. T. Huang arXiv:1308.1697.

[11] S. Ananth and S. Theisen, *Phys. Lett.* **B652**, 128 (2007).

[12] H. Kawai, D. C. Lewellen and S. H. H. Tye, *Nucl. Phys.* **B269**, 001 (1986).

[13] Y. S. Akshay, S. Ananth and M. Mali, *Nucl. Phys.* **B884**, 66 (2014), arXiv:1401.5933.
S. Ananth and M. Mali, *Phys. Lett. B* **745**, 48 (2015), arXiv:1504.02839.

[14] E. Cremmer, B. Julia and J. Scherk, *Phys. Lett. B* **76**, 409 (1978).

[15] S. Ananth, L. Brink and P. Ramond, *JHEP* **0407**, 082 (2004), arXiv:hep-th/0405150.
S. Ananth, L. Brink and P. Ramond, *JHEP* **0505**, 003 (2005), arXiv:hep-th/0501079.

[16] S. Mandelstam, *Nucl. Phys.* **B213**, 149 (1983).

[17] Y. S. Akshay and Sudarshan Ananth, *Phys. Rev. D* **91**, 085029 (2015).

[18] L. Brink, O. Lindgren and B. E. W. Nilsson, *Nucl. Phys.* **B212**, 401 (1983).
L. Brink, O. Lindgren and B. E. W. Nilsson, *Phys. Lett. B* **123**, 323 (1983).

[19] S. Ananth, L. Brink, S. S. Kim and P. Ramond, *Nucl. Phys. B* **722**, 166 (2005).

[20] E. Cremmer and B. Julia, *Nucl. Phys. B* **159**, 141 (1979).
E. Cremmer, B. Julia and J. Scherk, *Phys. Lett. B* **76**, 409 (1978).
L. Brink and P. S. Howe, *Phys. Lett. B* **88**, 268 (1979).
B. de Wit and H. Nicolai, *Phys. Lett. B* **155**, 47 (1985).
B. de Wit and H. Nicolai, *Nucl. Phys. B* **274**, 363 (1986).
H. Nicolai, *Phys. Lett. B* **187**, 316 (1987).

[21] S. Ananth, L. Brink, R. Heise and H. G. Svendsen, *Nucl. Phys. B* **753**, 195 (2006).

S. Ananth, L. Brink and M. Mali, *JHEP* **1508**, 153 (2015), arXiv:1507.01068.
[22] L. Brink, S. Kim and P. Ramond, *JHEP* **0806**, 034 (2008).
[23] S. Ananth, L. Brink, S. Majumdar, arXiv:1601.02836.
[24] P. C. West, *Class. Quant. Grav.* **18**, 4443 (2001).
T. Damour, M. Henneaux and H. Nicolai, *Phys. Rev. Lett.* **89**, 221601 (2002).
T. Damour, A. Kleinschmidt and H. Nicolai, *Phys. Lett. B* **634**, 319 (2006).
[25] G. Bossard and H. Nicolai, *JHEP* **1108**, 074 (2011).
[26] P. Ramond, Unpublished.

Reducible Higher-Spin Systems in Metric-like and Frame-like Formulation

Alessandro Agugliaro[a,*], Francesco Azzurli[b,†] and Dmitri Sorokin[c,‡]

[a]*Dipartimento di Fisica, Universitá di Firenze and INFN Sezione di Firenze, Via G. Sansone 1, 50019 Sesto Fiorentino, Italy*

[b]*Dipartimento di Fisica, Università di Milano Bicocca, Piazza della Scienza 3, I-20126 Milano, Italy*

[c]*INFN, Sezione di Padova, Via F. Marzolo 8, 35131 Padova, Italy*

[*]*alessandro.agugliaro@fi.infn.it*

[†]*francesco.azzurli@mib.infn.it*

[‡]*dmitri.sorokin@pd.infn.it*

We briefly review main properties of systems of massless higher-spin fields which form the spectrum of string theory in a tensionless limit and present a new result regarding the construction of the metric-like action and equations of motion of fermionic higher-spin triplet systems in anti-de-Sitter space.

1. Introduction

Understanding a detailed structure of higher-spin excitations in string theory and the role they play in making this theory consistent, as well as their relation to existing higher-spin field theories, are long standing problems that theorists have addressed for decades.

String excitations give rise to an infinite number of fields of increasing spin, whose mass squared is proportional to the string tension and grows linearly with the spin. The infinite tower of the higher-spin states ensures a smooth ultraviolet behavior (or even UV finiteness) of superstring theory, thus making it a consistent theory of quantum gravity. One of the conceptual issues still waiting for a full comprehension is the fate of string theory in a limit in which the string tension goes to zero and the higher-spin states become massless. Generic arguments suggest [1–7] that this limit results in a huge enhancement of gauge (higher-spin) symmetries pointing at an alluring hypothesis that string theory may be a spontaneously broken (tension generating) phase of an even more fundamental gauge theory of massless higher-spin fields[1].

One can try to deduce the structure of higher-spin field interactions from the

[1]For recent developments in studying these issues in the context of AdS/CFT correspondence for superstrings on *AdS* backgrounds see e.g.[8, 9] and references therein.

action of String Field Theory. However, because of the complexity of the problem, only partial results have been obtained in this way so far. In particular back in 1986, by looking at a tensionless limit of string theory in flat space-time, Ouvry and Stern [10], and Bengtsson [11] obtained Lagrangians and equations of motion of massless bosonic higher-spin systems of fields which formed reducible moduls of the Lorentz group [2]. The simplest of these systems consists of three symmetric tensor fields of rank s, $s-1$ and $s-2$. This motivated Francia and Sagnotti [12] to call them *higher-spin triplets.* Fermionic triplets of tensor-spinor fields were introduced in [13] and independently in [12]. Their origin from a tensionless limit of a Ramond-Neveu-Schwarz string was studied in detail in [14]. The simplest bosonic triplets form reducible multiplets of physical fields with even or odd spins running, respectively, from 0 or 1 to s, while the fermionic triplets consist of physical fields with half-integer spins running from 1/2 to s. In general, the class of higher-spin "triplets" includes more complicated systems consisting of mixed symmetry tensor fields, as was considered already in [10, 11].

Since 1986 higher-spin triplets have been studied from different perspectives (see e.g.[10–34]). In this contribution we will briefly review the derivation of the equations of motion of bosonic triplets by taking the tensionless limit of the BRST-quantized free string in flat space, describe main properties of these systems in metric-like and frame-like formulation and present our recent results [35] on the derivation of the metric-like action and equations of motion of the fermionic higher-spin triplets in anti-de-Sitter space from their frame-like counterparts. Our main notation and conventions are given in the Appendix.

2. Bosonic Higher-Spin Triplets in the Tensionless Limit of String Theory

2.1. *BRST quantization of the open bosonic string*

As it has been pointed out in the Introduction, massless bosonic triplets naturally arise in a limit of vanishing string tension $T \sim \frac{1}{\alpha'} \to 0$. To show this, we first sketch the BRST quantization of the open bosonic string. The string oscillator operators α_n^μ $(n = 0, \pm 1, \pm 2, \ldots, \pm\infty)$ satisfying the commutation relations

$$[\alpha_k^\mu, \alpha_l^\nu] = k\, \delta_{k+l}\, \eta^{\mu\nu}, \tag{2.1}$$

are used to construct the Virasoro generators

$$L_m = \frac{1}{2} \sum_{n=-\infty}^{+\infty} \alpha_{m-n} \cdot \alpha_n\,, \qquad \alpha_0^\mu = \sqrt{2\alpha'} p^\mu\,, \tag{2.2}$$

[2]As was pointed out in [11], the tensionless string limit in flat space-time is actually well defined not for all quantities of the theory. For instance, the Veneziano amplitude may diverge or become zero depending on values of its parameters. However, at least in the free theory, the limit does allow for a consistent truncation of the Virasoro algebra and the BRST charge, thus allowing to get the corresponding spectrum of quantized states, as we will review below.

where p^μ is the momentum of the string centre of mass. In a $D-$dimensional space-time the generators L_m form the centrally extended Virasoro algebra

$$[L_m, L_n] = (m-n)\, L_{m+n} + \frac{D}{12}\left(m^3 - m\right)\delta_{m+n}\,. \tag{2.3}$$

The Virasoro constraints on the string states are implemented with the use of the BRST operator of the form

$$Q = \sum_{-\infty}^{+\infty} L_{-m} C_m - \frac{1}{2}\sum_{-\infty}^{+\infty}(m-n) : C_{-m} C_{-n} B_{m+n} : -C_0\,, \tag{2.4}$$

where C_k and B_k are, respectively, the ghost and anti-ghost fields satisfying the anti-commutation relations

$$\{C_k, B_l\} = \delta_{k+l}. \tag{2.5}$$

The BRST charge is nilpotent, i.e. $Q^2 = 0$, if the string moves in a space-time of the critical dimension $D = 26$.

The string field equations are obtained by requiring that any physical state $|\Phi\rangle$ be annihilated by the BRST operator, namely

$$Q|\Phi\rangle = 0. \tag{2.6}$$

In view of the nilpotency of the BRST operator, the equation (2.6) is invariant under the gauge transformation

$$|\Phi'\rangle = |\Phi\rangle + Q|\Lambda\rangle. \tag{2.7}$$

2.2. *The tensionless limit*

In order to perform the tensionless limit, one rescales the Virasoro generators (2.2) as follows

$$l_k = \frac{1}{\sqrt{2\alpha'}} L_k, \quad k \neq 0\,; \qquad l_0 = \frac{1}{\alpha'} L_0 \tag{2.8}$$

Then, in the limit $\alpha' \to \infty$, one is left with

$$l_k = p_\mu \cdot \alpha_k^\mu, \quad k \neq 0; \qquad l_0 = p_\mu p^\mu. \tag{2.9}$$

These operators satisfy the following algebra which has no central charge

$$[l_k, l_j] = k\,\delta_{j+k}\, l_0\,. \tag{2.10}$$

To have, in this limit, a non-degenerate BRST operator one defines the rescaled ghost and anti-ghost operators as follows

$$c_k = \sqrt{2\alpha'} C_k, \qquad b_k = \frac{1}{\sqrt{2\alpha'}} B_k, \tag{2.11}$$

for $k \neq 0$, and

$$c_0 = \alpha' C_0, \qquad b_0 = \frac{1}{\alpha'} B_0. \tag{2.12}$$

After such redefinitions, at $\alpha' \to \infty$ the BRST operator takes the following form

$$Q = \sum_k \left[c_{-k} l_k - \frac{k}{2} b_0\, c_{-k} c_k \right] \tag{2.13}$$

which is now nilpotent in a space-time of any dimension.

Let us consider now a generic string state, generated by acting on the vacuum with the creation operators α^{μ}_{-k} $(k = 1, \ldots, \infty)$ and the ghost fields c_0, c_{-n} and b_{-n} $(n = 1, \ldots, \infty)$. This can be written schematically as

$$|\Phi\rangle = \sum_{s=1}^{\infty} \sum_{\{k_1,\ldots,k_s\}} \phi_{\mu_1\mu_2\ldots\mu_s}(x)\, \alpha^{\mu_1}_{-k_1}\, \alpha^{\mu_2}_{-k_2} \cdots \alpha^{\mu_s}_{-k_s}\, c \ldots b \ldots b|0\rangle \tag{2.14}$$

One may wonder whether there exist finite sets of states that satisfy equation (2.6) independently of the presence of other sets. It turns out that such systems of fields do exist. For instance, in the simplest case, picking up in (2.14) a single Regge trajectory (e.g. with $k = 1$), for each value of s we find that the following triplet state independently satisfies (2.6)

$$\begin{aligned} |\Phi\rangle_{triplet} =& \frac{1}{s!}\varphi_{\mu_1\ldots\mu_s}(x)\alpha^{\mu_1}_{-1}\ldots\alpha^{\mu_s}_{-1}|0\rangle \\ &- \frac{i}{(s-1)!}C_{\mu_1\ldots\mu_{s-1}}(x)\alpha^{\mu_1}_{-1}\ldots\alpha^{\mu_{s-1}}_{-1}c_0 b_{-1}|0\rangle \\ &+ \frac{1}{(s-2)!}D_{\mu_1\ldots\mu_{s-2}}(x)\alpha^{\mu_1}_{-1}\ldots\alpha^{\mu_{s-2}}_{-1}c_{-1}b_{-1}|0\rangle\,. \end{aligned} \tag{2.15}$$

The triplet consists of totally symmetric tensor fields φ_s, C_{s-1} and D_{s-2} of rank s, $s-1$ and $s-2$, respectively. Their equations of motion, which are deduced by inserting (2.15) into (2.6) and using $p_\mu = -i\partial_\mu$, have the following form

$$\begin{aligned} \Box\varphi_{\mu_1\ldots\mu_s} - \partial_{(\mu_1}C_{\mu_2\ldots\mu_s)} &= 0\,, \\ \Box D_{\mu_1\ldots\mu_{s-2}} - \partial^\nu C_{\nu\mu_1\ldots\mu_{s-2}} &= 0\,, \\ \partial^\nu\varphi_{\nu\mu_1\ldots\mu_{s-1}} - \partial_{(\mu_1}D_{\mu_2\ldots\mu_{s-1})} - C_{\mu_1\ldots\mu_{s-1}} &= 0. \end{aligned} \tag{2.16}$$

Finally, since equation (2.6) is invariant under the gauge transformations (2.7), we find taking $|\Lambda\rangle = \frac{i}{(s-1)!}\Lambda_{\mu_1\ldots\mu_{s-1}}(x)\alpha^{\mu_1}_{-1}\ldots\alpha^{\mu_{s-1}}_{-1}b_{-1}|0\rangle$ that the triplet equations (2.16) are invariant under the following gauge transformations with a local totally symmetric parameter $\Lambda_{\mu_1\ldots\mu_{s-1}}(x)$

$$\begin{aligned} \delta\varphi_{\mu_1\ldots\mu_s} &= \partial_{(\mu_1}\Lambda_{\mu_2\ldots\mu_s)}\,, \\ \delta C_{\mu_1\ldots\mu_{s-1}} &= \Box\Lambda_{\mu_1\ldots\mu_{s-1}}\,, \\ \delta D_{\mu_1\ldots\mu_{s-2}} &= \partial^\nu\Lambda_{\nu\mu_1\ldots\mu_{s-2}}. \end{aligned} \tag{2.17}$$

The gauge transformation (2.17) of the bosonic field $\varphi_{\mu_1\ldots\mu_s}$ can be regarded as a direct generalization of the (linearized) transformation of a metric under diffeomorphisms

$$\delta g_{\mu\nu} = \partial_{(\mu}\Lambda_{\nu)}.$$

For this reason, the field $\varphi_{\mu_1\ldots\mu_s}$ and the corresponding formalism are called metric-like.

3. Frame-like Formulation of Bosonic Triplets

One may wonder: what is the geometrical nature of the other two fields, C and D, of the triplet? The answer to this question is provided by the frame-like formulation of higher-spin fields generalized to the case of triplets in [24].

A fundamental field in this formulation is the "frame-like" 1-form, or the higher-spin vielbein, with $s-1$ symmetrized tangent-space indices a_i

$$e^{a_1\dots a_{s-1}} = dx^\mu e_\mu^{a_1\dots a_{s-1}}. \tag{3.1}$$

Note that, to describe the triplet system, we do not require this field to satisfy any traceless condition which means that $e^{a_1\dots a_{s-1}}$ transforms under an irreducible representation of the $GL(D)$ group rather than of $SO(1, D-1)$. The irreducibility under $SO(1, D-1)$ would require $e^{a_1\dots a_{s-1}}$ be traceless, which would correspond to the case of the Fronsdal field of spin s (see e.g.[24] for a review).

The gauge transformations of the higher-spin vielbein have the following form in flat space-time

$$\delta e^{a_1\dots a_{s-1}} = d\xi^{a_1\dots a_{s-1}} - dx^a \eta_{ab}\xi^{a_1\dots a_{s-1},b}, \tag{3.2}$$

where $\xi^{a_1\dots a_{s-1}}$ and $\xi^{a_1\dots a_{s-1},b}$ are zero-form parameters. The parameter $\xi^{a_1\dots a_{s-1},b}$ satisfies the $GL(D)$ irreducibility property

$$\xi^{(a_1\dots a_{s-1},b)} = 0\,. \tag{3.3}$$

The representation of the $GL(D)$ group in which $e_\mu^{a_1\dots a_{s-1}}$ sits can be decomposed, using the Young tableaux, as follows:

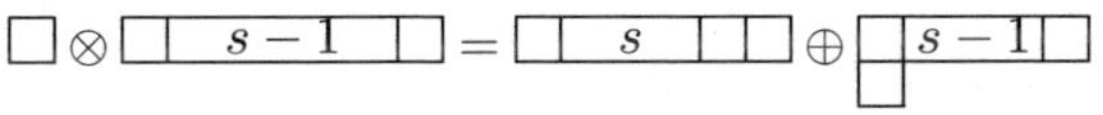

(3.4)

If the parameter $\xi^{a_1\dots a_{s-1},b}$ is not subject to any constraint, it could be used to gauge away the second term in the right hand side of (3.4). Then we would end up with a single totally symmetric trace-full field which can be associated with the field $\varphi^{\mu_1\dots\mu_s}$ of the triplet. If, however, we impose the condition

$$\eta_{a_1 b}\xi^{a_1\dots a_{s-1},b} = 0, \tag{3.5}$$

then this parameter can be used to gauge away only the corresponding part of the second term of the decomposition (3.4), and one is left with the two symmetric fields of rank s and $s-2$. Comparing their gauge transformations (3.2) with those of the triplet (2.17) allows us to make the following identification :

$$\varphi^{a_1\dots a_s} \equiv \eta^{\mu(a_s} e_\mu^{a_1\dots a_{s-1})}, \quad D^{a_1\dots a_{s-2}} \equiv \delta^\mu_{a_{s-1}} e_\mu^{a_1\dots a_{s-1}}, \quad \Lambda^{a_1\dots a_{s-1}} \equiv \xi^{a_1\dots a_{s-1}}. \tag{3.6}$$

Further on, he gauge parameter $\xi^{a_1\dots a_{s-1},b}$ is associated to a higher spin "connection" $\omega_\mu^{a_1\dots a_{s-1},b}$, which is auxiliary, subject to the traceless condition similar to (3.5)

$$\eta_{bc}\omega^{a_1\dots a_{s-2}b,c} = 0\,.$$

The connection can be expressed in terms of $e_\mu^{a_1\dots a_{s-1}}$ by means of a torsion-like constraint [3]

$$de^{a_1\dots a_{s-1}} - dx_b \wedge \omega^{a_1\dots a_{s-1},b} = 0. \tag{3.7}$$

This equation is gauge invariant under (3.2) and

$$\delta\omega^{a_1\dots a_{s-1},b} = d\xi^{a_1\dots a_{s-1},b} - dx_c\xi^{a_1\dots a_{s-1},bc}. \tag{3.8}$$

The new gauge parameter $\xi^{a_1\dots a_{s-1},bc}$, satisfying the condition $\eta_{a_1 b}\xi^{a_1\dots a_{s-1},bc} = 0$, is necessary for the two terms in (3.7) to have the same number of degrees of freedom and, ultimately, to ensure that $\omega_\mu^{a_1\dots a_{s-1},b}$ is completely auxiliary.

Looking at the transformations (3.2) and (3.8) we notice that (in view of the constraint $\eta_{bc}\xi^{a_1\dots a_{s-1},bc} = 0$, which is a consequence of $\eta_{a_1 b}\xi^{a_1\dots a_{s-1},bc} = 0$ and the Young tableau property $\xi^{(a_1\dots a_{s-1},b)c} = 0$) the following relation holds

$$\delta\left(\partial^\mu e_\mu^{a_1\dots a_{s-1}} + \delta_b^\mu \omega_\mu^{a_1\dots a_{s-1},b}\right) = \Box\xi^{a_1\dots a_{s-1}}$$

and so, in virtue of (3.6), the third triplet field is identified as the combination of the divergence of the higher-spin vielbein and a trace of the higher-spin connection

$$C^{a_1\dots a_{s-1}} \equiv \partial^\mu e_\mu^{a_1\dots a_{s-1}} + \delta_b^\mu \omega_\mu^{a_1\dots a_{s-1},b}.$$

The following remark is in order. For s even the lower-spin component of the metric-like triplet φ, C, D is a dynamical spin-0 field, while for s odd the lowest component is a spin-1 field. These physical fields are not contained in the higher-spin vielbein (3.1). To see this, consider the case of an even s. Upon setting $D = 0$ by fixing a gauge, the scalar field in the metric-like formalism is given by

$$\phi \equiv \eta_{\mu_1\mu_2}\cdots\eta_{\mu_{s-1}\mu_s}\varphi^{\mu_1\dots\mu_s},$$

which by (2.16) obeys the massless Klein-Gordon equation. On the other hand, in the frame-like formalism the D and φ fields both come from $e_\mu^{a_1\dots a_{s-1}}$ and are related, for s even, by the fact that taking their complete trace and using the identifications (3.6) one finds

$$\eta_{a_1 a_2}\cdots\eta_{a_{s-1}a_s}\varphi^{a_1\dots a_s} \propto \eta_{a_1 a_2}\cdots\eta_{a_{s-3}a_{s-2}}D^{a_1\dots a_{s-2}}.$$

This means that when D is gauge fixed to be zero, also $\phi = 0$ and thus no physical scalar field is present in the spectrum. By the same reasoning one can see that in the case of odd s the field $e_\mu^{a_1\dots a_{s-1}}$ does not contain the physical field of spin 1.

So, the metric-like and frame-like formulations are not completely equivalent. However, one can always separately add the scalar and the vector field to the frame-like description of the higher-spin triplets so that the field content in both formulations will be the same.

For further details on the structure of the frame-like Lagrangian formulation of the bosonic higher-spin triplets, their action and equations of motion in flat and AdS space we refer the reader to [24] and pass to a more detailed consideration of the fermionic higher-spin triplets.

[3] As was shown in [24], the necessity to impose the weak traceless constraint (3.5) on the higher-spin connections and the corresponding gauge parameters originates from the requirement of the gauge invariance of the actions for these reducible higher-spin systems.

4. Fermionic Triplets in Flat Space-time

4.1. *Metric-like formulation*

In the metric-like formulation a fermionic higher-spin triplet consists of three symmetric tensor-spinor fields $\Psi_\alpha^{\mu_1\ldots\mu_r}$, $\chi_\alpha^{\mu_1\ldots\mu_{r-1}}$ and $\lambda_\alpha^{\mu_1\ldots\mu_{r-2}}$, where α is a spinor index which we will usually skip, and $r = s - \frac{1}{2}$ with s denoting the highest spin in the spectrum of the triplet fields. In flat space-time they satisfy the following equations of motion [12]

$$\begin{aligned} \partial\!\!\!/ \Psi^{\mu_1\ldots\mu_r} + i\partial^{(\mu_1}\chi^{\mu_2\ldots\mu_r)} &= 0, \\ \partial_\nu \Psi^{\nu\mu_1\ldots\mu_{r-1}} - \partial^{(\mu_1}\lambda^{\mu_2\ldots\mu_{r-1})} + i\partial\!\!\!/ \chi^{\mu_1\ldots\mu_{r-1}} &= 0, \\ \partial\!\!\!/ \lambda^{\mu_1\ldots\mu_{r-2}} + i\partial_\nu\chi^{\nu\mu_1\ldots\mu_{r-2}} &= 0. \end{aligned} \tag{4.1}$$

which are invariant under the gauge transformations

$$\begin{aligned} \delta\Psi^{\mu_1\ldots\mu_r} &= \partial^{(\mu_1}\Lambda^{\mu_2\ldots\mu_r)}, \\ \delta\chi^{\mu_1\ldots\mu_{r-1}} &= i\partial\!\!\!/ \Lambda^{\mu_1\ldots\mu_{r-1}}, \\ \delta\lambda^{\mu_1\ldots\mu_{r-2}} &= \partial_\nu\Lambda^{\nu\mu_1\ldots\mu_{r-2}}. \end{aligned} \tag{4.2}$$

The field equations (4.1) follow from the flat-space action whose explicit form was given in [12]. However, the construction of the metric-like Lagrangian formulation of fermionic higher-spin triplets in AdS spaces encountered difficulties [14, 20] and has not been accomplished by now. In what follows we will show how this puzzle is resolved by deriving the metric-like action for the fermionic higher-spin triplets in AdS from their frame-like Lagrangian formulation.

4.2. *Frame-like formulation*

In the frame-like formulation [24] the fermionic triplet is associated with a 1-form that takes values in the space of symmetric tensor-spinors:

$$\psi_\alpha^{a_1\ldots a_{r-1}} = dx^\mu \psi_{\alpha\,\mu}^{a_1\ldots a_{r-1}}.$$

This form is not subject to any gamma-trace conditions[4].

In flat space the gauge transformation of $\psi^{a_1\ldots a_{r-1}}$ has the following form

$$\delta\psi^{a_1\ldots a_{r-1}} = d\xi^{a_1\ldots a_{r-1}} - dx_b\xi^{a_1\ldots a_{r-1},b}, \tag{4.3}$$

and the tensor-spinor gauge parameters $\xi^{a_1\ldots a_{r-1}}$ and $\xi^{a_1\ldots a_{r-1},b}$ are zero forms that remove from $\psi_\mu^{a_1\ldots a_{r-1}}$ the unphysical degrees of freedom. The second parameter is

[4]Remember that the frame-like fermionic field transforming under an irreducible representation of the Lorentz group of a spin $s = r + \frac{1}{2}$ is gamma-traceless

$$\gamma_b\psi_\mu^{a_1\ldots a_{r-2}b} = 0 \Longrightarrow \eta_{bc}\psi_\mu^{a_1\ldots a_{r-3}bc} = 0.$$

associated with a connection-like 1-form $\psi_\mu^{a_1\dots a_{r-1},b}$. The connection field is auxiliary and expressed (modulo the pure gauge degrees of freedom) as a function of $\psi_\mu^{a_1\dots a_{r-1}}$ via the torsion-like constraint

$$\mathcal{T}^{a_1\dots a_{r-1}} = d\psi^{a_1\dots a_{r-1}} - dx_c \wedge \psi^{a_1\dots a_{r-1},b} = 0, \tag{4.4}$$

The torsion is invariant under the gauge transformations (4.3) and

$$\delta\psi^{a_1\dots a_{r-1},b} = d\xi^{a_1\dots a_{r-1},b} - dx_c\xi^{a_1\dots a_{r-1},bc}. \tag{4.5}$$

For the consistency of the construction (see Section 4.3), while the higher-spin vielbein is unconstrained, the connection $\psi^{a_1\dots a_{r-1},b}$ and the gauge parameters $\xi^{a_1\dots a_{r-1},b}$ and $\xi^{a_1\dots a_{r-1},bc}$ should satisfy the following (gamma-)trace constraints

$$\gamma_b\psi^{a_1\dots a_{r-1},b} = 0, \quad \eta_{bc}\psi^{a_1\dots a_{r-2}b,c} = 0, \tag{4.6}$$

$$\gamma_b\xi^{a_1\dots a_{r-1},b} = 0, \quad \eta_{bc}\xi^{a_1\dots a_{r-2}b,c} = 0, \tag{4.7}$$

$$\gamma_b\xi^{a_1\dots a_{r-1},bc} = 0, \quad \eta_{db}\xi^{a_1\dots a_{r-2}d,bc} = 0.$$

The metric-like triplet fields Ψ, χ and λ introduced in Section 4.1 are related to $\psi_\mu^{a_1\dots a_{r-1}}$ by the following identifications motivated by the form of the gauge transformations (4.2)

$$\hat{\Psi}^{a_1\dots a_r} \equiv \delta^{\mu(a_r}\psi_\mu^{a_1\dots a_{r-1})}, \quad \hat{\chi}^{a_1\dots a_{r-1}} \equiv i\gamma^\mu\psi_\mu^{a_1\dots a_{r-1}}, \quad \hat{\lambda}^{a_1\dots a_{r-2}} \equiv \delta^\mu_{a_{r-1}}\psi_\mu^{a_1\dots a_{r-1}}, \tag{4.8}$$

where we introduced "hatted" quantities which differ from Ψ, χ and λ by a total trace, as we will explain now.

The splitting (4.8) manifests the fact that the representation in which $\psi^{a_1\dots a_{r-1}}$ sits is not irreducible, as it may have γ_a- and η_{ab}-traces. As a result, this field contains physical states with spins going down from $s = r + \frac{1}{2}$ to 3/2, while its spin 1/2 state is a pure gauge. Similarly to the bosonic case, this is due to the fact that the three fields (4.8), being all derived from $\psi_\mu^{a_1\dots a_{r-1}}$, are not completely independent. Indeed, let us define the complete trace $\tilde{\mathbb{T}}$ of a tensor-spinor $\mathbb{T}^{a_1\dots a_r}$ as

$$\tilde{\mathbb{T}} = \begin{cases} \eta_{a_1a_2}\cdots\eta_{a_{r-1}a_r}\mathbb{T}^{a_1\dots a_r} & \text{if } r \text{ is even} \\ \eta_{a_1a_2}\cdots\eta_{a_{r-2}a_{r-1}}i\gamma_{a_r}\mathbb{T}^{a_1\dots a_r} & \text{if } r \text{ is odd.} \end{cases} \tag{4.9}$$

Then, in the metric-like description, the equations of motion imply that for $s = r+\frac{1}{2}$

$$\begin{cases} \not{\partial}\left(\tilde{\Psi} - r\tilde{\lambda}\right) = 0 & \text{if } r \text{ is even} \\ \not{\partial}\left(\tilde{\Psi} - \tilde{\chi} - (r-1)\tilde{\lambda}\right) = 0 & \text{if } r \text{ is odd}. \end{cases} \tag{4.10}$$

Their form allows us to identify the spin $\frac{1}{2}$ field as

$$\rho \equiv \begin{cases} \tilde{\Psi} - r\tilde{\lambda} & \text{if } r \text{ is even} \\ \tilde{\Psi} - \tilde{\chi} - (r-1)\tilde{\lambda} & \text{if } r \text{ is odd.} \end{cases} \tag{4.11}$$

In the frame-like formulation, by virtue of (4.8), we find that $\rho \equiv 0$.

This analysis tells us that the triplet fields (4.8) obtained from $\psi_\mu^{a_1\dots a_{r-1}}$ are not quite the same as those appearing in the equations (4.1): their complete traces do not match. Of course, we can fix this issue by simply adding a spin-$\frac{1}{2}$ massless field ρ to the frame-like action of the theory, which we will introduce in the next section. Then the original metric-like triplet fields Ψ, χ and λ are related to the fields (4.8) as follows

$$\begin{aligned}\Psi^{a_1\dots a_r} &\equiv \hat{\Psi}^{a_1\dots a_r} + \eta^{(a_1a_2}\cdots\eta^{a_{r-1}a_r)}\rho\,,\\ \lambda^{a_1\dots a_{r-2}} &\equiv \hat{\lambda}^{a_1\dots a_{r-2}} + \eta^{(a_1a_2}\cdots\eta^{a_{r-3}a_{r-2})}\rho \quad \text{for } r \text{ even}\end{aligned} \tag{4.12}$$

and

$$\begin{aligned}\Psi^{a_1\dots a_r} &\equiv \hat{\Psi}^{a_1\dots a_r} - i\eta^{(a_1a_2}\cdots\eta^{a_{r-2}a_{r-1}}\gamma^{a_r)}\rho\\ \lambda^{a_1\dots a_{r-2}} &\equiv \hat{\lambda}^{a_1\dots a_{r-2}} - i\eta^{(a_1a_2}\cdots\eta^{a_{r-4}a_{r-3}}\gamma^{a_{r-2})}\rho\,,\\ \chi^{a_1\dots a_{r-1}} &\equiv \hat{\chi}^{a_1\dots a_{r-1}} + 2\eta^{(a_1a_2}\cdots\eta^{a_{r-2}a_{r-1})}\rho \quad \text{for } r \text{ odd.}\end{aligned} \tag{4.13}$$

4.3. *Frame-like action for fermionic triplets in flat space*

The frame-like action that reproduces, upon making the identifications (4.8), (4.12) and (4.13), the equations (4.1) can be found by an ansatz motivated by some simple requirements. It should be gauge-invariant under (4.5) and have a schematic form of a free fermion action $i\bar{\psi}\gamma\partial\psi$, where γ stays for a product of gamma-matrices and $\bar{\psi}$ is the Dirac conjugate of ψ. Therefore, to construct the action we use the gauge invariant higher-spin torsion 2-form

$$\mathcal{T}^{a_1\dots a_{r-1}} \equiv d\psi^{a_1\dots a_{r-1}} - dx_b \wedge \psi^{a_1\dots a_{r-1},b}, \tag{4.14}$$

which is considered to be non-zero off the mass shell (compare with (4.4)). Then, terms like $i\bar{\psi}\gamma \wedge \mathcal{T}$ are gauge invariant under the transformations of ψ but not $\bar{\psi}$. The most general Lorentz-invariant action constructed using such 3-forms is [24]

$$\begin{aligned}S_{\mathcal{T}} = i\int dx^{a_1}\wedge\cdots\wedge dx^{a_{D-3}}\wedge\varepsilon_{a_1\dots a_{D-3}cdf}\,\big(&\bar{\psi}^{b_1\dots b_{r-1}}\wedge\gamma^{cdf}\mathcal{T}_{b_1\dots b_{r-1}}\\ &+c\bar{\psi}^{b_1\dots b_{r-2}c}\wedge\gamma^{d}\mathcal{T}^{f}_{b_1\dots b_{r-2}}\big)\,,\end{aligned} \tag{4.15}$$

where $\varepsilon_{a_1\dots a_D}$ is the D-dimensional completely antisymmetric Levi-Civita tensor and c is an arbitrary constat.

This action has three issues to tackle. It is not invariant under the gauge transformations of $\bar{\psi}$, it is not real and contains the auxiliary field $\psi^{a_1\dots a_{r-1},b}$, that should be completely determined by $\psi^{a_1\dots a_{r-1}}$ through (4.4). These three issues are fixed by choosing a proper c. Indeed, it is possible to show that for $c = -6(r-1)$ and if $\psi^{a_1\dots a_{r-1},b}$ is constrained as in (4.6), all the terms proportional to this auxiliary field disappear. Then (4.15) becomes simply

$$\begin{aligned}S = i\int dx^{a_1}\wedge\cdots\wedge dx^{a_{D-3}}\wedge\varepsilon_{a_1\dots a_{D-3}cdf}\,\big(&\bar{\psi}^{b_1\dots b_{r-1}}\wedge\gamma^{cdf}d\psi_{b_1\dots b_{r-1}}\\ &-6(r-1)\bar{\psi}^{b_1\dots b_{r-2}c}\wedge\gamma^{d}d\psi^{f}_{b_1\dots b_{r-2}}\big)\,.\end{aligned} \tag{4.16}$$

Integrating (4.16) by parts we can turn it (modulo total derivatives) into its complex conjugate, so (4.16) satisfies the reality condition. Moreover, due to our choice of c, in the Hermitian conjugate version of (4.16) we can restore the (vanishing) terms proportional to $\bar{\psi}^{a_1\ldots a_{r-1},b}$ and rewrite the action as

$$S_{\bar{\mathcal{T}}} = i \int dx^{a_1} \wedge \cdots \wedge dx^{a_{D-3}} \wedge \varepsilon_{a_1\ldots a_{D-3}cdf} \left(\bar{\mathcal{T}}^{b_1\ldots b_{r-1}} \wedge \gamma^{cdf} \psi_{b_1\ldots b_{r-1}} \right.$$
$$\left. -6(r-1)\bar{\mathcal{T}}^{b_1\ldots b_{r-2}c} \wedge \gamma^d \psi^f_{\ b_1\ldots b_{r-2}}\right),$$

which is therefore equivalent to (4.15) up to total derivatives. The variation of (4.16) under gauge transformations can be then schematically rewritten as

$$\delta S = \frac{\delta S_{\mathcal{T}}}{\delta \psi}\delta\psi + \delta\bar{\psi}\frac{\delta S_{\bar{\mathcal{T}}}}{\delta\bar{\psi}} = 0.$$

It vanishes because of the manifest gauge invariance of $\mathcal{T}^{a_1\ldots a_{r-1}}$ and $\bar{\mathcal{T}}^{a_1\ldots a_{r-1}}$ and provided that the gauge parameters $\xi^{a_1\ldots a_{r-1},b}$ satisfy the constraints (4.7).

One can show [24] that the equations of motion derived from the frame-like action (4.16) are equivalent to the metric-like ones (4.1) modulo the subtlety with the spin-$\frac{1}{2}$ field ρ, which can be included into the frame-like formulation by simply adding to the action the massless Dirac Lagrangian $L_{\frac{1}{2}} = i\bar{\rho}\gamma^m\partial_m\rho$.

5. Fermionic Higher-Spin Triplets in Anti-de-Sitter Spaces

5.1. *The frame-like formalism in AdS*

The frame-like description of the field $\psi^{a_1\ldots a_{r-1}}$ can be generalized to the anti-de-Sitter space by employing a proper *AdS* covariant derivative in place of the flat-space derivative and by using a local basis for the *AdS* tangent space given by the vielbein $e^a = dx^m\, e^a_m(x)$. Actually, to construct the covariant derivatives we will employ two different kinds of connections. The one associated with the invariance under $SO(1, D-1)$ local Lorentz transformations and containing the spin–connection 1–form $\omega^{ab} = -\omega^{ba}$ will be denoted by

$$\nabla T^a = dT^a + \omega^a{}_b \wedge T^b \tag{5.1}$$

and will act on the D-dimensional vectorial tangent-space indices. For the spinorial indices one can also use the connection associated to the whole symmetry group of AdS_D, namely the isometry group $SO(2, D-1)$, which includes, in addition to the Lorentz transformations generated by J^{ab}, the non-commuting translations generated by P^a:

$$\left[P^a, P^b\right] = \Lambda J^{ab},$$

where Λ is a negative cosmological constant which defines the *AdS* curvature

$$R^{ab} = d\omega^{ab} + \omega^{ac} \wedge \omega_c{}^b = -\Lambda e^a \wedge e^b\,. \tag{5.2}$$

We denote this covariant derivative by

$$\mathcal{D}\psi_\alpha = d\psi_\alpha + \frac{1}{2}\omega^{ab}\,(J_{ab})_\alpha{}^\beta \wedge \psi_\beta + e^b\,(P_b)_\alpha{}^\beta \wedge \psi_\beta\,, \tag{5.3}$$

where in the spinorial representation

$$J^{ab} = \frac{1}{4}\left[\gamma^a, \gamma^b\right], \quad P^a = -\frac{i}{2}\sqrt{-\Lambda}\gamma^a.$$

Note that, in view of (5.2), when acting on the spinors the product of the two external differentials $\mathcal{D}$ vanishes (5.3)

$$\mathcal{D}^2\psi = 0\,. \tag{5.4}$$

When dealing with tensor-spinors we will assume that $\mathcal{D}$ acts as a covariant differential on vector indices and as (5.3) on spinorial ones. In particular, the matrices γ^a are annihilated by ∇ but not by $\mathcal{D}$

$$\mathcal{D}\gamma^a = -\frac{i}{2}\sqrt{-\Lambda}e_b\left[\gamma^b, \gamma^a\right] = -i\sqrt{-\Lambda}e_b\gamma^{ba}. \tag{5.5}$$

In view of (5.2) and (5.4), the following identity holds for the symmetric tensor-spinors

$$\mathcal{D}^2\psi^{a_1\ldots a_r} = \nabla^2\psi^{a_1\ldots a_r} = -\Lambda e^{(a_1} \wedge e_b \wedge \psi^{a_2\ldots a_r)b}\,. \tag{5.6}$$

The AdS counterpart of the fermionic higher-spin torsion (4.14) is defined as follows

$$\mathcal{T}^{a_1\ldots a_{r-1}} \equiv \mathcal{D}\psi^{a_1\ldots a_{r-1}} - e_b \wedge \psi^{a_1\ldots a_{r-1},b}. \tag{5.7}$$

In virtue of (5.6), it is gauge invariant under the following AdS-deformation of the gauge transformations (4.5)

$$\begin{aligned}\delta\psi^{a_1\ldots a_{r-1}} &= \mathcal{D}\xi^{a_1\ldots a_{r-1}} - e_b\xi^{a_1\ldots a_{r-1},b}\,,\\ \delta\psi^{a_1\ldots a_{r-1},b} &= \mathcal{D}\xi^{a_1\ldots a_{r-1},b} - e_c\xi^{a_1\ldots a_{r-1},bc}\\ &\quad+\Lambda\left(e^{(a_{r-1}}\xi^{a_1\ldots a_{r-2})b} - (r-1)e^b\xi^{a_1\ldots a_{r-1}}\right),\end{aligned} \tag{5.8}$$

which coincide with (4.5) in the flat limit $\Lambda \to 0$.

In flat space, for consistency, we required the gauge parameters to satisfy the constraints (4.7), which guarantee that the γ-trace of the higher-spin vielbein of rank $r = s - \frac{1}{2}$ transforms under the gauge transformations in the same way as a higher-spin vielbein of rank $r = s - \frac{3}{2}$. In the AdS space, in view of (5.5), the same requirement leads to the following constraints on the gauge parameters

$$\delta\left(\gamma_b\psi^{a_1\ldots a_{r-2}b}\right) = \mathcal{D}\Xi^{a_1\ldots a_{r-2}} - e_c\Xi^{a_1\ldots a_{r-2},c}\,.$$

where

$$\Xi^{a_1\ldots a_{r-2}} \equiv \gamma_b\xi^{a_1\ldots a_{r-2}b}, \quad \Xi^{a_1\ldots a_{r-2},c} \equiv -i\sqrt{-\Lambda}\gamma^c{}_b\xi^{a_1\ldots a_{r-2}b} + \gamma_b\xi^{a_1\ldots a_{r-2}b,c}.$$

The symmetrization properties of the parameter $\Xi^{a_1\ldots a_{r-2},c}$ impose the following constraint on $\xi^{a_1\ldots a_{r-1},b}$ which is the AdS generalization of (4.7)

$$\gamma_b\xi^{a_1\ldots a_{r-1},b} = -i\sqrt{-\Lambda}\gamma^{(a_1}{}_b\,\xi^{a_2\ldots a_{r-1})b}.$$

Correspondingly, one imposes the analogous constraint on the auxiliary higher-spin connection

$$\gamma_b\psi^{a_1\ldots a_{r-1},b} = -i\sqrt{-\Lambda}\gamma^{(a_1}{}_b\,\psi^{a_2\ldots a_{r-2})b}. \tag{5.9}$$

The relations involving contractions with the metric η_{ab} in (4.6) and (4.7) do not change.

5.2. *The frame-like action in AdS*

We construct the frame-like action for the reducible higher-spin fermionic field $\psi^{a_1\dots a_{r-1}}$ in the anti-de Sitter space in the same way as in flat space, i.e. with the use of the higher-spin torsion (5.7) and fix the coefficient by requiring that the action reduces to (4.15) in the $\Lambda\to 0$ limit. We thus get

$$S_{AdS}=i\int e^{a_1}\wedge\cdots\wedge e^{a_{D-3}}\wedge\varepsilon_{a_1\dots a_{D-3}cdf}\Big(\bar\psi^{b_1\dots b_{r-1}}\wedge\gamma^{cdf}\mathcal{T}_{b_1\dots b_{r-1}} \\ -6(r-1)\bar\psi^{b_1\dots b_{r-2}c}\wedge\gamma^{d}\mathcal{T}^{f}_{b_1\dots b_{r-2}}\Big). \tag{5.10}$$

This time the terms in this action containing the auxiliary field $\psi^{a_1\dots a_{r-1},b}$ cancel each other due to the deformed constraint (5.9). So, in comparison with (4.16) the action (5.10) contains more terms than in the flat space, namely

$$S_{AdS}=i\int\varepsilon_{a_1,\dots,a_{D-3}cdf}e^{a_1}\cdots e^{a_{D-3}}\Big\{\bar\psi^{b_1\dots b_{r-1}}\gamma^{cdf}\mathcal{D}\psi_{b_1\dots b_{r-1}} \\ -6(r-1)\bar\psi^{b_1\dots b_{r-2}c}\gamma^{d}\mathcal{D}\psi^{f}_{b_1\dots b_{r-2}}+i\frac{\sqrt{-\Lambda}}{D-2}e^{c}\Big[6(r-1)\bar\psi^{b_1\dots b_{r-2}d}\gamma^{f}\gamma^{g}\psi_{b_1\dots b_{r-2}g} \\ +3(r-1)\left(\bar\psi^{b_1\dots b_{r-1}}\gamma^{df}\psi_{b_1\dots b_{r-1}}-\bar\psi^{b_1\dots b_{r-2}e}\gamma_e\gamma^{df}\gamma^{g}\psi_{b_1\dots b_{r-2}g}\right) \\ -6(r-1)\Big((r-2)\bar\psi^{b_1\dots b_{r-3}ed}\gamma_e\gamma^{g}\psi^{k}_{gb_1\dots b_{r-3}}-(r-1)\bar\psi^{b_1\dots b_{r-2}d}\psi^{f}_{b_1\dots b_{r-2}}\Big)\Big]\Big\}, \tag{5.11}$$

where to simplify the appearance of the above expression we have skipped the wedge products of the differential forms. In the form (5.11) the AdS action was constructed in [24] by a "brute force", i.e. without the help of the gauge-invariant higher-spin torsion (5.7).

The total gauge invariance and reality of (5.10) (and therefore of (5.11)) is proven in the same way as in Section 4.3 by showing that the action is equivalent (modulo total derivatives) to the action constructed with the use of the Dirac conjugate torsion $\bar{\mathcal{T}}^{a_1\dots a_{r-1}}$.

As in the flat case, to include in the consideration the spin-$\frac{1}{2}$ field ρ we add to the action (5.11) the Dirac action

$$S_{\frac{1}{2}}=\int d^Dx\bar\rho\left(i\gamma^a\nabla_a+\frac{D-4}{2}\sqrt{-\Lambda}\right)\rho. \tag{5.12}$$

5.3. *The metric-like action and equations of motion for fermionic triplets in AdS*

Having at hand the frame-like action (5.11)+(5.12) we are now ready to derive its metric-like counterpart by replacing in the former the higher-spin vielbein $\psi_\mu^{a_1\dots a_{r-1}}$ with the fermionic higher-spin triplet fields defined in (4.8), (4.12) and (4.13).

Somewhat tedious but direct calculations results in the following metric-like action

$$
\begin{aligned}
S^{AdS}_{ML} = \int d^D x\sqrt{-g}\,\Big[& i\bar{\chi}^{b_1\dots b_{r-1}}\not{\nabla}\chi_{b_1\dots b_{r-1}} + \bar{\chi}_{b_1\dots b_{r-1}}\nabla_a\Psi^{ab_1\dots b_{r-1}} \\
& +\nabla_a\bar{\Psi}^{ab_1\dots b_{r-1}}\chi_{b_1\dots b_{r-1}} + \frac{i}{r}\bar{\Psi}^{b_1\dots b_r}\not{\nabla}\Psi_{b_1\dots b_r} \\
& -(r-1)\big(i\bar{\lambda}^{b_1\dots b_{r-2}}\not{\nabla}\lambda_{b_1\dots b_{r-2}} - \nabla_a\bar{\chi}^{ab_1\dots b_{r-2}}\lambda_{b_1\dots b_{r-2}} - \bar{\lambda}_{b_1\dots b_{r-2}}\nabla_a\chi^{ab_1\dots b_{r-2}}\big) \\
& -\sqrt{-\Lambda}\Big(\frac{D+2r}{2}\bar{\chi}^{b_1\dots b_{r-1}}\chi_{b_1\dots b_{r-1}} - \frac{D+2r-4}{2r}\bar{\Psi}^{b_1\dots b_r}\Psi_{b_1\dots b_r} \\
& +(r-1)\frac{D+2r-8}{2}\bar{\lambda}^{b_1\dots b_{r-2}}\lambda_{b_1\dots b_{r-2}} + \frac{3}{2}i(r-1)\bar{\not{\chi}}^{b_1\dots b_{r-2}}\lambda_{b_1\dots b_{r-2}} \\
& -\frac{3}{2}i(r-1)\bar{\lambda}^{b_1\dots b_{r-2}}\not{\chi}_{b_1\dots b_{r-2}} + \frac{3}{2}i\bar{\not{\Psi}}^{b_1\dots b_{r-1}}\chi_{b_1\dots b_{r-1}} \\
& -\frac{3}{2}i\bar{\chi}^{b_1\dots b_{r-1}}\not{\Psi}_{b_1\dots b_{r-1}} + \bar{\not{\Psi}}^{b_1\dots b_{r-1}}\not{\Psi}_{b_1\dots b_{r-1}} \\
& -(r-1)\bar{\not{\chi}}^{b_1\dots b_{r-2}}\not{\chi}_{b_1\dots b_{r-2}} - (r-1)(r-2)\bar{\not{\chi}}^{b_1\dots b_{r-3}}\not{\chi}_{b_1\dots b_{r-3}}\Big)\Big]
\end{aligned} \tag{5.13}
$$

This action is gauge invariant under the AdS version of the gauge transformations (4.5)

$$
\begin{aligned}
\delta\Psi^{a_1\dots a_r} &= \nabla^{(a_1}\xi^{a_2\dots a_r)} - \frac{i}{2}\sqrt{-\Lambda}\gamma^{(a_1}\xi^{a_2\dots a_r)} \\
\delta\chi^{a_1\dots a_{r-1}} &= i\not{\nabla}\xi^{a_1\dots a_{r-1}} + \frac{D+2r-2}{2}\sqrt{-\Lambda}\xi^{a_1\dots a_{r-1}} - \sqrt{-\Lambda}\gamma^{(a_1}\not{\xi}^{a_2\dots a_{r-1})} \\
\delta\lambda^{a_1\dots a_{r-2}} &= \nabla_b\xi^{a_1\dots a_{r-2}b} - \frac{i}{2}\sqrt{-\Lambda}\not{\xi}^{a_1\dots a_{r-2}},
\end{aligned}
$$

which follow from the first expression in (5.8).

The fermionic triplet equations of motion obtained by extremising this action have the following form

$$
\begin{aligned}
&\left(i\not{\nabla} + \frac{D+2r-4}{2}\sqrt{-\Lambda}\right)\Psi_{b_1\dots b_r} - \sqrt{-\Lambda}\gamma_{(b_1}\not{\Psi}_{b_2\dots b_r)} \\
&\qquad\qquad = \left(\nabla_{(b_1} + \frac{3}{2}i\sqrt{-\Lambda}\gamma_{(b_1}\right)\chi_{b_2\dots b_r)}\,, \\
&\left(i\not{\nabla} + \frac{D+2r-8}{2}\sqrt{-\Lambda}\right)\lambda_{b_1\dots b_{r-2}} - \sqrt{-\Lambda}\gamma_{(b_1}\not{\lambda}_{b_2\dots b_{r-2})} \\
&\qquad\qquad = \nabla^a\chi_{ab_1\dots b_{r-2}} + \frac{3}{2}i\sqrt{-\Lambda}\not{\chi}_{b_1\dots b_{r-2}}\,, \\
&\left(i\not{\nabla} - \frac{D+2r}{2}\sqrt{-\Lambda}\right)\chi_{b_1\dots b_{r-1}} + \sqrt{-\Lambda}\gamma_{(b_1}\not{\chi}_{b_2\dots b_{r-1})} \\
&\qquad = -\nabla^a\Psi_{ab_1\dots b_{r-1}} - \frac{3}{2}i\sqrt{-\Lambda}\not{\Psi}_{b_1\dots b_{r-1}} + \left(\nabla_{(b_1} + \frac{3}{2}i\sqrt{-\Lambda}\gamma_{(b_1}\right)\lambda_{b_2\dots b_{r-1})}\,.
\end{aligned} \tag{5.14}
$$

5.4. $(\frac{1}{2},\frac{3}{2})$ *doublet*

As the simplest example demonstrating basic properties of the above metric-like Lagrangian systems of fermionic fields in AdS, let us consider a doublet of fields χ

and Ψ_a propagating a spin $\frac{1}{2}$ and $\frac{3}{2}$. For this system the action is and the equations of motion (5.14) reduce to

$$\left(i\nabla\!\!\!/ + \frac{D-2}{2}\sqrt{-\Lambda}\right)\Psi_a - \sqrt{-\Lambda}\gamma_a\gamma^b\Psi_b = \left(\nabla_a + \frac{3}{2}i\sqrt{-\Lambda}\gamma_a\right)\chi\,,$$
$$i\nabla^a\Psi_a - \frac{3}{2}\sqrt{-\Lambda}\gamma^a\Psi_a = \left(\nabla\!\!\!/ + i\frac{D+2}{2}\sqrt{-\Lambda}\right)\chi. \qquad (5.15)$$

These are invariant under the gauge transformations

$$\delta\Psi_a = \nabla_a\xi - \frac{i}{2}\sqrt{-\Lambda}\gamma_a\xi\,, \qquad (5.16)$$
$$\delta\chi = i\gamma^a\nabla_a\xi + \frac{D}{2}\sqrt{-\Lambda}\xi\,.$$

By taking linear combinations of the equations (5.15) we get the disentangled conventional equations of motion for the gauge-invariant dynamical spin-1/2 field $\rho = \chi - i\gamma^a\Psi_a$ and the dynamical spin-3/2 field $\psi_a = \Psi_a - \frac{i}{D-2}\gamma_a\rho$

$$\left(i\gamma^a\nabla_a + \tfrac{D-4}{2}\sqrt{-\Lambda}\right)\rho = 0\,,$$
$$\gamma^b\mathcal{D}_{[b}\psi_{a]} = \gamma^b(\nabla_{[b} - \tfrac{i}{2}\sqrt{-\Lambda}\gamma_{[b})\psi_{a]} = 0. \qquad (5.17)$$

Let us now compare the equations (5.15) with equations for a $(\frac{1}{2},\frac{3}{2})$ doublet proposed in [14]. The latter have the following form

$$\left(i\nabla\!\!\!/ + \tfrac{D-2}{2}\sqrt{-\Lambda}\right)\Psi_a + \tfrac{\sqrt{-\Lambda}}{2}\gamma_a\gamma^b\Psi_b = \nabla_a\chi\,,$$
$$i\nabla^a\Psi_a + \tfrac{D-1}{2}\sqrt{-\Lambda}\gamma^a\Psi_a = \nabla\!\!\!/\chi. \qquad (5.18)$$

These equations are also invariant under the gauge transformations (5.16) but, as one can see, they differ from (5.15). As was shown in [14], the consistency of (5.18) requires that

$$\chi = i\gamma^a\Psi_a,$$

which means that the system (5.18) does not contain the physical spin-1/2 field, the issue which is solved by the properly modified equations (5.15).

6. Conclusion and Outlook

As we have reviewed, triplet systems of higher-spin fields naturally show up in a truncated free action for String Field Theory in flat space. Their symmetry properties and Lagrangian structure are governed by the tensionless limit of the string BRST charge.

As was shown in [24], in the frame-like formulation, the triplet fields are endowed with a geometrical meaning of higher-spin vielbeins and connections transforming under higher-spin local symmetries. This allows one to determine in a conventional way gauge-invariant higher-spin torsion and curvatures and use them for the construction of simply-looking frame-like actions for these systems both in flat and AdS spaces.

We have then demonstrated how, starting from the frame-like action for the unconstrained fermionic higher-spin vielbein and the spin-1/2 field in AdS space, and using the splitting of this vielbein into the metric-like triplet of fermionic fields, one resolves a long-standing issue of the construction of the metric-like Lagrangian description of the fermionic triplets in AdS spaces.

Having now at our disposal the metric-like Lagrangian formulation for bosonic and fermionic triplets in AdS, one can analyze the BRST structure associated with their gauge transformations and equations of motion, and use the obtained form of the BRST charge for studying whether and how the triplets in AdS may arise from the quantization of strings in AdS in a tensionless limit. Since in AdS one may play with two parameters, the string tension and the AdS radius, the tensionless limit of AdS strings may avoid singularities of its flat space counterpart. In this respect it will be interesting to revise the procedure and results of taking a tensionless limit of a bosonic AdS string considered in [5].

As another direction of research, one can proceed with studying interactions of higher-spin triplet fields (cubic and quartic vertices, current exchanges, etc.) along lines put forward in [23, 25, 31]. One of the advantages here is that reducibility of triplet systems can make things simpler, since a single triplet vertex contains a number of vertices of irreducible higher-spin fields.

Finally, one may speculate on the observation that minimal Vasiliev theory [36], describing fields of the even spins from 0 to infinity, is a theory of a single higher-spin "triplet" with $s = \infty$ and wonder whether it can be extended to an interacting theory of infinite sets of "triplets", as might happen in string theory.

Acknowledgments

The authors are grateful to Dario Francia, Ruslan Metsaev, Bo Sundborg, Mirian Tsulaia and Mikhail Vasiliev for useful discussions and comments. The work of D. S. was partially supported by the Russian Science Foundation grant 14-42-00047 in association with Lebedev Physical Institute and by the Australian Research Council (ARF) Discovery Project grant DP160103633. The work of F. A. was supported by INFN and Scuola Galileiana di Studi Superiori of Padua.

Appendix. Notation and conventions

The signature of the D-dimensional space-time metric is chosen to be almost minus $(+,-,\dots,-)$. The Greek letters $\mu,\nu,...$ denote world indices associated with space-time coordinates x^μ. The Latin letters $a,b,c...$ label the components of tangent-space tensors. The world indices are converted into the tangent space ones by means of the vielbein $e^a_\mu(x)$, which is just the unit matrix δ^a_μ in flat space-time.

Different groups of symmetric indices are separated by commas. Each group corresponds to a row in the Young tableau associated to the representation the tensor sits in. For example

$$\psi^{abc,de}$$

is a tensor whose symmetry properties are defined by the Young tableau

$$\begin{array}{|c|c|c|}\hline a & b & c \\ \hline d & e \\ \cline{1-2}\end{array}\,.$$

Symmetrizations of indices are not weighted and are denoted with round brackets, e.g.

$$A^{(a_1}B^{a_2a_3)} \equiv A^{a_1}B^{a_2a_3} + A^{a_2}B^{a_1a_3} + A^{a_3}B^{a_2a_1}\,.$$

We also use the short-hand notation for contractions involving γ matrices, e.g.

$$\not{\psi}^{a_1\ldots a_n} \equiv \gamma_{a_{n+1}}\psi^{a_1\ldots a_n a_{n+1}}, \qquad \not{\partial} \equiv \gamma_\mu \partial^\mu\,.$$

The gamma–matrices obey the Clifford algebra

$$\gamma^a\gamma^b + \gamma^b\gamma^a = 2\eta^{ab}\,.$$

In $D = 4$ the gamma-matrices are purely imaginary in the Majorana representation.

References

1. D. J. Gross, High-Energy Symmetries of String Theory, *Phys. Rev. Lett.* **60**, 1229, (1988). doi: 10.1103/PhysRevLett.60.1229.
2. D. J. Gross and P. F. Mende, String Theory Beyond the Planck Scale, *Nucl. Phys.* **B303**, 407, (1988). doi: 10.1016/0550-3213(88)90390-2.
3. G. W. Moore. Symmetries and symmetry breaking in string theory. In *International Workshop on Supersymmetry and Unification of Fundamental Interactions (SUSY 93) Boston, Massachusetts, March 29-April 1, 1993*, (1993).
4. B. Sundborg, Stringy gravity, interacting tensionless strings and massless higher spins, *Nucl. Phys. Proc. Suppl.* **102**, 113–119, (2001). doi: 10.1016/S0920-5632(01)01545-6.
5. G. Bonelli, On the tensionless limit of bosonic strings, infinite symmetries and higher spins, *Nucl. Phys.* **B669**, 159–172, (2003). doi: 10.1016/j.nuclphysb.2003.07.002.
6. U. Lindstrom and M. Zabzine, Tensionless strings, WZW models at critical level and massless higher spin fields, *Phys. Lett.* **B584**, 178–185, (2004). doi: 10.1016/j.physletb.2004.01.035.
7. A. Sagnotti and M. Taronna, String Lessons for Higher-Spin Interactions, *Nucl. Phys.* **B842**, 299–361, (2011). doi: 10.1016/j.nuclphysb.2010.08.019.
8. S. Giombi, S. Minwalla, S. Prakash, S. P. Trivedi, S. R. Wadia, and X. Yin, Chern-Simons Theory with Vector Fermion Matter, *Eur. Phys. J.* **C72**, 2112, (2012). doi: 10.1140/epjc/s10052-012-2112-0.
9. M. R. Gaberdiel and R. Gopakumar, String Theory as a Higher Spin Theory. (2015).
10. S. Ouvry and J. Stern, Gauge fields of any spin and symmetry, *Phys. Lett.* **B177**, 335, (1986). doi: 10.1016/0370-2693(86)90763-X.
11. A. K. H. Bengtsson, A unified action for higher spin gauge bosons from covariant string theory, *Phys. Lett.* **B182**, 321, (1986). doi: 10.1016/0370-2693(86)90100-0.
12. D. Francia and A. Sagnotti, On the geometry of higher-spin gauge fields, *Class. Quant. Grav.* **20**, S473–S486, (2003).
13. M. P. Bellon and S. Ouvry, $D = 4$ Supersymmetry for Gauge Fields of Any Spin, *Phys. Lett.* **B187**, 93, (1987). doi: 10.1016/0370-2693(87)90078-5.

14. A. Sagnotti and M. Tsulaia, On higher spins and the tensionless limit of string theory, *Nucl. Phys.* **B682**, 83–116, (2004). doi: 10.1016/j.nuclphysb.2004.01.024.
15. M. Henneaux and C. Teitelboim, In: Quantum Mechanics of Fundamental Systems, 2 , *eds. C. Teitelboim and J. Zanelli (Plenum Press, New York)*. p. p. 113, (1988).
16. A. I. Pashnev, Composite systems and field theory for a free Regge trajectory, *Theor. Math. Phys.* **78**, 272–277, (1989). doi: 10.1007/BF01017664.
17. G. Barnich, G. Bonelli, and M. Grigoriev. From BRST to light-cone description of higher spin gauge fields. In *Proceedings, 4th International Spring School and Workshop on Quantum Field Theory & Hamiltonian Systems*, (2005).
18. R. Metsaev, Cubic interaction vertices of massive and massless higher spin fields, *Nucl. Phys.* **B759**, 147–201, (2006). doi: 10.1016/j.nuclphysb.2006.10.002.
19. I. L. Buchbinder, A. Fotopoulos, A. C. Petkou, and M. Tsulaia, Constructing the cubic interaction vertex of higher spin gauge fields, *Phys. Rev.* **D74**, 105018, (2006). doi: 10.1103/PhysRevD.74.105018.
20. I. L. Buchbinder, A. V. Galajinsky, and V. A. Krykhtin, Quartet unconstrained formulation for massless higher spin fields, *Nucl. Phys.* **B779**, 155–177, (2007). doi: 10.1016/j.nuclphysb.2007.03.032.
21. A. Fotopoulos, N. Irges, A. C. Petkou, and M. Tsulaia, Higher-Spin Gauge Fields Interacting with Scalars: The Lagrangian Cubic Vertex, *JHEP.* **10**, 021, (2007). doi: 10.1088/1126-6708/2007/10/021.
22. R. R. Metsaev, Cubic interaction vertices for fermionic and bosonic arbitrary spin fields, *Nucl. Phys.* **B859**, 13–69, (2012). doi: 10.1016/j.nuclphysb.2012.01.022.
23. A. Fotopoulos and M. Tsulaia, Gauge Invariant Lagrangians for Free and Interacting Higher Spin Fields. A Review of the BRST formulation, *Int. J. Mod. Phys.* **A24**, 1–60, (2009). doi: 10.1142/S0217751X09043134.
24. D. P. Sorokin and M. A. Vasiliev, Reducible higher-spin multiplets in flat and AdS spaces and their geometric frame-like formulation, *Nucl. Phys.* **B809**, 110–157, (2009). doi: 10.1016/j.nuclphysb.2008.09.042.
25. A. Fotopoulos and M. Tsulaia, Current Exchanges for Reducible Higher Spin Multiplets and Gauge Fixing, *JHEP.* **10**, 050, (2009). doi: 10.1088/1126-6708/2009/10/050.
26. A. Fotopoulos and M. Tsulaia. Current Exchanges for Reducible Higher Spin Modes on AdS. In *19th International Colloquium on Integrable Systems and Quantum Symmetries (ISQS-19) Prague, Czech Republic, June 17-19, 2010.*
27. D. Francia, String theory triplets and higher-spin curvatures, *Phys. Lett.* **B690**, 90–95, (2010). doi: 10.1016/j.physletb.2010.05.006.
28. X. Bekaert and E. Meunier, Higher spin interactions with scalar matter on constant curvature spacetimes: conserved current and cubic coupling generating functions, *JHEP.* **11**, 116, (2010). doi: 10.1007/JHEP11(2010)116.
29. E. D. Skvortsov and Yu. M. Zinoviev, Frame-like Actions for Massless Mixed-Symmetry Fields in Minkowski space. Fermions, *Nucl. Phys.* **B843**, 559–569, (2011). doi: 10.1016/j.nuclphysb.2010.10.012.
30. A. Campoleoni and D. Francia, Maxwell-like Lagrangians for higher spins, *JHEP.* **03**, 168, (2013). doi: 10.1007/JHEP03(2013)168.
31. R. R. Metsaev, BRST-BV approach to cubic interaction vertices for massive and massless higher-spin fields, *Phys. Lett.* **B720**, 237–243, (2013). doi: 10.1016/j.physletb.2013.02.009.
32. M. Asano, Gauge invariant and gauge fixed actions for various higher-spin fields from string field theory, *Nucl. Phys.* **B868**, 75–101, (2013). doi: 10.1016/j.nuclphysb.2012.11.008.
33. M. Asano and M. Kato, Extended string field theory for massless higher-spin fields,

Nucl. Phys. **B877**, 1107–1128, (2013). doi: 10.1016/j.nuclphysb.2013.11.002.
34. X. Bekaert, N. Boulanger, and D. Francia, Mixed-symmetry multiplets and higher-spin curvatures, *J. Phys.* **A48**(22), 225401, (2015). doi: 10.1088/1751-8113/48/22/225401.
35. A. Agugliaro, F. Azzurli, and D. Sorokin, Fermionic higher-spin triplets in AdS, arXiv:1603.02251. (2016).
36. M. A. Vasiliev, Consistent equation for interacting gauge fields of all spins in (3+1)-dimensions, *Phys. Lett.* **B243**, 378–382, (1990). doi: 10.1016/0370-2693(90)91400-6.

Bulk Quartic Vertices from Boundary Four-Point Correlators

X. Bekaert[a], J. Erdmenger[b], D. Ponomarev[c] and C. Sleight[b]

[a]*Laboratoire de Mathématiques et Physique Théorique, Unité Mixte de Recherche 7350 du CNRS, Fédération de Recherche 2964 Denis Poisson, Université François Rabelais, Parc de Grandmont, 37200 Tours, France*

[b]*Max-Planck-Institut für Physik (Werner-Heisenberg-Institut), Föhringer Ring 6, D-80805 Munich, Germany*

[c]*The Blackett Laboratory, Imperial College, London SW7 2AZ, United Kingdom*

We present arguments which suggest that the bulk higher-spin gravity duals of weakly-coupled conformal field theories obey some refined notion of locality. In particular, we discuss the Mellin amplitude programme in this context. We focus on the $O(N)$ vector model and minimal higher-spin gravity as a paradigmatic example of such holographic dual pairs. We restrict ourselves to three- and four-point functions of scalar primary operators, but the qualitative conclusions are expected to hold for the generic case.

1. Bulk Locality and Mellin Amplitudes

Bulk locality is one of the most remarkable features of the AdS/CFT correspondence. The concept of bulk locality refers to the fact that the usual conditions for locality in a quantum field theory also hold in the interior of anti de Sitter (AdS) space. It is expected to hold in the usual regime where the duality is tested, i.e. when the curvature radius R_{AdS} is large compared to the string length ℓ_s which, in turn, is large compared to the Planck length ℓ_P. On the gravity side, this corresponds to (I) a semi-classical limit: $R_{AdS} \gg \ell_P$ and (II) a higher-spin gap: all particles of spins s greater than two have large masses $M_{s>2} \gg M_{s'\leq 2}$. On the conformal field theory (CFT) side, this translates into:

(I) a semi-classical limit: a perturbative expansion around a generalised free field theory when $N \gg 1$,

(II) a higher-spin gap: all single-trace operators of higher-spins have large conformal dimensions $\Delta_{s>2} \gg \Delta_{s'\leq 2}$.

These two properties were argued to provide necessary and sufficient conditions for a CFT to possess a semiclassical and local bulk dual [1, 2]. Later, a third condition was added from the perspective of effective field theory [3]:

(III) polynomial boundedness of Mellin amplitudes.

Over the last years, significant progress has been made in the Mellin amplitude programme. Initiated in [4], the goal is to interpret the Mellin amplitudes [5] associated to CFT correlation functions as AdS scattering amplitudes (the corresponding dictionary is summarised in Table 1).

Table 1. Mellin amplitude dictionary.

Minkowski spacetime	Anti-de Sitter spacetime
Feynman diagram	Witten diagram
Mandelstam invariants	Mellin variables
Scattering amplitude	Mellin amplitude

For a given correlation function, the corresponding Mellin amplitude is a suitably normalised Mellin transform of the factor that is not fixed by conformal symmetry. Consider, for example, the simplest case of a four-point correlator of a scalar primary operator $\mathcal{O}$ with scaling dimension Δ,

$$\langle \mathcal{O}(y_1)\,\mathcal{O}(y_2)\,\mathcal{O}(y_3)\,\mathcal{O}(y_4)\rangle = \frac{1}{\left(y_{12}^2 y_{34}^2\right)^{\Delta}}\, F(u,v)\,, \tag{1}$$

with u and v the two cross ratios $u = \frac{y_{12}^2 y_{34}^2}{y_{13}^2 y_{24}^2}$, $v = \frac{y_{14}^2 y_{23}^2}{y_{13}^2 y_{24}^2}$, and $y_{ij} = y_i - y_j$. Then the *reduced Mellin amplitude* [5] is the function $M(s,t)$ of two scattering variables s and t defined via the relation

$$F(u,v) = \int \frac{ds}{2\pi i} \int \frac{dt}{2\pi i}\, u^s\, v^{\Delta - s - t}\, M(s,t), \tag{2}$$

providing a Mellin-Barnes representation of the conformal correlator (1). The *Mellin amplitude* itself is the function $\mathcal{M}(s,t)$ defined by

$$M(s,t) = \left[\Gamma(\Delta - s)\Gamma(\Delta - t)\Gamma(s + t - \Delta)\right]^2 \mathcal{M}(s,t), \tag{3}$$

where a numerical coefficient (irrelevant for the present discussion) has been dropped. The extraction of the corresponding gamma functions is important to ensure the following property: For a large-N CFT with a discrete set of primary operators, the Mellin amplitudes of single-trace primary-operator correlators are meromorphic functions with simple poles determined by the twists of the single-trace operators in the OPE [4, 5]. In particular, the reduced Mellin amplitude $M(s,t)$ possesses extra poles with respect to the Mellin amplitude $\mathcal{M}(s,t)$, due to the gamma functions in (3). These extra poles originate from the double-trace operators in the OPE.

Table 2. Amplitude properties.

Bulk process	Boundary amplitude
Local contact interaction	Polynomial
Particle exchange	Simple pole

For individual (Feynman or Witten) diagrams, scattering amplitudes of elementary particles and Mellin amplitudes for single-trace primary operators thus share common properties as functions of Mandestam invariants vs Mellin variables [4–7] (cf. Table 2) suggesting a common

Boundary criterion of bulk locality: *Interactions on flat/AdS spacetime are local iff the amplitudes of the corresponding contact Feynman/Witten diagram are polynomial functions of Mandelstam invariants / Mellin variables.*

In this sense, the condition (III) can be seen as the criterion of bulk locality itself, in the same way that the condition (I) stands for the criterion of existence of a semiclassical limit. Nevertheless, the condition (II) remains of interest because the presence of a higher-spin gap sets a higher-spin symmetry breaking scale in the bulk. Even if the gravity theory may be nonlocal in the sense of being an effective field theory, the previous scale controls the low-energy expansion of the theory and the validity range of its local truncations.

The Mellin amplitude programme of rewriting CFT correlators as Witten diagrams seemingly applies to the large class of strongly-coupled CFTs obeying the criteria (I)-(III). Under these hypotheses, the corresponding bulk dual would possess a simultaneously weakly-coupled & weakly-curved regime. Indeed, the Mellin amplitude programme seems well adapted (but presently restricted) to the holographic reconstruction of bulk theories (or individual scattering processes) possessing a weakly-coupled & weakly-curved limit. However it does not apply directly to the simplest example of CFTs: free ones (or weakly coupled ones). In fact, the Mellin transform of correlators of free CFTs is often not even well defined (see section 4). The bulk duals of free CFTs are conjectured to be higher-spin gravity theories which are indeed non-local in the restricted sense of locality (see section 2).

What could be a mild replacement of the criteria (II) and (III), that could provide necessary and sufficient conditions for CFTs (including weakly-coupled ones) to possess (mildly non-local) bulk duals (including higher-spin gravity theories)? A tentative answer could be:

(II') a finite number of single-trace primary operators with conformal dimension below any fixed dimension, in the $N \to \infty$ limit,

(III') analyticity of Mellin amplitudes.

The criterion (II') is inspired from the second basic property in [8]. In spirit, it is analogous to the second assumption of the Coleman-Mandula theorem [9] since the conformal dimension of a single-trace operator translates into the energy of the corresponding elementary particle in the bulk. In contrast to flat spacetime, in AdS this condition does not rule out a tower of massless higher spins as their dimensions grow with spin. Another motivation for criterion (II') is that bulk locality would be obscure if an infinite number of fields were relevant in a scattering process at a given energy. The criterion (III') relaxes strict locality and replaces it with the milder requirement that the coefficients in the Taylor series expansion of the amplitude decrease fast enough in order to have an infinite radius of convergence. In other words, the amplitude can be approximated, for any fixed accuracy, by a polynomial of sufficiently high degree, i.e. by a local interaction of sufficiently high order. More precisely, the criterion (III') amounts to the following

boundary criterion of weak locality:[10] *Interactions on flat/AdS spacetime are weakly local iff the amplitudes of the corresponding contact Feynman/Witten diagram are entire functions of Mandelstam invariants / Mellin variables.*

2. Bulk Locality and Higher-Spin Holography

The higher-spin holographic duality arose shortly after the birth of AdS/CFT correspondence and was initially motivated by the semiclassical but stringy regime ($\ell_s \gg R_{AdS} \gg \ell_P$) in the strong version of Maldacena conjecture. A more general picture progressively emerged in a long series of paper (see [11–17] for some early steps) but the basic idea underlying higher-spin holography is easy to summarise.

Free (or integrable) CFTs have an infinite number of global higher (sometimes called "hidden") symmetries (including conformal symmetry). Applying Noether's theorem, one deduces that their spectrum must contain an infinite tower of traceless conserved currents with unbounded spin (including spin two). Therefore, the AdS/CFT dictionary suggests that free (or some integrable) CFTs should be dual to "higher-spin gravity" theories, in the sense of theories whose spectra contain an infinite tower of gauge fields with unbounded spin (including spin two). The simplest example of such a scenario states that the singlet sector of the free and critical vector models should be holographically dual to minimal higher-spin gravity [16, 17].

Turning back to the issue of bulk locality, higher-spin interactions in four (and higher) dimension appear to be generically:

(A) **quasi-local** in the sense that they possess a perturbative expansion (in powers of fields and their derivatives) where each individual term in the Lagrangian is local (effective field theories are typically quasilocal).

(B) **non-local** in the sense that the total number of derivatives in the complete Lagrangian is unbounded (as in string field theory). This is a corollary of:

   (a) *Metsaev bounds:* The number of derivatives appearing in a non-trivial cubic vertex evaluated on the free mass-shell is bounded from [18]

      i. below by the highest spin involved,
      ii. above by the sum of the spins involved.

   (b) *Higher-spin algebra structure:* The Jacobi identity requires a spectrum with an infinite tower of fields with unbounded spin [19, 20].

   The number of derivatives is bounded from below by the highest spin involved by virtue of Metsaev's lower bound (a.i). However, there cannot be any upper bound on the spin in a consistent theory due to point (b). Hence, the number of derivatives is necessarily unbounded.

(C) **weakly-local** in the sense that the Mellin amplitude of contact Witten diagrams are entire functions of the Mellin variables.

   This weak-locality holds at cubic level due to Metsaev upper bound and should hold for the quartic self-interactions of the AdS scalar field due to general facts about Mellin amplitudes.

   (a) *Locality of cubic vertices:* Individual cubic higher-spin interactions are indeed local in the sense that any 3-point contact Witten diagram with fixed external legs is a polynomial function of the Mellin variables. This follows as a corollary from Metsaev upper bound (a.ii): For any triplet of spins, the number of derivatives in any relevant cubic vertex is bounded from above by the sum of the spins.
   (b) *Weak-locality of quartic vertices:* The quartic self-interactions of the AdS scalar field dual to the single-trace scalar primary operator in the $O(N)$ model (or, more generally, in a large-N CFT with discrete spectrum) appear to be weakly local, in the sense that the 4-point contact Witten diagram with four scalar external legs is an entire function of the Mellin variables.

   This follows from the following general facts about the decomposition of single-trace scalar primary operator 4-point function in conformal blocks at leading order in $1/N$:

   (i) This decomposition contains both single-trace and double-trace conformal blocks.
   (ii) Any single-trace conformal block can be accounted by a Witten exchange diagram.
   (iii) The Mellin amplitude of a 4-point function is an entire function of the Mellin variables iff its conformal block decomposition does *not* contain any single-trace contribution.

   The point (i) is expected; the coinciding point limit of two single-trace operators will produce double-trace operators as well. The point (ii) is AdS/CFT standard lore. The correspondence between single-trace conformal blocks and

Witten exchange diagrams is one-to-one: It is essentially an identity up to double-trace conformal blocks (see e.g. the recent discussion in [8]). The point (iii) holds because Mellin amplitudes are meromorphic functions with simple poles arising only from single-trace operators [4].

The proof of (C.b) goes as follows: Consider a single-trace scalar primary operator 4-point function in a CFT with discrete spectrum and at large N. First, decompose this 4-point function into conformal blocks. Second, using facts (i)-(ii) associate to each single-trace conformal block its corresponding Witten exchange diagram. Third, substract from the 4-point function all these Witten exchange diagrams. By construction, the difference that remains is a 4-point function whose decomposition contains only double-trace conformal blocks. Fourth, conclude from (iii) that this remainder is an entire function of the Mellin variables since it does not contain any single-trace contribution. This allows to interpret the remainder as a 4-point contact Witten diagram associated to a weakly local quartic vertex.

As one can see, the boundary criterion of weak locality (see Section 1) tantamounts to the existence of a neat separation in the amplitude between the exchange and contact contributions: Poles in the scattering amplitude are accounted by particle exchanges and what remains is interpreted as the contact amplitude.

Remarks:

- It would be interesting to compare the criterion of weak locality for higher-spin gravity advocated here with the recent proposals [21] and [22] based, respectively, on functional classes of star-product elements and on classes of field redefinitions leaving Witten diagrams invariant.
- A caveat of our argument that the bulk dual of the $O(N)$ model is a weakly local higher-spin gravity is that it implicitly assumes that Mellin amplitudes are well defined functions while they actually require some regularisation. They may also be thought as distributions (c.f. Section 4) but then it is their analyticity properties which are somewhat elusive.
- Another subtlety is that the third step in the proof (substraction of all Witten exchange diagrams) can bring infinities in the double-trace conformal block decomposition of the remainder. Taking an optimistic standpoint, this step might actually regularise the Mellin amplitudes alluded above.

3. Holographic Reconstruction of Higher-Spin Gravity Vertices

The line of arguments presented in section 2 is expected to generalise (to all spins and to higher numbers of points) and indicates that the higher-spin interactions for the bulk dual of the $O(N)$ model might be weakly local. This picture is very suggestive but remains somewhat qualitative. The explicit holographic reconstruction of some quartic vertices in higher-spin gravity, to which we now turn, provides a more concrete playground to test bulk locality.

More specifically, we will consider the holographic reconstruction of quartic AdS interations from a free CFT. In practice, this task amounts to:

- Compute the 3 and 4 point conformal correlators via Wick contraction,
- Write the most general ansatz for the relevant cubic and quartic vertices,
- Compute the corresponding exchange and contact amplitudes,
- Fix the coefficients of vertices by matching correlators with total amplitudes.

A priori, it is not guaranteed that such a purely holographic reconstruction produces interactions compatible with the Noether procedure. However, it is natural to expect that these two perturbative procedures are compatible with each other since Ward identities of the boundary CFT should be dual to Noether identities of the AdS theory.

The holographic reconstruction was performed in [10, 23] for the simplest non-trivial case: The quartic self-interaction of the AdS_4 scalar field in the higher-spin multiplet dual to the $d = 3$ free $O(N)$ model. The relevant four-point Witten diagrams are displayed in figure 1. Diagrams (a)-(c) are exchanges of massless spin-s fields between two pairs of the real scalar. The contact diagram (d) is the amplitude associated to the quartic vertex.

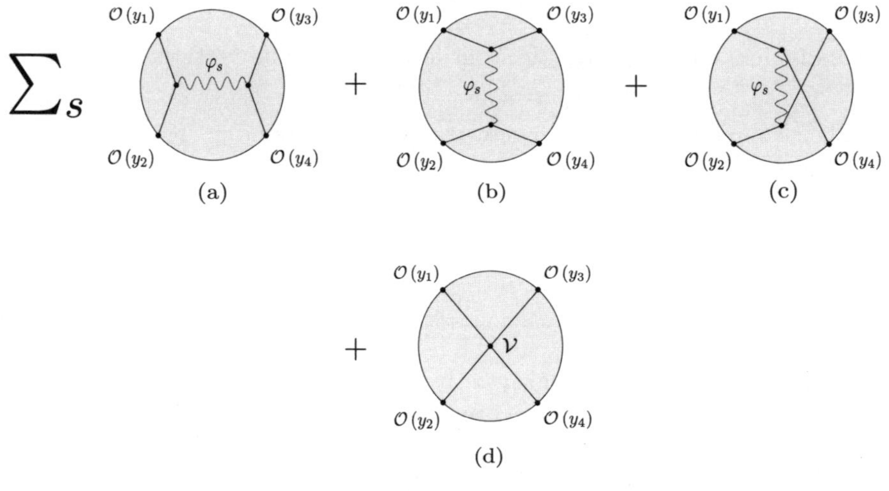

Fig. 1. Four-point Witten diagrams contributing to the holographic reconstruction.

Among the important technical simplifications in this example are the structure of relevant vertices: On the one hand, the bulk cubic vertices $s - 0 - 0$ are of Noether type $\varphi_s J_s$ (see [24, 25]), i.e. a gauge field φ_s times a conserved current $J_s = \varphi_0(\nabla)^s\varphi_0 + ...$ bilinear in the scalar field φ_0, and traceless on-shell. Therefore, the diagrams (a)-(c) correspond to current exchange $J_s\mathcal{P}_sJ_s$, where $\mathcal{P}_s$ is the spin-s gauge field propagator. On the other hand, all bulk quartic vertices $0 - 0 - 0 - 0$ are also of current exchange type $J_s\,\Box^m J_s$ (see [1, 10]). Consequently, the exchange and contact Witten diagrams are of the same type and can be easily compared for each spin s and in each channel (see figure 2).

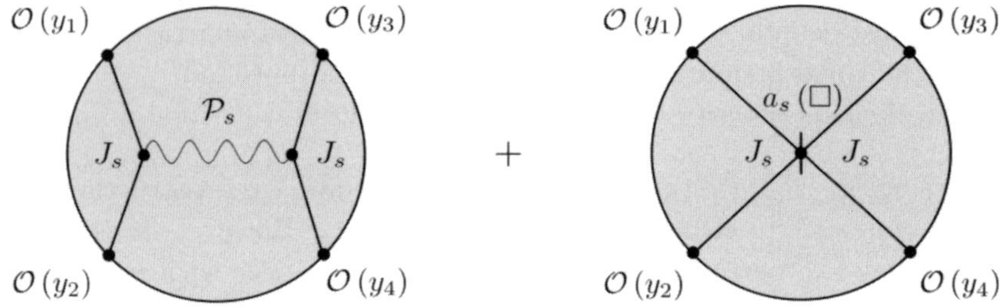

Fig. 2. Comparison of exchange and contact Witten diagrams in a single channel.

The form of the quartic vertex can be summarised as follows:

$$\mathcal{V} = \sum_{s\in 2\mathbb{N}} \mathcal{V}_s\,, \qquad \mathcal{V}_s = J_{\mu_1\cdots\mu_s}\, a_s(\Box)\, J^{\mu_1\cdots\mu_s}, \tag{4}$$

where $a_s(\Box)$ are generating functions

$$a_s(\Box) = \sum_{m=0}^{\infty} a_{m,s}\,\Box^m, \tag{5}$$

for the individual couplings $a_{m,s}$ in front of the individual vertices

$$\mathcal{V}_{m,s} = J_{\mu_1\cdots\mu_s}\,\Box^m\, J^{\mu_1\cdots\mu_s}\,. \tag{6}$$

The explicit expression of the quartic vertex $\mathcal{V}$ can be found in [10]. The generating functions are defined by

$$a_s\left(\nu^2+s+\tfrac{9}{4}\right) \propto \frac{2^{8-s}}{\nu^2+(s-\frac{1}{2})^2}\left[\frac{\pi}{\Gamma\left(\frac{2s-2i\nu+1}{4}\right)^2\Gamma\left(\frac{2s+2i\nu+1}{4}\right)^2}-\frac{1}{\Gamma(s)^2}\right]$$
$$-\frac{(-1)^{\frac{s}{2}}\,\pi^{\frac{3}{2}}\,2^{s+5}\,\Gamma\left(s+\frac{3}{2}\right)\Gamma\left(\frac{s}{2}+\frac{1}{2}\right)}{\sqrt{2}\,\Gamma\left(\frac{s}{2}+1\right)\Gamma(s+1)\,\Gamma\left(\frac{3}{4}-\frac{i\nu}{2}\right)\Gamma\left(\frac{3}{4}+\frac{i\nu}{2}\right)\Gamma\left(s+\frac{1}{2}+i\nu\right)\Gamma\left(s+\frac{1}{2}-i\nu\right)}\,,$$

which are entire functions, though it is not manifest. Heuristically, this indicates that each contact diagram associated to the cubic vertex $\mathcal{V}_s$ and depicted in figure 2 looks weakly local.

A subtlety is that the quartic vertices (6) are linearly independent on the free mass shell only for $2m \leqslant s$. However, the generating function includes infinitely many terms with $2m > s$. The latter should be reexpressed in terms of the previous independent ones. Since this involves infinitely many couplings and since we are on AdS_4 background where covariant derivatives do not commute, this procedure can a priori produce infinities, c.f. the examples in [22, 26] at cubic level (see also [27–30] where this possibility was discussed). In other words, the couplings in front of the independent quartic vertices might be infinite. However, one may expect this phenomenon not to happen in the present context of holographic reconstruction because infinite couplings in front of the independent quartic vertices would correspond to infinite coefficients in the double-trace conformal block decomposition.

4. Mellin Amplitudes of the Free and Critical $O(N)$ Model

As mentioned above, Mellin amplitudes of weakly-coupled theories are sometimes ill-defined as genuine functions, so they may either require some regularisation procedure or to be treated as generalised functions (i.e. distributions). We will investigate the latter option and focus on the $O(N)$ model.

The scalar single-trace primary operator $\mathcal{O}$ has conformal dimension $\Delta = d-2$ for the d-dimensional free $O(N)$ model. Its 4-point correlator (1) has a factor not fixed by conformal symmetry

$$F_{\text{free}}(u,v) = F_{\text{dis.}}(u,v) + \frac{1}{N}\,F_{\text{con.}}(u,v) \tag{7}$$

where the first term $F_{\text{dis.}}(u,v)$ arise from disconnected diagrams and the second term $F_{\text{con.}}(u,v)$ from connected diagrams. Both terms are power functions of the two cross ratios. This is true for the disconnected piece

$$F_{\text{dis.}}(u,v) = 1 + u^{\Delta} + \left(\frac{u}{v}\right)^{\Delta} \tag{8}$$

and for the connected piece

$$F_{\text{con.}}(u,v) = u^{\frac{\Delta}{2}} + \left(\frac{u}{v}\right)^{\frac{\Delta}{2}} + u^{\frac{\Delta}{2}}\left(\frac{u}{v}\right)^{\frac{\Delta}{2}}, \tag{9}$$

where, again, front numerical factors have been neglected for the sake of simplicity.

A power function, say

$$f(x) = x^{\alpha} \qquad (\alpha \in \mathbb{R}), \tag{10}$$

does not have a well defined Mellin transform [31]

$$M(z) = \int_0^{\infty} x^z\, f(x)\, \frac{dx}{x}, \tag{11}$$

or, rather, its fundamental strip of convergence degenerates to the vertical line $z = -\alpha + i\,\mathbb{R}$ in the complex plane. Actually, the Mellin transform (11) of the power function (10) may be understood[a] as a distribution [32]: $M(z) = \delta(z+\alpha)$. With this interpretation, the Mack ampitudes corresponding to (8) and (9) are, respectively,

$$M_{\text{dis.}}(s,t) = \delta(s)\,\delta(t-\Delta) + \delta(s-\Delta)\,\delta(t) + \delta(s-\Delta)\,\delta(t-\Delta) \tag{12}$$

and

$$M_{\text{con.}}(s,t) = \delta(s-\tfrac{\Delta}{2})\,\delta(t-\tfrac{\Delta}{2}) + \delta(s-\tfrac{\Delta}{2})\,\delta(t-\Delta) + \delta(s-\Delta)\,\delta(t-\tfrac{\Delta}{2})\,. \tag{13}$$

Due to the distributional nature of these reduced Mellin amplitudes $M_{\text{dis.}}(s,t)$ and $M_{\text{con.}}(s,t)$, the definition (3) for the corresponding Mellin amplitudes $\mathcal{M}_{\text{dis.}}(s,t)$ and $\mathcal{M}_{\text{con.}}(s,t)$ is difficult to apply. The problem is that each value in the support

[a] We thank S. Rychkov and M. A. Vasiliev for pointing out to us this possibility, which we later traced in the mathematical litterature.

of the left-hand-side in (3) corresponds to a pole at zero of one (even two for the disconnected piece) gamma function on the right-hand-side. Formally, treating the Mellin amplitudes $\mathcal{M}_{\text{dis.}}(s,t)$ and $\mathcal{M}_{\text{con.}}(s,t)$ as genuine functions, one would conclude that they must vanish identically. In other words, although the reduced Mellin amplitudes (12) and (13) are well defined distributions, the nature of the corresponding Mellin amplitudes remains to be clarified.

Let us stress that the above observations are not restricted to free CFTs but even holds for some integrable CFTs. A good example is the large-N limit of the $O(N)$ model at large N whose integrability has been recognised a long time ago [33]. From a modern perspective, the integrability of the large-N limit of the $O(N)$ model is intimately related to its equivalence to the free $O(N)$ model, up to a Legendre transformation. This latter fact lies at the heart of the Klebanov-Polyakov conjecture [16]. In the $N \to \infty$ limit, the scalar single-trace primary operator $\mathcal{O}$ for the critical $O(N)$ model has conformal dimension $\Delta = 2$ and its 4-point correlator (1) has a factor not fixed by conformal symmetry

$$F_{\text{int}}(u,v) = G_{\text{dis.}}(u,v) + \frac{1}{N}\,G_{\text{con.}}(u,v) + O\Big(\frac{1}{N^2}\Big) \tag{14}$$

which is also a sum of power functions of the two cross ratios. More precisely, the disconnected part is given by formula (8) with $\Delta = 2$ and the connected part reads for $d = 3$ [34]

$$G_{\text{con.}}(u,v) = u^2\left[u^{-\frac{3}{2}}\,(1+u-v) + v^{-\frac{3}{2}}\,(1-u+v) - (uv)^{-\frac{3}{2}}\,(1-u-v)\right]. \tag{15}$$

The reduced Mellin amplitude of the connected part is thus

$$\begin{aligned} M_{\text{con.}}(s,t) = {} & \delta(s-\tfrac{1}{2})\,\delta(t-\tfrac{3}{2}) + \delta(s-\tfrac{3}{2})\,\delta(t-\tfrac{1}{2}) - \delta(s-\tfrac{1}{2})\,\delta(t-\tfrac{1}{2}) \\ & + \delta(s-2)\,\delta(t-\tfrac{3}{2}) - \delta(s-3)\,\delta(t-\tfrac{1}{2}) + \delta(s-2)\,\delta(t-\tfrac{1}{2}) \\ & - \delta(s-\tfrac{1}{2})\,\delta(t-3) + \delta(s-\tfrac{3}{2})\,\delta(t-2) + \delta(s-\tfrac{1}{2})\,\delta(t-2)\,. \end{aligned} \tag{16}$$

Again the support of the Dirac delta functions coincide with poles of the gamma functions in (3). The presence of a product of two Dirac distributions in the reduced Mellin amplitudes (12), (13) and (16) seem to preclude their interpretation as scattering amplitudes since trivial scattering (i.e. alignment of pairs of momenta) should correspond to fixing a single Mandelstam variable (and not all of them). This may be a hint that Mellin amplitudes of free (or integrable) CFTs do not admit a sensible flat limit, a reasonable expectation since unbroken higher-spin gravity theories do not admit a weakly-coupled flat regime.

The distributional nature of tree-level amplitudes of four scalar fields and their absence of scattering has also been recently observed in conformal higher-spin gravity [35]. The triviality of scattering amplitudes may look surprising at first sight since higher-spin gravity theory has nonvanishing bulk interaction vertices. Nevertheless, retrospectively it seems reasonable to expect that the algebra of asymptotic higher-spin symmetries is so huge as to constrain the scattering of particles to be essentially trivial.

5. Conclusion and Open Directions

Characterising the degree of (non)locality of the bulk duals of weakly-coupled CFTs is an important issue for understanding holographic duality beyond the regime where the bulk spacetime is weakly curved. Looking at higher-spin gravity suggests to broaden the definition of locality in order to encompass interactions on flat/AdS spacetime for which the Feynman/Witten amplitudes of the corresponding contact diagrams are not polynomial but entire functions. At the level of four-point conformal correlators and quartic bulk vertices, this weak locality criterion is equivalent to the property that Witten exchange diagrams account for all single-trace conformal blocks in the decomposition of correlators. General arguments and explicit computations indicate that, for the higher-spin gravity dual of the free $O(N)$ model, the quartic self-interaction of the bulk scalar field is weakly (non)local.

Extending the holographic reconstruction of quartic vertices to spin $s \neq 0$ using twistor techniques, or to boundary dimension $d \neq 3$, are interesting challenges, as well as comparing the quartic vertices obtained purely from holographic reconstruction with the corresponding ones that can be extracted from Vasiliev equations.

Acknowledgments

We thank E. Joung, E. Skvortsov, M. Taronna and M. A. Vasiliev for useful discussions.

X.B. is very grateful to the Institute of Advanced Studies from Nanyang Technological University in Singapore for hospitality, and to the organisers of the workshop "Higher Spin Gauge Theories" for providing the opportunity to present this work and contribute to the proceedings. He also acknowledges the Asia Pacific Center for Theoretical Physics (APCTP) in Pohang and the Institute for Studies in Theoretical Physics and Mathematics (IPM) in Tehran for hospitality where part of this work was done during, respectively, the program "Duality and Novel Geometry in M-theory" and the "IPM school on Higher Spin Theory". D.P. thanks E. Joung and Seoul National University for their kind hospitality.

The research of X.B. was supported by the Russian Science Foundation grant 14-42-00047 in association with Lebedev Physical Institute. The work of J.E. and C.S. was partially supported by the European Science Foundation Holograv network (Holographic methods for strongly coupled systems). The work of D.P. was supported by the ERC Advanced grant No.290456.

References

[1] I. Heemskerk, J. Penedones, J. Polchinski, and J. Sully, Holography from Conformal Field Theory, *JHEP.* **0910**, 079, (2009). doi: 10.1088/1126-6708/2009/10/079.

[2] A. L. Fitzpatrick, E. Katz, D. Poland, and D. Simmons-Duffin, Effective Conformal Theory and the Flat-Space Limit of AdS, *JHEP.* **07**, 023, (2011). doi: 10.1007/JHEP07(2011)023.

[3] A. L. Fitzpatrick and J. Kaplan, AdS Field Theory from Conformal Field Theory, *JHEP.* **02**, 054, (2013). doi: 10.1007/JHEP02(2013)054.
[4] J. Penedones, Writing CFT correlation functions as AdS scattering amplitudes, *JHEP.* **03**, 025, (2011). doi: 10.1007/JHEP03(2011)025.
[5] G. Mack, D-independent representation of Conformal Field Theories in D dimensions via transformation to auxiliary Dual Resonance Models. Scalar amplitudes. (2009).
[6] M. F. Paulos, Towards Feynman rules for Mellin amplitudes, *JHEP.* **10**, 074, (2011). doi: 10.1007/JHEP10(2011)074.
[7] A. L. Fitzpatrick, J. Kaplan, J. Penedones, S. Raju, and B. C. van Rees, A Natural Language for AdS/CFT Correlators, *JHEP.* **11**, 095, (2011). doi: 10.1007/JHEP11(2011)095.
[8] S. El-Showk and K. Papadodimas, Emergent Spacetime and Holographic CFTs, *JHEP.* **1210**, 106, (2012). doi: 10.1007/JHEP10(2012)106.
[9] S. R. Coleman and J. Mandula, All Possible Symmetries of the S Matrix, *Phys. Rev.* **159**, 1251–1256, (1967). doi: 10.1103/PhysRev.159.1251.
[10] X. Bekaert, J. Erdmenger, D. Ponomarev, and C. Sleight, Quartic AdS Interactions in Higher-Spin Gravity from Conformal Field Theory, *JHEP.* **11**, 149, (2015). doi: 10.1007/JHEP11(2015)149.
[11] B. Sundborg, Stringy gravity, interacting tensionless strings and massless higher spins, *Nucl. Phys. Proc. Suppl.* **102**, 113–119, (2001). doi: 10.1016/S0920-5632(01)01545-6. [,113(2000)].
[12] S. E. Konstein, M. A. Vasiliev, and V. N. Zaikin, Conformal higher spin currents in any dimension and AdS / CFT correspondence, *JHEP.* **12**, 018, (2000). doi: 10.1088/1126-6708/2000/12/018.
[13] E. Sezgin and P. Sundell, Doubletons and 5-D higher spin gauge theory, *JHEP.* **09**, 036, (2001). doi: 10.1088/1126-6708/2001/09/036.
[14] A. Mikhailov, Notes on higher spin symmetries. (2002).
[15] E. Sezgin and P. Sundell, Massless higher spins and holography, *Nucl. Phys.* **B644**, 303–370, (2002). doi: 10.1016/S0550-3213(02)00739-3. [Erratum: Nucl. Phys.B660,403(2003)].
[16] I. R. Klebanov and A. M. Polyakov, AdS dual of the critical O(N) vector model, *Phys. Lett.* **B550**, 213–219, (2002). doi: 10.1016/S0370-2693(02)02980-5.
[17] E. Sezgin and P. Sundell, Holography in 4D (super) higher spin theories and a test via cubic scalar couplings, *JHEP.* **0507**, 044, (2005). doi: 10.1088/1126-6708/2005/07/044.
[18] R. R. Metsaev, Cubic interaction vertices of massive and massless higher spin fields, *Nucl. Phys.* **B759**, 147–201, (2006). doi: 10.1016/j.nuclphysb.2006.10.002.
[19] E. S. Fradkin and M. A. Vasiliev, Candidate to the Role of Higher Spin Symmetry, *Annals Phys.* **177**, 63, (1987). doi: 10.1016/S0003-4916(87)80025-8.
[20] N. Boulanger, D. Ponomarev, E. Skvortsov, and M. Taronna, On the uniqueness of higher-spin symmetries in AdS and CFT, *Int. J. Mod. Phys.* **A28**, 1350162, (2013). doi: 10.1142/S0217751X13501625.
[21] M. A. Vasiliev, Invariant Functionals in Higher-Spin Theory. (2015).
[22] E. D. Skvortsov and M. Taronna, On Locality, Holography and Unfolding, *JHEP.* **11**, 044, (2015). doi: 10.1007/JHEP11(2015)044.
[23] X. Bekaert, J. Erdmenger, D. Ponomarev, and C. Sleight, Towards holographic higher-spin interactions: Four-point functions and higher-spin exchange, *JHEP.* **1503**, 170, (2015). doi: 10.1007/JHEP03(2015)170.
[24] F. A. Berends, G. J. H. Burgers, and H. van Dam, Explicit Construction of Conserved Currents for Massless Fields of Arbitrary Spin, *Nucl. Phys.* **B271**, 429, (1986).

[25] X. Bekaert and E. Meunier, Higher spin interactions with scalar matter on constant curvature spacetimes: conserved current and cubic coupling generating functions, *JHEP.* **1011**, 116, (2010). doi: 10.1007/JHEP11(2010)116.

[26] N. Boulanger, P. Kessel, E. D. Skvortsov, and M. Taronna, Higher Spin Interactions in Four Dimensions: Vasiliev vs. Fronsdal, *J. Phys.* **A49**(9), 095402, (2016). doi: 10.1088/1751-8113/49/9/095402.

[27] S. Giombi and X. Yin, Higher Spins in AdS and Twistorial Holography, *JHEP.* **04**, 086, (2011). doi: 10.1007/JHEP04(2011)086.

[28] S. Giombi and X. Yin, Higher Spin Gauge Theory and Holography: The Three-Point Functions, *JHEP.* **09**, 115, (2010). doi: 10.1007/JHEP09(2010)115.

[29] N. Boulanger, S. Leclercq, and P. Sundell, On The Uniqueness of Minimal Coupling in Higher-Spin Gauge Theory, *JHEP.* **08**, 056, (2008). doi: 10.1088/1126-6708/2008/08/056.

[30] X. Bekaert, N. Boulanger, and P. Sundell, How higher-spin gravity surpasses the spin two barrier: no-go theorems versus yes-go examples, *Rev. Mod. Phys.* **84**, 987–1009, (2012). doi: 10.1103/RevModPhys.84.987.

[31] P. Flajolet, X. Gourdon, and P. Dumas, Mellin transforms and asymptotics: Harmonic sums, *Theor. Comp. Sc.* **144**, 3, (1995). doi: http://dx.doi.org/10.1016/0304-3975(95)00002-E.

[32] J. Bertrand, P. Bertrand, and J.-P. Ovarlez. The Mellin transform. In ed. A. D. Poularikas, *Transforms and applications handbook*, chapter 11. CRC Press, (2000).

[33] H. E. Stanley, Spherical model as the limit of infinite spin dimensionality, *Phys. Rev.* **176**, 718–722, (1968). doi: 10.1103/PhysRev.176.718.

[34] T. Leonhardt and W. Rühl, The minimal conformal O(N) vector sigma model at d = 3, *J. Phys.* **A37**, 1403–1413, (2004). doi: 10.1088/0305-4470/37/4/023.

[35] E. Joung, S. Nakach, and A. A. Tseytlin, Scalar scattering via conformal higher spin exchange, *JHEP.* **02**, 125, (2016). doi: 10.1007/JHEP02(2016)125.

The Sky is the Limit: Free Boundary Conditions in AdS_3 Chern-Simons Theory

Luis Apolo and Bo Sundborg

Department of Physics & The Oskar Klein Centre, Stockholm University, AlbaNova University Centre, SE-106 91 Stockholm, Sweden

We test the effects of new diffeomorphism invariant boundary terms in $SL(2, R) \times SL(2, R)$ Chern-Simons theory. The gravitational interpretation corresponds to free AdS_3 boundary conditions, without restrictions on the boundary geometry. The boundary theory is the theory of a string in a target AdS_3. Its Virasoro conditions can eliminate ghosts. Generalisations to $SL(N, R) \times SL(N, R)$ higher spin theories and many other questions are still unexplored.

1. The Sky is the Limit

By the sky is meant the imagined surface (at infinity) from which light can reach us. Under certain circumstances this surface is the boundary of spacetime. We do not wish to assume anything about the geometry of the boundary, in other words we impose no prior geometry.

AdS gravity depends on the boundaries. Physicists are often cavalier about boundary conditions: our main interest lies in equations of motion and their consequences. AdS/CFT duality is an exception to this rule since the coupling of boundary operators to bulk degrees of freedom constitutes an important part of the dictionary relating two theories. In an early precursor to this duality Brown and Henneaux[1] related 2d conformal symmetries to asymptotic symmetries of 3d gravity. Three-dimensional gravity being a theory without local degrees of freedom, boundary conditions are all important in this case. It has been argued to be very closely related (if not defined by) 3d Chern-Simons theory with gauge group identified with the local spacetime symmetry group.[2,3] For the case of a negative cosmological constant, the symmetry can be written as $SL(2, R) \times SL(2, R)$.

New asymptotic boundary conditions are potentially important in both gravity and higher spin theory. Notably, the conditions at boundaries of spacetime are not uniquely determined. Several choices are possible. The main requirement is that a unique time evolution can be ascertained, at least locally. Famous examples in the present context are the Brown-Henneaux boundary conditions[1] which entail specific fall off conditions on components of the metric. This is standard and entirely consistent, even if it involves a choice of coordinates in the bulk and a specific boundary geometry. Since the boundaries are crucial in 3d gravity, and diffeomorphism invariance is a time-honoured principle, it could be worthwhile to formulate boundary conditions in a purely geometric way and without prescribing the boundary geometry. This could be achieved by not imposing any structure on the sky, i.e. by employing free boundary conditions[a]. We studied this approach in,[9] and discuss this work and some further consequences of it here. Because $SL(2,R) \times SL(2,R)$ Chern-Simons theory can be regarded as the simplest 3d $SL(N,R) \times SL(N,R)$ higher spin theory, and because higher spin theories in general can be formulated as generalised Chern-Simons theories (see for instance Ref. 10), we believe that our ideas may be useful also in this extended context.

Pure three-dimensional gravity is notorious by not disclosing its microscopic degrees of freedom,[11,12] and its existence as a quantum theory is in doubt.[13] Before settling this question the crucial boundary conditions should be re-examined, because boundary conditions may lead to new degrees of freedom without introducing matter. Indeed, an effective string description of three-dimensional gravity emerges from this examination[b]. A generally covariant formulation typically jeopardises unitarity, but the connections to strings is likely to resolve this issue for AdS_3 gravity.

2. Introduction

AdS_3 Chern-Simons theory has been advocated[3] as a soluble quantum theory of gravity, and by the discovery of black hole solutions of the classical theory[17,18] the stakes were raised: one could envision calculating the entropy of black holes of varying sizes from first principles and study their formation from collisions. This could be feasible because Chern-Simons theory is in a sense a simple theory: it is essentially topological and contains no local degrees of freedom. Indeed, black hole formation from two colliding massless particles has been described exactly in classical AdS_3 gravity.[19]

Furthermore, the same theory is the simplest in an $SL(N,R) \times SL(N,R)$ series of Chern-Simons theories. All the others are simple higher spin theories without matter. A host of questions on conical singularities,[20,21] black holes[22–25] and gauge invariances vs. geometry in higher spin gravity find their simplest manifestations in the $SL(2,R)$ case. Spin 2 is in a sense the lowest high spin.

Unfortunately, hopes about this approach to gravity has dwindled.[12,26] There does not seem to be solutions to the physical questions that could be addressed in the theory, and worse, there does not even seem to be a consensus on what the problems are. One obvious problem, the unitarity of the Chern-Simons theory, is probably solved by the construction explained below, when taken over to the

[a]For previous work in this direction, see Refs. 4–8.
[b]Previous connections to strings include Refs. 14–16.

quantum theory. The unitarity issue of higher spin versions of the theory could have similar resolutions.

The confusing state of affairs in AdS3 gravity challenges us to reconsider how to approach it. We take the prominence of boundaries as a clue in this effort: the boundary is truly "where the action is". We adapt to higher spin generalisations by using the Chern-Simons formulation, but we maintain contact with standard metric gravity. The strategy is to use Chern-Simons theory as a theory of 3d gravity classically, test coordinate invariant boundary conditions and derive a boundary theory. In fact, we demonstrate that it is the world-sheet theory of a string propagating in AdS3!

3. Diffeomorphism Invariant Boundaries

3.1. *The Chern-Simons formulation of gravity*

First, a review of the Chern-Simons description. The vector potentials

$$A = (w^a + e^a)\, T_a, \qquad \bar{A} = (w^a - e^a)\, \bar{T}_a, \tag{1}$$

are the fundamental fields in the Achúcarro-Townsend-Witten gauge theory description of AdS_3 gravity. They are expressed in terms of the vielbein e and the dual ω to the spin connection. T_a and $\bar{T}_a$ represent generators of the two $SL(2,R)$ factors in the AdS_3 group. The metric

$$g_{\mu\nu} = \frac{1}{4}\mathrm{tr}\left\{(A-\bar{A})_\mu (A-\bar{A})_\nu\right\} \tag{2}$$

is now a composite field. As a consequence there is nothing in gauge theory that forbids locally degenerate metrics. The degeneration locus is generically of codimension 1. Degenerate metrics are potentially important in the interpretation of the theory, and appear naturally in the description of BTZ black holes, as described later. The action that reproduces gravity is the difference of integrals over two Chern-Simons three forms:

$$I_{CS;\mathrm{bulk}}[A,\bar{A}] = I_{CS}[A] - I_{CS}[\bar{A}] \tag{3}$$

$$I_{CS}[A] = \int_\Sigma \mathrm{tr}\Big(A \wedge dA + \frac{2}{3} A \wedge A \wedge A\Big). \tag{4}$$

Diffeomorphisms can be represented in a special way in the Chern-Simons formulation. They are equivalent on shell to particular gauge transformations

$$\delta A = -Du, \qquad \delta \bar{A} = -Du \tag{5}$$

with parameter

$$u = \rho^a P_a = v^\mu e^a_\mu P_a \tag{6}$$

where P_a generates AdS translations and v^μ parametrises diffeomorphisms. The map between gauge transformations and diffeomorphisms degenerates if and only if the metric degenerates.

We may ask if there are interesting solutions with degenerate metrics. This question is somewhat out of line with the rest of this presentation, but it is potentially important for interpreting the Chern-Simons formulation. The equations of motion are flatness conditions $F = 0 = \bar{F}$ on the field strengths corresponding to the potentials A and $\bar{A}$. The solutions

$$A = 0 = \bar{A}, \tag{7}$$

certainly give rise to a degenerate metric but they are a bit trivial, and do not agree with the gravitational boundary conditions we want to impose.

A more interesting example is the BTZ black hole metric in the form

$$ds^2 = -\sinh^2\rho[r_+dt - r_-d\phi]^2 + \cosh^2\rho[r_-dt - r_+d\phi]^2 \tag{8}$$

which degenerates at the outer horizon joining two exterior solutions. But the corresponding gauge potential

$$\begin{aligned} A^{0+} &= -\tfrac{1}{\ell}\left(r_+ \mp r_-\right)\sinh\rho\left(\tfrac{dt}{\ell} \pm d\phi\right) \\ A^{1\pm} &= \tfrac{1}{\ell}\left(r_+ \mp r_-\right)\cosh\rho\left(\tfrac{dt}{\ell} \pm d\phi\right) \\ A^{2\pm} &= \pm d\rho, \end{aligned} \tag{9}$$

is perfectly regular. In metric gravity we are supposed to only use this solution away from the coordinate singularity at the horizon, but the Chern-Simons gauge theory description does not discriminate against this solution at the horizon. It is thus more radical than a simple reformulation. If we try to interpret the above solution in a metric formulation we need energy momentum from a domain wall (string) at the horizon for consistency with Einstein's equations. Perhaps such strings are the missing ingredients in Ref. 12.

3.2. *Chern-Simons theory for double gauge groups $G \times G$*

The special product structure of the AdS_3 isometry group $SL(2,R) \times SL(2,R)$ has consequences for the description of the boundary degrees of freedom which are always associated with Chern-Simons theory. The same structure applies to higher spin generalisations $SL(N,R) \times SL(N,R)$ and it makes sense to use it.

The Einstein-Hilbert action can be rewritten in terms of Chern-Simons fields,

$$\int_\Sigma \sqrt{|g|}\,(R+2) = I_{CS}[A] - I_{CS}[\bar{A}] - \int_{\partial\Sigma} \mathrm{tr}\big(A \wedge \bar{A}\big), \tag{10}$$

leads to the equations of motion

$$0 = F = dA + A \wedge A, \qquad 0 = \bar{F} = d\bar{A} + \bar{A} \wedge \bar{A}, \tag{11}$$

implying that all solutions are locally of the form,

$$A = g^{-1}dg, \qquad \bar{A} = \bar{g}^{-1}d\bar{g}. \tag{12}$$

They are locally pure gauge, but not necessarily globally, due to topology and boundary conditions.

Thus, there are no local degrees of freedom, but there can be boundary degrees of freedom. Their action can be found to be a non-chiral Wess-Zumino-Witten action

$$S_{WZW}[G] = -\frac{k}{4\pi}\left\{\frac{1}{2}\int_{\partial\Sigma}\eta^{\mu\nu}\mathrm{tr}[(G^{-1}\partial_\mu G)(G^{-1}\partial_\nu G)] + \frac{1}{3}\int_\Sigma \mathrm{tr}\left(G^{-1}dG\right)^3\right\}, \tag{13}$$

with

$$G = g\bar{g}^{-1}. \tag{14}$$

There are many different derivations, typically using boundary equations of motion[c] to combine two chiral WZW actions into one non-chiral action. The procedure involves specification of boundary terms breaking boundary diffeomorphisms (by explicit use of a boundary Minkowski metric $\eta_{\mu\nu}$).

For the purposes of eventually finding a complete boundary quantum description we wish to derive an off shell boundary theory, and we also wish to keep boundary general covariance. The first requirement was satisfied in a derivation by Arcioni, Blau and O'Loughlin.[28] To restore boundary diffeomorphisms ($\eta_{\mu\nu}$) we just replace $\eta^{\mu\nu} \to \sqrt{|\gamma|}\gamma^{\mu\nu}$ and let $\gamma_{\mu\nu}$ be dynamical.

Given solutions to the Chern-Simons equations

$$A = g^{-1}dg, \qquad \bar{A} = \bar{g}^{-1}d\bar{g}, \tag{15}$$

using a Wess-Zumino-Witten action

$$W[g] = -\frac{k}{4\pi}\left\{\frac{1}{2}\int_{\partial\Sigma}\sqrt{|\gamma|}\gamma^{\mu\nu}\left(g^{-1}\partial_\mu g\right)\left(g^{-1}\partial_\nu g\right) + \frac{1}{3}\int_\Sigma \left(g^{-1}dg\right)^3\right\}, \tag{16}$$

and the notation $\Gamma^{\mu\nu}_{\pm} = \epsilon^{\mu\nu} \pm \sqrt{|\gamma|}\gamma^{\mu\nu}$, the gravitational action *with appropriate boundary terms* becomes

$$W[g] + W[\bar{g}^{-1}] - \frac{k}{4\pi}\int_{\partial\Sigma}\Gamma^{\mu\nu}_{-}\left(g^{-1}\partial_\mu g\right)\left(\bar{g}^{-1}\partial_\nu \bar{g}\right) = W[g\bar{g}^{-1}] \tag{17}$$

$$= S_{WZW}[G](\eta^{\mu\nu} \to \sqrt{|\gamma|}\gamma^{\mu\nu}) \tag{18}$$

from a Polyakov-Wiegmann identity. A generally covariant non-chiral WZW action for $G = g\bar{g}^{-1}$ results.

This agrees with our intent. The Metric gravity ↔ Chern-Simons formulation is not a completely off shell relation. Thus, to start with a bulk off shell formulation would be unnecessarily accurate. However, to capture physical degrees of freedom in the boundary a boundary path integral is needed and therefore an off shell boundary theory.

One piece is still missing, the form of the appropriate boundary terms.

[c] A benchmark reference is Ref. 27.

3.3. *The Chern-Simons route to the string*

Now write the Cherns-Simons formulation with appropriate boundary terms, which do not specify a particular boundary geometry:

$$S_{CS} = \frac{k}{4\pi}\left\{I_{CS}[A] - I_{CS}[\bar{A}] + J_{\alpha,\beta}[A,\bar{A}]\right\}, \tag{19}$$

where $k = 1/4G_N$ and the boundary term $J_{\alpha,\beta}[A,\bar{A}]$ reads[d]

$$J_{\alpha,\beta}[A,\bar{A}] = (2\alpha-1)\int_{\partial\Sigma}\mathrm{tr}(A\wedge\bar{A}) \pm \frac{\beta}{2}\int_{\partial\Sigma}\sqrt{|\gamma|}\gamma^{\mu\nu}\mathrm{tr}\left[\left(A-\bar{A}\right)_\mu\left(A-\bar{A}\right)_\nu\right]. \tag{20}$$

Arcioni, Blau and O'Loughlin take

- $\alpha = 0$ to get local Lorentz invariance at the boundary.
- Also $\beta = 1 - 2\alpha = 1$ for a regular action in the metric formalism.
- $\gamma_{\mu\nu} = \eta_{\mu\nu}$.

We let the 2d "world-sheet" metric $\gamma_{\mu\nu}$ be arbitrary. We fix no prior geometry. This differs from previous approaches.

The modified term in the action is

$$\int_{\partial\Sigma}\sqrt{|\gamma|}\gamma^{\mu\nu}\mathrm{tr}\left[\left(A-\bar{A}\right)_\mu\left(A-\bar{A}\right)_\nu\right] = 4\int_{\partial\Sigma}\sqrt{|\gamma|}\gamma^{\mu\nu}g_{\mu\nu} \tag{21}$$

by equation (2). We now integrate out $\gamma_{\mu\nu}$. This will favour no boundary geometry, and give two constraints

$$g_{\mu\nu} = \frac{1}{2}\gamma_{\mu\nu}\gamma^{\rho\sigma}g_{\rho\sigma}. \tag{22}$$

Compare the string action, $\int\sqrt{|\gamma|}\gamma^{\mu\nu}\partial_\mu X\partial_\nu X$, with the same dependence on the 2d metric $\gamma_{\mu\nu}$ and only two Virasoro constraints due to Weyl invariance ($\gamma_{\mu\nu}\to e^{\phi}\gamma_{\mu\nu}$).

Gauge fix $\gamma_{\mu\nu}\to e^{\phi}\eta_{\mu\nu}$, and vary the action:

$$\delta S_{CS} = \frac{k}{2\pi}\left\{\int_\Sigma \mathrm{tr}\left(\delta A\wedge F + \delta\bar{A}\wedge\bar{F}\right) - 2\int_{\partial\Sigma}\mathrm{tr}\left(e_-\delta A_+ - e_+\delta\bar{A}_-\right)\right\}. \tag{23}$$

A well-defined action principle is obtained for $\delta A_+ = \delta\bar{A}_- = 0$, which together with factorisation $G = g\bar{g}^{-1}$ reproduce the (standard) boundary conditions

$$A_+ = \bar{A}_- = 0, \tag{24}$$

but now after a gauge fixing from a generally covariant boundary action.

There are consequences of starting from a geometric action: there are constraints to consider. As alluded to above, they appear as equations of motion from variation of $\gamma_{\mu\nu}$:

$$0 = \frac{\delta}{\delta\gamma_{\alpha\beta}}\int_{\partial\Sigma}\sqrt{|\gamma|}\gamma^{\mu\nu}\mathrm{tr}\left[\left(A-\bar{A}\right)_\mu\left(A-\bar{A}\right)_\nu\right] = 4\frac{\delta}{\delta\gamma_{\alpha\beta}}\int_{\partial\Sigma}\sqrt{|\gamma|}\gamma^{\mu\nu}g_{\mu\nu}. \tag{25}$$

[d] In the presence of higher rank symmetric tensors, and higher rank gauge groups, it would be natural to include boundary terms of higher order than quadratic.

Choosing a $\gamma_{\mu\nu} = e^{\phi}\eta_{\mu\nu}$ gauge, and applying the $A_+ = \bar{A}_- = 0$ boundary conditions, two constraints affect the metric

$$0 = 4g_{--} = \mathrm{tr}\big[A_- A_-\big], \quad 4g_{-+} = -\mathrm{tr}\big[A_- \bar{A}_+\big], \quad 0 = 4g_{++} = \mathrm{tr}\big[\bar{A}_+ \bar{A}_+\big]. \tag{26}$$

Inserting the solutions $A = g^{-1}dg$, $\bar{A} = \bar{g}^{-1}d\bar{g}$ we find

$$0 = \mathrm{tr}\big[g^{-1}\partial_- g g^{-1}\partial_- g\big] = \mathrm{tr}\big[\partial_- g g^{-1}\partial_- g g^{-1}\big] = \mathrm{tr}\big[J_- J_-\big], \quad 0 = \mathrm{tr}\big[\bar{J}_+ \bar{J}_+\big], \tag{27}$$

where J_- and $\bar{J}_+$ are chiral conserved SL(2,R) currents. We recognise the stress tensor components T_{--} and T_{++} in a WZW CFT, and their vanishing is familiar from the world sheet string as Virasoro constraints.

Since a string is more or less defined as a 2d CFT with Virasoro constraints, we conclude that we have a string interpretation of 3d gravity.

Some important observations:

- g and $\bar{g}$ depend on all bulk coordinates, but lose dependence on one chiral coordinate each close to the boundary, due to the $A_+ = \bar{A}_- = 0$ boundary conditions.
- $G = g\bar{g}^{-1}$ at the boundary describes the embedding of the string world sheet into $SL(2,R)$ which is geometrically AdS_3.
- Gauge potentials are *not* simply related to currents at the boundary: $A_- = g^{-1}\partial_- g \not\equiv \partial_- g g^{-1} = J_-$, in contrast to other approaches, e.g. Ref. 27.
- For a string in a conformal gauge, the bulk metric is conformally flat at the boundary with the conformal factor determined by the string solution: $g_{++} = g_{--} = 0, \quad 4g_{-+} = -\mathrm{tr}\big[g^{-1}\partial_- g\bar{g}^{-1}\partial_+ \bar{g}\big]$.

3.4. *Metric AdS3 gravity and diffeomorphism invariant boundaries*

Metric gravity seems to be much more difficult to formulate than the Chern-Simons version in geometries with boundaries. It has been done for non-compact spacetimes with boundaries at infinity. Asymptotic fall off conditions like

$$g_{\mu\nu} \equiv h_{\mu\nu} = r^2 h^{(0)}_{\mu\nu} + \mathcal{O}(r^0), \quad g_{r\mu} = \mathcal{O}(r^{-3}), \quad g_{rr} = r^{-2} + \mathcal{O}(r^{-4}), \tag{28}$$

permit some terms to be dropped, while other conditions follow from requiring finiteness of charges. Here r is a radial coordinate, and $h^{(0)}_{\mu\nu}$ is the metric at the conformal boundary. If the boundary metric is fixed, we have standard Brown-Henneaux type boundary conditions. If the boundary metric permitted to vary we have *free* boundary conditions.

To determine whether boundary conditions make sense it is necessary to check the action principle. Free boundary are only compatible with a well-defined action principle if the variation with respect to the boundary metric vanishes:[4,8]

$$0 = T^{BY}_{\mu\nu} \equiv -\frac{2}{\sqrt{|h^{(0)}|}}\frac{\delta S}{\delta h^{(0)\mu\nu}} = -\frac{1}{\kappa}\big(K_{\mu\nu} - Kh_{\mu\nu} + h_{\mu\nu}\big), \tag{29}$$

where $T^{BY}_{\mu\nu}$ is the Brown-York stress tensor and $K_{\mu\nu}$ is the extrinsic curvature of the boundary. Note that we need the correct boundary terms for this to work.

To compare with the Chern-Simons discussion boundary conditions which do not specify a prior boundary geometry should be studied. It is no big surprise that free boundary conditions do the job. But

- The boundary terms come with non-standard coefficients.
- The requirement of a vanishing Brown-York tensor $T_{\mu\nu}^{BY}$ for free boundary conditions gives 3 equations rather than 2 expected from Virasoro constraints.

Boundary terms are constrained by demanding a well-defined variational principle of the generalised action

$$S_{GR} = \frac{1}{\kappa}\left\{\frac{1}{2}\int_{\Sigma}\sqrt{|g|}\,(R+2) + \alpha\int_{\partial\Sigma}\sqrt{|h|}K \pm \frac{\beta}{2}\int_{\partial\Sigma}\sqrt{|\gamma|}\gamma^{\mu\nu}h_{\mu\nu}\right\}. \tag{30}$$

This action is not well defined for arbitrary coefficients of the boundary terms[e]. However, if $2\alpha+\beta=1$ the bulk equations of motion and the vanishing of $T_{\mu\nu}^{BY}$ are enough.[9,29] The standard gravity action corresponds to $\alpha=1, \beta=-1$, while our Chern-Simons case corresponds to $\alpha=0, \beta=1$.

Given that the Chern-Simons approach leads to two Virasoro constraints[f] it not surprising that the traceless part of a two dimensional stress tensor defined from metric gravity vanishes. This tensor is the Brown-York tensor. It is less clear what causes the vanishing of the trace part in the Chern-Simons picture. In the boundary string we get a vanishing trace of the stress tensor due to Weyl invariance, but there is not even an expression for the trace part which can be studied off shell. It can however be shown that the the vanishing follows from the chirality of the group elements g and $\bar{g}$ on the approach to the boundary.

Returning to the more general framework we can choose an arbitrary world-sheet metric on the boundary, modulo topological obstructions. The boundary metric is conformal to the world-sheet metric,

$$h_{\mu\nu}^{(0)} = e^{2\psi}\gamma_{\mu\nu}, \tag{31}$$

and the subleading terms of the bulk metric are constrained by Virasoro constraints. They imply that the 2d metrics in the surfaces orthogonal to the radial direction are conformally flat[g]. The conformal factor is constrained by vanishing Brown-York tensor (or chirality of g and $\bar{g}$), cf. Ref. 8. For a boundary metric with vanishing scalar curvature (and $\gamma_{\mu\nu}=\eta_{\mu\nu}$) the flat wave equation is obeyed, $\partial_+\partial_-\psi=0$. Such boundary geometries are even insensitive to quantum corrections from the boundary trace anomaly.

It remains to connect the metric gravity picture with the Chern-Simons/string picture. An explicit extension of world-sheet string solutions to corresponding

[e]Note that a conventional term proportional to $\int\sqrt{|h|}$ has been replaced by the last term containing the world sheet metric $\gamma_{\mu\nu}$. Varying with respect to γ and reinserting in the action reproduces the standard form of the action.
[f]Vanishing of the traceless part of a 2d stress tensor.
[g]In Fefferman-Graham coordinates.[30]

Chern-Simons solutions provides such a connection if it contains an asymptotic region and is formulated in terms of coordinates that obey the appropriate fall off conditions. In the general case this is work in progress,[31] but some known facts are reviewed below.

3.5. *The string and the boundary theory of gravity*

We propose a string interpretation of 3d gravity, inspired by the analogous structures we have found in the string and in 3d gravity. First, a rough sketch of what such an interpretation would entail. If we consider spacetime boundaries with vanishing curvature scalar, the trace anomaly will not ruin conformal invariance and we get the entries without question marks in Table 1. The last two lines are tempting but much more tentative. So far it seems that one string is enough to describe the geometries (AdS and BTZ with boundary waves) which are typically considered in AdS_3 gravity, with the possible exception of solutions with additional 2π excess angle conical singularities.[20]

String	Gravity
world sheet conformal field theory	boundary CFT
$\rightarrow$ vanishing stress tensor trace	$\rightarrow$ vanishing Brown-York tensor trace
2d diffeomorphism	free boundary conditions
$\rightarrow$ Virasoro constraints	$\rightarrow$ vanishing Brown-York tensor
target space AdS_3 metric	boundary action
antisymmetric tensor	bulk (Chern-Simons) action
multiple strings?	multiple boundaries?
string interactions?	boundary topology change?

More concretely any string solution in terms of g and $\bar{g}$ can be be extended to Chern-Simons solution in several ways. A simple choice is obtained by introducing a radial coordinate ρ and taking $g \rightarrow gM(\rho)$ and $\bar{g} \rightarrow M(\rho)^{-1}\bar{g}$, which solve the equations of motion and the Virasoro constraints for any ρ (cf. Eq. (26)). The conformal factor is proportional to $\mathrm{tr}\left[g^{-1}\partial_- g\bar{g}^{-1}\partial_+\bar{g}\right]$, and then depends on ρ. This is a proof of concept: string solutions generate 3d geometries. These geometries are however not automatically in a useful form and even those that are in such form are in an unfamiliar conformal gauge, not immediately recognisable in terms of standard parametrisations like that of Bañados.[32] Still the asymptotic charges can be evaluated.[6,8]

4. Comments and Outlook

We have associated a boundary string theory to AdS_3 gravity. The Virasoro algebra that comes with string theory plays the role of gauge generators and constraints, in contrast to the role played by the Brown-Henneaux Virasoro algebra, which is more familiar in the gravitational context. This role prohibits any classical central extension like the one of Brown and Henneaux,[1] and one may ask how our result connects to theirs. In fact, their algebra should not be expected to be apparent until the gauge freedom has been fixed and the constraint solved. A kind of light-cone gauge can be used for this purpose.[15]

Strings in AdS_3 have an interesting zoo of solutions and we already have simple examples[9] of a degeneracy: several string solutions map to the same 3d geometry. It would be extremely interesting to find more examples of this behaviour. The typical situation in 3d gravity, where there are too few states compared to the Bekenstein-Hawking entropy could be improved. One expression of this state of affairs is the belief that gravitation is a macroscopic theory and not a microscopic one. Could it be enough to have the Chern-Simons formulation and the resulting string as a microscopic description?

A next step could be to quantise the boundary theory. Since the string in AdS_3 is known, the quantum theory is almost within reach. However, we do not have a completely off shell derivation, since the bulk equations of motion have been used in our approach. The precise quantum theory could receive corrections. Furthermore, unless we add extra degrees of freedom we do not have a critical string, which is the closest quantum theory that is known.[33] If we were extremely lucky, these caveats would cancel, and we could extract information on quantum properties directly.

Finally, a few words about the potential applications to higher spin theory. We have argued that 3d gravity is effectively a simple example of such a theory, at least in the Chern-Simons formulation. All the questions that plague 3d gravity are then likely to show up and require a solution. The main higher spin role of our work is to prepare for such a discussion. In practice, one might ask how our approach can be extended. We have given some clues. Higher symmetric tensors can be included in higher order boundary terms[h] in $SL(N,R) \times SL(N,R)$ theories, and we expect higher order constraints of W algebra type, which would lead to W strings. All of this remains to be worked out.

References

1. J. D. Brown and M. Henneaux, Central Charges in the Canonical Realization of Asymptotic Symmetries: An Example from Three-Dimensional Gravity, *Commun.Math.Phys.* **104**, 207–226, (1986). doi: 10.1007/BF01211590.
2. A. Achucarro and P. Townsend, A Chern-Simons Action for Three-Dimensional anti-De Sitter Supergravity Theories, *Phys.Lett.* **B180**, 89, (1986). doi: 10.1016/0370-2693(86)90140-1.
3. E. Witten, (2+1)-Dimensional Gravity as an Exactly Soluble System, *Nucl.Phys.* **B311**, 46, (1988). doi: 10.1016/0550-3213(88)90143-5.
4. G. Compère and D. Marolf, Setting the boundary free in AdS/CFT, *Class.Quant.Grav.* **25**, 195014, (2008). doi: 10.1088/0264-9381/25/19/195014.

[h] A discussion of some of the likely terms necessary for such extensions were studied in Ref. 34.

5. G. Compère, W. Song, and A. Strominger, New Boundary Conditions for AdS_3, *JHEP.* **1305**, 152, (2013). doi: 10.1007/JHEP05(2013)152.
6. C. Troessaert, Enhanced asymptotic symmetry algebra of AdS_3, *JHEP.* **1308**, 044, (2013). doi: 10.1007/JHEP08(2013)044.
7. S. G. Avery, R. R. Poojary, and N. V. Suryanarayana, An sl(2,$\mathbb{R}$) current algebra from AdS_3 gravity, *JHEP.* **1401**, 144, (2014). doi: 10.1007/JHEP01(2014)144.
8. L. Apolo and M. Porrati, Free boundary conditions and the AdS_3/CFT_2 correspondence, *JHEP.* **1403**, 116, (2014). doi: 10.1007/JHEP03(2014)116.
9. L. Apolo and B. Sundborg, Strings from 3D gravity: asymptotic dynamics of AdS_3 gravity with free boundary conditions, *JHEP.* **06**, 171, (2015). doi: 10.1007/JHEP06(2015)171.
10. M. A. Vasiliev, Higher spin gauge theories: Star product and AdS space. (1999).
11. S. Carlip, Conformal field theory, (2+1)-dimensional gravity, and the BTZ black hole, *Class. Quant. Grav.* **22**, R85–R124, (2005). doi: 10.1088/0264-9381/22/12/R01.
12. A. Maloney and E. Witten, Quantum Gravity Partition Functions in Three Dimensions, *JHEP.* **1002**, 029, (2010). doi: 10.1007/JHEP02(2010)029.
13. E. J. Martinec, Conformal field theory, geometry, and entropy. (1998).
14. M. Bañados and A. Ritz, A Note on classical string dynamics on AdS(3), *Phys.Rev.* **D60**, 126004, (1999). doi: 10.1103/PhysRevD.60.126004.
15. B. Sundborg, Mapping pure gravity to strings in three-dimensional anti-de Sitter geometry. (2013).
16. J. Kim and M. Porrati, Long String Dynamics in Pure Gravity on AdS_3. (2014).
17. M. Bañados, C. Teitelboim, and J. Zanelli, The Black hole in three-dimensional spacetime, *Phys.Rev.Lett.* **69**, 1849–1851, (1992). doi: 10.1103/PhysRevLett.69.1849.
18. M. Bañados, M. Henneaux, C. Teitelboim, and J. Zanelli, Geometry of the (2+1) black hole, *Phys.Rev.* **D48**(6), 1506–1525, (1993). doi: 10.1103/PhysRevD.48.1506, 10.1103/PhysRevD.88.069902.
19. S. Holst and H.-J. Matschull, The Anti-de Sitter Gott universe: A Rotating BTZ wormhole, *Class. Quant. Grav.* **16**, 3095–3131, (1999). doi: 10.1088/0264-9381/16/10/305.
20. T. Mansson and B. Sundborg, Multi - black hole sectors of AdS(3) gravity, *Phys.Rev.* **D65**, 024025, (2002). doi: 10.1103/PhysRevD.65.024025.
21. A. Castro, R. Gopakumar, M. Gutperle, and J. Raeymaekers, Conical Defects in Higher Spin Theories, *JHEP.* **1202**, 096, (2012). doi: 10.1007/JHEP02(2012)096.
22. M. Gutperle and P. Kraus, Higher Spin Black Holes, *JHEP.* **05**, 022, (2011). doi: 10.1007/JHEP05(2011)022.
23. M. Ammon, M. Gutperle, P. Kraus, and E. Perlmutter, Spacetime Geometry in Higher Spin Gravity, *JHEP.* **10**, 053, (2011). doi: 10.1007/JHEP10(2011)053.
24. A. Castro, E. Hijano, A. Lepage-Jutier, and A. Maloney, Black Holes and Singularity Resolution in Higher Spin Gravity, *JHEP.* **01**, 031, (2012). doi: 10.1007/JHEP01(2012)031.
25. A. Perez, D. Tempo, and R. Troncoso, Higher Spin Black Holes, *Lect. Notes Phys.* **892**, 265–288, (2015). doi: 10.1007/978-3-319-10070-8_10.
26. E. Witten, Three-Dimensional Gravity Revisited. (2007).
27. O. Coussaert, M. Henneaux, and P. van Driel, The Asymptotic dynamics of three-dimensional Einstein gravity with a negative cosmological constant, *Class.Quant.Grav.* **12**, 2961–2966, (1995). doi: 10.1088/0264-9381/12/12/012.
28. G. Arcioni, M. Blau, and M. O'Loughlin, On the boundary dynamics of Chern-Simons gravity, *JHEP.* **0301**, 067, (2003). doi: 10.1088/1126-6708/2003/01/067.
29. S. Detournay, D. Grumiller, F. Schöller, and J. Simón, Variational principle and

one-point functions in three-dimensional flat space Einstein gravity, *Phys.Rev.* **D89** (8), 084061, (2014). doi: 10.1103/PhysRevD.89.084061.
30. C. Fefferman and C. R. Graham. Conformal invariants. In *Elie Cartan et les Mathématiques d'aujourd'hui.* Asterisque, (1985).
31. L. Apolo, S. Sjors, and B. Sundborg, In preparation.
32. M. Bañados, Three-dimensional quantum geometry and black holes, *AIP Conf.Proc.* **484**, 147–169, (1999). doi: 10.1063/1.59661.
33. J. M. Maldacena and H. Ooguri, Strings in AdS(3) and SL(2,R) WZW model 1.: The Spectrum, *J.Math.Phys.* **42**, 2929–2960, (2001). doi: 10.1063/1.1377273.
34. L. Apolo, The covariant action of higher spin black holes in three dimensions. (2015).

Higgs Mechanism and Holography of Partially Massless Higher Spin Fields

Seungho Gwak,* Jaewon Kim* and Soo-Jong Rey*,†

**School of Physics and Astronomy & Center for Theoretical Physics, Seoul National University, Seoul 151-747, Korea*

†Fields, Gravity & Strings, Center for Theoretical Physics of the Universe, Institute for Basic Sciences, Daejeon 305-811, Korea

We present a geometric approach via Kaluza-Klein compactification to trigger Higgs mechanism for massive and partially massless higher-spin gauge fields. Starting from a free massless higher-spin field in five-dimensional anti-de Sitter space and compactifying on Janus geometry, we obtain an infinite tower of heavy, light and massless higher-spin fields in four-dimensional anti-de Sitter space. In this talk, we explain how the Higgs mechanism is triggered for spin-one and spin-two fields. All massive, partially massless fields are described manifest gauge invariantly in terms of Stüeckelberg coupling to lower-spin Goldstone fields. The mass spectrum depends on the boundary conditions. For Robin boundary conditions, we get unitary representations of massive and massless fields. For higher-derivative boundary conditions, equivalently, spectrum-dependent boundary conditions, we get either unitary massive spin-one field or non-unitary partially-massless spin-two fields. We also discuss how the pattern continues to higher spin fields and holographic dual description in terms of interface conformal field theory in three dimensions.

"Things without all remedy should be without regard: What is done, is done."
Shakespeare, Macbeth (Act 5, scene 1, 68)

1. Introduction

Consider a higher-spin field of spin-s in four dimensions. If massless, it has two polarization states. If massive, it has in general $(2s+1)$ polarization states. Intuitively, it can be viewed as a sum of massless fields of spin $0, 1, \ldots, s$, as can be gleaned from

$$1 + \sum_{\text{spin}=1}^{s} 2 = (2s+1). \tag{1}$$

Is it possible for a spin-s field to have polarization states less than $(2s+1)$? In Minkowski space-time, this is impossible as, according to finite-dimensional unitary representation of Poincare group, a massive spin-s field ought to encompass all of $(2s+1)$ polarization states, while a massless spin-s field encompasses only 2 helicity states. Perhaps less known, in case the cosmological constant Λ is non-zero, the unitary representations of (anti-)de Sitter group indicate that a spin-s field can encompass less than $(2s+1)$ but larger than 2 polarization states,

$$-s, -s+1, \ldots, -k-1, -k, [\quad], k, k+1 \cdots s-1, s \quad (k \leq s). \tag{2}$$

It is referred to as "partially massless"[1] spin-s field $\psi^{(k|s)}$ of depth $t := s-k$. The massless spin-s field corresponds to partially massless spin-s field of depth-zero $(k = s)$, while the massive spin corresponds to partially massless spin-s field of depth-s $(k = 0)$. It has the partial gauge invariance

$$\delta\psi^{(k|s)} = (\nabla \cdots \nabla)_{(s-k)}\xi^{(n)} + \cdots, \tag{3}$$

where the ellipses denote contributions involving fewer number of derivatives, and, for Ricci scalar $\mathcal{R}$, has the mass-squared

$$m^2(k|s) = \frac{\mathcal{R}}{12}[s(s-1) - k(k-1)] \quad (k = 0, 1, \ldots, s). \tag{4}$$

The partially massless fields are peculiar and mysterious. For example, consider them in the AdS (anti-de Sitter) space, $\mathcal{R} < 0$, for which rules of the AdS/CFT correspondence for matching bulk fields and boundary operators, which is just the Dirichlet boundary-value problem, are well established. Firstly, the partial gauge invariance under Eq. (3) is nontrivial for higher spins, $s > 1$. Secondly, as the mass formula Eq. (4) indicates, the partially massless spin-s has mass-squared lower than that of spin-s gauge field $(k = s)$, but still above massive spin-s field $(k = 0)$. So, the massive spin-s field exists even for negative value of mass-squared and partially massless spin-s field exists at discrete values inside the continuum of mass-squared. Thirdly, in the dual CFT, they are associated with a partially conserved current $\mathcal{O}^{(k|s)}$ of rank s and depth $t = s-k$ obeying the conformally invariant operator conservation law

$$(\nabla \cdots \nabla)_{(s-k)}\mathcal{O}^{(k|s)} + \cdots \big|_{\text{boundary}} = 0. \tag{5}$$

The Higgs mechanism for rendering mass to higher spin fields were better understood in recent years. Can we similarly understand the Higgs mechanism of turning massless spin-s field to partially massless spin-s field? In this talk, we will present a geometric approach toward the partially massless Higgs mechanism. The idea is to utilize Kaluza-Klein[2] compactification of massless higher-spin field in AdS_5 space and compactify on so-called Janus wedge to AdS_5 space. Recall that the Janus geometry[3] is a domain-wall configuration of Type IIB string theory on $AdS_5 \times \mathbb{S}^5$ space. From the viewpoint of underlying AdS_5 space, the domain-wall configuration is an open space with two boundaries preserving AdS_4 isometries. As such, higher spin field in this background is completely determined only after self-adjoint boundary conditions are specified. With detailed analysis explained below, we found the following results:

- Compactification spectrum is rich due to the presence of two length scales, the Janus wedge and the anti-de Sitter curvature scale. The spectrum comprises of Kaluza-Klein modes whose masses depend on the size of Janus wedge and ground modes whose masses are controlled by radius of the anti-de Sitter space.
- Generally, boundary conditions involve higher-derivatives. This leads in particular to existence of partially massless fields as part of ground modes. They are (non-)unitary in (anti-)de Sitter space, precisely because of the higher-derivative nature of boundary conditions.

We organize this paper as follows. In Sec. 2, we will recall various aspects of the partially massless spin-two field, both from the AdS space viewpoint and the CFT viewpoint. In Sec. 3, we first consider spectral analysis of the flat space waveguide for spin-one field. In Sec. 4, we describe so-called Janus wedge in AdS_4 space. In Sec. 5, we study waveguide spectrum of spin-one field. In Sec. 6, we describe how partially massless spin-two field can arise from massless spin-two field via the straightforward generalization of the Stueckelberg mechanism.[4] As recalled in Sec. 2, one mysterious aspect of partially massless field is that descendant of it has a negative norm. We will show that this arises from second-order boundary conditions at the wedge boundaries. As such, Higgs mechanism for partially massless spin-two field is achieved by choosing appropriate boundary conditions. The mechanism is straightforwardly generalizable to higher spins, which I relegate to a separate paper.[5] In Sec. 7, I consider the partially massless higher spin field in cosmological background and show that the field supports a conserved charge if the cosmological background is de Sitter. In Sec. 8, I discuss various generalizations of the results I report in this talk.

2. Views on Partially Massless Spin-Two Field

We first recall salient features of partially massless spin-two field from various viewpoints. We follow closely earlier treatments in Ref. 6 and, in particular, Refs. 7–10. These fields exist only in de Sitter (dS) or anti-de Sitter (AdS) space, so the (A)dS/CFT correspondence provides an integral part for understanding their nature. To make results compact and concrete, we shall limit our discussions to partially massless spin-two field (possibly interacting with massless spin-two field, viz. gravity) in AdS_4 space. Throughout this talk, we will restrict the analysis to the linear level of equations of motion.

2.1. *AdS viewpoint*

Suppose the gravity in AdS_4 space contains a massless or partially massless or massive spin-2 fields. The AdS_4 space is a metric space $g_{\mu\nu}$ that solves the Einstein equation

$$\mathcal{R}_{\mu\alpha\nu\beta} = \frac{\Lambda}{3}(g_{\mu\nu}g_{\alpha\beta} - g_{\mu\beta}g_{\nu\alpha}), \quad W_{\mu\alpha\nu\beta} = 0, \quad \mathcal{R}_{\mu\nu}(g) = \Lambda g_{\mu\nu}, \quad \mathcal{R} = 4\Lambda. \quad (6)$$

In this background, the linearized field equation of partially massless spin-two field $h_{\mu\nu}$ is given by

$$\mathcal{K}h_{\mu\nu} - \mathcal{K}_{\mu\nu}h = 0 \quad \text{where } \mathcal{K}_{\mu\nu} := \left[\nabla_\mu\nabla_\nu - \frac{\Lambda}{3}g_{\mu\nu}\right] \quad \text{and } \mathcal{K} := g^{\mu\nu}\mathcal{K}_{\mu\nu}. \quad (7)$$

Here, ∇ is the covariant derivative with Christoffel connection of the background metric $g_{\mu\nu}$ and h is the trace of $h_{\mu\nu}$. Taking covariant divergence of the equation of motion, we find the constraint

$$\nabla^\mu h_{\mu\nu} - \nabla_\nu h = 0. \quad (8)$$

Unlike massive Pauli-Fierz spin-two field, the partially massless spin-two field has gauge invariance. We see this readily from that the linearized equation of motion does not change under the partial gauge transformation

$$h_{\mu\nu} \quad \rightarrow \quad h_{\mu\nu} + \mathcal{K}_{\mu\nu}\xi. \quad (9)$$

As the gauge parameter ξ is a scalar field, we conclude that the partially massless field $h_{\mu\nu}$ has one less degrees of freedom than the massive Pauli-Fierz field, viz. $5 - 1 = 4$ polarizations.

2.2. *Bulk-boundary viewpoint*

One expects the bulk-boundary map of AdS/CFT correspondence, originally formulated for lower spin fields, continues to hold for partially massless higher-spin fields. Again, we shall focus on partially massless spin-two field in AdS_4 space. Recall that the massless spin-two field in the bulk is dual to source and expectation value of symmetric rank-two energy-momentum tensor operator. Recall also that the gauge invariance of the bulk spin-two field implies conservation of the boundary energy-momentum tensor. By extrapolation, we expect that partially massless spin-two field in the bulk is dual to source and expectation value of symmetric rank-two tensor operator and that the partially massless gauge invariance in the bulk implies partial conservation of the boundary tensor.

To solve for the linearized equation of partially massless spin-two field, we take the covariant gauge conditions, $\nabla^\mu h_{\mu\nu} = 0$ and $h^\mu_\mu = 0$. This leads to

$$\mathcal{K} h_{\mu\nu} = 0. \tag{10}$$

We take the metric near the boundary of AdS$_4$ space expressed in Fefferman-Graham coordinates

$$\mathrm{d}s_4^2 = \frac{R^2_{\mathrm{AdS}}}{u^2}\left[\mathrm{d}u^2 + g_{ab}(u, \mathbf{x})\mathrm{d}\mathbf{x}^i \mathrm{d}\mathbf{x}^j\right] \quad \text{where } R^2_{\mathrm{AdS}} = \frac{12}{|\mathcal{R}|}. \tag{11}$$

Near the conformal boundary $u = 0$, the Einstein equation can be solved in power series or Frobenius series. The solution takes the form

$$g_{ab}(u, \mathbf{x}) = \bar{g}_{ab}(\mathbf{x}) - u^2\big(\bar{R}^{(3)}_{ab}(\mathbf{x}) - \frac{1}{4}\bar{R}^{(3)}(\mathbf{x})\bar{g}_{ab}(\mathbf{x})\big) + \cdots \tag{12}$$

where $\bar{g}_{ab}$ is the conformal metric of the boundary at $u = 0$ and the ellipses denote cubic or higher-order terms in u.

Using this, we can derive the covariant form of partial conservation law Eq. (5). Taking $(3+1)$-decomposition Eq. (11), the linearized field equation for the fluctuation $h_{\mu\nu}$ around the background Eq. (11) becomes with gauge conditions $\nabla^\mu h_{\mu\nu} = 0, h^\mu_\mu = 0$ to

$$\mathcal{K} h_{\mu\nu} = \left[-\frac{3}{u}\partial_u - \frac{3}{u^2} + \big(\bar{\nabla}^2_{(3)} - \frac{1}{8}\bar{R}_{(3)}\big)\right] h_{\mu\nu} = 0. \tag{13}$$

We solve this for the transverse-traceless fluctuation $h_{ab}(u, \mathbf{x})$ near the boundary $u = 0$. In the unitary gauge, the solution is given by

$$h_{ab}(u, \mathbf{x}) = \frac{1}{u} J_{ab}(\mathbf{x}) + V_{ab}(\mathbf{x}) \quad \text{and} \quad h^a_b(u, \mathbf{x}) = u\, J^a_b(\mathbf{x}) + u^2\, V^a_b(\mathbf{x}), \tag{14}$$

where the fields on the right-hand side are with respect to the conformal boundary metric $\bar{g}_{ab}(\mathbf{x})$. The field h^a_b is the conformal tensor of weight two, the bulk graviton couples to boundary operator $\mathcal{O}^a_b(\mathbf{x})$ in Weyl invariant manner

$$S_H = \int \mathrm{d}^3\mathbf{x}\, \sqrt{\bar{g}(\mathbf{x})}\, J^a_b(\mathbf{x})\mathcal{O}^b_a(\mathbf{x}). \tag{15}$$

The operator $\mathcal{O}^a_b(\mathbf{x})$ is precisely the partially conserved rank $(1,1)$ tensor operator. The partially massless gauge transformations at the boundary $u = 0$ is given by

$$\delta J_{ab}(\mathbf{x}) = \left[\bar{\nabla}_a\bar{\nabla}_b + \left(\bar{R}^{(3)}_{ab} - \frac{1}{4}\bar{R}^{(3)}\bar{g}_{ab}\right)\right]\zeta(\mathbf{x}). \tag{16}$$

Demanding that S_J is invariant under this gauge transformation and tracelessness of $\mathcal{O}_{ab}$,

$$\left[\bar{\nabla}_a\bar{\nabla}_b + \bar{R}^{(3)}_{ab}\right]\mathcal{O}^{ab}(\mathbf{x}) = 0. \tag{17}$$

This is the covariant form of conformally invariant equation for partially conserved tensor operator $\mathcal{O}_{ab}(\mathbf{x})$ in CFT_3. For tensors of more general rank, the corresponding conformally invariant equations are known.[11–13]

2.3. *Group theoretic viewpoint*

The isometry algebra of AdS_4 is $\mathfrak{so}(3,2)$ and equals to the conformal algebra of $\mathbb{R}^{2,1}$. Recall that the Dirac singletons represent AdS_4 fields whose physical degrees of freedom reside at its conformal infinity, equivalently, free conformal fields defined in $\mathbb{R}^{2,1}$. According to the Flato-Fronsdal theorem, tensor product of two Dirac's singletons can be decomposed to an infinite sum of irreducible representations of $\mathfrak{so}(3,2)$. These representations are interpretable as on-shell higher-spin fields in AdS_4, equivalently, as conserved current of CFT_3. The infinite sum corresponds to an infinite tower of higher-spin gauge fields, equivalently, to the one-particle states of CFT_3.

An extension of the Flato-Fronsdal theorem incorporates so-called higher-order singletons. They are non-unitary, ultra-short irreducible representations of $\mathfrak{so}(3,2)$. It turns out that tensor product of two higher-order singletons gives rise to partially massless fields having the higher-order gauge invariance. If the higher-order singletons are bosonic, the corresponding partially massless fields have odd depths. If the higher-order singletons are fermionic, the corresponding partially massless fields have even depths.

We first recall the definition of the Verma $\mathfrak{so}(d,2)$-module for general d. Consider a finite dimensional module $\mathcal{Y}(\Delta, Y)$ of sub-algebra $\mathfrak{so}(2)\oplus\mathfrak{so}(d)$. We use Δ to denote conformal dimension and Y to denote Young diagram of $\mathfrak{so}(d)$. For the analysis of symmetric higher spin, we limit ourself to the Young diagram of a single row of length s. The Verma $\mathfrak{so}(d,2)$-module $\mathcal{V}(\Delta,\, s)$ is the space generated by action of the raising operators to the module $\mathcal{Y}(\Delta,\, Y)$. We will also denote $\mathcal{D}(\Delta, s)$ for the irreducible quotient of Verma module $\mathcal{V}(\Delta,\, s)$. For generic value, Verma module $\mathcal{V}(\Delta,\, s)$ is irreducible and therefore coincides with $\mathcal{D}(\Delta, s)$. However, for specific values, it becomes reducible with a non-trivial submodule. For instance, $\Delta = d+k-1$ with an integer $0\le k\le s-1$, there is a submodule $\mathcal{D}\,(d+s-1,\, k)$. Therefore, $\mathcal{D}\,(d+k-1,\, s)$ is not equal to Verma module but is to the quotient of Verma module:

$$\begin{aligned}\mathcal{V}\,(d+k-1,\, s) &\simeq \mathcal{D}\,(d+k-1,\, s)\oplus\mathcal{D}\,(d+s-1,\, k),\\ \mathcal{D}\,(d+k-1,\, s) &\simeq \frac{\mathcal{V}\,(d+k-1,\, s)}{\mathcal{D}\,(d+s-1,\, k)}.\end{aligned}\tag{18}$$

For $k=s-1$, $\mathcal{D}(d+(s-1)-1,s)$ is unitary and its field theoretical realization is the massless spin-s field propagating in AdS_{d+1}. For $0\le k<s-1$, $\mathcal{D}(d+k-1,s)$ is non-unitary and their field theoretical realizations are partially massless fields with depth $t=(s-k-1)$. (For more general cases, see Refs. 9 and 14.) The action for partially massless field has the partial gauge symmetry which contains

covariant derivatives up to order $t-1$. This can be derived by the Stueckelberg form of partially massless gauge transformation

$$\delta\,\phi_{\mu_1\mu_2\cdots\mu_s} = \nabla_{(\mu_1}\cdots\nabla_{\mu_{t+1}}\,\xi_{\mu_{t+2}\cdots\mu_s)} + \cdots \tag{19}$$

We thus collect the properties of partially massless fields as

Partially Massless	Δ_+	m^2	gauge variation: $\delta\,\phi_{\mu_1\mu_2\cdots\mu_s}$
Depth-t	$d+s-t-2$	$-\frac{\sigma}{\ell^2}\,t\,(d+2s-t-4)$	$\nabla_{(\mu_1}\cdots\nabla_{\mu_{t+1}}\,\xi_{\mu_{t+2}\cdots\mu_s)}+\cdots$

Here, the mass m is defined by the following convention. By the mass of a field, we refer to the mass in flat limit. Therefore, it is zero when the higher spin gauge symmetry exist. In this convention, the relation between mass-squared and the conformal dimension is given by

$$m^2\,R^2 = (\Delta - d/2)^2 - (s-2+d/2)^2\,. \tag{20}$$

Note that this is different from the mass-squared which appears in Fierz-Pauli equation in AdS$_d$ space,[15] $\left(\nabla^2+\kappa^2\right)\phi_{\mu_1\,\mu_2\,\cdot\mu_s} = 0$, which is given by $\kappa^2\,R^2 = \Delta\,(\Delta-d)-s$.

The $\mathfrak{so}(d+1,2)$-module for massless spin-s can be decomposed into $\mathfrak{so}(d,2)$-modules by the following branching rules:[20]

$$\begin{aligned}&\mathcal{D}(d+s-1,s)_{\mathfrak{so}(d+1,2)}\\ &\quad=\bigoplus_{n=0}^{\infty}\mathcal{D}(d+n+s-1,s)_{\mathfrak{so}(d,2)}\oplus\bigoplus_{l=0}^{s-1}\mathcal{D}(d+s-1,l)_{\mathfrak{so}(d,2)}.\end{aligned} \tag{21}$$

For the isometry algebra of AdS$_4$ space, $\mathfrak{so}(3,2)$, the Verma module and the irreducible module $\mathcal{D}$ are related each other by

$$\mathcal{D}(\Delta,s)\simeq\frac{\mathcal{V}(\Delta,s)}{\mathcal{D}(3-\Delta,s)}\simeq\frac{\mathcal{V}(\Delta,s)}{\mathcal{V}(3-\Delta,s)}. \tag{22}$$

The AdS$_4$ Di and Rac singletons, first introduced by Dirac,[16] are associated with the irreducible modules

$$\begin{aligned}\text{Di}:\ \mathcal{D}\,(1,1/2)&\simeq\frac{\mathcal{V}\,(1,1/2)}{\mathcal{D}\,(2,1/2)},\\ \text{Rac}:\ \mathcal{D}\,(1/2,0)&\simeq\frac{\mathcal{V}\,(1/2,0)}{\mathcal{D}\,(5/2,0)}.\end{aligned} \tag{23}$$

The characters for the corresponding modules are given by

$$\chi_{\text{Di}}(q,x)=q^{3/2}\left(\sqrt{q}-\frac{1}{\sqrt{q}}\right)\chi(x)\,\mathcal{P}(q,x),\quad \chi_{\text{Rac}}(q,x)=q^{3/2}\left(q-\frac{1}{q}\right)\mathcal{P}(q,x), \tag{24}$$

where

$$\mathcal{P}(q,x) = \frac{1}{(1-q)}\frac{1}{(1-qx)(1-qx^{-1})}, \quad \chi(x) = (x-x^{-1})(x^{1/2}-x^{-1/2}) \quad (25)$$

are so(3) invariant partition function and so(3) irreducible character, respectively.

Likewise, the $(k+1)$-th order Di and Rac singletons are associated with the irreducible modules

$$\begin{aligned} (k+1)\text{Di}: \quad & \mathcal{D}\,(1-k,1/2) \simeq \frac{\mathcal{V}\,(1-k,1/2)}{\mathcal{D}\,(2+k,1/2)}, \\ (k+1)\text{Rac}: \quad & \mathcal{D}\,(1/2-k,0) \simeq \frac{\mathcal{V}\,(1/2-k,0)}{\mathcal{D}\,(5/2+k,0)}. \end{aligned} \quad (26)$$

The characters for these $(k+1)$-th order modules are given by

$$\begin{aligned} \chi_{(k+1)\text{Di}}(q,x) &= q^{3/2}\left(q^{k+1/2} - \frac{1}{q^{k+1/2}}\right)\chi(x)\,\mathcal{P}(q,x), \\ \chi_{(k+1)\text{Rac}}(q,x) &= q^{3/2}\left(q^{k+1} - \frac{1}{q^{k+1}}\right)\mathcal{P}(q,x). \end{aligned} \quad (27)$$

The singletons are precisely the massless scalar and spinor fields in AdS_4 boundary. They are the lowest $k = 0$ of the tower of higher-order singletons. These higher-order $k \geq 1$ singletons are precisely the partially massless scalar and spinor fields, as can be seen from the agreement of their scaling weight Δ.

More generally, the modules corresponding to partially massless, totally symmetric AdS_4 fields are given by

$$\begin{aligned} \mathcal{D}_{1/2}(1+s-k,s) &\simeq \frac{\mathcal{V}_{1/2}(1+s-k,s)}{\mathcal{D}_{1/2}(2+s,s-k-1)} \quad (0 \leq k \leq s-1/2), \\ \mathcal{D}_0(1+s-k,s) &\simeq \frac{\mathcal{V}_0(1+s-k,s)}{\mathcal{D}_0(2+s,s-k-1)} \quad (0 \leq k \leq s), \end{aligned} \quad (28)$$

where the subscripts 0, 1/2 are Di and Rac refer to trajectories, respectively, and $k = 0$ corresponds to the massless, totally symmetric bosonic or fermionic fields.

We finally quote the fusion rules of higher-order singletons. The tensor product of two $(k+1)$-th singletons in AdS_4 space decomposes into

$$\mathcal{D}_0(1/2-k,0) \otimes \mathcal{D}_0(1/2-k,0) \simeq \bigoplus_{\ell=0}^{k}\bigoplus_{s=0}^{\infty}\mathcal{D}_0(s+1-2\ell,0)$$

$$\mathcal{D}_{1/2}(1-k,1/2) \otimes \mathcal{D}_0(1/2-k,0) \simeq \bigoplus_{t=0}^{2k}\bigoplus_{s=0}^{\infty}\mathcal{D}_{1/2}(3/2+s-t,s+1/2)$$

$$\mathcal{D}_{1/2}(1-k,1/2)\otimes\mathcal{D}_{1/2}(1-k,1/2)\simeq\bigoplus_{t=-2k}^{2k}\mathcal{D}_0(2-t,0)$$
$$\oplus\bigoplus_{s=0}^{\infty}\mathcal{D}_1(2+s-2k,s+1)$$
$$\oplus\, 2\bigoplus_{t=1}^{2k}\bigoplus_{s=0}^{\infty}\mathcal{D}_1(3+s-t,s). \quad (29)$$

In words, the tensor product of higher-order scalar singletons gives rise to partially massless, totally symmetric boson tensor fields, the tensor product of higher-order scalar and spinor singletons gives rise to partially massless, totally symmetric fermionic tensor fields, while the tensor product of higher-order spinor singletons gives massive and partially massless bosonic tensor fields. Interestingly, the last involves tensor fields of not only totally symmetric type but also mixed symmetric type.

2.4. *Conformal field theory viewpoint*

We next study the partially massless field from the representation theoretic viewpoint. Suppose a 3-dimensional conformal field theory, living at spacelike infinity of AdS_4 space, contains a partially conserved spin-s conformal tensor $\mathcal{O}_\Delta^{(k|s)}$ of depth $t=s-k$. We are primarily interested in understanding the tensor from the representation theory of conformal algebra. So, we conveniently take the AdS_4 space Euclidean, foliate it by flat $\mathbb{R}^3$ leaves, and adopt the radial quantization $\mathbb{R}^3\simeq\mathbb{S}^2\times\mathbb{R}_+$ up to a conformal equivalence class. The Weyl invariance in turn ensures invariance of the CFT_3 under the Euclidean conformal group $SO(4,1)$ acting on $\mathbb{R}^3$. This Euclidean conformal algebra is spanned by the generators

$$\mathfrak{so}(4,1)=\{M_{ab},P_a,K^a,D,\,a,b=1,\dots,3\}, \quad (30)$$

with commutation relations

$$[M_{[ab]},M_{[cd]}]=\delta_{ac}M_{bd}-\delta_{ad}M_{bc}+\delta_{bd}M_{ac}-\delta_{bc}M_{ad}$$
$$[M_{ab},K_c]=\delta_{ac}K_b-\delta_{bc}K_a,\quad [M_{ab},P_c]=\delta_{ac}P_b-\delta_{bc}P_a$$
$$[D,K_a]=-K_a,\quad [D,P_a]=+P_a,\quad [K_a,P_b]=\delta_{ab}D+M_{ab}. \quad (31)$$

Using this algebra, we shall extract conformal dimension of the operator $\mathcal{O}^{(k|s)}$ by requiring that it obeys partial conservation law.

In the radial quantization of Euclidean CFT_3, one labels operators and associated states (obtained by the action of operator on a vacuum) according to the decomposition

$$\mathfrak{so}(3)\oplus\mathfrak{so}(1,1)\subset\mathfrak{so}(4,1), \quad (32)$$

In this decomposition, the dilatation generator D plays the role of Hamiltonian and measures the $\mathfrak{so}(1,1)$ scaling dimension Δ of its eigenstates. Then, from the conformal algebra Eq. (31), generators P_m, K_m act as raising and lowering operators, respectively, for the tower of eigenstates. Finally, the generator M_{mn} measures the $\mathfrak{so}(3)$ spin s of eigenstates. So, every state of the conformal algebra is labelled by (Δ, s). It should be noted that, in Euclidean space, these generators obey the hermiticity property given by

$$D^\dagger = D, \quad P_m^\dagger = K_m, \quad K_m^\dagger = P_m, \quad M_{mn}^\dagger = -M_{mn}. \tag{33}$$

We assume that a vacuum $|0\rangle$ exists as a unique state which belongs to the null space of D, P_m, K_m and M_{mn}. The fixed point of $\mathfrak{so}(1,1)$ action is $x = 0$ (and its conformal inversion $x = \infty$). Therefore, to each state $|\mathcal{O}\rangle$ labelled by (Δ, s), we can associate a unique operator $\mathcal{O}_\Delta^{(s)}(x)$ at the fixed point

$$|\mathcal{O}; \Delta, s\rangle = \lim_{x\to 0} \mathcal{O}_\Delta^{(s)}|0\rangle \tag{34}$$

and its Hermitian conjugate in the dual Hilbert space.

Our goal is to derive the representation of partially conserved conformal tensor $\mathcal{O}_\Delta^{(k|s)}$ under the $\mathfrak{so}(4,1)$. In radial quantization around the origin $x = 0$, repeating the above operator-state map Eq. (34), we should be able to associate the partially conserved conformal operator $\mathcal{O}_\Delta^{(k|s)}$ with a state excited by the operator above the vacuum,

$$|\mathcal{O}; \Delta, s, k\rangle = \lim_{x\to 0} \mathcal{O}_\Delta^{(k|s)}(x)|0\rangle. \tag{35}$$

This state forms an irreducible representation in terms of the $\mathfrak{so}(1,1)$ scaling dimension and $\mathfrak{so}(3)$ spin. Making tensor indices somewhat explicit,

$$\begin{aligned} &D|\mathcal{O}_{c(\cdot)}; \Delta, s, k\rangle = \Delta|\mathcal{O}_{p(\cdot)}; \Delta, s, k\rangle \\ &L_{ab}|\mathcal{O}_{c(\cdot)}; \Delta, s, k\rangle = \delta_{ac}|\mathcal{O}_{b(\cdot)}; \Delta, s, k\rangle - \delta_{bc}|\mathcal{O}_{a(\cdot)}; \Delta, s, k\rangle + (\text{other indices}). \end{aligned} \tag{36}$$

An irreducible representation of the conformal algebra is specified by a primary operator $\mathcal{O}_\Delta^{(k|s)}$ belonging to the null space of $[K_a, \cdot]$. Other operators in this representation is spanned by the conformal descendants formed by action of raising operator P_a. For example, at level-ℓ, the conformal descendant is spanned by the image of $[P_{a_1}, [P_{a_2}, \ldots, [P_{a_\ell}, \cdot] \cdots]]$ acting on $\mathcal{O}_\Delta^{(k|s)}$. They have conformal dimension $\Delta_J + \ell$ and spins $s - \ell, s - \ell + 1, \ldots, s + \ell - 1, s + \ell$.

Suppose first the conformal tensor $\mathcal{O}^{a(\cdot)}$ is a conformal primary ($[K_s, \mathcal{O}^{a(\cdot)}] = 0$) obeying conservation law, $\partial_a \mathcal{O}^{a(\cdot)} = 0$. The gauge current and the energy-momentum tensor are two lower-spin examples. The conservation law $\partial_a \mathcal{O}^{a(\cdot)}(x) = 0$ is now read as the condition on the operator, equivalently, on the state of the representation whose primary is labelled by (Δ, s):

$$[P_a, \mathcal{O}^{a(\cdot)}(0)] = 0, \quad \text{equivalently,} \quad P_a|\mathcal{O}^{a(\cdot)}\rangle = 0 \quad \text{and hence } \|P_a|\mathcal{O}^{a(\cdot)}\rangle\|^2 = 0. \tag{37}$$

Thus, the conservation law is equivalent to the statement that the level-one descendant of the primary state $|\mathcal{O}^{a(\cdot)}\rangle$ is null.

For the null state to appear, however, the conformal dimension of primary state $|\mathcal{O}\rangle$ needs to take a discrete value. We can see this from the conformal algebra and the hermiticity property. Take for example spin-two current operator $\mathcal{O}^{ab}$ obeying conservation law $\partial_a \mathcal{O}^{ab} = 0$. We have

$$\begin{aligned} \|P_a|\mathcal{O}^{ab}\rangle\|^2 &:= \langle P_a\mathcal{O}^{ab}|\delta_{bd}|P_c\mathcal{O}^{cd}\rangle = \langle\mathcal{O}^{ab}|K_a\delta_{bd}P_c|\mathcal{O}^{cd}\rangle \\ &= \delta_{bd}\langle\mathcal{O}^{ab}|[K_a, P_c]|\mathcal{O}^{cd}\rangle = \langle\mathcal{O}^{ab}|\delta_{ac}\delta_{bd}D + M_{ac}\delta_{bd}|\mathcal{O}^{cd}\rangle \\ &= (\Delta - 3)\|\,|\mathcal{O}^{ab}\rangle\|^2. \end{aligned} \tag{38}$$

If the operator $\mathcal{O}^{ab}$ corresponds to a conserved current tensor, this must vanish. Since the state associated with the primary operator must have nonzero norm, it follows that the conformal dimension Δ should equal to 3.

What if the spin-two current operator $\mathcal{O}^{ab}$ obeys a partial conservation law, $\partial_m\partial_n\mathcal{O}^{ab} = 0$? In this case, we have

$$\begin{aligned} &[P_a, [P_b, \mathcal{O}^{ab}(0)]] = 0, \quad \text{equivalently} \quad P_aP_b|\mathcal{O}^{ab}\rangle = 0 \\ &\text{and hence } \|P_aP_b|\mathcal{O}^{ab}\rangle\|^2 = 0. \end{aligned} \tag{39}$$

Thus, the partial conservation law is equivalent to the statement that the level-two descendant is null. We first start from the level-one descendant whose non-zero inner product is given by

$$\|P_a|\mathcal{O}^{ab}\rangle\|^2 = (\Delta - 3)\|\,|\mathcal{O}^{ab}\rangle\|^2 \quad \text{and hence } \Delta \neq 3. \tag{40}$$

The level-two inner product then follows

$$\begin{aligned} \|\,P_aP_b|\mathcal{O}^{ab}\rangle\|^2 &= \langle\mathcal{O}^{ab}|K_aK_bP_cP_d|\mathcal{O}^{cd}\rangle \\ &= \langle\mathcal{O}^{ab}|K_a[K_b, P_c]P_d|\mathcal{O}_{cd}\rangle + \langle\mathcal{O}^{ab}|K_aP_c[K_b, P_d]|\mathcal{O}^{cd}\rangle \\ &= 2(\Delta - 2)\delta_{bd}\langle\mathcal{O}^{ab}|K_aP_c|\mathcal{O}^{cd}\rangle \\ &= 2(\Delta - 2)(\Delta - 3)\,\|\,|\mathcal{O}^{ab}\rangle\|^2. \end{aligned} \tag{41}$$

Taking account of Eq. (40), the partial conservation law then implies that the scaling dimension of the partially conserved current is given by $\Delta = 2$. The pattern of conformal dimensions, $\Delta = 3, 2$, for conventionally conserved and partially conserved current operators matches perfectly with the pattern of masses for massless and partially massless spin-two fields in AdS_4 space.

We see that partially massless states are special states with partially restored gauge invariance. The mass formula indicates that partially massless states exists only for nonzero cosmological constant, for masses below the massless states, and that they are Higgsed to fully massive states away from the special values. This then brings us to the question how the Higgs mechanism may be triggered for the partially massless higher spin in AdS_4 space. We now turn to this issue for the rest of this talk.

3. Waveguide and Boundary Conditions: Flat Space

To achieve the Higgs mechanism of higher-spin fields in AdS_4 space, the salient feature of our approach is to compactify AdS_5 to AdS_4 times an angular wedge whose size is tunable, thus forming an "AdS waveguide". At each boundary of the wedge, we must impose suitable boundary conditions, which in turn uniquely determine the mass spectrum in AdS_4. The effect of boundary condition to the mass spectrum is best illustrated by the flat space counterpart. Here, we digress to this situation and gain intuition for relation between boundary conditions and spectra.

3.1. *Kalauza-Klein mode expansion*

For a start, we study the electromagnetic — massless spin-one — waveguide in five-dimensional flat spacetime with boundaries, paying particular attention to relations between boundary conditions and spectra for fields of different spins. The flat spacetime is $\mathbb{R}^{1,3} \times I_L$, where interval $I_L \equiv \{0 \leq z \leq L\}$. The five-dimensional coordinates can be decomposed into parallel and perpendicular directions: $x^M = (x^\mu, z)$. The five-dimensional spin-one field is decomposed in four dimensions to a spin-one field and a spin-zero field: $A_M = (A_\mu, \phi)$. The equations of motions are decomposed as

$$\partial^M F_{M\nu} = \partial^\mu F_{\mu\nu} - \partial_z(\partial_\nu \phi - \partial_z A_\nu) = 0, \tag{42}$$

$$\partial^M F_{Mz} = \partial^\mu (\partial_\mu \phi - \partial_z A_\mu) = 0, \tag{43}$$

while the gauge transformations are decomposed as

$$\delta A_\mu = \partial_\mu \Lambda, \quad \delta \phi = \partial_z \Lambda. \tag{44}$$

We note that both the equations of motion and the gauge transformations manifest the structure of Stüeckelberg system.[4] Recall that the Stüeckelberg Lagrangian of massive spin-one vector field is given by

$$\mathcal{L} = -\frac{1}{4} F_{\mu\nu} F^{\mu\nu} - \frac{1}{2} \partial_\mu \phi \, \partial^\mu \phi + m A_\mu \left(\frac{m}{2} A^\mu - \partial^\mu \phi \right), \tag{45}$$

which is invariant under Stüeckelberg gauge transformations

$$\delta A_\mu = \partial_\mu \lambda \quad \text{and} \quad \delta \phi = m \lambda. \tag{46}$$

The field ϕ is referred to as the Stueckelberg spin-zero field. This field is redundant for $m \neq 0$ because it can be eliminated by a suitable gauge transformation. In the massless limit, $m \to 0$, the Stüeckelberg system breaks into a spin-one gauge system and a massless spin-zero system.

Inside the waveguide, the five-dimensional spin-one field A_M is excited along the z-direction. The field can be mode-expanded, and expansion coefficients are four-dimensional spin-one and spin-zero fields of varying masses. Importantly, mode functions can be chosen from any complete set of basis functions. It is natural to choose them by the eigenfunctions of $\Box := -(\partial_z)^2$ with a prescribed boundary condition. Mode functions of the gauge parameter Λ should be chosen compatible with the mode function of spin-one field A_M. Combining the two gauge variations Eq. (44), we learn that the mode functions ought to be related to each other as

$$\begin{aligned}&\partial_z\,(\text{mode function of spin-one field } A_\mu(x,z))\\&\qquad\propto (\text{mode function of spin-zero field } \phi(x,z)).\end{aligned} \tag{47}$$

Being a local relation, this relation must hold at each boundaries as well.

It would be instructive to understand, instead of the required Eq. (47), what might go wrong if one imposes the same boundary conditions for both A_μ and ϕ, such as zero-derivative (Dirichlet) or one-derivative (Neumann) boundary conditions. Suppose one adopts the zero-derivative (Dirichlet) boundary condition for both fields. From $A_\mu(z)|_{z=0,\,L} = 0$, $\phi(z)|_{z=0,L} = 0$ and from the field equation of ϕ, Eq. (43), it follows that

$$\left(\partial^\mu\,\partial_\mu\,\phi(z) - \partial^\mu\,\partial_z\,A_\mu(z)\right)\Big|_{z=0,\,L} = -\partial^\mu\,\partial_z\,A_\mu(z)|_{z=0,\,L} = 0, \tag{48}$$

and hence $\partial_z\,A_\mu(z)|_{z=0,L} = 0$. But A_μ satisfies second-order partial differential equation, so these two sets of boundary conditions — $A_\mu(z)|_{z=0,\,L} = 0$ and $\partial_z\,A_\mu(z)|_{z=0,L} = 0$ — imply that $A_\mu(z)$ must vanish everywhere. Likewise, ϕ satisfies a first-order differential equation Eq. (42), so the two sets of boundary conditions imply that $\phi(z)$ vanishes everywhere as well. One concludes that there is no non-trivial field excitations satisfying such boundary conditions. We remind that this conclusion follows from the fact that these boundary conditions do not preserve the relation Eq. (47).

The most general boundary conditions compatible with the relation Eq. (47) restricts the form of boundary conditions for spin-one and spin-zero fields. For example, if we impose the Robin boundary condition for the spin-zero field, $\mathcal{M}(\partial_z)\phi|_{z=0,L} := (a\partial_z + b)\phi|_{z=0,L} = 0$ where a, b are arbitrary constants, the relation Eq. (47) imposes the boundary condition for the spin-one field as $\mathcal{M}\,\partial_z\,A_\mu|_{z=0,\,L} = 0$. Modulo higher-derivative generalizations, we have two possible boundary conditions: $a = 0, b \neq 0$ corresponding to the vector boundary condition and $a \neq 0, b = 0$ corresponding to the scalar boundary condition. Hereafter, we analyze each of them explicitly.

3.2. *Vector boundary condition*

We may impose one-derivative (Neumann) boundary condition on the spin-one field $A_\mu(x,z)$ field and zero-derivative (Dirichlet) boundary condition on spin-zero field

$\phi(x,z)$ at $z=0,\ L$. The corresponding mode expansion for A_μ and ϕ reads

$$A_\mu(z) = \sum_{n=0}^{\infty} A_\mu^{(n)} \cos\left(\frac{n\,\pi}{L}\,z\right) \quad \text{and} \quad \phi(z) = \sum_{n=1}^{\infty} \phi^{(n)} \sin\left(\frac{n\,\pi}{L}\,z\right), \tag{49}$$

so we mode-expand the field equations Eqs. (42) and (43) in a suggestive form

$$\sum_{n=0}^{\infty} \cos\left(\frac{n\,\pi}{L}\,z\right)\left[\partial^\mu\, F^{(n)}{}_{\mu\nu} - \left(\frac{n\,\pi}{L}\right)\left(\frac{n\pi}{L}\,A_\nu^{(n)} + \partial_\nu\,\phi^{(n)}\right)\right] = 0, \tag{50}$$

$$\sum_{n=1}^{\infty} \sin\left(\frac{n\,\pi}{L}\,z\right)\partial^\mu\left(\frac{n\,\pi}{L}\,A_\mu^{(n)} + \partial_\mu\,\phi^{(n)}\right) = 0. \tag{51}$$

The standing-wave mode functions for $n = 0, 1, \ldots$ form a complete set of the orthogonal basis for square-integrable functions over I_L, so individual coefficient in the above equation ought to vanish. The zero-mode $n = 0$ is special, as only the first equation is nonempty and gives the equation of motion for massless spin-one field. All Kaluza-Klein modes, $n \geq 1$, satisfies the Stüeckelberg equation of motion for massive spin-one field[a] with mass $m_n = n\,\pi/L$. The second equation follows from divergence of the first equation, so just confirms consistency of the prescribed boundary conditions. In the limit $L \to 0$, all Stüeckelberg fields become infinitely massive. As such, there only remains the massless spin-one field $A_\mu^{(0)}$ with associated gauge invariance. Also, there is no spin-zero field $\phi^{(0)}$, an important result that follows from the prescribed boundary conditions. Intuitively, $A_\mu^{(0)}$ remains massless and gauge invariant, so Stüeckelberg spin-zero field $\phi^{(0)}$ is not needed. Moreover, the spectrum is consistent with the fact that this boundary condition ensures no energy flow across the boundary $z = 0, L$.

The key observation crucial for foregoing discussion is that the same result is obtainable from Kaluza-Klein compactification of gauge transformations Eq. (44). The gauge transformations that preserve the vector boundary conditions can be expanded by the Fourier modes:

$$\Lambda = \sum_{n=0}^{\infty} \Lambda^{(n)} \cos\left(\frac{n\,\pi}{L}\,z\right). \tag{52}$$

The gauge transformations of four-dimensional fields read

$$\delta\,A_\mu^{(n)} = \partial_\mu\,\Lambda^{(n)} \quad (n \geq 0) \quad \text{and} \quad \delta\,\phi^{(n)} = -\frac{n\,\pi}{L}\,\Lambda^{(n)} \quad (n \geq 1). \tag{53}$$

We note that the $n = 0$ mode is present only for the gauge transformation of spin-one field. This is the gauge transformation of a massless gauge vector field. We also note that gauge transformations of all higher $n = 1, 2, \ldots$ modes take precisely the form of Stüeckelberg gauge transformations. Importantly, the Stüeckelberg gauge

[a] For flat higher-spin Kaluza-Klein compactification, the Stüeckelberg structure were observed at Ref. 17.

invariance fixes quadratic part of action as the Stüeckelberg action for a tower of Proca fields with masses $m_n = n\,\pi/L$, $(n = 1, 2, \ldots)$.

We can turn the logic reversed. Suppose we want to retain massless spin-one field $A_\mu^{(0)}$ in four dimensions, along with associated gauge invariance. This requirement then singles out one-derivative (Neumann) boundary condition for A_μ. This and the divergence for A_μ, in turn, single out zero-derivative (Dirichlet) boundary condition for ϕ. Clearly, the massless fields are associated with gauge or global symmetries (as invariances under inhomogeneous transformations). So, this argument shows that proper boundary conditions for free field equations can be extracted just from gauge transformations.

3.3. *Scalar boundary condition*

Alternatively, one might impose no-derivative (Dirichlet) boundary condition to the spin-one field A_μ and one-derivative (Neumann) boundary condition to the spin-zero ϕ. In this case, the equations of motion, when mode-expanded, take exactly the same form as above except that the standing-wave mode functions are interchanged:

$$\sum_{n=1}^{\infty} \sin\left(\frac{n\,\pi}{L}\,z\right)\left[\partial^\mu\, {F^{(n)}}_{\mu\nu} - \left(\frac{n\,\pi}{L}\right)\left(\frac{n\pi}{L}\,A_\nu^{(n)} - \partial_\nu\,\phi^{(n)}\right)\right] = 0, \tag{54}$$

$$\sum_{n=0}^{\infty} \cos\left(\frac{n\,\pi}{L}\,z\right)\partial^\mu\left(\frac{n\,\pi}{L}\,A_\mu^{(n)} - \partial_\mu\,\phi^{(n)}\right) = 0. \tag{55}$$

Consequently, the zero-mode $n = 0$ consists of massless spin-zero field $\phi^{(0)}$ only ($A_\mu^{(0)}$ is absent from the outset). All Kaluza-Klein modes $n \neq 0$ are again Stüeckelberg massive spin-one fields with mass $m_n = n\pi/L$. In the limit $L \to 0$, these Stüeckelberg field becomes infinitely massive. Below the Kaluza-Klein scale $1/L$, there only remains the massless spin-zero field $\phi^{(0)}$. Once again, this is consistent with the fact that this boundary condition ensures no energy flow across the boundary.

Once again, the key idea is that the above results are obtainable from the Kaluza-Klein compactification of the gauge transformations. For a gauge transformation that preserves the scalar boundary condition, the gauge function can be expanded as

$$\Lambda(x, z) = \sum_{n=1}^{\infty} \Lambda^{(n)}(x)\sin\left(\frac{n\,\pi}{L}\,z\right). \tag{56}$$

With these modes, the gauge transformations of fields are

$$\delta\,A_\mu^{(n)} = \partial_\mu\,\Lambda^{(n)} \quad (n \geq 1) \quad \text{and} \quad \delta\,\phi^{(n)} = \frac{n\,\pi}{L}\,\Lambda^{(n)} \quad (n \geq 0). \tag{57}$$

There is no $n = 0$ zero-mode gauge transformation, and so no massless gauge spin-one field. The spin-zero zero-mode $\phi^{(0)}$ is invariant under the gauge transformations. We also note that the gauge transformations take the form of the Stüeckelberg gauge symmetries with masses $m_n = n\,\pi/L$.

Once again, we can turn the logic reversed. Suppose we want to retain massless spin-zero field $\phi^{(0)}$ in four dimensions. This then singles out one-derivative (Neumann) boundary condition for ϕ. This and the divergence of A_μ equation of motion, in turn, put zero-derivative (Dirichlet) boundary condition for the spin-one field A_μ.

Summarizing,

- Kaluza-Klein spectrum is obtainable either from field equations or from gauge transformations.
- Stüeckelberg structure naturally arises from Kaluza-Klein compactification.
- Boundary conditions of lower-dimensional component fields (for example, A_μ and ϕ from A_M) are correlated to each other (for example as in Eq. (47)).

4. Waveguide in Anti-de Sitter Space

One might try to extend the spin-one example of the previous section to spin-two and higher in flat spacetime. However, this poses an issue. The free field system we study should be defined in the context of fully interacting theory of higher spin fields. It is by now well known that the higher spin gauge theory is well defined only if it is put on (A)dS space. Therefore, if I want to understand Higgs mechanism of higher spin gauge fields via the waveguide compactification, I must perform the Kaluza-Klein reduction not for flat space (as was previously done in Refs. 17 and 18) but for (A)dS space.

We now turn to waveguide in (A)dS space. Here, we first explain how, starting from AdS_5 space, we can construct a "tunable" AdS_4 waveguide — a waveguide which retains $\mathfrak{so}(3,2)$ isometry within $\mathfrak{so}(4,2)$ and which has a tunable size of internal space.

Consider the AdS_5 in the Poincaré patch with coordinates $(t,\, \mathbf{x}_{d-1},\, y,\, z) \in \mathbb{R}^{1,d} \times \mathbb{R}^+$:

$$\mathrm{d}s(\text{AdS}_5)^2 = \frac{\ell^2}{z^2}\left(-\mathrm{d}t^2 + \mathrm{d}\mathbf{x}_2^2 + \mathrm{d}z^2\right) + \frac{\ell^2}{z^2}\mathrm{d}y^2 = \mathrm{d}s(\text{AdS}_4)^2 + g_{yy}\mathrm{d}y^2. \tag{58}$$

The five-dimensional Poincaré metric is independent of y, and remaining four-dimensional space is again Poincaré patch. Therefore, it appears that this foliation of AdS metric would work well for compactification. Actually, it is not. The reason is as follows. Locally at each y, the isometry $\mathfrak{so}(3,2)$ is part of the original isometry $\mathfrak{so}(4,2)$. However, globally, this does not hold in the Poincaré patch. The reason is that $\mathfrak{so}(3,2)$ isometry transformation does not commute with translation along y direction. Moreover, when compactifying along the y-direction, the five-dimensional tensor does *not* give rise to four-dimensional tensors. Consider, for example, a small fluctuation of the metric. The tensor $\nabla_\mu h_{\nu y}$ is dimensionally reduced to $\nabla_\mu A_\nu + \delta_{\mu z} \frac{1}{z} A_\nu$, where $A_\mu \equiv h_{\mu y}$. The second term is a manifestation of non-tensorial transformation in four dimensions,[b]

In fact, any attempt of compactifying along an isometry direction faces the same difficulties. As such, we shall instead foliate AdS_5 space into a semi-direct product of AdS_4 hypersurface and an angular coordinate θ and Kaluza-Klein compactify along the θ-direction over a finite interval:

$$\mathrm{d}s(\mathrm{AdS}_5)^2 = \frac{1}{\cos^2\theta}\left[\mathrm{d}s(\mathrm{AdS}_4)^2 + \ell^2\,\mathrm{d}\theta^2\right]. \tag{59}$$

Here, the conformal factor arises because we compactified the internal space along a direction which is globally non-isometric. This compactification bypasses the issues that arose in the compactification Eq. (58). In particular, five-dimensional tensors continue to be four-dimensional tensors. For instance, $\nabla_\mu h_{\nu\theta}$ becomes $\nabla_\mu A_\nu - \tan\theta\, h_{\mu\nu} + \tan\theta \frac{1}{\ell^2} g_{\mu\nu}\,\phi$. Under mild assumptions, we were also able to show that the semi-direct product waveguide Eq. (59) is the unique compactification that preserves covariance of tensors.

We can explicitly construct the semi-direct product metric from appropriate foliations of AdS_5 space. We start from Poincaré patch of AdS_5 space and change bulk radial coordinate z and another spatial coordinate y to polar coordinates, $z = \rho\cos\theta$, $y = \rho\sin\theta$.[c] With this parametrization, the AdS_5 space can be represented as a fibration of AdS_4 space over the interval, $\theta \in [-\frac{\pi}{2}, \frac{\pi}{2}]$:

$$\begin{aligned}\mathrm{d}s_5{}^2 &= \frac{\ell^2}{z^2}\left(-\mathrm{d}t^2 + \mathrm{d}\vec{x}^{\,2} + \mathrm{d}y^2 + \mathrm{d}z^2\right) = \frac{\ell^2}{\rho^2\cos^2\theta}\left(-\mathrm{d}t^2 + \mathrm{d}\vec{x}^{\,2} + \mathrm{d}\rho^2 + \rho^2\,\mathrm{d}\theta^2\right)\\ &= \frac{1}{\cos^2\theta}(\mathrm{d}s_4{}^2 + \ell^2\,\mathrm{d}\theta^2).\end{aligned} \tag{60}$$

The boundary of AdS_5 space is at $\theta = \pm\frac{\pi}{2}$. We now construct the AdS waveguide by taking the wedge $-\alpha \le \theta \le \alpha$. See Fig. 1.

An important consequence of compactifying along non-isometry direction is the appearance of the conformal factor $\frac{1}{\cos^2\theta}$. One might try to conceive an alternative compactification scheme based on anti-de Sitter tube by putting periodic boundary condition that identifies the two boundaries at $\theta = \pm\alpha$. This is not possible. The vector ∂_θ is not a Killing vector, so although the metric at hyper-surfaces $\theta = \pm\alpha$ are equal, their first derivatives differ each other. We reiterate that anti-de Sitter waveguide is the unique option for tunable compactification.

[b]The mismatch of the Kaluza-Klein tower with massive fields in AdS space was first pointed out by Metsaev.[19]

[c]The choice of spatial Poincaré direction "y" does not play a special role. It can be chosen from any direction of the $SO(3)/SO(2)$ coset. The semi-direct product structure can be straightforwardly generalized to other descriptions of the (A)dS space.

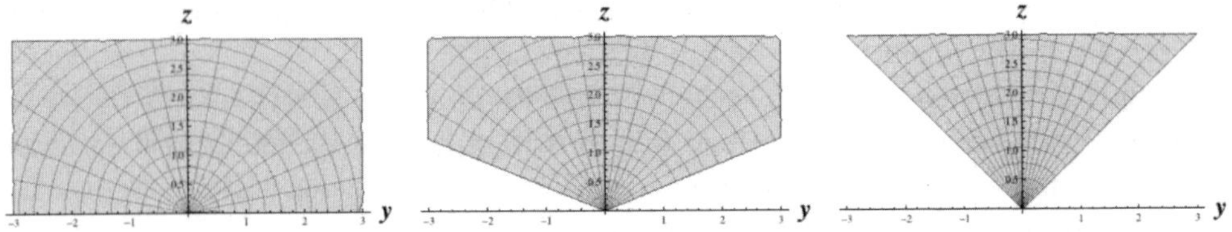

Fig. 1. Anti-de Sitter waveguide: The left depicts a slice of AdS space in Poincaré coordinates (y, z). In polar coordinates (ρ, θ), the AdS boundary is located at $\theta = \pm\pi/2 (z = 0)$. The waveguide is constructed by taking the angular domain $-\alpha \leq \theta \leq +\alpha$ for $\alpha < \pi/2$, as in the middle figure. For example, the waveguide for $\alpha = \pi/4$ is given in the right figure.

Before concluding this section, we introduce the notations that will be extensively used in later sections. We introduce the mode functions as follows:

$$\Theta_n^{s|S}(\theta) = n\text{-th mode function for four-dimensional spin-}s\text{ component that arise from five-dimensional spin-}S\text{ field upon waveguide compactification.} \tag{61}$$

We also introduce the first-order differential operators Ł_n $(n \in \mathbb{Z})$ of Weyl weight n in the Hilbert space $\mathcal{L}^2[-\alpha, \alpha]$ spanned by the above mode functions:

$$\text{Ł}_n = \partial_\theta + n \tan\theta. \tag{62}$$

All of ∂_θ derivatives in the (A)dS waveguide are in the combination of these operators, so we will use Ł_n's to express the Kaluza-Klein equations, gauge transformations and boundary conditions. As we will see, a Sturm-Liouville differential operator acting on spin-s field will carry the Weyl weight $(3 - 2s)$.

Several remarks are in order. Firstly, Metsaev[19] proposed a consideration similar to ours but only to scalar field. Secondly, Vasiliev[20] studied so-called 'dimensional digression' of decomposing AdS_5 higher spin field into towers of AdS_4 higher spin fields. Their idea is not related to ours; in particular, the reduction does not give rise to massless higher-spin fields in lower dimensions. Thirdly, we can generalize the above results to the de Sitter space counterpart. The only technical difference is to replace $(\tan\theta)$ to $(-\tanh\theta)$. The results in de Sitter space are obtainable by changing tangent functions in Eq. (62).

5. AdS Waveguide Spectrum of Spin-One Field

In this section, we focus on the lowest spin field, spin-one, in AdS space and systematically work out Kaluza-Klein compactification. In Sec. 3, we learned that boundary conditions of different polarization fields are related. In our AdS compactification, where the semi-direct product structure is the key feature, the choice of boundary conditions were not clear a priori. Here, we develop methods for identifying proper boundary conditions. This will be extended to higher-spin fields in later sections.

Through lower-spin examples, spin-one in this section and spin-two and spin-three in later sections, we shall compare two methods. One method is using equations of motions, while the other method is using gauge transformations. As the equation of motion contains many terms, the first method is less practical for applying to general higher-spin fields. However, the second method is relatively easy to deal and can be applied to general spin fields. The second method has one more important advantage: the dimensionally reduced equations of motion can be derived by the second method. After compactification, the gauge transformations become Stüeckelberg transformation. The point is that the Stüeckelberg symmetries are as restrictive as the gauge symmetries (because the latter follows from the former, as we will show below), so it completely fixes the equations of motion for all massive higher-spin fields[7] to the same extent that higher-spin gauge symmetries fix the equations of motion of higher-spin gauge fields. Therefore, it suffices to use the gauge transformations for obtaining information about the mass spectra of dimensionally reduced higher-spin fields.

For foregoing analysis, we use the following notations and conventions. The capital letters $M, N, \ldots$ will be used to represent the indices of AdS_5 space: they run from 0 to 4. The greek letters $\mu, \nu, \ldots$ are the indices of AdS_4: they run from 0 to 3. For the waveguide, index for the internal direction is θ. Therefore, $M = \{\mu, \theta\}$. The barred quantities represent tensors in AdS_{d+2} space, while unbarred quantities are tensors of AdS_4 space. The AdS radius is denoted by ℓ.

5.1. *Mode functions of spin-one waveguide*

We first consider the method using the equation of motion. The spin-one field equation in AdS_5 decompose into two polarization components:

$$\sec^2\theta\, \bar{g}^{MN}\, \overline{\nabla}_M\, \bar{F}_{\mu N} = \nabla^\nu\, F_{\mu\nu} - \text{Ł}_{d-2}\,(\text{Ł}_0\, A_\mu - \partial_\mu\, \phi) = 0, \tag{63}$$

$$\sec^2\theta\, \bar{g}^{MN}\, \overline{\nabla}_M\, \bar{F}_{\theta N} = \nabla^\mu\,(\text{Ł}_0\, A_\mu - \partial_\mu\, \phi) = 0, \tag{64}$$

where $\bar{A}_M = (\bar{A}_\mu, \bar{A}_\theta) := (A_\mu, \phi)$. The four-dimensional fields A_μ, ϕ can be expanded in terms of a complete set of mode functions on the interval $\theta \in [-\alpha, \alpha]$:

$$A_\mu = \sum_{n=0}^{\infty} A_\mu^{(n)}\, \Theta_n^{1|1}(\theta) \quad \text{and} \quad \phi = \sum_{n=0}^{\infty} \phi^{(n)}\, \Theta_n^{0|1}(\theta). \tag{65}$$

Mode functions are determined once proper boundary conditions are prescribed. As stated above, our key strategy is not to specify some boundary conditions at the outset but to require gauge invariance of various higher-spin fields and then classify possible boundary conditions that are compatible with such gauge invariances.

What we learned in Sec. 3 is that boundary condition or mode function for A_μ and ϕ must be related such that each term of Eqs. (63) and (64) obey the same boundary condition. Otherwise, as we saw already in Sec. 3, equations of motion are accompanied with independent boundary conditions for each field and there would be no degree of freedom left after dimensional reduction. Therefore, each term of Eqs. (63) and (64) need to be expanded by the same set of mode functions:

$$Ł_1\, Ł_0\, \Theta_n^{1|1} = c_n^{11}\, \Theta_n^{1|1} \quad \text{and} \quad \begin{pmatrix} 0 & Ł_1 \\ Ł_0 & 0 \end{pmatrix} \begin{pmatrix} \Theta_n^{1|1} \\ \Theta_n^{0|1} \end{pmatrix} = \begin{pmatrix} c_n^{01}\Theta_n^{1|1} \\ c_n^{10}\Theta_n^{0|1} \end{pmatrix}. \tag{66}$$

Here, c_n's are complex-valued coefficients. Equation (63) indicates that the coefficient $-c_n^{11}$ is the mass-squared of spin-one field in four dimensions. These equations display the Sturm-Liouville(SL) problem at hand involves factorized differential operators, leading to

$$Ł_1\big(Ł_0\, \Theta_n^{1|1}\big) = c_n^{10}\, c_n^{01}\, \Theta_n^{1|1} \quad \text{such that } c_n^{11} = c_n^{10} c_n^{01} \tag{67}$$

and also

$$Ł_0\big(Ł_{d-2}\, \Theta_n^{0|1}\big) = c_n^{01}\, Ł_0\, \Theta_n^{1|1} = c_n^{01}\, c_n^{10}\, \Theta_n^{0|1} \quad \text{and hence } c_n^{00} = c_n^{11}. \tag{68}$$

By these relations, the field equations become

$$\begin{aligned} \sum_n \Big[\nabla^\mu F_{\mu\nu}^{(n)} + c_n^{01}(c_n^{10}\, A_\nu^{(n)} - \partial_\nu\, \phi^{(n)})\Big]\, \Theta_n^{1|1} &= 0, \\ \sum_n \nabla^\mu \Big[c_n^{10}\, A_\mu^{(n)} - \partial_\mu \phi^{(n)}\Big]\, \Theta_n^{0|1} &= 0. \end{aligned} \tag{69}$$

From these equations, we draw two important results. Firstly, the n-th Kaluza-Klein mode of spin-one and spin-zero fields are related to each other so as to have their eigenvalues equal

$$c_n^{11} = c_n^{10} c_n^{01} = c_n^{00} = -M_n^2. \tag{70}$$

Secondly, these spin-one and spin-zero states combine together to trigger the Stüeckelberg mechanism. Note that central to this conclusion is the factorization property that the Sturm-Liouville second-order differential operator determining the Kaluza-Klein spectrum is a product of two first-order differential operators:

$$Ł_1 \;\; \underset{\Theta_n^{0|1}}{\overset{\Theta_n^{1|1}}{\upharpoonleft\!\downharpoonright}} \;\; Ł_0 \quad \text{with } -M_n^2 = c_n^{11} = c_n^{00} = c_n^{10} c_n^{01}. \tag{71}$$

This shows that Ł_1, Ł_0 are raising and lowering operators between spin-one and spin-zero polarization components. As such, we refer to the above relations as "spectrum generating complex" for spin-one field in AdS_5 space. Using the factorization property, we can study spectrum in AdS_4 space. For nonzero M_n, the spectrum generating complex shows that there are two modes: spin-one and spin-zero. This pairing is precisely the structure needed for the Stüeckelberg system that realizes the Higgs mechanism. The spin-zero field provides the Goldstone mode to the massive spin-one (Proca) field when the mass of spin-one is zero. A novelty of this system is that the scalar field, despite being a Goldstone mode, is massive. There are two special cases of the Stüeckelberg system which lead to massless modes:

$$\begin{aligned} c_0^{10} &= 0: \text{ no spin-zero mode, massless spin-one} \\ c_0^{01} &= 0: \text{ no spin-one mode, massless spin-zero.} \end{aligned}$$

Each case can be realized as the lowest Kaluza-Klein mode of specific boundary conditions. Attentive reader might have noticed the same underlying mathematical structure as the supersymmetry:

$$\begin{aligned} c_0^{10} &= 0: \text{ dim Ker } \text{Ł}_1 \neq 0 \rightarrow \text{ Dirichlet BC on } \Theta^{1|1} \\ c_0^{01} &= 0: \text{ dim Ker } \text{Ł}_0 \neq 0 \rightarrow \text{Neumann BC on } \Theta^{1|1}. \end{aligned} \tag{72}$$

Note that the modes which belong to the image of each Ł-operators always combine together and form massive modes via the Stüeckelberg mechanism.

We next consider the method using the gauge transformation. In the waveguide, only those gauge transformations that do not change the boundary condition would make sense, viz. gauge fields and gauge transformation parameters should obey the same boundary conditions and hence the same mode functions. So, we have

$$\sum_n \delta A_\mu^{(n)}\, \Theta_n^{1|1}(\theta) = \sum_n \partial_\mu\, \Lambda^{(n)}\, \Theta_n^{1|1}(\theta), \tag{73}$$

$$\sum_n \delta\phi^{(n)}\, \Theta_n^{0|1}(\theta) = \sum_n \text{Ł}_0\, \Lambda^{(n)}\, \Theta_n^{1|1}(\theta) = \sum_n c_n^{10}\, \Lambda^{(n)}\, \Theta_n^{0|1}(\theta). \tag{74}$$

Importantly, we note that the mode function relations Eq. (71), where it was obtained by the method using the equation of motion, can now be derived by the variations Eq. (74) and the Sturm-Liouville equation, Eq. (66).

Putting together field equations and gauge transformations of n-th Kaluza-Klein modes, we get

$$\begin{aligned} &\nabla^\mu F_{\mu\nu}^{(n)} + c_n^{01}\left[c_n^{10}\, A_\nu^{(n)} - \partial_\nu\, \phi^{(n)}\right] = 0 \\ &\nabla^\mu\left[c_n^{10}\, A_\mu^{(n)} - \partial_\mu \phi^{(n)}\right] = 0 \\ &\delta A_\mu^{(n)} = \partial_\mu\, \Lambda^{(n)} \\ &\delta\phi^{(n)} = c_n^{10}\, \Lambda^{(n)}. \end{aligned} \tag{75}$$

We recognize these equations as the Stüeckelberg equations of motion and Stüeckelberg gauge transformations that describe a massive spin-one gauge field (Proca field) in AdS_4 space. Comparing them with the standard form of Stüeckelberg system, we also identify the coefficients c_n's with mass of the Proca field, $c_n^{10} = -c_n^{01} = M_n$.

5.2. *Waveguide boundary conditions for spin-one field*

Having identified structure of mode functions as well as raising and lowering operators mapping among them, we are now ready for examine boundary conditions and modes. To simplify and systematize the analysis, we shall first concentrate on boundary conditions which do not contain derivatives higher than first-order.[d] In this case, all possible boundary conditions is identifiable with all possible choice of mode functions to nontrivial kernel of raising and lowering operators, as in this case these zero modes trivially satisfy the Dirichlet boundary condition. So, we have two possibilities:

$$\begin{cases} \Theta^{1|1}\,|_{\theta=\pm\alpha} = 0, \quad Ł_1\,\Theta^{0|1}\,|_{\theta=\pm\alpha} = 0 & \text{Dirichlet} \\ Ł_0\,\Theta^{1|1}\,|_{\theta=\pm\alpha} = 0, \; \Theta^{0|1}\,|_{\theta=\pm\alpha} = 0 & \text{Neumann.} \end{cases} \tag{76}$$

The first case corresponds to the situation that $\Theta^{0|1}$ is a zero mode, and the second case corresponds to the situation that $\Theta^{1|1}$ is a zero mode.

For each of the above two boundary conditions, the mass spectrum is determined by the Sturm-Liouville problem Eq. (68). We emphasize again that the above choice of boundary conditions put all Kaluza-Klein modes form automatically the Stüeckelberg system. The massless spin-one and spin-zero fields arise from the ground mode. The ground mode of Neumann boundary condition, $\Theta_0^{1|1}(\theta) =$ (constant) has vanishing mass, $M_0 = 0$, for spin-one field and there is no spin-zero ground mode. The ground mode of Dirichlet boundary condition, $\Theta_0^{0|1}(\theta) \propto (\sec\theta)^{2-d}$, has vanishing mass, $M_0 = 0$, for spin-zero field and there is no spin-one ground mode. So, we see that the two possible boundary conditions Eq. (76) are precisely the AdS counterparts of "vector" and "scalar" boundary conditions for flat space waveguide studied in Sec. 3. We summarize the spectrum of each boundary conditions in Fig. 2.

Our result for the ground modes, which comprises of massless spin-one or spin-zero fields, fits perfectly to the $\mathfrak{so}(3,2)$ representation theory of AdS_4 space. The set of normalizable solutions to the free field equation form a $\mathfrak{so}(3,2)$-module. Consider an irreducible representation $\mathcal{D}(\Delta, s)$ of $\mathfrak{so}(3,2)$. The conformal weight Δ (the Casimir of $\mathfrak{so}(2)$ subalgebra) is related to the mass-squared of spin-s field by

$$m^2_{\text{spin}-0}\,\ell^2 = \Delta\,(\Delta - d), \quad m^2_{\text{spin}-1}\,\ell^2 = \Delta\,(\Delta - d) + (d-1)\,. \tag{77}$$

[d] For $s \geq 2$, boundary conditions necessarily involve higher derivative terms in order to accommodate all possible mass spectra of higher-spin fields.

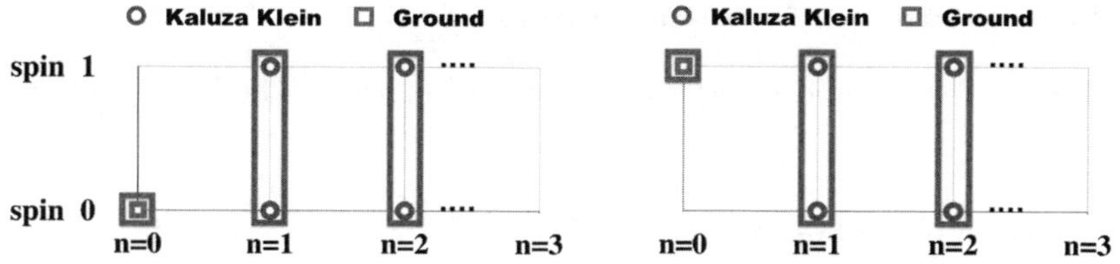

Fig. 2. Mass spectrum (M_n, s) for spin-one field in AdS waveguide. The left is for Dirichlet boundary condition to spin-one component, and the right is for Dirichlet boundary condition to spin-zero component. The red squares are the ground modes, while the blue circles are the Kaluza-Klein modes. Circles inside the same rectangle have the same eigenvalue and form a Stüeckelberg system of massive spin-one field.

The ground modes of Dirichlet and Neumann boundary conditions are massless spin-zero and spin-one fields, respectively. We see from Eq. (77) that each of them corresponds to the $\mathfrak{so}(3,2)$ representations, $\mathcal{D}(3,\,0)$ and $\mathcal{D}(2,\,1)$, respectively. Moreover, both are irreducible parts of the reducible Verma $\mathfrak{so}(3,2)$-module $\mathcal{V}(2,\,1)$, viz.

$$\mathcal{V}(2,\,1) = \underbrace{\mathcal{D}(2,\,1)}_{\text{Neumann}} \oplus \underbrace{\mathcal{D}(3,\,0)}_{\text{Dirichlet}}. \tag{78}$$

The pattern that ground mode comes from the irreducible representations of reducible Verma module continues to hold for higher-spin fields as well, and is an integral part of our main results in this paper.

Summarizing, from Kaluz-Klein compactification of spin-one field in AdS_5 space, we take following lessons.

- The mode functions of different spins in AdS_4 space are related to each other, which permits Stüeckelberg structure. For spin-one, this relation is shown in Eq. (71).
- It is known that free part of Stüeckelberg equation and action are uniquely determined by Stüeckelberg gauge transformations. Therefore, we could derive the lower-dimensional equations of motion just from consideration of the lower-dimensional gauge transformations.

6. Waveguide Spectrum of Spin-Two Field

In this section, we extend the analysis to spin-two field in AdS waveguide. The idea is basically the same as the spin-one case, but the result becomes more interesting for the ground modes. We shall present the analysis as closely parallel to the spin-one as possible and highlight salient differences that begin to show up for spin two and higher.

6.1. *Mode functions of spin-two waveguide*

We begin with the method using the equation of motion. The Pauli-Fierz equation of motion for a massive spin-two field in AdS_5 space is given by

$$\mathcal{K}_{MN}(\bar{h}) - 4\,(2\,\bar{h}_{MN} - \bar{g}_{MN}\,\bar{h}) - M^2\,(\bar{h}_{MN} - \bar{g}_{MN}\,\bar{h}) = 0, \tag{79}$$

where M^2 is the mass-squared, $\bar{g}_{MN}$ is the metric of AdS_5 space, and $\mathcal{K}_{MN}(\bar{h})$ is the spin-two Lichnerowicz operator:

$$\begin{aligned}\mathcal{K}_{MN}(\bar{h}) = \Box\,\bar{h}_{MN} - (\overline{\nabla}^L\,\overline{\nabla}_N\,\bar{h}_{ML} + \overline{\nabla}^L\,\overline{\nabla}_M\,\bar{h}_{NL}) + \bar{g}_{MN}\,\overline{\nabla}_K\,\overline{\nabla}_L\,\bar{h}^{KL} \\ + \overline{\nabla}_M\,\overline{\nabla}_N\,\bar{h} - \bar{g}_{MN}\,\overline{\Box}\,\bar{h},\end{aligned} \tag{80}$$

where ($\bar{h}$ denotes for the trace part, $\bar{g}^{MN}\bar{h}_{MN}$). After the compactification, the five-dimensional spin-two field is decomposed to four-dimensional spin-two, spin-one, and spin-zero component fields:

$$h_{\mu\nu} = \bar{h}_{\mu\nu} + \frac{1}{2}\,g_{\mu\nu}\,\bar{h}_{\theta\theta}, \quad \bar{h}_{\mu\theta} = A_\mu, \quad \bar{h}_{\theta\theta} = \phi. \tag{81}$$

Note that the spin-two field $h_{\mu\nu}$ is defined by the linear combination of $\bar{h}_{\mu\nu}$ and $\bar{h}_{\theta\theta}$.[e]

The massless spin-two equation of motion in AdS_5 space decomposes into equations of motion for component fields ($h_{\mu\nu}$, A_μ, ϕ) in AdS_4 space ($d = 3$):

$$\begin{aligned}&\mathcal{K}_{\mu\nu}(h) - 3\,(2\,h_{\mu\nu} - g_{\mu\nu}\,h) + \text{Ł}_1\,\text{Ł}_{-2}\,(h_{\mu\nu} - g_{\mu\nu}\,h) \\ &\qquad - \text{Ł}_1\,(\nabla_\mu\,A_\nu + \nabla_\nu\,A_\mu - 2\,g_{\mu\nu}\,\nabla^\rho\,A_\rho) + \frac{3}{2}\,g_{\mu\nu}\,\text{Ł}_1\,\text{Ł}_0\,\phi = 0,\end{aligned} \tag{82}$$

$$\nabla^\mu\,F_{\mu\nu} - 6\,A_\nu - \text{Ł}_{-2}\,(\nabla^\mu\,h_{\mu\nu} - \nabla_\nu\,h) - \frac{3}{2}\,\text{Ł}_0\,\nabla_\nu\,\phi = 0, \tag{83}$$

$$\Box\,\phi - (2\,\text{Ł}_{-1}\,\text{Ł}_0 + 4)\,\phi - 2\,\text{Ł}_{-1}\,\nabla^\mu\,A_\mu + \text{Ł}_{-1}\,\text{Ł}_{-2}\,h = 0\,, \tag{84}$$

where h is the trace part, $g^{\mu\nu}h_{\mu\nu}$. The mode expansion of four-dimensional spin-two, spin-one and spin-zero component fields reads

$$\begin{aligned}h_{\mu\nu} = \sum_n h^{(n)}{}_{\mu\nu}\,\Theta_n^{2|2}(\theta), \quad A_\mu = \sum_n A^{(n)}{}_\mu\,\Theta_n^{1|2}(\theta), \\ \phi = \sum_n \phi^{(n)}\,\Theta_n^{0|2}(\theta).\end{aligned} \tag{85}$$

[e] The equations of motion have cross terms between $\bar{h}$ and $\nabla^2\phi$. This linear combination removes these cross terms. This specific combination is also the linear part of diagonalized metric in the original Kaluza-Klein compactification, $\bar{g}_{\mu\nu} = e^{\phi/2}\,g_{\mu\nu}$.

From Eqs. (82), (83) and (84), we can expect the relations between mode-functions which can be summarized by following two matrix equations:

$$\begin{pmatrix} 0 & \text{Ł}_1 \\ \text{Ł}_{-2} & 0 \end{pmatrix} \begin{pmatrix} \Theta_n^{2|2} \\ \Theta_n^{1|2} \end{pmatrix} = \begin{pmatrix} c_n^{12}\Theta_n^{2|2} \\ c_n^{21}\Theta_n^{1|2} \end{pmatrix} \tag{86}$$

$$\begin{pmatrix} 0 & \text{Ł}_0 \\ \text{Ł}_{-1} & 0 \end{pmatrix} \begin{pmatrix} \Theta_n^{1|2} \\ \Theta_n^{0|2} \end{pmatrix} = \begin{pmatrix} c_n^{01}\Theta_n^{1|2} \\ c_n^{10}\Theta_n^{0|2} \end{pmatrix} \tag{87}$$

where c_n's are coefficients. We now have two sets of raising and lowering operators, connecting spin-zero and spin-one and spin-one and spin-two, respectively. Accordingly, we have two pairs of Sturm-Liouville problems. Equation (86) leads to the first set of Sturm-Liouville problems for spin-two and spin-one, respectively:

$$\begin{aligned} \text{Ł}_1\,\text{Ł}_{-2}\,\Theta_n^{2|2} &= c_n^{21}\,c_n^{12}\,\Theta_n^{2|2} = -M_n^2\,\Theta_n^{1|2}, \\ \text{Ł}_{-2}\,\text{Ł}_1\,\Theta_n^{1|2} &= c_n^{12}\,c_n^{21}\,\Theta_n^{1|2} = -M_n^2\,\Theta_n^{1|2}. \end{aligned} \tag{88}$$

The Eq. (87) leads to the second set of Sturm-Liouville problems for spin-one and spin-zero, respectively:

$$\begin{aligned} \text{Ł}_0\,\text{Ł}_{-1}\,\Theta_n^{1|2} &= c_n^{10}\,c_n^{01}\,\Theta_n^{1|2}, \\ \text{Ł}_{-1}\,\text{Ł}_0\,\Theta_n^{0|2} &= c_n^{01}\,c_n^{10}\,\Theta_n^{0|2}. \end{aligned} \tag{89}$$

The two sets of equations appear overdetermined, as the spin-one mode function $\Theta_n^{1|2}$ is the eigenfunction of two separate Sturm-Liouville problems. However, it can be shown that the two Sturm-Liouville problems are actually one and the same problem by using the identity

$$\text{Ł}_m\,\text{Ł}_n - \text{Ł}_{n-1}\,\text{Ł}_{m+1} = (n - m - 1). \tag{90}$$

This also leads to eigenvalues relations

$$c_n^{10}c_n^{01} = c_n^{21}c_n^{12} - 2. \tag{91}$$

So, the Sturm-Liouville problems can be summarized by the relations

$$\begin{array}{ccl} & \Theta_n^{2|2} & \\ \text{Ł}_1 & \uparrow\downarrow\ \text{Ł}_{-2} & : \quad -M_{n,2|2}^2 = -M_n^2 = c_n^{21}c_n^{12} \\ & \Theta_n^{1|2} & \\ \text{Ł}_0 & \uparrow\downarrow\ \text{Ł}_{-1} & : \quad -M_{n,1|2}^2 = -(M_n^2 + 2) = c_n^{10}c_n^{01} \\ & \Theta_n^{0|2} & \end{array} \tag{92}$$

We note that these relations, defined by raising and lowering operators between four-dimensional fields of adjacent spins, is precisely the structure required for Stüeckelberg mechanism.[f] If $M_{n,\,2|2}$ and $M_{n,\,1|2}$ were nonzero, the corresponding modes among different spin fields combine and become the Stüeckelberg spin-two system. There are two special cases, vanishing $M_{n,\,2|2}$ or vanishing $M_{n,\,1|2}$. As these are important exceptional situations, leading to so-called partially massless spin-two fields, we will analyze them separately in Subsec. 6.2 with examples.

We can also obtain Eq. (92) from the method using gauge transformations. The gauge transformations in AdS$_4$ space, with the gauge parameter $\bar{\xi}_M = \{\xi_\mu,\, \xi_\theta\}$, are decomposed into components

$$\begin{aligned} \delta h_{\mu\nu} &= \nabla_{(\mu}\, \xi_{\nu)} + \frac{1}{2}\, g_{\mu\nu}\, \mathrm{Ł}_{d-2}\, \xi_\theta, \\ \delta A_\mu &= \frac{1}{2}\, \partial_\mu\, \xi_\theta + \frac{1}{2}\, \mathrm{Ł}_{-2}\, \xi_\mu, \\ \delta\phi &= \mathrm{Ł}_{-1}\, \xi_\theta. \end{aligned} \tag{93}$$

Again, to retain the gauge invariances, the mode functions of gauge parameter are set proportional to mode functions of the fields:

$$\xi_\mu = \sum_n \xi_\mu^{(n)}\, \Theta_n^{2|2}(\theta), \quad \xi_\theta = \sum_n \xi_\theta^{(n)}\, \Theta_n^{1|2}(\theta). \tag{94}$$

By substituting these to Eq. (93) and comparing mode expansion terms in the gauge variations, we see we can recover precisely the same raising and lowering operators as in Eq. (92), which was previously derived from the field equations Eqs. (82), (83) and (84).

After the mode expansion, the component field equations read

$$\begin{aligned} &\mathcal{K}_{\mu\nu}(h^{(n)}) - 3\left[2\, h^{(n)}{}_{\mu\nu} - g_{\mu\nu}\, h^{(n)}\right] + c_n^{21}\, c_n^{12}\left[h^{(n)}{}_{\mu\nu} - g_{\mu\nu} h^{(n)}\right] \\ &\qquad - c_n^{12}\left[\nabla_\mu\, A^{(n)}{}_\nu + \nabla_\nu\, A^{(n)}{}_\mu - 2\, g_{\mu\nu}\, \nabla^\rho\, A^{(n)}{}_\rho\right] + c_n^{01}\, c_n^{12}\, \frac{3}{2}\, g_{\mu\nu}\, \phi^{(n)} = 0, \end{aligned} \tag{95}$$

$$\nabla^\mu\, F^{(n)}{}_{\mu\nu} - 6\, A^{(n)}{}_\nu - c_n^{21}\, \nabla^\mu\left[h^{(n)}{}_{\mu\nu} - g_{\mu\nu}\, h^{(n)}\right] - c_n^{01}\, \frac{3}{2}\, \nabla_\nu\, \phi^{(n)} = 0, \tag{96}$$

$$\Box\, \phi^{(n)} - \left[2\, c_n^{01}\, c_n^{10} + 4\right]\phi^{(n)} - 2\, c_n^{10}\, \nabla^\mu\, A^{(n)}{}_\mu + c_n^{21}\, c_n^{10}\, h^{(n)} = 0. \tag{97}$$

Their gauge transformations read

$$\begin{aligned} \delta h_{\mu\nu}^{(n)} &= \nabla_{(\mu}\, \xi_{\nu)}^{(n)} + \frac{c_n^{12}}{2}\, g_{\mu\nu}\, \xi^{(n)}, \\ \delta A_\mu^{(n)} &= \frac{1}{2}\, \partial_\mu\, \xi^{(n)} + \frac{c_n^{21}}{2}\, \xi_\mu^{(n)}, \quad \delta\phi^{(n)} = c_n^{10}\, \xi^{(n)}. \end{aligned} \tag{98}$$

[f]Note, however, $M_{n,1|2}$ is not related with mass-like term of spin-one field in Eq. (83). $M_n = M_{n,2|2}$ is the mass of the spin-two field in the Eq. (82).

We see that this system, Eqs. (95), (96), (97), (98), coincides precisely with the spin-two Stüeckelberg system on AdS_4 space, once we redefine c_n's as

$$c_n^{12} = -\sqrt{2}M_n, \quad c_n^{21} = \frac{M_n}{\sqrt{2}}, \quad c_n^{01} = -\sqrt{\frac{3}{4}({M_n}^2+2)}, \quad c_n^{10} = \sqrt{\frac{4}{3}({M_n}^2+2)}.$$

It is also known that the Stüeckelberg gauge symmetries can uniquely fix free parts in the field equations or equivalently in the action. Therefore, from the knowledge of the gauge transformations Eq. (98), we can fully reconstruct the field equations Eqs. (95), (96) and (97). In practice, the gauge transformations are much simpler to handle than the field equations. Note that the modes which are neither in the kernel of raising operators nor in the kernel of lowering operators always combine together and undergo the Stüeckelberg mechanism for massive spin-two fields.

Before classifying possible boundary conditions, we summarize Stüeckelberg spin-two system and Goldstone mode decomposition pattern of it. For general values of the masses, Stüeckelberg spin-two system describes the same physical degree of freedom as a massive spin-two field (having maximal number of longitudinal polarizations). This is because spin-one and spin-zero fields can be algebraically removed by the gauge symmetries Eq. (98), corresponding to the unitary gauge fixing. However, such gauge fixing is not possible were if the masses take special values:

$$M_n^2 = 0 \quad \text{and} \quad M_n^2 = -\frac{2}{\ell^2}. \tag{99}$$

At these special values of the mass parameters, the Stüeckelberg system breaks into subsystems which can be deduced just from the gauge transformation.

For the situation that $M_n = 0$, the gauge transformations are

$$\delta\, h_{\mu\nu} = \nabla_{(\mu}\, \xi_{\nu)}, \quad \delta\, A_\mu = \frac{1}{2}\, \partial_\mu\, \xi, \quad \delta\, \phi = \frac{1}{\ell}\sqrt{\frac{8}{3}}\, \xi. \tag{100}$$

We see from the first equation that the spin-two field ought to be massless as it has the spin-two gauge symmetry. We also see that the remaining two equations are precisely the spin-one Stüeckelberg system with $m^2 = 6/\ell^2$. This implies that the Goldstone field of the massive spin-two is given by the massive spin-one system, which in turn was formed by the Stüeckelberg system of massless spin-one and massless spin-zero fields. It should be noted that the normalization of each field is not standard.

For the situation that $M_n^2 = -2/\ell^2$, a subtlety arises as the coefficients c_n^{12} and c_n^{21} are pure imaginary. Specifically, the relation Eq. (86) implies that one of the two mode functions $\Theta_n^{1|2}$, $\Theta_n^{2|2}$ and corresponding field become pure imaginary. We are thus led to redefine the mode functions $\tilde{\Theta}_n^{1|2} = \pm i\, \Theta_n^{1|2}$ and the fields $\tilde{A}_\mu = \pm i\, A_\mu$.[g]

[g]In the path integral formulation, this amounts to choosing that the integration contour purely imaginary.

The gauge transformations now become

$$\delta h_{\mu\nu} = \nabla_{(\mu}\,\xi_{\nu)} + \frac{1}{\ell}\,g_{\mu\nu}\,\xi, \quad \delta A_\mu = \frac{1}{2}\,\partial_\mu\,\xi + \frac{1}{2\,\ell}\,\xi_\mu, \quad \delta\phi = 0. \tag{101}$$

The spin-two gauge transformations and spin-one gauge transformations are coupled each other. In fact, they are precisely the Stüeckelberg system of partially massless (PM) spin-two field.[7] We can always gauge-fix the spin-one field to zero, and the remanent gauge symmetry coincides with the partially-massless (PM) spin-two gauge symmetry:[7,8]

$$\delta h_{\mu\nu} = \nabla_\mu\,\nabla_\nu\,\lambda - \frac{1}{\ell^2}\,g_{\mu\nu}\,\lambda, \quad \text{where } \lambda = \ell\,\xi. \tag{102}$$

Therefore, when the mass-squared hits the special value $M_n^2 = -2/\ell^2$, the Stüeckelberg system breaks into a spin-two partially-massless (PM) Stüeckelberg system and a massive spin-zero field of mass-squared $m^2 = 4/\ell^2$, as given above in Eq. (97).

This spectrum decomposition pattern perfectly fits to the reducibility structure of the Verma $\mathfrak{so}(3,2)$-module $\mathcal{V}(\Delta, 2)$ for spin-two field. For the special values of conformal weights, $\Delta = 3$ and $\Delta = 2$, the Verma module becomes reducible and break into

$$\begin{aligned}
\mathcal{V}\,(3,\,2) &= \underbrace{\mathcal{D}\,(3,\,2)}_{\text{massless } s=2} \;\oplus\; \underbrace{\mathcal{D}\,(4,\,1)}_{\text{massive } s=1}\,,\\
\mathcal{V}\,(2,\,2) &= \underbrace{\mathcal{D}\,(2,\,2)}_{\text{PM } s=2} \oplus\; \underbrace{\mathcal{D}\,(4,\,0)}_{\text{massive } s=0}\,.
\end{aligned} \tag{103}$$

Here, $\mathcal{D}\,(3,\,2)$ and $\mathcal{D}\,(2,\,2)$ are irreducible representations of massless and partially massless states, respectively. Using the relation between the mass-squared and the conformal weights[h]

$$m^2_{\text{spin}-1}\,\ell^2 = \Delta\,(\Delta - 3) + 2 \quad \text{and} \quad m^2_{\text{spin}-0,\,2}\,\ell^2 = \Delta\,(\Delta - 3), \tag{104}$$

one finds that $\mathcal{D}\,(4,\,1)$ corresponds to spin-one field with $m^2 = 2\,d/\ell^2$, and $\mathcal{D}\,(4,\,0)$ corresponds to spin-zero field with $m^2 = (4)\,/\ell^2$. This result exactly matches with the spectrum decomposition patterns we analyzed above.

Here, we tabulate the four types of fields that appear at special values of masses (the four irreducible representations that appear in Eqs. (103), as they will be shown to arise as the ground modes of the Sturm-Liouville problems with appropriate boundary conditions in Subsec. 6.2.

[h]We define the mass-squared equal to the mass-squared in flat space limit. Therefore, it differs from the mass-squared values dictated by the Fierz-Pauli equations.

Table 1. The types of field involved in the inverse Higgs mechanism when spin-two Stüeckelberg systems decompose into spin-two gauge field and Goldstone field. Type I and II are Goldstone fields of spin-zero and spin-one. In AdS_4 space, these Goldstone fields are massive. Type III is massless, spin-two gauge field. Type IV is partially massless, spin-two gauge field.

Type	$\mathcal{D}(\Delta, s)_{\mathfrak{so}(d,2)}$	Field	Mass-squared
Type I	$\mathcal{D}(4,1)$	Massive Stüeckelberg spin-one	$m^2 = 6/\ell^2$
Type II	$\mathcal{D}(4,0)$	Massive spin-zero field	$m^2 = 4/\ell^2$
Type III	$\mathcal{D}(3,2)$	Massless spin-two	$m^2 = 0$
Type IV	$\mathcal{D}(2,2)$	Partially-massless Stüeckelberg spin-two	$m^2 = -2/\ell^2$

6.2. *Waveguide boundary conditions for spin-two field*

With mode expansions at hand, we now classify possible boundary conditions. In the spin-one situation, boundary conditions of different component fields (spin-one and spin-zero in that case) were related. This property continues to hold for spin-two situation. For instance, suppose we impose Dirichlet boundary condition for the spin-one component field in AdS_4, $\Theta^{1|2}|_{\theta=\pm\alpha} = 0$. Then, the spectrum generating complex Eq. (92) immediately imposes unique boundary conditions for other component fields:

$$\begin{aligned} Ł_{-2}\,\Theta^{2|2}_n &\sim \Theta^{1|2}_n, \qquad Ł_{-2}\,\Theta^{2|2}|_{\theta=\pm\alpha} = 0, \\ Ł_0\,\Theta^{0|2}_n &\sim \Theta^{1|2}_n, \qquad Ł_0\,\Theta^{0|2}|_{\theta=\pm\alpha} = 0. \end{aligned} \tag{105}$$

Likewise, if we impose a boundary condition to a component field, the spectrum generating complex Eq. (92) uniquely fixes boundary conditions for all other component fields. The minimal choice is imposing the Dirichlet boundary condition to one of the component fields. As there are $s + 1 = 3$ component fields (spin-two, spin-one and spin-zero), there are three possible minimal boundary conditions:[i]

$$\begin{aligned} &\textbf{B.C. 1:} \quad \{ \quad \Theta^{2|2}| = 0, \qquad Ł_1\,\Theta^{1|2}| = 0, \quad Ł_1\,Ł_0\,\Theta^{0|2}| = 0 \ \} \\ &\textbf{B.C. 2:} \quad \{ \ Ł_{-2}\,\Theta^{2|2}| = 0, \qquad \Theta^{1|2}| = 0, \qquad Ł_0\,\Theta^{0|2}| = 0 \ \}, \\ &\textbf{B.C. 3:} \quad \{ Ł_{-1}\,Ł_{-2}\,\Theta^{2|2}| = 0, \ Ł_{-1}\,\Theta^{1|2}| = 0, \qquad \Theta^{0|2}| = 0 \quad \} \end{aligned} \tag{106}$$

where $\Theta|$ is a shorthand notation for the boundary values, $\Theta|_{\theta=\pm\alpha}$. We reiterate that the boundary conditions on each set are automatically fixed by the spectrum generating complex Eq. (92). We now examine mass spectra and mode functions for each of the three types of boundary conditions, Eq. (106).

[i]Note that the first and the third conditions are higher-derivative boundary conditions (HD BC). HD BC is not self-adjoint in the functional space L^2, but is self-adjoint in expanded functional space.

To deliver our exposition clear and explicit, we shall perform the analysis for compactification of AdS_4 to AdS_3 times angular wedge, where the mode solutions of the Strum-Liouville prpblem, Eq. (92), are elementary:

$$\Theta^{2|2} = \begin{cases} \sec\theta\,(\tan\theta\,\cos(z_n\theta) - z_n\,\sin(z_n\theta)), & \text{odd parity} \\ \sec\theta\,(\tan\theta\,\sin(z_n\theta) + z_n\,\cos(z_n\theta)), & \text{even parity} \end{cases} \tag{107}$$

$$\Theta^{1|2} = \begin{cases} \sec\theta\,\sin(z_n\theta), & \text{odd parity} \\ \sec\theta\,\cos(z_n\theta), & \text{even parity} \end{cases} \tag{108}$$

$$\Theta^{0|2} = \begin{cases} \sec\theta\,\sin(z_n\theta), & \text{odd parity} \\ \sec\theta\,\cos(z_n\theta), & \text{even parity} \end{cases} \tag{109}$$

with $z_n^2 = M_n^2 + 1$. Note that the Sturm-Liouville equation and the boundary condition are symmetric under the parity $\theta \to -\theta$, so the solutions are also labelled as either odd or even parity of θ.

We begin our analysis with **B.C. 1**. Substituting the above mode functions to the **B.C. 1**, we get the same expression for spin-two and spin-one component fields except the condition that the parity of the mode functions must take opposite values:

$$\begin{cases} \sec\theta\,(\tan\theta\,\cos(z_n\theta) - z_n\,\sin(z_n\theta))\,|_{\theta=\pm\alpha}, & \text{odd } \Theta^{2|2} \text{ and even } \Theta^{1|2} \\ \sec\theta\,(\tan\theta\,\sin(z_n\theta) + z_n\,\cos(z_n\theta))\,|_{\theta=\pm\alpha}, & \text{even } \Theta^{2|2} \text{ and odd } \Theta^{1|2} \end{cases}. \tag{110}$$

We also get the boundary condition for spin-zero component $\Theta^{0|2}$ as

$$\begin{cases} z_n\,\sec\theta\,(\tan\theta\,\cos(z_n\theta) - z_n\,\sin(z_n\theta))\,|_{\theta=\pm\alpha}, & \text{odd } \Theta^{(0|2)} \\ z_n\,\sec\theta\,(\tan\theta\,\sin(z_n\theta) + z_n\,\cos(z_n\theta))\,|_{\theta=\pm\alpha}, & \text{even } \Theta^{(0|2)} \end{cases}. \tag{111}$$

We note that, modulo the overall spectral factor z_n, this spin-zero boundary condition is the same as the boundary conditions Eq. (110). This agreement is not accidental. Once again, they are consequences of the spectrum generating complex Eq. (92) and the boundary condition Eq. (105).

In general, solutions of each boundary condition, z_n, depend on the domain of angular wedge, α. They are the AdS-counterpart of flat space Kaluza-Klein compactification, and so z_n and M_n blow up as α is sent to zero. They correspond to the "Kaluza-Klein modes". For these modes, mode functions of each component spin fields combine and form spin-two Stüeckelberg system with mass-squared, $M_n^2 = z_n^2 - 1$.

There are, however, two special solutions that are independent of α, $z_n = 1$ and $z_n = 0$. They correspond to "ground modes" and have interesting features that are not shared with the Kaluza-Klein modes. Firstly, masses of the ground modes are equal to the special masses Eq. (99) at which the unitary gauge-fixing ceases to work and the Stüeckelberg system decomposes into subsystems. Secondly, mode function of some spin components are absent. For $z_n = 1$, the spin-two field is absent as $\Theta^{2|2} = 0$ in this case. The spin-one and spin-zero fields combine and form the Stüeckelberg spin-one system of **Type I**. For $z_n = 0$, only massive spin-zero field is present because $z_n = 0$ is not a solution of boundary conditions Eq. (110) or corresponding mode function is 0. This spin-zero field is of **Type II**.

By completing the analysis to other boundary conditions, we find the following spectrums of ground modes:

$$\begin{array}{l} \textbf{B.C. 1: Type I} \quad \text{and} \quad \textbf{Type II} \\ \textbf{B.C. 2: Type II} \quad \text{and} \quad \textbf{Type III} \\ \textbf{B.C. 3: Type III} \quad \text{and} \quad \textbf{Type IV} \end{array} \tag{112}$$

We see that **B.C. 1** keeps mostly spin-zero, **B.C. 3** keeps mostly spin-two, while **B.C. 2** keeps spin-zero and spin-two even. The complete spectrum of each set of boundary conditions is summarized in Fig. 3.

The ground mode spectra associated with **B.C. 3** deserves further elaboration, as they in fact describe non-unitary system. Firstly, it is non-unitary because the mass-squared is below the Breitenlohner-Freedman bound of spin-two field in AdS_{d+1} space. Secondly, norms of some mode functions are negative-definite, implying that the Hilbert space has the structure of indefinite metric, leading classically to spectrum unboundedness and instability classically and quantum mechanically to negative probability. Explicitly, for the mode functions

$$\begin{cases} \Theta_1^{2|2} = N_5 \sec^2\theta & \textbf{Type III} \\ \Theta_0^{2|2} = N_3 \sec\theta \tan\theta, \quad \Theta_0^{1|2} = N_4 \tan\theta & \textbf{Type IV}, \end{cases} \tag{113}$$

the norms of $\Theta_0^{2|2}$ and $\Theta_1^{2|2}$ are $-2\alpha\, {N_3}^2$ and $-\frac{2}{\tan\alpha}{N_5}^2$ and hence negative-definite for all choices of α. Such negative norms indicate that the higher-spin fields associated with these ground modes in **B.C. 3** have wrong sign kinetic term.

As the waveguide size α tends to $\pi/2$, the boundary surfaces of the waveguide approach to the timelike asymptotic boundary of the AdS_{d+2} space. In other words,

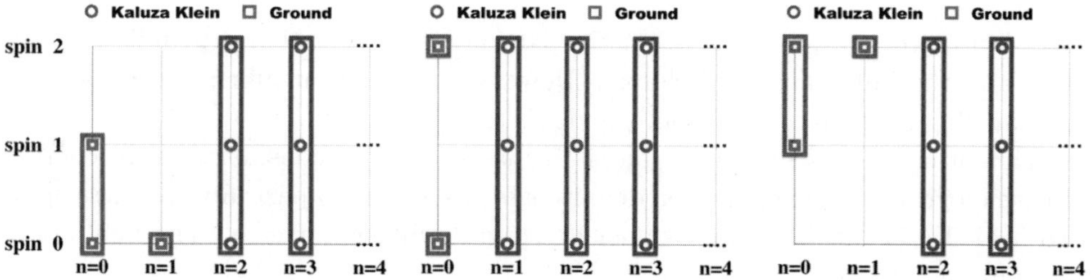

Fig. 3. Spectral pattern for three types of Dirichlet conditions, **B.C. 1**, **B.C. 2** and **B.C. 3** from left to right. The spin contents of each excitation level $n = 0, 1, 2, \ldots$ is depicted. Each point represents one mode: squares are from ground modes, while circles are from Kaluza-Klein modes. Points inside the same rectangle have the same eigenvalues and form Stüeckelberg system. Compare this with the pattern for arbitrary higher-spin in Fig. ??.

the waveguide decompactifies to the AdS_{d+2} space. In this limit, though, the mass spectrum for each boundary conditions does not necessarily gets to the spectrum of massless spin-two field in AdS_{d+2} space. The reason is that some of the boundary conditions we choose are singular in this limit in the sense that mode functions are ill-defined. Take for instance the mass spectrum for **B.C. 2**. It contains the massless spin-two ground mode as well as spin-zero ground mode whose normalized mode functions are

$$\begin{cases} \Theta_0^{0|2} = N_1 \sec\theta, & \textbf{Type II}, \\ \Theta_0^{2|2} = N_2 \sec^2\theta, & \textbf{Type III}, \end{cases} \qquad N_1 = \frac{1}{\sqrt{2\alpha}}, \quad N_2 = \sqrt{\frac{1}{2\tan\alpha}}. \tag{114}$$

These ground-mode functions are not normalizable in AdS_{d+2} space: the normalized mode functions vanish as N_2 vanishes in the decompactification limit. This explains why there is no massless spin-two field in the "dimensional degression" method.[20] We also checked that, for arbitrary spacetime dimension and spin of higher-spin field, the mass spectrum of "dimensional degression" spectrum is the spectrum of **B.C. 1** in the decompactification limit.

Summarizing,

- The mode functions of different spins in AdS_4 are related to each other by the spectrum generating complex Eq. (92), whose structure is uniquely fixed by consideration of Kaluza-Klein compactification of higher-spin gauge transformations.
- At special values of masses, the Stüeckelberg spin-two system decomposes into irreducible representations of massless or partially massless spin-two fields and massive Goldstone fields. The ground modes of Dirichlet boundary conditions Eq. (106) are precisely these irreducible representations in Table 1 at the special mass values.

7. Partially Massless Cosmology

We ask if there is any situation that the partially conserved charge is indistinguishable from the fully conserved charge. Here, we answer that affirmatively positive by providing a concrete situation, the de Sitter space.

This situation arises when the partially massless field is considered in the context of cosmological background. For definiteness, we consider partially massless spin-two field $K^{\mu\nu}(x)$ in four-dimensional Einstein-de Sitter universe. The field obeys the equation of motion

$$\nabla_\mu \nabla_\nu K^{\mu\nu} = 0, \quad g_{\mu\nu} K^{\mu\nu}(x) = 0. \tag{115}$$

In Friedmann-Robertson-Walker foliation, the spacetime background is given by

$$\mathrm{d}s_4^2 = \mathrm{d}t^2 - R^2(t)\mathrm{d}\mathbf{x}^2 = R^2(t)\left[\mathrm{d}\eta^2 - \mathrm{d}\mathbf{x}^2\right], \tag{116}$$

where $R(t)$ is the scale factor of comoving frame, and $\eta(t)$ is the conformal time defined by $\mathrm{d}\eta = \mathrm{d}t/R(t)$. With homogeneity and isotropy of the universe, the most general ansatz of partially conserved spin-two field $K^{\mu\nu}$, which has the scaling dimension two, is given by

$$K^{tt}(t) = \sigma(t), \quad K^{ti}(t) = 0, \quad K^{ij}(t) = \frac{\delta^{ij}}{3}\frac{\sigma(t)}{R^2(t)}. \tag{117}$$

The equation of motion $\nabla_\mu \nabla_\nu K^{\mu\nu} = 0$ is then reduced to

$$(\partial_t + 4H(t))(\partial_t + 3H(t))\sigma(t) = 0 \quad \text{where } H(t) \equiv \frac{\dot{R}(t)}{R(t)}. \tag{118}$$

The general solution is given by

$$\sigma(t) = \frac{1}{R^3(t)}\left(C_1 + C_2 \int_0^t \frac{\mathrm{d}s}{R(s)}\right), \tag{119}$$

where C_1, C_2 are constants that are fixed by the initial condition.

Rewriting the equation of motion as

$$\nabla_t(\nabla_\mu K^{\mu t}) + \nabla_i(\nabla_\mu K^{\mu i}) = 0 \tag{120}$$

and integrating over the spatial slice, we can define the partially conserved charge as

$$\begin{aligned} Q(t) &= \int \mathrm{d}^3x\sqrt{g_3}\nabla_\mu K^{\mu t} = \int \mathrm{d}^3 R^3(t)(\partial_t K^{tt} + 4H(t)K^{tt}) \\ &= \mathrm{Vol}_3\left[C_1 H(t) + C_2\left(\frac{1}{R(t)} + H(t)\int^t \frac{\mathrm{d}s}{R(s)}\right)\right]. \end{aligned} \tag{121}$$

In general, the partially conserved charge is not conserved and evolves in time. However, from the explicit form

$$\dot{Q}(t) = \mathrm{Vol}_3\left[C_1 + C_2\int^t \frac{\mathrm{d}s}{R(s)}\right]\dot{H}(t), \tag{122}$$

we see that, when the Universe is de Sitter and $\dot{H}(t) = 0$, the charge $Q(t)$ is conserved much the same way as the conventional conserved charges. If the accelerating Universe at present epoch is interpreted as a de Sitter space, this then implies that the number of partially massless quanta does not change with time. If the near-Planck scale inflation was again of de Sitter type, the number was conserved at that epoch. The two numbers are not equal, as the Universe evolved through decelerating acceleration phase between the two epochs. During radiation dominated ear,

$$\dot{Q}(t) = \mathrm{Vol}_3[A_1 R^{-2}(t) + A_2 R^{-1}(t)]. \tag{123}$$

During matter dominated era,

$$\dot{Q}(t) = \mathrm{Vol}_3[B_1 R^{-3/2}(t) + B_2 R^{-1}(t)]. \tag{124}$$

The constants A_1, A_2, B_1, B_2 are determined by initial conditions of each era. It is evident that, in both cases, the partially conserved charge decreases monotonically.

8. Generalizations

In this talk, I presented a geometric description of Higgs mechanism for (partially) massless spin-two fields in AdS_4 space. Our approach was based on the Kaluza-Klein compactification on a wedge of Janus geometry. We found that the Higgs mechanism for partially massless spin-two field is provided by the higher-derivative boundary conditions at the AdS_3 boundaries of Janus wedge. By taking the boundary condition to be of the Robin type and tuning its coefficients, we can trigger the (inverse) Higgs mechanism, (de)coupling a single scalar degrees of freedom to a massive AdS_4 graviton.

The geometric Higgs mechanism I explained can be extended. This was done for dimensions higher than four and for arbitrary spin in my recent work.[5] The Higgs mechanism is much richer than the spin-two since partially massless higher spin fields come with various depths. The mechanism can also be studied in lower dimensions.[21] Spin-two and higher spin fields in AdS_3 space are non-dynamical. Nevertheless, they have residual gauge invariances at asymptotic boundary[22] and the Higgs mechanism can still be formulated in terms of these boundary degrees of freedom. Using the Chern-Simons formulation, I recently found that the Higgs mechanism is nicely achieved[23] by color-decorating the spin-two or higher spin fields.[24,25]

The Higgs mechanism of (partially) massless higher spin fields is actually more general than the gauge theory of higher spin fields. Higgs mechanism of tensionless string is certainly the picture of generating tension for string. One situation in which the physics is clean and AdS/CFT correspondence can also be formulated in terms of well-developed CFTs was studied in the context of holographic dual of Gaiotto's $\mathcal{N} = 2$ superconformal quiver gauge theories in four dimensions,[26] based on the earlier hint that tension of Wilson loop can become arbitrarily light.[27]

Acknowledgments

We are grateful to Xavier Bekaert, Nicholas Boulanger, Marc Henneaux, Kurt Hinterbichler, Dima Polyakov, Evgeny Skvotsov, Augusto Sagnotti and Yurii Zinoviev for very helpful discussions. SHG and JWK acknowledge the APCTP Focus Program "Liouville, Integrability and Branes (11)" for providing stimulating scientific environment. SJR acknowledges warm hospitality and stimulating atmosphere of Tessaloniki Workshop "Aspects of Conformal Field Theory" (September 21–25, 2015) and Singapore Workshop "Higher Spin Gauge Theories" (November 4–6, 2015), where part of this work was presented. This work was supported in part by the National Research Foundation Grants 2005-0093843, 2010-220-C00003 and 2012K2A1A9055280.

References

1. S. Deser and A. Waldron, Gauge invariances and phases of massive higher spins in (A)dS, *Phys. Rev. Lett.* **87**, 031601 (2001) [hep-th/0102166]; Null propagation of partially massless higher spins in (A)dS and cosmological constant speculations, *Phys. Lett. B* **513**, 137 (2001) [hep-th/0105181].
2. T. Kaluza, Zum Unitätsproblem der Physik, *Sitz. Preuss. Akad. Wiss. Phys. Berlin.* (Math. Phys.): 966–972 (1921);
   O. Klein, Quantentheorie und fünfdimensionale Relativitätstheorie, *Zeits. Phys.* **37**, 895 (1926).
3. D. Bak, M. Gutperle and S. Hirano, A Dilatonic deformation of AdS(5) and its field theory dual, *JHEP* **0305**, 072 (2003) [hep-th/0304129].
4. E. C. G. Stüeckelberg, D̈ie Wechselwirkungskräfte in der Elektrodynamik und in der Feldtheorie der Kräfte, *Helv. Phys. Acta* **11**, 225 (1938).
5. S. Gwak, J. Kim and S. J. Rey, Massless and massive higher spins from Anti-de Sitter space waveguide, arXiv:1605.06526 [hep-th].
6. L. Dolan, C. R. Nappi and E. Witten, Conformal operators for partially massless states, *JHEP* **0110**, 016 (2001) [hep-th/0109096].
7. Y. M. Zinoviev, *On massive high spin particles in AdS,* [hep-th/0108192].
8. Y. M. Zinoviev, On massive spin 2 interactions, *Nucl. Phys. B* **770**, 83 (2007) [hep-th/0609170].
9. X. Bekaert and M. Grigoriev, Higher order singletons, partially massless fields and their boundary values in the ambient approach, *Nucl. Phys. B* **876**, 667 (2013) [hep-th/1305.0162].
10. T. Basile, X. Bekaert and N. Boulanger, Flato-Fronsdal theorem for higher-order singletons, *JHEP* **1411**, 131 (2014) [arXiv:1410.7668 [hep-th]].
11. M. G. Eastwood and J. W. Rice, Conformally invariant differential operators on Minkowski space and their curved analogues, *Commun. Math. Phys.* **109**, 207 (1987) [Erratum: *ibid* **144**, 213 (1992)].
12. M. G. Eastwood, Notes on conformal differential geometry, *Suppl. Rendi. Circ. Mat. Palermo* **43**, 57 (1996).
13. M. G. Eastwood, Higher symmetries of the Laplacian, *Annals Math.* **161**, 1645 (2005) [hep-th/0206233].
14. O. V. Shaynkman, I. Y. Tipunin and M. A. Vasiliev, Unfolded form of conformal equations in M dimensions and $o(M+2)$ modules, *Rev. Math. Phys.* **18**, 823 (2006) [hep-th/0401086].
15. R. Camporesi and A. Higuchi, Arbitrary spin effective potentials in anti-de Sitter space-time, *Phys. Rev. D* **47**, 3339 (1993); Spectral functions and zeta functions in hyperbolic spaces, *J. Math. Phys.* **35**, 4217 (1994).
16. P. A. M. Dirac, A remarkable representation of the $3+2$ de Sitter group, *J. Math. Phys.* **4**, 901 (1963). doi:10.1063/1.1704016
17. C. Aragone, S. Deser and Z. Yang, Massive higher spin from dimensional reduction of gauge fields, *Annals Phys.* **179**, 76 (1987);
   S. D. Rindani and M. Sivakumar, Gauge — invariant description of massive higher — spin particles by dimensional reduction, *Phys. Rev. D* **32**, 3238 (1985).

18. T. Biswas and W. Siegel, Radial dimensional reduction: Anti-de Sitter theories from flat, *JHEP* **0207**, 005 (2002) [hep-th/0203115];
K. Hallowell and A. Waldron, Constant curvature algebras and higher spin action generating functions, *Nucl. Phys. B* **724**, 453 (2005) [hep-th/0505255];
R. Manvelyan, R. Mkrtchyan and W. Ruehl, Radial Reduction and Cubic Interaction for Higher Spins in (A)dS space, *Nucl. Phys. B* **872**, 265 (2013) [hep-th/1210.7227].
19. R. R. Metsaev, Massive fields in AdS(3) and compactification in AdS space time, *Nucl. Phys. Proc. Suppl.* **102**, 100 (2001) [hep-th/0103088].
20. A. Y. Artsukevich and M. A. Vasiliev, On dimensional degression in AdS(d), *Phys. Rev. D* **79**, 045007 (2009) [hep-th/0810.2065].
21. M. A. Vasiliev, Higher spin gauge theories in four-dimensions, three-dimensions, and two-dimensions, *Int. J. Mod. Phys. D* **5**, 763 (1996) [hep-th/9611024];
Higher spin gauge theories: Star product and AdS space, in *The many faces of the superworld*, M. A. Shifman (ed.), 533–610 [hep-th/9910096].
22. M. Henneaux and S. J. Rey, Nonlinear W_∞ as asymptotic symmetry of three-dimensional higher spin Anti-de Sitter gravity, *JHEP* **1012**, 007 (2010) [hep-th/1008.4579].
23. S. Gwak *et al.*, Rainbow valley of colored (Anti) de Sitter gravity in three dimensions, *JHEP* **1604**, 055 (2016) [arXiv:1511.05220 [hep-th]];
Rainbow vacua of colored higher spin gravity in three dimensions, *JHEP* **1605**, 150 (2016) [arXiv:1511.05975 [hep-th]].
24. M. A. Vasiliev, Extended higher spin superalgebras and their realizations in terms of quantum operators, *Fortsch. Phys.* **36**, 33 (1988);
S. E. Konstein and M. A. Vasiliev, Extended higher spin superalgebras and their massless representations, *Nucl. Phys. B* **331**, 475 (1990);
M. A. Vasiliev, Higher spin superalgebras in any dimension and their representations, *JHEP* **0412**, 046 (2004) [hep-th/0404124].
25. M. Henneaux *et al.*, Super- W(infinity) asymptotic symmetry of higher-spin AdS_3 supergravity, *JHEP* **1206**, 037 (2012) [hep-th/1203.5152].
26. O. Aharony, M. Berkooz and S. J. Rey, Rigid holography and six-dimensional $\mathcal{N} = (2,0)$ theories on AdS_5 $\mathbb{S}^1$, *JHEP* **1503**, 121 (2015) [hep-th/1501.02904].
27. S. J. Rey and T. Suyama, Exact results and holography of Wilson loops in $N = 2$ superconformal (Quiver) gauge theories, *JHEP* **1101**, 136 (2011) [arXiv:1001.0016 [hep-th]].

Quartic Amplitudes for Minkowski Higher Spin

Anders K. H. Bengtsson
Academy of Textiles, Engineering and Economics
University of Borås, Allégatan 1, SE-50190 Borås, Sweden

The old problem of finding general quartic interaction terms between fields of higher helicities on the light-front is discussed from the point of view of calculating the corresponding amplitudes directly from the cubic vertices using BCFW recursion. Amplitude based no-go results that has appeared in the literature are reviewed and discussed and it is pointed out how they may perhaps be circumvented.

1. Introduction

An old and still unsolved problem in light-front higher spin gauge field theory in Minkowski space-time is the computation of general quartic interactions. The problem was addressed quite a while ago by Metsaev[1,2] and some partial results were obtained but it seems that explicit interaction terms were not arrived at. Furthermore, the success of the Vasiliev theory of higher spin – which is background independent but do not seem to have a consistent Minkowski limit – has all the time cast doubts on the very existence of consistent interactions in Minkowski space-time. Adding to this list of negative indications is the computation of quartic amplitudes[3,4] which yield still more restrictions on quartic interactions in Minkowski background. This paper is an attempt to review and address this problem by pointing out some still open access routes to circumventing the mounting no-go results.

Cubic self-interactions (in four spacetime dimensions) were first found in 1983[5] and a few years later all possible cubic interactions between three fields of different spins was written down.[6] These results have been generalised to higher dimensions, half-integer helicities and mixed symmetry and massive fields by Metsaev.[7–9] The spin 1 and 2 quartic interactions are certainly known as they can be derived from the covariant theories through light-front gauge fixing. Spin 1 is rather trivial, whereas for spin 2 there is some clever algebra to perform.[10,11] Even the spin 3 cubic self-interaction[12] can be gauge fixed to the light-front from the BBvD covariant spin 3 interaction from 1984.[13] It hasn't been possible to discern any obvious pattern in the spin 1 and 2 light-front quartic interactions that lend itself to a generalisation to higher spin s interactions. The problem of pattern recognition is also confounded by the possibility of making field redefinitions.

Anyway, we know that beyond spin 2 and cubic interactions we need a theory containing an infinite tower of higher spins in order to have any hope of finding a fully consistent interacting theory. Even at the cubic level we have interactions connecting fields of three different spin s_1, s_2, s_3 (with certain restrictions).[6] Therefore what is called for – at the very least – is a general quartic interaction term for four different spin s_1, s_2, s_3, s_4 (possibly with certain restrictions).

To address this problem head on using the configuration space techniques of the original 1983 paper doesn't seem to be tractable. Beyond the cubic level we need a formalism that can maintain a spectrum of all helicities. Such a formalism becomes most natural in momentum space together with Fock space basis in which a higher spin field $|\Phi\rangle$ can be expanded over some internal parameter, for example excitations of oscillators. Such a formalism was set up by Bengtsson et. al.[6] and has been elaborated by Metsaev[7] and further systematised by myself.[14] Cubic interaction data is then maintained by vertex operators that are themselves expansions in oscillators and momenta. In this way the confusing details of individual interaction terms are subsumed into the vertex operators that store the data with no particular reference to individual spin states of the interacting fields. One would hope that such vertex operators should exist also at the quartic level and above.

This said, it seems that even with the systematisation provided by a momentum space and vertex operator formalism, the quartic vertex is still elusive to calculate. This is so because there appear certain technical problems the solution of which probably requires further understanding of the cubic vertex and higher order kinematics as well as development of the formalism itself.

As already mentioned, the consistency of Minkowski spacetime higher spin theory beyond the cubics is questioned on the basis of various no-go results. If it is the case that higher spin theory in a Minkowski background cannot be extended beyond spin 2, one would perhaps have thought that these problems should occur already at the cubic level. Instead the free theory and the cubic theory are perfectly seamless generalisations of lower spin. However, the light-front cubic interactions are only restricted by kinematics[14] and do not really probe interactions. In terms of amplitudes this appears in the fact that the cubic amplitudes are determined by little group scaling and dimensional analysis.

A few years ago, quartic amplitudes for massless fields were investigated by Benincasa and Cachazo[3] and Benincasa and Conde,[4,15] in order to map out the consistent interactions among higher and lower spin particles. They use BCFW[16,17] recursion to build quartic tree amplitudes out of cubic amplitudes. As they show, the method – which is a generalisation of now quite standard amplitude methods – is very general and can be set up to test higher order consistency of gauge field theories without knowing the underlying Lagrangian. The method is based on a modern version of S-matrix theory in which five basic assumption are made: (i) analyticity,

(ii) unitarity, (iii) Poincaré invariance, (iv) existence of one-particle states and (v) locality. I will try to review parts of this work below.

On the face of it, higher spin gauge interactions turn out to be very severely restricted in Minkowski spacetime. In the cited work, the focus was on single spin theories. The four-particle test (to be described in more detail below) rules out all quartic higher spin interactions except what would be generalisations of spin one F^3 and spin two R^3 interactions.

On the other hand we know that once spin 2 is passed we need an infinite tower of higher spin excitations in order to have any hope at all at finding interactions. I will discuss how this property of higher spin field theories might escape the four-particle no-go results. There are two assumptions that can be questioned: the fifth one on locality of the S-matrix, and the fourth one on the existence of one-particle states. Of course, questioning these two basic assumption more or less amounts to questioning the S-matrix programme itself.

The first three assumptions are harder to question. Giving up Poincaré invariance is the same as bidding farewell to Minkowski higher spin which we don't want to do just as yet.

It is customary in higher spin theory to collect an infinite spectrum (or tower) of higher spin fields into some bilocal object $\Phi(p, \xi)$ where the parameter ξ works as an expansion variable over which, for instance, symmetric tensor fields can be expanded. Sometimes this is just a convenient calculational device and no particular dynamics is associated to the internal parameter. But more often ξ is considered to be more of a physical variable, perhaps in connection to underlying particle mechanics constraints. In my opinion, this last viewpoint is the more physical interesting and the one more likely to lead to an eventual understanding of higher spin theory.

Now it is – at least heuristically – possible to see how such a viewpoint might lead to a rejection of S-matrix assumptions (iv) and (v). A state $\Phi(p, \xi)$ depending on two variables – a momentum p, and some internal coordinate ξ – is indeed the most simple non-local object one can imagine. If the expansion over ξ is discrete (as in a power series) then the coefficients can be identified (at least in principle) with one-particle states. However, allowing for a more generous function space, we can (and perhaps must) drop assumption (iv) on one-particle states. Perhaps the most natural implementation of such a scenario is to connect to the continuous spin representations of the Poincaré group.

2. The Light-Front and Amplitudes

It is interesting to discuss the light-front cubic vertices and their relation to cubic amplitudes.

2.1. *Cubic light-front vertices*

Cubic interaction vertices for arbitrary helicities $\lambda_1, \lambda_2, \lambda_3$ have a remarkably simple structure when written in a light-front momentum frame.[6] Taking all λ as positive integers (denoting a negative helicity explicitly as $-\lambda$), there is essentially just one vertex for each helicity assignment. For instance a $(-\lambda_1, \lambda_2, \lambda_3)$ vertex takes the form

$$\frac{\gamma_1^{\lambda_1}}{\gamma_2^{\lambda_2}\gamma_3^{\lambda_3}}\,\bar{\mathbb{P}}^{\lambda_2+\lambda_3-\lambda_1}\,. \tag{1}$$

This term should be multiplied by the momentum conservation delta function and the appropriate coupling factor. In the light-front action this term is supplemented with its complex conjugate to make the action real.

In the particular case of three equal helicities $\lambda_1 = \lambda_2 = \lambda_3 = \lambda$ we have

$$\left(\frac{\gamma_1}{\gamma_2\gamma_3}\,\bar{\mathbb{P}}\right)^{\lambda}. \tag{2}$$

This formula can be interpreted as a generalised KLT-relation[18] with the higher spin λ cubic vertex being the Yang-Mills vertex raised to the power λ as pointed out by Ananth[19] (see also our paper[6]). In these formulas, γ is equal to the momentum component p^+ and $\mathbb{P}$ is

$$\mathbb{P} = \frac{1}{3}\big((\gamma_1 p_2 - \gamma_2 p_1) + (\gamma_2 p_3 - \gamma_3 p_2) + (\gamma_3 p_1 - \gamma_1 p_3)\big) \equiv \frac{1}{3}(\mathbb{P}_{12} + \mathbb{P}_{23} + \mathbb{P}_{31}) \tag{3}$$

and correspondingly for the complex conjugate transverse momenta $\bar{\mathbb{P}}$. Due to momentum conservation, all three $\mathbb{P}_{12}$, $\mathbb{P}_{23}$ and $\mathbb{P}_{31}$ are equal.

The form of the most general cubic interaction vertex on the light-front can be controlled in the following simple way, based on Poincaré analysis.[6,14]

Consider the cubic interaction of three fields of helicities s_1, s_2 and s_3 (positive or negative). As is customary we write the field as ϕ_i if the helicity s_i is positive and $\bar{\phi}_i$ if the helicity s_i is negative. The transverse momentum dependence allowed by Poincaré invariance is $\bar{\mathbb{P}}^n\mathbb{P}^m$ where n and m are restricted by $n - m = s_1 + s_2 + s_3$. Factors of $\bar{\mathbb{P}}\mathbb{P}$ correspond to field redefinitions of the free theory so we can set either $n = 0$ or $m = 0$.

Treating $m = 0$ first so that $n = s_1 + s_2 + s_3 \geq 0$ we get the cases

(a) All $s_i \geq 0$. Put all $s_i = \lambda_i$. Field and momentum structure become

$$\phi_1\phi_2\phi_3\frac{1}{\gamma_1^{\lambda_1}\gamma_2^{\lambda_2}\gamma_3^{\lambda_3}}\bar{\mathbb{P}}^{\lambda_1+\lambda_2+\lambda_3}$$

(b) Two $s_i \geq 0$, one $s_i \leq 0$. Say $s_2 = \lambda_2 \geq 0$, $s_3 = \lambda_3 \geq 0$ while $s_1 = -\lambda_1 \leq 0$. Field and momentum structure become

$$\bar{\phi}_1\phi_2\phi_3\frac{\gamma_1^{\lambda_1}}{\gamma_2^{\lambda_2}\gamma_3^{\lambda_3}}\bar{\mathbb{P}}^{\lambda_2+\lambda_3-\lambda_1} \quad \text{where} \quad \lambda_1 \leq \lambda_2 + \lambda_3$$

(c) One $s_i \geq 0$, two $s_i \leq 0$. Say $s_3 = \lambda_3 \geq 0$ while $s_1 = -\lambda_1 \leq 0$, $s_2 = -\lambda_2 \leq 0$. Field and momentum structure become

$$\bar{\phi}_1\bar{\phi}_2\phi_3\frac{\gamma_1^{\lambda_1}\gamma_2^{\lambda_2}}{\gamma_3^{\lambda_3}}\bar{\mathbb{P}}^{\lambda_3-\lambda_1-\lambda_2} \quad \text{where} \quad \lambda_1+\lambda_2 \leq \lambda_3.$$

Next treating $n = 0$ so that $m = -(s_1 + s_2 + s_3) \geq 0$ we get the cases

(a') All $s_i \leq 0$. Put all $s_i = -\lambda_i$. Field and momentum structure become

$$\bar{\phi}_1\bar{\phi}_2\bar{\phi}_3\frac{1}{\gamma_1^{\lambda_1}\gamma_2^{\lambda_2}\gamma_3^{\lambda_3}}\mathbb{P}^{\lambda_1+\lambda_2+\lambda_3}$$

(b') Two $s_i \leq 0$, one $s_i \geq 0$. Say $s_2 = -\lambda_2 \leq 0$, $s_3 = -\lambda_3 \leq 0$ while $s_1 = \lambda_1 \geq 0$. Field and momentum structure become

$$\phi_1\bar{\phi}_2\bar{\phi}_3\frac{\gamma_1^{\lambda_1}}{\gamma_2^{\lambda_2}\gamma_3^{\lambda_3}}\mathbb{P}^{\lambda_2+\lambda_3-\lambda_1} \quad \text{where} \quad \lambda_1 \leq \lambda_2+\lambda_3$$

(c') One $s_i \leq 0$, two $s_i \geq 0$. Say $s_3 = -\lambda_3 \leq 0$ while $s_1 = \lambda_1 \geq 0$, $s_2 = \lambda_2 \geq 0$. Field and momentum structure become

$$\phi_1\phi_2\bar{\phi}_3\frac{\gamma_1^{\lambda_1}\gamma_2^{\lambda_2}}{\gamma_3^{\lambda_3}}\mathbb{P}^{\lambda_3-\lambda_1-\lambda_2} \quad \text{where} \quad \lambda_1+\lambda_2 \leq \lambda_3.$$

As expected, cases (a') - (c') are the complex conjugates of cases (a) - (c) and they can all occur in the action. Cases (a) and (a') correspond to the F^3 (spin 1) and R^3 (two-loop pure gravity counterterm) interactions (and their higher spin generalisations) mentioned above.

The cases (b), (c), (b') and (c') contain, among many other types of interactions, the pure spin s interactions. Just to show how it works, consider $\lambda_1 = \lambda_2 = \lambda_3 = \lambda$. Then (b) and (b') yield the interaction term

$$\bar{\phi}_1\phi_2\phi_3\left(\frac{\gamma_1}{\gamma_2\gamma_3}\right)^{\lambda}\bar{\mathbb{P}}^{\lambda} + \phi_1\bar{\phi}_2\bar{\phi}_3\left(\frac{\gamma_1}{\gamma_2\gamma_3}\right)^{\lambda}\mathbb{P}^{\lambda}. \tag{4}$$

Cases (c) and (c') are not allowed for this helicity configuration.

2.2. *Spinor helicity formalism*

The light-front formalism is very close to amplitude formalism as has been noted by Ananth.[19] Indeed, working out $\not{p}$ we find (with σ^0 the unit matrix)

$$p_{a\dot{a}} = p_\mu\sigma^\mu_{a\dot{a}} = \sqrt{2}\begin{pmatrix} -p^- & \bar{p} \\ p & -p^+ \end{pmatrix} = /\text{on shell}/ = \sqrt{2}\begin{pmatrix} -\frac{p\bar{p}}{\gamma} & \bar{p} \\ p & -\gamma \end{pmatrix}. \tag{5}$$

Since the determinant of this two-by-two matrix is zero it can be written as a product of two-component angle and bracket spinors. Bracketing two such spinors for different momenta p_1 and p_2 we get

$$\begin{aligned}\langle p_1 p_2\rangle &= \langle p_1|_{\dot a}|p_2\rangle^{\dot a} = \frac{\sqrt[4]{2}}{\sqrt{\gamma_1}}(p_1, -\gamma_1)\frac{\sqrt[4]{2}}{\sqrt{\gamma_2}}\begin{pmatrix}-\gamma_2\\ -p_2\end{pmatrix} = \frac{\sqrt{2}}{\sqrt{\gamma_1\gamma_2}}\mathbb{P}_{12}\\ [p_1 p_2] &= [p_1|^a|p_2]_a = -\frac{\sqrt{2}}{\sqrt{\gamma_1\gamma_2}}\bar{\mathbb{P}}_{12}\end{aligned} \tag{6}$$

where the translation into light-front notation is explicit. We may thus expect an almost (modulo light-front energy conservation – see below) seamless translation between cubic amplitudes written in terms of spinor helicity variables and the light-front cubic vertices irrespective of the helicities of the external massless particles. It is of course special for the cubic interactions that the on-shell amplitude coincides up to numerical factors with the Feynman diagram vertex.

The question is now if we can exploit this dictionary between light-front variables and amplitude variables to compute quartic higher spin amplitudes directly from the cubic amplitudes. We will try to do it using BCFW recursion. But first let us spell out a little more of the translation between the formalisms. In the following the shorthand notation $|p_j\rangle = |j\rangle$ and $|p_j] = |j]$ will be used for the spinor variables.

Going back to the formulas (6) we should note that the left hand side inner products are $SL(2, \mathbf{C})$ invariants, while the right hand sides are explicit light-front constructs. Let us delve into this in a little more detail. The explicit translation can be done in the following way. Start with $p_{a\dot a} = -\lambda_a\tilde\lambda_{\dot a}$ in $(-+++)$ metric and introduce bra and kets according to

$$\begin{aligned}\lambda_a &= |p]_a \quad \text{with} \quad [p|^a = \epsilon^{ab}|p]_b\\ \tilde\lambda_{\dot a} &= \langle p|_{\dot a} \quad \text{with} \quad |p\rangle^{\dot a} = \epsilon^{\dot a\dot b}\langle p|_{\dot b}.\end{aligned} \tag{7}$$

Then the equation $p_{a\dot a} = -\lambda_a\tilde\lambda_{\dot a}$ is implemented by the assignments

$$|p]_a = \frac{\sqrt[4]{2}}{\sqrt{\gamma}}\begin{pmatrix}\bar p\\ -\gamma\end{pmatrix} \quad \text{and} \quad \langle p|_{\dot a} = \frac{\sqrt[4]{2}}{\sqrt{\gamma}}\begin{pmatrix}p & -\gamma\end{pmatrix} \tag{8}$$

from which formulas (6) follows.

With the covariant angle and square bracket notation comes a set of very useful identities.

$$\text{Anti-symmetry}: \quad \langle p\,q\rangle = -\langle q\,p\rangle, \quad [p\,q] = -[q\,p], \quad \langle p\,p\rangle = [p\,p] = 0 \tag{9}$$

$$\text{Schouten identities}: \quad \langle i\,j\rangle\langle k\,l\rangle + \langle i\,k\rangle\langle l\,j\rangle + \langle i\,l\rangle\langle j\,k\rangle = 0 \tag{10}$$

$$[i\,j][k\,l] + [i\,k][l\,j] + [i\,l][j\,k] = 0 \tag{11}$$

$$\text{Momentum conservation}: \quad \sum_{i=0}^{n}|i\rangle[i| = 0 \quad \Rightarrow \quad \sum_{i=0}^{n}\langle p\,i\rangle[i\,q] = 0 \tag{12}$$

$$\text{Squaring}: \quad \langle p\,q\rangle[p\,q] = 2p\cdot q = 2(p+q)^2 = -\mathbf{s}_{pq} \tag{13}$$

Of these equations, the anti-symmetry and Schouten identities are satisfied by the light-front form of the brackets. However, momentum conservation must be treated with some care when expressed in terms of the light-front form. Explicitly we get

$$\sum_{i=0}^{n} |i\rangle[i| = \sum_{i=0}^{n} \frac{\sqrt[4]{2}}{\sqrt{\gamma}_i}\begin{pmatrix}\gamma_i\\ p_i\end{pmatrix}\frac{\sqrt[4]{2}}{\sqrt{\gamma}_i}\begin{pmatrix}\gamma_i & \bar{p}_i\end{pmatrix} = \sqrt{2}\sum_{i=0}^{n}\begin{pmatrix}\gamma_i & \bar{p}_i\\ p_i & \frac{p_i\bar{p}_i}{\gamma_i}\end{pmatrix}. \tag{14}$$

The terms $\frac{p_i\bar{p}_i}{\gamma_i}$ do not sum to zero. Rather we have for $n = 3$

$$\sum_{i=0}^{3} \frac{p_i\bar{p}_i}{\gamma_i} = -\frac{\mathbb{P}\,\bar{\mathbb{P}}}{\gamma_1\gamma_2\gamma_3} \tag{15}$$

and for $n = 4$, for instance

$$\sum_{i=0}^{4} \frac{p_i\bar{p}_i}{\gamma_i} = -\frac{\mathbb{P}_{12}\,\bar{\mathbb{P}}_{23}}{\gamma_1\gamma_2\gamma_3} - \frac{\mathbb{P}_{14}\,\bar{\mathbb{P}}_{43}}{\gamma_1\gamma_4\gamma_3}. \tag{16}$$

It is a little bit interesting to see how the algebra works out. From the momentum conservation equation (12) we get, bracketing with (for instance) $\langle 1|$ and $|3]$

$$\langle 1|\left(\sum_{i=0}^{n}|i\rangle[i|\right)|3] = \langle 12\rangle[23] + \langle 14\rangle[43] = -2\frac{\mathbb{P}_{12}\,\bar{\mathbb{P}}_{23}}{\sqrt{\gamma_1\gamma_2}\sqrt{\gamma_2\gamma_3}} - 2\frac{\mathbb{P}_{14}\,\bar{\mathbb{P}}_{43}}{\sqrt{\gamma_1\gamma_4}\sqrt{\gamma_4\gamma_3}}$$

where we have used (6). On the other hand bracketing (14) we get

$$\frac{\sqrt[4]{2}}{\sqrt{\gamma}_1}\begin{pmatrix}p_1 & -\gamma_1\end{pmatrix}\sqrt{2}\begin{pmatrix}0 & 0\\ 0 & \sum_{i=0}^{4}\frac{p_i\bar{p}_i}{\gamma_i}\end{pmatrix}\frac{\sqrt[4]{2}}{\sqrt{\gamma}_3}\begin{pmatrix}\bar{p}_3\\ -\gamma_3\end{pmatrix} =$$

$$2\sqrt{\gamma_1\gamma_3}\sum_{i=0}^{4}\frac{p_i\bar{p}_i}{\gamma_i} = -2\frac{\mathbb{P}_{12}\,\bar{\mathbb{P}}_{23}}{\sqrt{\gamma_1\gamma_2}\sqrt{\gamma_2\gamma_3}} - 2\frac{\mathbb{P}_{14}\,\bar{\mathbb{P}}_{43}}{\sqrt{\gamma_1\gamma_4}\sqrt{\gamma_4\gamma_3}}$$

where the equality (16) is used.

The cubic formula (15) is used when deriving the light-front cubic vertices. It can be expected that the corresponding quartic formula (16) will play a similar role for the quartic vertex. This may be one of the technical pieces missing in earlier attempts at the full quartic vertex.

As regards the squaring equations (13) we get, for instance for the Mandelstam invariant $\mathbf{s}$

$$\mathbf{s} = -\langle 12\rangle[12] = 2\frac{\mathbb{P}_{12}\,\bar{\mathbb{P}}_{12}}{\gamma_1\gamma_2}. \tag{17}$$

2.3. *Cubic amplitudes*

We will only consider tree amplitudes and the notation $M_n[1^{s_1}2^{s_2}\dots n^{s_n}]$ will be used for a partial (colour stripped) amplitude with n outgoing on-shell massless particles with helicities $s_1, s_2, \dots s_n$. The s_i are integers (as above, we use $\lambda_i > 0$ when we want to be explicit about the sign of the helicity).

It is well-known that the cubic amplitudes are determined (up to numerical factors) by little group scaling and dimensional analysis to be either of

$$\begin{aligned} M_3(1^{s_1}2^{s_2}3^{s_3}) &\sim [12]^{s_1+s_2-s_3}[23]^{s_2+s_3-s_1}[31]^{s_3+s_1-s_2} \quad \text{for } s_1+s_2+s_3>0\\ M_3(1^{s_1}2^{s_2}3^{s_3}) &\sim \langle 12\rangle^{s_3-s_1-s_2}\langle 23\rangle^{s_1-s_2-s_3}\langle 31\rangle^{s_2-s_3-s_1} \quad \text{for } s_1+s_2+s_3<0 \end{aligned} \tag{18}$$

These amplitudes are complex conjugates of each other and for real momenta, due to momentum conservation and the on-shell conditions, they actually vanish. Indeed, $p_1 + p_2 + p_3 = 0$ and $p_i^2 = 0$ imply $p_i \cdot p_j = 0$ for any i and j. This gives $\langle 12\rangle[12] = \langle 23\rangle[23] = \langle 31\rangle[31] = 0$ which can only be satisfied either with all $\langle ij\rangle = 0$ or all $[ij] = 0$ since the spinors belong to a two-dimensional vector space.

However, for complex momenta, $\langle ij\rangle$ and $[ij]$ are independent of each other, and the cubic amplitudes in (18) can be used as a basis for constructing higher order amplitudes recursively.

These amplitudes are exactly the same as the ones written above in light-front notation using the translation (6). This close correspondence is interesting because it could have been expected that covariantising the light-front vertices would be an awkward calculation. However, we don't know if this is special for the cubic level.

2.4. *Coupling constants*

The light-front higher spin cubic interactions discussed above should be supplied by a coupling factors $\rho(s_1, s_2, s_3)$ of mass dimension $-|s_1 + s_2 + s_3|$. For instance, a pure spin $s = \lambda$ interaction comes with the couplings $\rho(\lambda, \lambda, -\lambda)$ and $\rho(\lambda, -\lambda, -\lambda)$ of mass dimension $-\lambda$ as it should (to compensate the dimension of the momenta at the vertex). It is convenient to let the dimension be carried by a parameter ρ of dimension -1 and write

$$\rho(s_1, s_2, s_3) = \rho^{|s_1+s_2+s_3|} y_{s_1,s_2,s_3} \tag{19}$$

where y_{s_1,s_2,s_3} is a numerical factor. It is obvious that y_{s_1,s_2,s_3} has full permutational symmetry in s_1, s_2 and s_3 but note that the signs of the s are significant. $y_{\lambda,\lambda,\lambda}$ need not be equal to $y_{\lambda,\lambda,-\lambda}$ (for $\lambda = 2$ corresponding to the cubic R^3 term and the spin 2 basic cubic interaction respectively). However, we do have $y_{\lambda,\lambda,-\lambda} = y_{\lambda,-\lambda,-\lambda}$ ensuring hermiticity of the light-front action.

3. Higher Spin Quartic Amplitudes Through BCFW Recursion

We will attempt a computation of the general pure quartic higher spin amplitude $M_4[1^{s_1}, 2^{s_2}, 3^{s_3}, 4^{s_4}]$ using BCFW recursion closely following the procedure in Benincasa and Cachazo[3] specialising to the pure spin s MHV amplitude with $s_1 = -s_2 = s_3 = -s_4 = s$. However, we will consider what happens when particles of all allowed helicity are summed over in the factorisation channels. It may therefore be useful to briefly discuss the BCFW method itself.

3.1. *BCFW recursion*

On-shell tree level scattering amplitudes are determined by the momenta of the external particles i and their types, in the case of higher spin theory, by the helicities s_i encoded by polarisations ϵ_i, and colour indices $\{a_i b_i\}$ for odd spin. For our purposes, it is enough to consider four-particle amplitudes with $p_1+p_2+p_3+p_4 = 0$ and

$p_i^2 = 0$ for all the particles. A tree amplitude (under the assumptions of S-matrix theory) is therefore a rational function of the Lorentz invariant quantities $p_i \cdot p_j$ and $\epsilon_i \cdot p_j$ and poles can only come from internal particle propagators. For quartic amplitudes this means poles in **s**, **t** and **u**. There are many reviews treating amplitude methods and BCFW recursion.[20–22] Here we will just outline the procedure in the special case of four-particle amplitudes.

In practice the recursion is done by deforming two momenta, say p_1 and p_2 into $\hat{p}_1 = p_1 - zq$ and $\hat{p}_2 = p_2 + zq$ with complex z and $q \cdot q = p_1 \cdot q = p_2 \cdot q = 0$ so that momentum conservation and the on-shell conditions still hold. Then one studies the meromorphic function $M_n(z)/z$. Consider a certain internal momenta in the amplitude (that gets shifted), say $\hat{P}_k(z) = \hat{p}_1 + p_3$. It goes on-shell when

$$\hat{P}_k(z)^2 = P_k^2 - 2zP_k \cdot q \to 0 \quad \Rightarrow \quad z \to z_k = \frac{P_k^2}{2P_k \cdot q} \tag{20}$$

or

$$z - z_k = \frac{2zP_k \cdot q - P_k^2}{2P_k \cdot q}. \tag{21}$$

This means that a pole in the amplitude will occur as

$$\frac{1}{\hat{P}_k(z)^2} = \frac{1}{2P_k \cdot q} \frac{1}{z - z_k} = \frac{z_k}{P_k^2} \frac{1}{z - z_k} \tag{22}$$

which – by the way – explains why one studies $M_n(z)/z$.

When an internal propagator goes on-shell – corresponding to z approaching the corresponding pole – the amplitude becomes dominated by the pole and factorises into a product of two subamplitudes and the propagator.

Consider the Riemann sphere $\mathcal{R} = \mathbb{C} \cup \{\infty\}$. If $f(z)$ is a meromorphic function on the sphere (only a finite number of singularities) then the sum of the residues at the singularities is zero.[23] Applying this theorem to the function $M_n^{(i,j)}(z)$ (where the superscript (i,j) denotes which momenta are shifted) yields

$$\frac{1}{2\pi i} \int_{\mathcal{R}} \frac{M_n^{(i,j)}(z)}{z} dz = M_n^{(i,j)}(0) + \sum_k \frac{c_k^{(i,j)}}{z_k} - \mathcal{C}_n^{(i,j)} = 0 \tag{23}$$

where $M_n^{(i,j)}(0)$ is the residue at $z = 0$ to be identified by the physical amplitude. The $c_k^{(i,j)}$ are the residues at the poles z_k and $\mathcal{C}_n^{(i,j)}$ is boundary term at infinity. Different deformations (i,j) yield shifted amplitudes $M_n^{(i,j)}(z)$ with different locations of the poles z_k, all however containing the physical amplitude at $z = 0$.

Consider a certain pole z_k corresponding to a certain channel $z \to z_k$. As seen above, the momenta $\hat{P}_k(z)$ goes on-shell and this channel dominates. The amplitude factorises into

$$M_n^{(i,j)}(z) \to M_L^{(i,j)}(z_k) \frac{z_k}{P_k^2} \frac{1}{z - z_k} M_R^{(i,j)}(z_k) \tag{24}$$

and the residue is given by

$$-\frac{c_k^{(i,j)}}{z_k} = M_L^{(i,j)}(z_k) \frac{1}{P_k^2} M_R^{(i,j)}(z_k). \tag{25}$$

The subscripts L and R denote the subamplitudes, in the case of four-particle these are cubic amplitudes. The physical amplitude is given by

$$M_n = M_n^{(i,j)}(0) = M_L^{(i,j)}(z_k)\frac{1}{P_k^2}M_R^{(i,j)}(z_k) + \mathcal{C}_n^{(i,j)}. \tag{26}$$

The boundary contribution vanishes if $\lim_{z\to\infty} M_n^{(i,j)}(z)$.

In this paper I will only consider amplitudes under the assumption that deformed amplitudes go to zero as $z \to \infty$. Benincasa and Conde[4,15] has studied BCFW deformations with a non-zero boundary term. They show that the factorisation of the amplitudes into subamplitudes remain the same but that the propagator factor gets multiplied by a "weight" factor $f_k^{\nu,n}$ such that

$$M_n = M_L^{(i,j)}(z_k)\frac{f_k^{\nu,n}}{P_k^2}M_R^{(i,j)}(z_k) \tag{27}$$

where ν denotes the power with which the boundary term diverges.

3.2. *Constructibility and the four-particle test*

Benincasa and Cachazo[3] introduce two concepts, *constructibility* and *the four-particle test*. A theory is constructible if the four-particle tree level amplitudes can completely computed from the three-particle amplitudes.

The four-particle test amounts to computing a certain amplitude using two different deformations and requiring the results to be equal. In the case of the four-particle amplitudes considered below we will deform particles 1 and 2 and particle 1 and 4 respectively using the notation $M_4^{(i,j)}(z)$ for a i and j deformation. Then the test amounts to requiring $M_4^{(1,2)}(0) = M_4^{(1,4)}(0)$. In practice, $M_4^{(1,4)}(0)$ is obtained from $M_4^{(1,2)}(0)$ simply by interchanging the labels 2 and 4 in the formula for $M_4^{(1,4)}(0)$. In terms of the kinematic invariants this means interchanging $\mathbf{s}$ and $\mathbf{t}$ (but of course polarisation factors may also change).

Benincasa and Cachazo[3] consider single spin theories. Here I will study what happens when one sums over an infinite spectrum of particles in the channels.

3.3. *Computing the four-particle amplitude*

In Figure 1 the $\mathbf{t}$-channel contribution to the amplitude we want to compute is depicted. The $\mathbf{u}$-channel should be added, but that we get by interchanging legs 3 and 4. The $\mathbf{s}$-channel do not contribute since we will be shifting momenta 1 and 2. The channel momentum is denoted by s_c. Eventually it will be summed over a certain range. Writing down the parts of this amplitude (before doing the

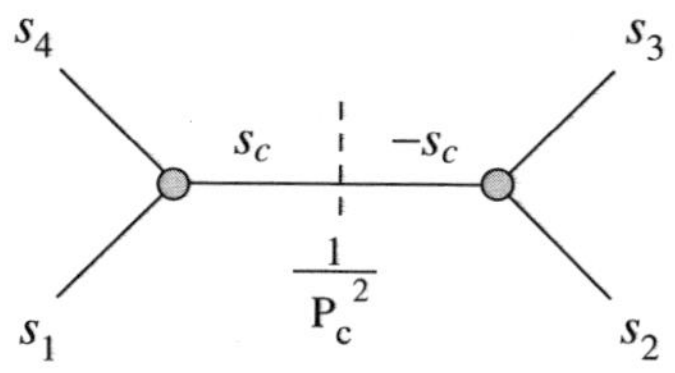

Figure 1. The **t**-channel contribution to the four-particle amplitude.

momentum shifts) using (18) for the cubic amplitudes we get

$$\begin{aligned}&\Big(\Theta_{s_1+s_4+s_c}[14]^{s_1+s_4-s_c}[4c]^{s_4-s_1+s_c}[c1]^{s_1-s_4+s_c}+\\&\Theta_{-(s_1+s_4+s_c)}\langle 14\rangle^{s_c-s_1-s_4}\langle 4c\rangle^{s_1-s_4-s_c}\langle c1\rangle^{s_4-s_1-s_c}\Big)\frac{1}{P_{14}^2}\times\\&\Big(\Theta_{s_2+s_3-s_c}[c3]^{s_3-s_2-s_c}[32]^{s_3+s_2+s_c}[2c]^{s_2-s_3-s_c}+\\&\Theta_{-(s_2+s_3-s_c)}\langle c3\rangle^{s_c+s_2-s_3}\langle 32\rangle^{-s_c-s_3-s_2}\langle 2c\rangle^{s_3-s_2+s_c}\Big)\end{aligned} \tag{28}$$

where Θ_x is 1 for $x > 0$ and 0 for $x < 0$. The channel momentum c is $P_{14} = p_1 + p_4$. The coupling factors $\rho(s_1, s_4, s_c)$ and $\rho(s_2, s_3, s_c)$ that should multiply the Θ factors were discussed in section 2.4. They will be included when needed simply by writing for instance $\Theta_{s_1+s_4+s_c} \to \rho(s_1, s_4, s_c)\Theta_{s_1+s_4+s_c} = \rho^{|s_1+s_4+s_c|}y_{s_1,s_4,s_c}\Theta_{s_1+s_4+s_c}$.

Now we perform the following shift

$$\begin{aligned}|\hat{1}\rangle &= |1\rangle + z|2\rangle \qquad & |\hat{1}] &= |1]\\|\hat{2}\rangle &= |2\rangle & |\hat{2}] &= |2] - z|1].\end{aligned} \tag{29}$$

In terms of momenta this works out as

$$\begin{aligned}\hat{p}_1 &= p_1 - z|1]\langle 2| = p_1 - z|2\rangle[1|\\\hat{p}_2 &= p_2 + z|1]\langle 2| = p_2 + z|2\rangle[1|\end{aligned} \tag{30}$$

depending on downstairs (first equality) or upstairs (second equality) dotted and undotted indices. Momentum conservation and the on-shell conditions are maintained since $p_1 = -|1]\langle 1| = -|1\rangle[1|$ and $p_2 = -|2]\langle 2| = -|2\rangle[2|$. It is also clear why the **s**-channel momentum doesn't get shifted.

Upon performing the shift we get from (28)

$$\begin{aligned}&\Big(\Theta_{s_1+s_4+s_c}[\hat{1}4]^{s_1+s_4-s_c}[4\hat{P}_{14}]^{s_4-s_1+s_c}[\hat{P}_{14}\hat{1}]^{s_1-s_4+s_c}+\\&\Theta_{-(s_1+s_4+s_c)}\langle\hat{1}4\rangle^{s_c-s_1-s_4}\langle 4\hat{P}_{14}\rangle^{s_1-s_4-s_c}\langle\hat{P}_{14}\hat{1}\rangle^{s_4-s_1-s_c}\Big)\frac{1}{\hat{P}_{14}^2}\times\\&\Big(\Theta_{s_2+s_3-s_c}[\hat{P}_{14}3]^{s_3-s_2-s_c}[3\hat{2}]^{s_3+s_2+s_c}[\hat{2}\hat{P}_{14}]^{s_2-s_3-s_c}+\\&\Theta_{-(s_2+s_3-s_c)}\langle\hat{P}_{14}3\rangle^{s_c+s_2-s_3}\langle 3\hat{2}\rangle^{-s_c-s_3-s_2}\langle\hat{2}\hat{P}_{14}\rangle^{s_3-s_2+s_c}\Big).\end{aligned} \tag{31}$$

Just to be clear: this is the **t**-channel part of the $M_4^{(1,2)}(z)$ amplitude. Before we are finished we shall have to add the **u**-channel part (interchanging legs 3 and 4) and sum over channel momentum. Next it should be compared to the same amplitude computed via a $(1,4)$ shift, i.e. $M_4^{(1,4)}(z)$ (done by interchanging labels 2 and 4).

Evaluation of the shifted spinors In order to compute the amplitude, the shifted spinors in this expression must be evaluated at the pole in z. From $\hat{P}_{14}^2(z) = 0$ we get $\hat{P}_{14}^2 = 2\hat{p}_1 \cdot p_4 = \langle\hat{1}4\rangle[14] = 0$. Therefore z must be chosen such that $\langle\hat{1}4\rangle = \langle 14\rangle + z\langle 14\rangle = 0$. Thus in the **t**-channel $z_t = -\langle 14\rangle/\langle 24\rangle$.

For $\hat{P}_{14}(z_t)$ we get

$$\hat{P}_{14}(z_t) = -|1]\langle 1| + \frac{\langle 14\rangle}{\langle 24\rangle}|1]\langle 2| - |4]\langle 4| = -\frac{[14]}{[13]}|3]\langle 4| \tag{32}$$

using Schouten identities and momentum conservation. But we also need the spinors $|\hat{P}_{14}]$ and $\langle\hat{P}_{14}|$. From (32) we get

$$\begin{aligned}\langle\hat{P}_{14}| &= \alpha\langle 4| \\ |\hat{P}_{14}] &= \beta|3]\end{aligned} \tag{33}$$

with $\alpha\beta = [14]/[13]$. Furthermore, again using Schouten identities and momentum conservation to rewrite the shifted spinors of (30), we have

$$\begin{aligned}|\hat{1}\rangle &= \frac{\langle 21\rangle}{\langle 24\rangle}|4\rangle \quad \text{and} \quad |\hat{1}] = |1] \\ |\hat{2}] &= \frac{[12]}{[13]}|3] \quad \text{and} \quad |\hat{2}\rangle = |2\rangle.\end{aligned} \tag{34}$$

Evaluation of the subamplitudes of expression (31)

L1:

$$\begin{aligned}&\Theta_{s_1+s_4+s_c}[\hat{1}4]^{s_1+s_4-s_c}[4\hat{P}_{14}]^{s_4-s_1+s_c}[\hat{P}_{14}\hat{1}]^{s_1-s_4+s_c} = \\ &\Theta_{s_1+s_4+s_c}[14]^{s_1+s_4-s_c}[43]^{s_4-s_1+s_c}[31]^{s_1-s_4+s_c}\beta^{2s_c}\end{aligned} \tag{35}$$

L2:

$$\begin{aligned}&\Theta_{-(s_1+s_4+s_c)}\langle\hat{1}4\rangle^{s_c-s_1-s_4}\langle 4\hat{P}_{14}\rangle^{s_1-s_4-s_c}\langle\hat{P}_{14}\hat{1}\rangle^{s_4-s_1-s_c} = \\ &\Theta_{-(s_1+s_4+s_c)}\left(\frac{\langle 21\rangle}{\langle 24\rangle}\right)^{-2s_1}\alpha^{-2s_c}\langle 44\rangle^{-(s_1+s_4+s_c)} = 0\end{aligned} \tag{36}$$

R1:

$$\begin{aligned}&\Theta_{s_2+s_3-s_c}[\hat{P}_{14}3]^{s_3-s_2-s_c}[3\hat{2}]^{s_3+s_2+s_c}[\hat{2}\hat{P}_{14}]^{s_2-s_3-s_c} = \\ &\Theta_{s_2+s_3-s_c}\left(\frac{[12]}{[13]}\right)^{2s_2}\beta^{-2s_2}[33]^{s_2+s_3-s_c} = 0\end{aligned} \tag{37}$$

R2:

$$\begin{aligned}&\Theta_{-(s_2+s_3-s_c)}\langle\hat{P}_{14}3\rangle^{s_c+s_2-s_3}\langle 3\hat{2}\rangle^{-s_c-s_3-s_2}\langle\hat{2}\hat{P}_{14}\rangle^{s_3-s_2+s_c} = \\ &\Theta_{-(s_2+s_3-s_c)}\langle 43\rangle^{s_c+s_2-s_3}\langle 32\rangle^{-s_c-s_3-s_2}\langle 24\rangle^{s_3-s_2+s_c}\alpha^{2s_c}\end{aligned} \tag{38}$$

We can now put together the **t**-channel contribution to $M_4^{(1,2)}(0)$. After some algebra we get[3]

$$M_{4,\mathbf{t}}^{(1,2)}(0) = \Theta_{s_1+s_4+s_c}\Theta_{-(s_2+s_3-s_c)}\frac{(-P_{3,4}^2)^{s_c}}{P_{1,4}^2}\times \left(\frac{[14][31]}{[43]}\right)^{s_1}\left(\frac{\langle 43\rangle}{\langle 32\rangle\langle 24\rangle}\right)^{s_2}\left(\frac{\langle 24\rangle}{\langle 43\rangle\langle 32\rangle}\right)^{s_3}\left(\frac{[14][43]}{[31]}\right)^{s_4}. \tag{39}$$

These terms contribute to the channel sum over s_c for $s_c > \max(-(s_1+s_4),(s_2+s_3))$. The **u**-channel contribution to $M_4^{(1,2)}(0)$ is obtained by interchanging labels 3 and 4.

$$M_{4,\mathbf{u}}^{(1,2)}(0) = \Theta_{s_1+s_3+s_c}\Theta_{-(s_2+s_4-s_c)}\frac{(-P_{4,3}^2)^{s_c}}{P_{1,3}^2}\times \left(\frac{[13][41]}{[34]}\right)^{s_1}\left(\frac{\langle 34\rangle}{\langle 42\rangle\langle 23\rangle}\right)^{s_2}\left(\frac{\langle 23\rangle}{\langle 34\rangle\langle 42\rangle}\right)^{s_4}\left(\frac{[13][34]}{[41]}\right)^{s_3}. \tag{40}$$

These terms contribute to the channel sum over s_c for $s_c > \max(-(s_1+s_3),(s_2+s_4))$.

3.4. *Pure spin s four-particle MHV amplitude*

We now have enough data to compute a pure spin s MHV four-particle amplitude with $s_1 = -s_2 = s_3 = -s_4 = s$. Then the channel sum runs $s_c > 0$ in the **t**-channel contribution and $s_c > -2s$ **u**-channel contribution. Also introduce the coupling factors according to (19)

$$\begin{aligned}\rho(s_1,s_4,s_c)\rho(s_2,s_3,-s_c) &= \rho(s,-s,s_c)\rho(-s,s,-s_c)\\ &= \rho^{2|s_c|}y_{s,-s,s_c}y_{-s,s,-s_c} = \rho^{2|s_c|}y_{s,-s,s_c}^2\end{aligned} \tag{41}$$

for the **t**-channel, and

$$\begin{aligned}\rho(s_1,s_3,s_c)\rho(s_2,s_4,-s_c) &= \rho(s,s,s_c)\rho(-s,-s,-s_c)\\ &= \rho^{2|s_c+2s|}y_{s,s,s_c}y_{-s,-s,-s_c} = \rho^{2|s_c+2s|}y_{s,s,s_c}^2.\end{aligned} \tag{42}$$

Specialising formulas (39) and (40) and submitting them to some further algebra we get for the **t**-channel

$$-\sum_{s_c>0}\rho^{2|s_c|}y_{s,-s,s_c}^2\frac{\mathbf{s}^{s_c-2s}}{\mathbf{t}}\left([31]\langle 24\rangle\right)^{2s} \tag{43}$$

and for the **u**-channel

$$-\sum_{s_c>-2s}\rho^{2|s_c+2s|}y_{s,s,s_c}^2\frac{\mathbf{s}^{s_c}}{\mathbf{u}}\left([31]\langle 24\rangle\right)^{2s}. \tag{44}$$

The sum of these two sums is a candidate for a pure spin s four-particle MHV amplitude computed by a $(1,2)$ BCFW shift. For a further simplification, let us only consider even spin (so that we need not take care of colour factors), for concreteness scattering of spin $s = 2$ and $s = 4$ respectively.

Pure spin 2 scattering

$$-\left([31]\langle 24\rangle\right)^4 \Big(\rho^4 y_{2,-2,2}^2 \frac{\mathbf{s}^{-2}}{\mathbf{t}} + \rho^8 y_{2,-2,4}^2 \frac{\mathbf{s}^{0}}{\mathbf{t}} + \rho^{12} y_{2,-2,6}^2 \frac{\mathbf{s}^{2}}{\mathbf{t}} + \ldots$$
$$\rho^4 y_{2,2,-2}^2 \frac{\mathbf{s}^{-2}}{\mathbf{u}} + \rho^8 y_{2,2,0}^2 \frac{\mathbf{s}^{0}}{\mathbf{u}} + \rho^{12} y_{2,2,2}^2 \frac{\mathbf{s}^{2}}{\mathbf{u}} + \ldots\Big) \tag{45}$$

A theory with only spin 2 in the channel would have only $y_{2,-2,2} = y_{2,2,-2}$ non-zero with an amplitude depending on the kinematic invariants as

$$\sim \mathbf{s}^{-2}\left(\frac{1}{\mathbf{t}} + \frac{1}{\mathbf{u}}\right) = \frac{1}{\mathbf{stu}}. \tag{46}$$

The amplitude computed using a $(1,4)$ shift can be obtained by interchanging labels 2 and 4 (as discussed above). The prefactor $[31]\langle 24\rangle$ and $\mathbf{u}$ stays unchanged while $\mathbf{s} \leftrightarrow \mathbf{t}$. The form of the prefactor depends on the helicity configuration $s_1 = s_3 = -s_2 = -s_4 = s$. Therefore the amplitudes that results from a $(1,2)$ shift and a $(1,4)$ shift are the same. But note that one is then actually discarding a pure spin 2 term in the $\mathbf{u}$ channel, namely

$$\rho^{12} y_{2,2,2}^2 \frac{\mathbf{s}^2}{\mathbf{u}}$$

corresponding to the (cubic level) R^3 two-loop counterterm.

Pure spin 4 scattering

$$-\left([31]\langle 24\rangle\right)^8 \Big(\rho^4 y_{4,-4,2}^2 \frac{\mathbf{s}^{-6}}{\mathbf{t}} + \rho^8 y_{4,-4,4}^2 \frac{\mathbf{s}^{-4}}{\mathbf{t}} + \rho^{12} y_{4,-4,6}^2 \frac{\mathbf{s}^{-2}}{\mathbf{t}} + \ldots$$
$$\rho^4 y_{4,4,-6}^2 \frac{\mathbf{s}^{-6}}{\mathbf{u}} + \rho^8 y_{4,4,-4}^2 \frac{\mathbf{s}^{-4}}{\mathbf{u}} + \rho^{12} y_{4,4,-2}^2 \frac{\mathbf{s}^{-2}}{\mathbf{u}} + \ldots\Big) \tag{47}$$

A theory with only spin 4 in the channel and just one coupling constant $\rho^4 y_{4,4,-4}$ (and thus discardaring a pure spin 4 term further up in the $\mathbf{u}$-channel) would have a dependence on the kinematic invariants

$$\sim \mathbf{s}^{-4}\left(\frac{1}{\mathbf{t}} + \frac{1}{\mathbf{u}}\right) = \frac{\mathbf{s}^{-2}}{\mathbf{stu}}. \tag{48}$$

Again, the amplitude computed using a $(1,4)$ shift can be obtained by interchanging labels 2 and 4, with the result for the dependence on kinematic invariants

$$\sim \mathbf{t}^{-4}\left(\frac{1}{\mathbf{s}} + \frac{1}{\mathbf{u}}\right) = \frac{\mathbf{t}^{-2}}{\mathbf{stu}}. \tag{49}$$

So, according to the four-particle test, pure spin 4 scattering is ruled out since $M_4^{(1,2)} \neq M_4^{(1,4)}$. Let us look at the general pure spin s scattering.

Pure even spin s scattering

$$-([31]\langle 24\rangle)^{2s}\Big(\rho^4 y^2_{s,-s,2}\frac{\mathbf{s}^{2-2s}}{\mathbf{t}}+\rho^8 y^2_{s,-s,4}\frac{\mathbf{s}^{4-2s}}{\mathbf{t}}+\rho^{12} y^2_{s,-s,6}\frac{\mathbf{s}^{6-2s}}{\mathbf{t}}+\dots$$
$$\rho^4 y^2_{s,s,2-2s}\frac{\mathbf{s}^{2-2s}}{\mathbf{u}}+\rho^8 y^2_{s,s,4-2s}\frac{\mathbf{s}^{4-2s}}{\mathbf{u}}+\rho^{12} y^2_{s,s,6-2s}\frac{\mathbf{s}^{6-2s}}{\mathbf{u}}+\dots\Big) \tag{50}$$

Picking out the terms corresponding to spin s in the channel and uniform dimension of the coupling constant, we get for the $(1,2)$ shift

$$M_4^{(1,2)}\sim([31]\langle 24\rangle)^{2s}\rho^{2s}y^2_{s,s,-s}\mathbf{s}^{-s}\left(\frac{1}{\mathbf{t}}+\frac{1}{\mathbf{u}}\right)\sim\frac{\mathbf{s}^{2-s}}{\mathbf{stu}} \tag{51}$$

and for the $(1,4)$ shift

$$M_4^{(1,4)}\sim([31]\langle 24\rangle)^{2s}\rho^{2s}y^2_{s,s,-s}\mathbf{t}^{-s}\left(\frac{1}{\mathbf{s}}+\frac{1}{\mathbf{u}}\right)\sim\frac{\mathbf{t}^{2-s}}{\mathbf{stu}}. \tag{52}$$

This essentially the four-particle test as derived by Benincasa and Cachazo.[3] Demanding the two expressions for the scattering amplitude to be the same forces $s=2$, ruling out higher spin scattering.

3.5. *End of story?*

Embarrasing as this result is for Minkowski higher spin, it is not yet the end of story. First of all, we really have to study the full series of terms i (43) and (44). A somewhat similar analysis has been reported by Dempster and Tsulaia.[24]

For concreteness, study the case of pure spin 2 scattering. Let us also assume that the numerical y coefficients are equal at each power $\mathbf{s}$: $y_{2,-2,n}=y_{2,-2,n-4}\equiv y_n$. This amounts to, for instance, that the spin $(2,2,2)$ coupling comes with the same numerical strength as the spin $(2,-2,6)$ coupling involving a spin 6 field (which of course need not be the case). We get

$$-([31]\langle 24\rangle)^{4}\Big(\rho^4 y^2_{2,-2,2}\frac{\mathbf{s}^{-2}}{\mathbf{t}}+\rho^8 y^2_{2,-2,4}\frac{\mathbf{s}^{0}}{\mathbf{t}}+\rho^{12} y^2_{2,-2,6}\frac{\mathbf{s}^{2}}{\mathbf{t}}+\dots$$
$$\rho^4 y^2_{2,2,-2}\frac{\mathbf{s}^{-2}}{\mathbf{u}}+\rho^8 y^2_{2,2,0}\frac{\mathbf{s}^{0}}{\mathbf{u}}+\rho^{12} y^2_{2,2,2}\frac{\mathbf{s}^{2}}{\mathbf{u}}+\dots\Big)= \tag{53}$$
$$-([31]\langle 24\rangle)^{4}\rho^4\frac{1}{\mathbf{stu}}\big(y_2^2+y_4^2\rho^4\mathbf{s}^2+y_6^2\rho^8\mathbf{s}^4+\dots\big).$$

The corresponding expression coming from the $(1,4)$ shift becomes

$$-([31]\langle 24\rangle)^{4}\rho^4\frac{1}{\mathbf{stu}}\big(y_2^2+y_4^2\rho^4\mathbf{t}^2+y_6^2\rho^8\mathbf{t}^4+\dots\big). \tag{54}$$

The expressions still disagree. Considered as power series they will have a certain radius of convergence within which they certainly differ and outside of which they both diverge. However, both results make some sense. We know that higher spin theory entails ever increasing powers of momenta in the interactions, cubic and higher order. In terms of amplitudes this most likely would manifest itself through increasing powers of kinematic invariants.

We see that computing the amplitude through a $(1, 2)$ shift, we pick up poles in $\mathbf{t}$ and $\mathbf{u}$ while we get power series in the $\mathbf{s}$ channel. Likewise, computing the amplitude through a $(1, 4)$ shift, we pick up poles in $\mathbf{s}$ and $\mathbf{u}$ while we get power series in the $\mathbf{t}$ channel. In both cases the missing pole also appears when adding the two series.

Had we computed the amplitude through a $(1, 3)$ shift, we would get poles in $\mathbf{s}$, $\mathbf{t}$ and $\mathbf{u}$ multiplying a power series in the $\mathbf{u}$ channel. Indeed, we would get by interchanging labels 2 and 3

$$-\left([21]\langle 34\rangle\right)^4 \rho^4 \frac{1}{\mathbf{stu}}\left(y_2^2 + y_4^2\rho^4\mathbf{u}^2 + y_6^2\rho^8\mathbf{u}^4 + \dots\right) \tag{55}$$

where the polarisation factor changes since the helicity configuration changes.

These circumstances throw some doubt on the combined power och constructibility and the four-particle test. Even though quartic amplitudes for lower spin theories can be fully reconstructed by BCFW recursion using just one type of shift, it is not clear to me that it should be sufficient for higher spin. Perhaps a certain shift only probes part of the full amplitude and that rather than requiring equality of the results of different shifts, perhaps we should combine them. The full amplitude then shows poles in all channels as well as power series in all invariants. However, this line of thought must be supplemented by a careful analysis of the $z \to \infty$ limit of the terms in the shifted amplitudes and the behaviour of the boundary term. This analysis remains to be done.

Furthermore, it may be that higher spin scattering amplitudes are not constructible, meaning that there could be an irreducible part to the four-particle amplitude that has to be added to the terms coming from cubic amplitudes. Which seems to return the question to computing the full quartic (Feynman) vertex.

4. Can Minkowski Higher Spin Be Rescued?

The answer to this question – if there is a positive one – must be sought in physical understanding as much as in technical calculations. There is one basic physical puzzle in all of higher spin theory. If there are higher spin excitations in nature: then where are they, what do they do and what role do they play?

That we are thinking about extreme high energy phenomena is obvious. Perhaps one could speak of "beyond quantum gravity" which is probably what people have in mind when they write about the subject in terms of "higher spin gravity". With this in mind, what sense does it make to think of scattering spin s particles on spin s' particles? Not very much it seems.

Exciting one higher spin particle must clearly excite a whole bunch of them – indeed all of them – if we take seriously what we already know. So, in my opinion, considering particular higher spin fields by themselves doesn't make much sense. What we should consider is the full spectrum of fields as one object. One should work with non-local objects such as $\Phi(x, \xi)$. I have no idea of how an S-matrix theory of such an object would look like. But it should be possible to – formally at least – join all cubic subamplitudes into one cubic amplitude maintaining them all. One way of doing that in a principled way could be to consider the component higher spin fields as components of a continuous spin field. The continuous spin representations naturally organize all helicities as well as a naturally involve a dimensionfull parameter. On a computational level that would mean not only summing over channel spin but also over external spin s_1, s_2, s_3, s_4 corresponding to scattering continuous spin fields on contiuous spin fields. This is clearly a little bit speculative, but not more so than that a calculation along these lines could be attempted.

Acknowledgment

This paper is an extended and modified version of a presentation I gave at the international workshop on Higher Spin Gauge Theories at the Institute of Advanced Studies at Nanyang Technological University in Singapore, November 4 - 6, 2015. I'd like to thank the organizers for the invititation and the opportunity it gave to look into this fascinating aspect of higher spin theory (of which I wasn't up to date when giving my talk). I also thank Slava Didenko for pointing out the work by Metsaev on quartic interactions that had slipped my memory.

References

1. R. R. Metsaev, Poincaré invariant dynamics of massless higher spins: Fourth order analysis on mass shell, *Mod. Phys. Lett.* **A6**, 359, (1991).
2. R. R. Metsaev, S-matrix approach to massless higher spins theory: Ii. the case of internal symmetry, *Mod. Phys. Lett.* **A6**, 2411, (1991).
3. P. Benincasa and F. Cachazo, Consistency conditions on the s-matrix of massless particles. (2008). arXiv:0705.4305.
4. P. Benincasa and E. Conde, Exploring the s matrix of massless particles, *Phys. Rev. D.* **86, 025007**, (2011). arXiv:1108.3078.
5. A. K. H. Bengtsson, I. Bengtsson, and L. Brink, Cubic interaction terms for arbitrary spin, *Nucl. Phys. B.* **227**, 31–40, (1983).
6. A. K. H. Bengtsson, I. Bengtsson, and N. Linden, Interacting higher-spin gauge fields on the light front, *Class. Quant. Grav.* **4**, 1333, (1987).
7. R. R. Metsaev, Generating function for cubic interaction vertices of higher spin fields in any dimension, *Modern Physics Letters A.* **8**(25), 2413–2426, (1993).
8. R. R. Metsaev, Cubic interaction vertices for massive and massless higher spin fields, *Nucl. Phys. B.* **759**, 147–201, (2006). arXiv:hep-th/0512342.
9. R. R. Metsaev, Cubic interaction vertices for fermionic and bosonic arbitrary spin fields, *Nucl. Phys. B.* **859**, 13–69, (2012). arXiv:0712.3526.
10. I. Bengtsson, M. Cederwall, and O. Lindgren, Light-cone actions for gravity and higher spins. (1983). Göteborg preprint 83-55.
11. S. Ananth, The quintic interaction vertex in light-cone gravity, *Phys. Lett.* **B664**, 219–223, (2008). arXiv:0803.1494.

12. F. A. Berends, G. J. H. Burgers, and H. van Dam, On spin three self interactions, *Z. Phys. C.* **24**, 247–254, (1984).
13. I. Bengtsson. Private communication.
14. A. K. H. Bengtsson, Systematics of higher-spin light-front interactions. (2012). arXiv:1205.6117.
15. P. Benincasa and E. Conde, On the tree-level structure of scattering amplitudes of massless particles, *JHEP.* **11(2011)074**, (2011). arXiv:1106.0166.
16. R. Britto, F. Cachazo, and B. Feng, New recursion relations for tree amplitudes of gluons, *Nucl. Phys. B.* **715**, 499, (2005). arXiv:hep/th-0412308.
17. R. Britto, F. Cachazo, B. Feng, and E. Witten, Direct proof of tree-level recursion relation in yang-mills theory, *Phys. Rev. Lett.* **94**, 181602, (2005). arXiv:hep/th-0501052.
18. H. Kawai, D. Lewellen, and S. Tye, A relation between tree amplitudes of closed and open strings, *Nucl. Phys. B.* **269**, 1–23, (1986).
19. S. Ananth, Spinor helicity structures in higher spin theories, *J. High Energy Phys.* **1211**, 089, (2012). arXiv:1209.4960.
20. L. Dixon, A brief introduction to modern amplitude methods, *CERN-2014-008, pp. 31-67.* (2013). arXiv:1302.3225.
21. B. Feng and M. Luo, An introduction to on-shell recursion relations, *Front. Phys.* **7(5)**, 533–575, (2012). arXiv:1111.5759.
22. H. Elvang and Y.-T. Huang, *Scattering Amplitudes in Gauge Theory and Gravity.* (Cambridge University Press, 2015).
23. H. Cohn, *Conformal Mapping on Riemann Surfaces.* Dover Books on Advanced Mathematics, (Dover Publications, 1967). Theorem 2-4.
24. P. Dempster and M. Tsulaia, On the structure of quartic vertices for massless higher spin fields on minkowski background, *Nucl. Phys. B.* **865**, 353–375, (2012). arXiv:1203.5597.

Solutions in Bosonic String Field Theory and AdS Higher Spin Algebras

Dimitri Polyakov
Center for Theoretical Physics
Sichuan University
29 Wangjiang Road, Chengdu 640064, China

1. Introduction. SFT vs Vasiliev's HS EOM: The Motivation

It is well known that the equations of motion of cubic string field theory:

$$Q\Psi + \Psi \star \Psi = 0$$

resemble the Vasiliev's equations in the unfolding formalism in higher-spin theories:

$$dW + W \wedge \star W = 0$$

(flatness condition for connections in infinite-dimensional higher-spin algebras).

Higher spin holography implies, however, that this resemblance may be much more than just a formal similarity. The generalized 1-form W contains all the higher-spin gauge fields components in AdS spaces which, by holography principle, are related to various multi-index composite operators in the dual CFT. Any of these CFT's, in turn, must be a low-energy limit of string theory in AdS_{d+1}, with the CFT_d correlators reproduced by the worldsheet correlation functions of the vertex operators in AdS string theory, with the space-time fields polarized along the boundary of the AdS space. On the other hand, the second-quantized string field Ψ, satisfying the SFT EOM is nothing but the expansion containing infinite number of modes determined by these vertex operators. Both string fields and higher spin gauge fields in the both equations are known to be complicated objects to work with. The structure of these interactions is, however, crucial for our understanding of higher-spin extensions of the holography principle and non-supersymmetric formulation of AdS/CFT.

At the same time, the string field theory still remains our best hope to advance towards background-independent formulation of string dynamics. Analytic solutions in string field theory appear to be one of the most crucial ingredients in order to approach such string theories in the SFT formalism, using the concept of background independence. To illustrate this, suppose a string field Ψ_0 is a solution of the SFT EOM. Then the form of the EOM is invariant under the shift

$$\Psi \to \tilde{\Psi} = \Psi + \Psi_0$$

with the simultaneous shift of the BRST charge $Q \to \tilde{Q}$, so that $Q^2 = \tilde{Q}^2 = 0$ and the new nilpotent charge $\tilde{Q}$ defined according to

$$\tilde{Q}\Psi = Q\Psi + \Psi_0 \star \Psi + \Psi \star \Psi_0$$

for any Ψ. Then the new BRST charge $\tilde{Q}$ defines the new cohomology, different from that of the original charge Q, corresponding to string theory in a new background, depending on the structure of Ψ_0. In general, identifying analytic solutions of the SFT EOM is hard because of the complexity of the star product.

The SFT star product is most naturally defined in terms of the worldsheet correlators according to

$$\begin{aligned}
\langle\langle \Psi, \Psi \star \Psi \rangle\rangle &= \langle F_1^3 \circ \Psi(0) F_2^3 \circ \Psi(0) F_3^3 \circ \Psi(0) \rangle, \\
F_k^3(z) &= g \circ f_k^3(z) \quad (k = 1, 2, 3), \\
g(z) &= i\frac{1-z}{1+z}, \\
f_k^3(z) &= e^{\frac{2i\pi(k-1)}{3}} \left(\frac{1-iz}{1+iz}\right)^{\frac{2}{3}}
\end{aligned} \tag{1}$$

with $f_k^3(z)$ conformally mapping the worldsheets of 3 interacting strings to 3 wedges of a single disc and $g(z)$ further mapping this disc to the half-plane.

Generic string field Ψ, containing infinite number of non-primary operators without normal ordering, transforms under (1) in a complicated manner, with calculations easily wobbling out of control beyond low truncation levels.

For this reason, there are not many known examples of analytic solutions having a clear physical interpretation. One of the most fascinating and well-known solutions, describing the nonperturbative tachyonic vacuum in string theory is the Schnabl's famous solution. This solution, however, is restricted to the subspace of the s.c. "wedge states" and their descendants for which the star product gets simplified and can be computed. Our aim is to describe the class of the SFT solutions that lead to the nonperturbative physics of interacting HS theories at all orders of the interaction. The structure of this class of solutions in bosonic SFT is determined by the enveloping of HS algebra in AdS_3 but extensions to higher-dimensional AdS can be considered as well. The solution is given by the generating function for products of Bell polynomial operators in bosonic target space coordinates, which, in turn, form a natural OPE realization of w_∞ and higher-spin algebra in AdS_3.

2. $c = 1$ Model and HS Algebra in SFT Solution: A Warm-up Example

One particularly simple and almost obvious example of a class of string fields solving (1) can be constructed as follows. Let $V_i(z,p)$ $(i = 1, \ldots)$ be the set of all physical vertex operators in string theory in the cohomology of the original BRST charge Q (primary fields of ghost number 1 and conformal dimension 0) and $\lambda^i(p)$ are the corresponding space-time fields (where p is the momentum in space-time and we suppress the space-time.

Then the string field

$$\Psi_0 = \sum_i \lambda^i V_i$$

is the solution provided that the zero β-function conditions:

$$\beta_{\lambda^i} = 0$$

are imposed on the space-time fields in the leading order of the perturbation theory. This statement is easy to check. The on-shell invariance conditions on V_i imply $\{Q, \lambda^i V_i = \hat{L}\lambda^i = 0$ where $\hat{L}$ is some differential operator (e.g. a Laplacian plus the square of mass) acting on λ^i. Since the operators are the dimension zero primaries, they are invariant under the transformations (1) and therefore the star product can be computed simply by using

$$\begin{aligned}\langle\langle \Psi, \Psi \star \Psi \rangle\rangle &= \left\langle \prod_{n=1}^{3} F_n^3 \circ \Psi(0) \right\rangle \\ &= \sum_{i,j,k} C_{ijk} \lambda^i \lambda^j \lambda^k,\end{aligned}$$

where $C_{ijk}(p_1, p_2)$ are the structure constants in front of the simple pole in the OPE of the vertex operators:

$$V_i(z_1, p_1) V_j(z_2, p_2) \sim (z_1 - z_2)^{-1} C_{ijk}(p_1, p_2) V^k \left(\frac{1}{2}(z_1 + z_2), p_1 + p_2 \right).$$

Substituting this into SFT EOM then leads to the constraints on λ^i space-time fields:

$$\hat{L}\lambda^i + C^i_{jk} \lambda^j \lambda^k = 0$$

which are nothing but $\beta_{\lambda^i} = 0$ equation in the leading order.

The practical significance of this solution class is rather limited: it defines the *perturbative* changes of the background caused by the RG flows from the original conformal point (e.g. corresponding to flat background) to the new fixed point (corresponding to a certain solution of the low-energy effective equations of motion in string perturbation theory).

Physically, far more interesting is the case when the operators entering the string field solution are no longer the primaries since those are related to *nonperturbative* background deformations. Nevertheless, this class of solutions includes an instructive example, leading to the appearance of the higher-spin algebra and giving a hint for the structure of the solution describing *nonperturbative* higher-spin backgrounds. Now consider the noncritical open one-dimensional bosonic string theory (also known as the $c = 1$ model). It is well-known that this string theory does not contain a photon in the massless spectrum, however, due to the $SU(2)$ symmetry at the self-dual point, it does contain the $SU(2)$ multiplet of the discrete states which are physical at integer or half-integer momentum values only and become massless upon the Liouville dressing. To obtain the vertex operators for these states, consider the $SU(2)$ algebra generated by

$$T_\pm = \oint dz e^{\pm iX\sqrt{2}}$$
$$T_0 = \frac{i}{\sqrt{2}}\partial X$$

where X is a single target space coordinate and the dressed BRST-invariant highest weight vectors

$$V_l = \int dz e^{(ilX+(l-1)\varphi)\sqrt{2}}$$

where φ is the Liouville field and l is integer or half-integer. The SU(2) multiplet of the physical operators is then obtained by acting on V_l with the lowering operator T_- of SU(2):

$$U_{l|m} = T_-^{l-m} V_l$$
$$-l \leq m \leq l.$$

The dressed $U_{l|m}$ operators are the physical operators (massless states) of the $c = 1$ model and are the worldsheet integrals of primary fields of dimension one (equivalently, the primaries of dimension 0 at the unintegrated ghost number 1 picture). Manifest expressions for $U_{l|m}$ vertex operators are complicated, however, their structure constants have been deduced by using symmetry arguments.[1,2] One has

$$U_{l_1|m_1}(z)U_{l_2|m_2}(w) \sim (z-w)^{-1}C(l_1,l_2,l_3|m_1,m_2,m_3)f(l_1,l_2)U_{l_3,m_3}$$

where the $SU(2)$ Clebsch-Gordan coefficients are fixed by the symmetry while the function of Casimir eigenvalues $f(l_1, l_2)$ is nontrivial and was deduced to be given by

$$f(l_1,l_2) = \frac{\sqrt{l_1+l_2}(2l_1+2l_2-2)!}{\sqrt{2l_1l_2}(2l_1-1)!(2l_2-1)!}.$$

Remarkably, these structure constants coincide (up to a simple field redefinition) exactly with those of chiral copy higher spin algebra in AdS_3 in a certain basis, computed in a completely different context.[3]

Thus the discrete state primaries form the simplest vertex operator realization of AdS_3 higher-spin algebra.

The related OSFT solution is then constructed similarly to the previous one. It is given simply by

$$\Psi = \sum_{l,m} \lambda^{l|m} U_{l|m}$$

with the constants $\lambda^{l|m}$ satisfying the β-function condition

$$S^{l_3|m_3}_{l_1 m_1|l_2 m_2} \lambda^{l_1 m_1} \lambda^{l_2 m_2} = 0,$$

where $S^{l_3|m_3}_{l_1 m_1|l_2 m_2} = C(l_1, l_2, l_3|m_1, m_2, m_3) f(l_1, l_2)$ are the AdS_3 higher spin algebra's structure constants. As previously, this solution describes the perturbative background's change

$$Q \to \tilde{Q} = Q + \sum_{l,m} \lambda^{l|m} U_{l|m}.$$

Our goal is thus to find the off-shell analogues of this string field. solving the SFT EOM, with λ-constants satisfying the relation with the structure $\hat{L}\lambda + C\lambda^2$ with C being related to the constants of the higher-spin algebras. An instructive and helpful hint comes from studying the manifest expressions for the $U_{l|m}$ vertex operators, although in original papers their structure constants were obtained by symmetry arguments without knowing their precise form (which was not discussed in the literature before). Explicit calculation of the operator products gives the answer

$$\begin{aligned} U_{l|l-m} &= m! \oint dw e^{(i(l-m)X + (l-1)\varphi)\sqrt{2}} \\ &\quad \times \sum_{m^2|k_1 \ldots k_m} (-1)^{\pi(k_1,\ldots,k_m)} B^{(2l-k_1)}_{-i\sqrt{2}X} B^{(2l-k_2)}_{-i\sqrt{2}X} \ldots B^{(2l-k_m)}_{-i\sqrt{2}X}. \end{aligned}$$

Here, for any function $f(z)$, $B^{(n)}_{f(z)} \equiv B^{(n)}(\partial_z f, \ldots, \partial_z^n f)$ are the inhomogenious rank n normalized Bell polynomials in the derivatives of f, defined according to

$$\begin{aligned} B^{(n)}(\partial_z f, \ldots, \partial_z^n f) &= B^{(n)}(x_1, \ldots, x_n)|_{x_k \equiv \partial^k f; 1 \leq k \leq n} \\ &= \frac{1}{n!} \sum_{k=1}^{n} B_{n|k}(x_1, \ldots, x_{n-k+1}), \end{aligned}$$

and $B_{n|k}(x_1, \ldots, x_{n-k+1})$ are the *partial* Bell polynomials defined according to

$$\begin{aligned} &B_{n|k}(x_1, \ldots, x_{n-k+1}) \\ &\quad = \sum_{p_1, \ldots p_{n-k+1}} \frac{n!}{p_1! \ldots p_{n-k+1}!} x_1^{p_1} \left(\frac{x_2}{2!}\right)^{p_2} \ldots \left(\frac{x_{n-k+1}}{(n-k+1)!}\right)^{p_{n-k+1}} \end{aligned}$$

with the sum taken over all the combinations of non-negative p_j satisfying

$$\sum_{j=1}^{n-k+1} p_j = k,$$

$$\sum_{j=1}^{n-k+1} jp_j = n.$$

The sum for $U_{l|m}$ is taken over all the ordered length m partitions of $m^2 = k_1 + \cdots + k_m$ with $1 \leq k_1 \leq \cdots \leq k_m \leq 2m-1$ and with the parity $\pi(k_1, \ldots, k_m)$ of each partition defined as follows.

Consider a particular partition of m^2 $k_1 \leq k_2 \cdots \leq k_m$. By redpermutation we shall call any exchange between two neighboring elements of the partition with one unit that does not break the order of the partition, e.g.

$$\{k_1 \leq \cdots k_{i-1} \leq k_i \leq k_{i+1} \leq k_{i+2} \leq \cdots \leq k_m\}$$
$$\to \{k_1 \leq \cdots k_{i-1} \leq (k_i \pm 1) \leq (k_{i+1} \mp 1) \leq k_{i+2} \leq \cdots \leq k_m\}.$$

Then $\pi(k_1, \ldots, k_m)$ for any length m partition of m^2 is the minimum number of permutations needed to obtain the partition $m^2 = k_1 + \cdots + k_m$ from the reference partition $m^2 = 1 + 3 + 5 + \cdots + (2m-1)$.

Note that, possibly up to an overall sign change of $U_{l|m}$, any partition can be chosen as a reference partition.

3. SFT Ansatz Solution: Bell Polynomials, Star Product Computation and Generating Function for Enveloping of HS Algebras

Based on the above observations, our ansatz for the solution in D-dimensional bosonic string field theory is the following. Define the generating functions for the normalized Bell polynomials:

$$H(B) = \sum_n h_n B^{(n)}_{\vec{\alpha}\vec{X}},$$

$$G(H(B)) = \sum_n \frac{g_n}{n!} H^n(B),$$
$$= \sum_{N|p_1 \ldots p_n} \frac{N!}{p_1! \ldots p_n! q_{p_1}! \ldots q_{p_n}!} h_{p_1} \ldots h_{p_n} B^{(p_1)}_{\vec{\alpha}\vec{X}} \ldots B^{(p_n)}_{\vec{\alpha}\vec{X}}$$

$$\tilde{G}(H(B)) = \sum_n \frac{g_n}{n!} H^n(B),$$
$$= \sum_{N|p_1 \ldots p_n} \frac{N!}{p_1! \ldots p_n! q_{p_1}! \ldots q_{p_n}!} (-1)^{\pi(p_1, \ldots p_n)} h_{p_1} \ldots h_{p_n} B^{(p_1)}_{\vec{\alpha}\vec{X}} \ldots B^{(p_n)}_{\vec{\alpha}\vec{X}},$$

where h_n, g_n are expansion coefficients of associate characteristic functions

$$H(x) = \sum_n \frac{h_n x^n}{n!} G(x) = \sum_n \frac{g_n x^n}{n!}. \tag{2}$$

Our main result is that $G(H(B))$ and $\tilde{G}(H(B))$ form two families of solutions of SFT EOM, parametrized by AdS_3 HS algebra structure constants and $SU(2)$ enveloping, with $\alpha^2 - 2$ being the parameter of the enveloping.

To compute the correlators, the first step is to determine the transformation law for the SFT ansatz under F_k^3 conformal transformations.

Introducing

$$R_N^{n_1 \dots n_q} = \prod_{j=1}^{q} B_\psi^{(n_j)}(w)$$
$$N = \sum_j n_j$$

the result is given by

$$\begin{aligned} cR_N^{n_1\dots n_q}(w) &\to_{z\to f(z)} \left(\frac{df}{dz}\right)^{N-1} cR_N^{n_1\dots n_q} \\ &- \sum_{j=1}^{n_j+1}\sum_{k_j=2} \frac{1}{k_j}\left(\frac{df}{dz}\right)^{N-k_j} B^{(k_j-1)}\left(-\left(n_j+1+\alpha^2+-\frac{1}{2}(\alpha^2+2)k_j\right)\right. \\ &\times \left.\log\left(\frac{df}{dz}\right)\right) cR_{N-n_j}^{n_1\dots n_q|_j} B_\psi^{(n_j-k_j+1)} \\ &+ \sum_{l,m=1;l<m}^{q} \frac{1}{(k_l+k_m-1)}\left(\frac{df}{dz}\right)^{N-k_l-k_m+1} B_{k_l+k_m-2}\left(-\alpha^2\log\left(\frac{df}{dz}\right)\right) \\ &\times cR_{N-n_l-n_m}^{n_1\dots n_q|_{l,m}} B_\psi^{(n_l-k_l+1)} B_\psi^{(n_m-k_m+1)}, \end{aligned}$$

where $n_1 \dots n_q|_j$ stands for the set of $q-1$ indices with n_j excluded, similarly for $n_1 \dots n_q|_{l,m}$.

That's the absolutely crucial point. Operators composed of the products of the Bell polynomials form a closed subspace of operators under conformal transformations, making it technically possible to compute the star products in this subspace (far larger then the subspace of the wedge states related operators).

Note that the Bell polynomials of the logs of derivatives of the conformal transformation at $z = 0$, shaping the form of the SFT solution, are the higher order generalizations of Schwarzian derivatives, well-known to appear in CFT in various conformal transformations of non-primaries, such as stress-energy tensor (conformal symmetry generator). E.g.

$$S(f(z)) = 2B_2\left(-\frac{1}{2}\log(f')\right).$$

Their values at $z=0$ for $I(z) = -\frac{1}{z}$ in the kinetic term and for $F_k^3(z)$ are (κ is an arbitrary number)

$$B^{(n)}(\kappa\log(I'(z))) = z^{-n}B^{(n)}(2\kappa, -2\kappa, \ldots (-1)^n 2(n-1)!\kappa)$$

and

$$B^{(n)}\left(\kappa\log\left(\frac{d}{dz}g\circ f_1^3(z)\right)\right)\Big|_{z=0}$$
$$= \sum_{k,l,m|k+l+m=n} \frac{e^{\frac{i\pi}{2}(k-l)}2^{-m}}{k!l!m!}\frac{\Gamma(1-\frac{5}{3}\kappa)\Gamma(1-\frac{1}{3}\kappa)\Gamma(1-2\kappa)}{\Gamma(1-\frac{5}{3}\kappa-k)\Gamma(1-\frac{1}{3}\kappa-l)\Gamma(1-2\kappa-m)}$$
$$B^{(n)}\left(\kappa\log\left(\frac{d}{dz}g\circ f_2^3(z)\right)\right)\Big|_{z=0}$$
$$= \sum_{k,l,m|k+l+m=n} \frac{e^{i\pi\left(\frac{1}{2}(k-l)+\frac{2m}{3}\right)}}{k!l!m!}\frac{\Gamma(1-\frac{5}{3}\kappa)\Gamma(1-\frac{1}{3}\kappa)\Gamma(1-2\kappa)}{\Gamma(1-\frac{5}{3}\kappa-k)\Gamma(1-\frac{1}{3}\kappa-l)\Gamma(1-2\kappa-m)}$$
$$B^{(n)}\left(\kappa log\left(\frac{d}{dz}g\circ f_3^3(z)\right)\right)\Big|_{z=0}$$
$$= \sum_{k,l,m|k+l+m=n} \frac{e^{i\pi\left(\frac{1}{2}(k-l)-\frac{2m}{3}\right)}}{k!l!m}\frac{\Gamma(1-\frac{5}{3}\kappa)\Gamma(1-\frac{1}{3}\kappa)\Gamma(1-2\kappa)}{\Gamma(1-\frac{5}{3}\kappa-k)\Gamma(1-\frac{1}{3}\kappa-l)\Gamma(1-2\kappa-m)}$$

with the sums taken over the unordered partitions of $n = k+l+m$. These relations altogether fully determine the transformation properties of our string field ansatz, including the star product.

The final absolutely crucial building block to identify the SFT solution defining the nonperturbative higher-spin background in *AdS* is the correlation functions and operator products of the Bell polynomial operators and their products. It is the structure of these OPE's which allows us to relate our SFT ansatz to envelopings of higher-spin algebras.

Despite the complexity of operators, the OPE turns out to have a remarkably simple structure:

$$B_{\vec{\alpha}\vec{X}}^{(N_1)}(z)B_{\vec{\alpha}\vec{X}}^{(N_2)}(w) = \sum_{n_1=0}^{N_1}\sum_{n_2=0}^{N_2}\left\langle B_{\vec{\alpha}\vec{X}}^{(n_1)}(z)B_{\vec{\alpha}\vec{X}}^{(n_2)}(w)\right\rangle$$
$$\times :B_{\vec{\alpha}\vec{X}}^{(N_1-n_1)}(z)B_{\vec{\alpha}\vec{X}}^{(N_2-n_2)}(w):$$

with the correlation functions of the Bell polynomial operators given by derivatives of the generating function:

$$\left\langle B^{(n_1)}_{\vec{\alpha}\vec{X}}(z)B^{(n_2)}_{\vec{\alpha}\vec{X}}(w)\right\rangle = \frac{\lambda_{n_1|n_2}}{(z-w)^{n_1+n_2}},$$

$$\lambda_{n_1|n_2} = \frac{1}{n_1!n_2!}\partial_x^{n_1}\partial_y^{n_2}R(x,y)|_{x,y=0},$$

$$R(x,y) = \left[\frac{(1+x)(1-y)}{1+x-y}\right]^{-\vec{\alpha}\vec{\beta}}.$$

Using the above OPE's, it is straightforward to compute the star product and the SFT correlators leading to the ansats described above.

The SU(2) enveloping is then realized by the operator algebra of

$$\Psi_{N|p} = \sum_{q=1}^{p}\sum_{N|n_1\ldots n_q}\left(\prod_{i=1}^{q}[\sigma(n_i)]^{-1}\right)B^{(n_1)}\ldots B^{(n_q)}$$

with the midpoint OPE given by

$$\begin{aligned}&\Psi_{N_1|p_1}(z)\Psi_{N_2|p_2}(z)\\ &\quad = (z-w)^{-1}(N_2p_1-N_1p_2)\Psi_{N_1+N_2-1|p_1+p_2}\left(\frac{z+w}{2}\right)\\ &\qquad + \sum_{n=2}^{N_1+N_2}(z-w)^{-n}\gamma_n(\alpha^2)(N_1,p_1|N_2,p_2)\Psi_{N_1+N_2-n|p_1+p_2}\left(\frac{z+w}{2}\right)\end{aligned} \tag{3}$$

with

$$\lambda_n \sim (\alpha^2-2)\frac{\Gamma(\alpha^2+n-2)}{\Gamma(\alpha^2-2)}.$$

The classical w_∞ commutator appears at the simple pole of this OPE while the full enveloping appears in the commutator of the charges $\oint dz\Psi_{N|n}$.

Using the OPE algebras described above and substituting the SFT ansatz into EOM leads to cumbersome defining functional relations for $G(H)$. These relations define the family of SFT solutions, parametrized by the structure constants of SU(2) enveloping and can be solved explicitly in some cases. Examples of the solutions:

$$h_n = \partial_x^n\partial_y^n(1-R(x,y))^{-1}|_{x=-y=\frac{1}{\sqrt{3}}},$$

$$g_n = \sum_{k=0}^{n}\frac{1}{k}\left(-\frac{8}{3}\right)^{n-k}\sum_{a,b,c|a+b+c=k-1}\frac{e^{(\frac{i\pi}{2}(a-b)+\frac{2i\pi c}{3})}2^{-c}}{a!b!c!}$$

$$\frac{\Gamma\left(1-\frac{5}{3}(1-n)\right)\Gamma\left(1-\frac{1}{3}(1-n)\right)\Gamma(1-2(1-n))}{\Gamma\left(1-\frac{5}{3}(1-n)-a\right)\Gamma\left(1-\frac{1}{3}(1-n)-b\right)\Gamma(1-2(1-n)-c)}$$

for the $G(F(B))$-type family of solutions and

$$h_n = \partial_x^n \partial_y^n e^{-R(x,y)}|_{x=y=0},$$

$$g_n = \frac{1}{k+l-1}\left(-\frac{8}{3}\right)^{n-k-l+1}$$

$$\times \sum_{k,l=0;k\leq l}^{k+l=n+1} \sum_{a,b,c|a+b+c=k+l-2} \frac{e^{\frac{i\pi}{2}(a-b)}2^{-c}}{a!b!c!} \frac{\Gamma(-\frac{7}{3})\Gamma\left(\frac{1}{3}\right)\Gamma(-3)}{\Gamma\left(-\frac{7}{3}-a\right)\Gamma\left(\frac{1}{3}-b\right)\Gamma(-3-c)}$$

for the $\tilde{G}(F(B))$-type family of solutions. This constitutes the main result. To summarize, the SFT solution, explained above, is given by generating function, which expansion coefficients define certain enveloping of AdS_3 HS algebra. As such, it must be related to background defined by nonperturbative HS configuration The physics of this background is defined by cohomologies of BRST operator, shifted by the solution It appears to be of importance to study these cohomologies, as well as to find the generalizations to HS algebras in higher dimensions. Does the BRST charge of the collective higher spin background admit a homotopy operator $\{\tilde{Q}, H\} = 1$? If yes, triviality of shifted HS BRST cohomology $\rightarrow$ "triviality" Beccaria-Tseytlin partition function (quantum HS analogue of Sen-Schnabl's mechanism). Another interesting question, certainly worth investigating is whether $G - 0\tilde{G}$-isomorphism between families of SFT solutions isomorphism in enveloping of HS algebras. We hope to address these issues in our future works.

References

1. E. Witten, *Nucl. Phys. B* **373**, 187–213 (1992).
2. I. Klebanov and A. M. Polyakov, *Mod. Phys. Lett. A* **6**, 3273–3281 (1991).
3. M. Henneaux and S.-J. Rey, *JHEP* **1012**, 007 (2010).

Comments on Higher-Spin Fields in Nontrivial Backgrounds

Rakibur Rahman[a] and Massimo Taronna[b]
[a]*Max-Planck-Institut für Gravitationsphysik (Albert-Einstein-Institut), Am Mühlenberg 1, D-14476 Potsdam-Golm, Germany*
rakibur.rahman@aei.mpg.de
[b]*Université Libre de Bruxelles, ULB-Campus Plaine C.P. 231, B-1050 Bruxelles, Belgium*
massimo.taronna@ulb.ac.de

We consider the free propagation of totally symmetric massive bosonic fields in nontrivial backgrounds. The mutual compatibility of the dynamical equations and constraints in flat space amounts to the existence of an Abelian algebra formed by the d'Alembertian, divergence and trace operators. The latter, along with the symmetrized gradient, symmetrized metric and spin operators, actually generate a bigger non-Abelian algebra, which we refer to as the "consistency" algebra. We argue that in nontrivial backgrounds, it is some deformed version of this algebra that governs the consistency of the system. This can be motivated, for example, from the theory of charged open strings in a background gauge field, where the Virasoro algebra ensures consistent propagation. For a gravitational background, we outline a systematic procedure of deforming the generators of the consistency algebra in order that their commutators close. We find that equal-radii $\mathrm{AdS}_p \times \mathrm{S}^q$ manifolds, for arbitrary p and q, admit consistent propagation of massive and massless fields, with deformations that include no higher-derivative terms but are non-analytic in the curvature. We argue that analyticity of the deformations for a generic manifold may call for the inclusion of mixed-symmetry tensor fields like in String Theory.

1. Introduction

It is a challenging task to construct consistent interacting theories of higher-spin (HS) fields. Generic interactions of massless fields in flat space are in tension with HS gauge invariance, and this leads to various no-go theorems.[1–5] Consistency issues arise even for the free propagation of HS fields in nontrivial backgrounds as the corresponding equations of motion and constraints may cease to remain mutually compatible, as noticed first by Fierz and Pauli.[6] A Lagrangian formulation is free from the latter kind of difficulties, but generically suffers from the Velo-Zwanziger problem: the resulting system of equations may allow superluminal propagation.[7–10]

A consistent Lagrangian description of free massive HS fields may be furnished by appropriate non-minimal terms, at least for backgrounds with constant curvature.[11–15] The resulting equations of motion are rather simple, but this simplicity is obscured at the Lagrangian level.[11] One may wonder whether consistency can be achieved without taking any recourse to the Lagrangian formulation. Indeed, it is possible to systematically deform the dynamical equations and constraints to render them mutually compatible in some nontrivial backgrounds.[16,17] Moreover, consistent interactions with other dynamical fields could be introduced more easily at the level of equations of motion. After all, nonlinear equations for interacting HS gauge fields in AdS space have been successfully constructed,[18] but their Lagrangian embedding is very difficult.[19]

In this article, we will restrict our attention only to totally symmetric massive bosonic fields. A spin-s boson is customarily represented by a rank-s symmetric traceless Lorentz tensor, say $\varphi_{\mu_1\cdots\mu_s}$. The dynamical equations and constraints that describe its free propagation in flat space—the Fierz-Pauli equations—are given by:

$$\left(\partial^2 - m^2\right)\varphi_{\mu_1\cdots\mu_s} = 0, \qquad \partial\cdot\varphi_{\mu_1\cdots\mu_{s-1}} = 0, \qquad \varphi'_{\mu_1\cdots\mu_{s-2}} = 0, \tag{1}$$

where a dot denotes contraction of indices w.r.t. the Minkowski metric and a prime denotes a trace. The first one of these equations is the Klein-Gordon equation for mass m, while the second and third are respectively the divergence and trace constraints. The constraints are crucial in the counting of propagating degrees of freedom. In D spacetime dimensions this number is given by $\binom{D-4+s}{s} + 2\binom{D-4+s}{s-1}$, which of course reduces to $2s+1$ in $D=4$.

The mutual compatibility of the Fierz-Pauli equations (1) is automatic, thanks to the commuting nature of ordinary derivatives. This is no longer true in a nontrivial background since covariant derivatives do not commute. In a constant electromagnetic background, for example, one may consider the minimal coupling by replacing the ordinary derivatives with covariant ones, $\partial_\mu \to \mathcal{D}_\mu$, to obtain

$$\left(\mathcal{D}^2 - m^2\right)\varphi_{\mu_1\cdots\mu_s} = 0, \qquad \mathcal{D}\cdot\varphi_{\mu_1\cdots\mu_{s-1}} = 0, \qquad \varphi'_{\mu_1\cdots\mu_{s-2}} = 0. \tag{2}$$

The mutual compatibility is lost as the equations imply an unwarranted constraint:

$$iqF^{\alpha}{}_{(\mu_1}\varphi_{\mu_2\cdots\mu_s)\alpha} = 0, \tag{3}$$

which disappears when the background is turned off, and so the system (2) does not describe the same number of degrees of freedom as in flat space.

However, the covariantization (2) of the Fierz-Pauli equations (1) is a naïve attempt. One may systematically incorporate non-minimal corrections to the minimally coupled equations (2) in order to restore consistency;[16,20] this systematics amounts to the closure of an algebra generated by the deformed d'Alembertian, divergence and trace operators. In this article, we argue that a bigger algebra (including more operators than just the d'Alembertian, divergence and trace) governs the consistency of propagation and interactions of HS fields. The trio therefore generates a subalgebra of the latter algebra.

The organization of this article is as follows. In the remaining of this section, we explain the operator formalism that will be used throughout this article. In Section 2, we elucidate what we mean by the "consistency" algebra for free HS fields in flat space and why it is relevant. To motivate the importance of such an algebra, we consider in Section 3 the theory of charged open strings in a constant electromagnetic background, where it is the Virasoro algebra whose closure ensures consistency. We devote Section 4 to the free propagation of HS fields in a gravitational background, where under some simplifying assumptions we outline a systematic procedure of deforming the generators of the consistency algebra in order that their commutators close. We show among others that equal-radii $\mathrm{AdS}_p \times \mathrm{S}^q$ manifolds, for arbitrary p and q, consistently propagate totally symmetric massive bosonic fields without invoking higher-derivative kinetic terms. Moreover, these manifolds admit the propagation of massless fields just like AdS space. We conclude in Section 5 with some remarks and open questions.

The Operator Formalism

In the operator formalism, contraction and symmetrization of indices are realized through auxiliary variables, so that tensor operations are much simplified in terms of operator calculus. Symmetric fields are represented by generating functions:

$$\varphi(x,u) = \frac{1}{s!}\,\varphi_{\mu_1\cdots\mu_s}(x)\, e^{\mu_1}_{a_1}(x)u^{a_1}\ \cdots\ e^{\mu_s}_{a_s}(x)u^{a_s}, \tag{4}$$

where e^μ_a is the vielbein and u^a is an auxiliary tangent variable. The action of the covariant derivative is defined as a differential operator involving both x and u:

$$\nabla_\mu = \bar{\nabla}_\mu + \omega_\mu{}^{ab} u_a \tfrac{\partial}{\partial u^b}, \tag{5}$$

with $\bar{\nabla}_\mu$ being the standard covariant derivative acting on naked tensorial indices, and $\omega_\mu{}^{ab}$ the spin connection. We will work only with the contracted auxiliary variable and the associated derivative:

$$u^\mu \equiv e^\mu_a(x)u^a, \quad d_\mu \equiv e^a_\mu(x)\tfrac{\partial}{\partial u^a}. \tag{6}$$

Then the vielbein postulate implies $[\nabla_\mu, u^\nu] = 0$ and $[\nabla_\mu, d_\nu] = 0$. The commutator of covariant derivatives on a scalar function of u and d will be given by:

$$[\nabla_\mu, \nabla_\nu] = R_{\mu\nu\rho\sigma}(x)u^\rho d^\sigma. \tag{7}$$

2. Consistency Algebra in Flat Space

Let us note that the isometries of D-dimensional Minkowski space are captured by the Poincaré group $ISO(D-1,1)$, which incorporates the momentum as generator of spacetime translations and the Lorentz generators. For flat space, where $e^\mu_a = \delta^\mu_a$, one can construct the following set of basic operators:[21]

$$\mathfrak{g} = \left\{\partial^2,\, d\cdot\partial,\, d^2,\, u\cdot\partial,\, u^2,\, u\cdot d\right\}, \tag{8}$$

which commute with the Poincaré generators. The set (8) comprises six operators: the d'Alembertian ∂^2, divergence $d \cdot \partial$, trace d^2, symmetrized gradient $u \cdot \partial$, symmetrized metric u^2, and spin $u \cdot d$. The first three appear in the Fierz-Pauli equations (1), which can now be rewritten as:

$$\partial^2 \varphi = m^2 \varphi, \qquad d \cdot \partial \, \varphi = 0, \qquad d^2 \varphi = 0. \tag{9}$$

The mutual compatibility of these equations can be reexpressed in terms of the following commutation relations:

$$\left[\, \partial^2, \, d \cdot \partial \,\right] = 0, \qquad \left[\, \partial^2, \, d^2 \,\right] = 0, \qquad \left[\, d \cdot \partial, \, d^2 \,\right] = 0, \tag{10}$$

which imply that the d'Alembertian, divergence and trace form an Abelian algebra.

The algebra (10) is in fact a subalgebra of a non-Abelian algebra generated by all the elements of $\mathfrak{g}$. In particular, the d'Alembertian operator by definition is the commutator of divergence and gradient:

$$[\, d \cdot \partial, \, u \cdot \partial \,] = \partial^2. \tag{11}$$

The other nontrivial commutators include:

$$\left[\, d \cdot \partial, \, u^2 \,\right] = 2u \cdot \partial, \tag{12.1}$$

$$\left[\, d^2, \, u \cdot \partial \,\right] = 2d \cdot \partial, \tag{12.2}$$

$$\left[\, d^2, \, u^2 \,\right] = 4u \cdot d + 2D, \tag{12.3}$$

$$[\, d \cdot \partial, \, u \cdot d \,] = d \cdot \partial, \tag{12.4}$$

$$[\, u \cdot d, \, u \cdot \partial \,] = u \cdot \partial, \tag{12.5}$$

$$\left[\, d^2, \, u \cdot d \,\right] = 2d^2, \tag{12.6}$$

$$\left[\, u \cdot d, \, u^2 \,\right] = 2u^2. \tag{12.7}$$

The "consistency" algebra we will consider is simply the set of operators $\mathfrak{g}$ enumerated in Eq. (8), given the nontrivial commutation relations (11)–(12). Note that this algebra commutes with the Poincaré generators. In a generic manifold, one may have a different set of isometry generators if any. Yet, it makes sense to talk about some deformed version of the operators (8), and require that they generate an algebra. Moreover, the deformed d'Alembertian, divergence and trace operators should generate a subalgebra (perhaps non-Abelian) to ensure that the deformed Fierz-Pauli equations remain mutually compatible.

3. Charged Open String in Electromagnetic Background

To motivate the important role played by such an algebra, we take recourse to String Theory. For a flat background, one can construct an infinite set of Virasoro generators that commute with the target-space isometry. Of course, Poincaré symmetry is broken by the presence of a constant electromagnetic (EM) background, but the Virasoro algebra prevails for charged open strings modulo deformations of the individual generators.[11] It is the Virasoro algebra whose closure ensures consistent propagation of the massive HS string excitations in a constant EM background.

The world-sheet sigma model for a charged open bosonic string in a constant EM background is exactly solvable.[11,22] Upon quantization, one finds the usual infinite set of creation and annihilation operators[a]:

$$[\, a_m^\mu, a_n^{\dagger\nu} \,] \;=\; \eta^{\mu\nu}\delta_{mn}, \qquad [\, a_m^\mu,\, a_n^\nu \,] \;=\; [\, a_m^{\dagger\mu},\, a_n^{\dagger\nu} \,] \;=\; 0 \qquad m, n \in \mathbb{N}_1, \tag{13}$$

along with the Virasoro algebra

$$[L_m, L_n] \;=\; (m-n)L_{m+n} + \tfrac{1}{12}\, D(m^3 - m)\delta_{m,-n}\,, \tag{14}$$

which is the same as in flat space. The Virasoro generators do get deformed in the EM background. For example, one has (with $\alpha' = \frac{1}{2}$)

$$L_0 = -\tfrac{1}{2}\mathfrak{D}^2 + \sum_{m=1}^{\infty}(m+iG)_{\mu\nu}a_m^{\dagger\mu}a_m^\nu + \tfrac{1}{4}\mathrm{Tr}G^2, \tag{15.1}$$

$$L_1 = -i\left[\sqrt{1+iG}\right]_{\mu\nu}\mathfrak{D}^\mu a_1^\nu + \sum_{m=2}^{\infty}\left[\sqrt{(m+iG)(m-1+iG)}\right]_{\mu\nu}a_{m-1}^{\dagger\mu}a_m^\nu, \tag{15.2}$$

$$\begin{aligned} L_2 = &-i\left[\sqrt{2+iG}\right]_{\mu\nu}\mathfrak{D}^\mu a_2^\nu + \tfrac{1}{2}\left[\sqrt{1+G^2}\right]_{\mu\nu}a_1^\mu a_1^\nu \\ &+ \sum_{m=3}^{\infty}\left[\sqrt{(m+iG)(m-2+iG)}\right]_{\mu\nu}a_{m-2}^{\dagger\mu}a_m^\nu, \end{aligned} \tag{15.3}$$

where $\mathfrak{D}^\mu$ is the covariant derivative up to a rotation,

$$\mathfrak{D}^\mu = \left(\sqrt{G/qF}\right)^\mu{}_\nu\,\mathcal{D}^\nu\,, \qquad [\,\mathcal{D}^\mu,\,\mathcal{D}^\nu\,] = iqF^{\mu\nu}, \tag{16}$$

with $q = q_0 + q_\pi$ being the total charge of the string, and

$$G = \frac{1}{\pi}\left[\,\tanh^{-1}(\pi q_0 F) + \tanh^{-1}(\pi q_\pi F)\,\right]. \tag{17}$$

The physical state conditions for string states translate into a set of Fierz-Pauli equations for the string fields. The dynamical equations and constraints are deformed, and their mutual compatibility is guaranteed by the Virasoro algebra (14).

Restricting attention to totally symmetric fields, as we do in this article, means that we consider only the first Regge trajectory of string excitations and exclude the subleading Regge trajectories. This is tantamount to switching off all the creation and annihilation operators but $a_1^{\dagger\mu}$ and a_1^μ, which leaves us with only five nontrivial Virasoro generators: L_0, $L_{\pm 1}$ and $L_{\pm 2}$. In flat space, this quintet combines with the number operator, $\mathcal{N} \equiv \sum_{n=1}^{\infty} n a_n^\dagger \cdot a_n$, to generate what we call the "consistency" algebra for symmetric fields. In a constant EM background, however, this smaller set of operators no longer constitute an algebra, and one needs to turn on all the creation and annihilation operators to construct a set of operators that do form a closed algebra—the Virasoro algebra.[24] In other words, consistency of string field theory in a constant EM background is achieved through the inclusion of mixed symmetry fields. Our simplified approach does not include this feature, and may therefore lead to possibilities that are not realized in String Theory.[23]

[a]These operators are well defined in the regimes of physical interest.

4. Consistency Algebra in Curved Manifolds

In a curved background, the ordinary derivatives will be replaced by covariant derivatives: $\partial_\mu \to \nabla_\mu$, which do not commute but follow Eq. (7). Therefore, when the operators (8) are naïvely covariantized, their commutators give rise to terms proportional to the curvature tensor and its derivatives, and so the consistency algebra ceases to close in general. In this section, we will outline a systematic procedure to deform these generators such that the consistency algebra closes.

First, let us note that the contracted auxiliary variable u^μ and the associated derivative d_μ can be considered as a pair of creation and annihilation operators:

$$[d_\mu, u^\nu] = \delta^\nu_\mu, \qquad [d_\mu, d_\nu] = [u^\mu, u^\nu] = 0. \tag{18}$$

We start the deformation procedure with the following ansatz for the divergence:

$$d * \nabla \equiv H^{\mu\nu} d_\mu \nabla_\nu, \tag{19}$$

where $H^{\mu\nu}$ is a function of the curvature tensor and its derivatives. For simplicity, we consider neither the possible dependency of $H^{\mu\nu}$ on the contracted auxiliary variable and its associated derivative nor the appearance of higher spacetime derivatives. By Hermiticity, we also have the deformed gradient:

$$u * \nabla \equiv H_\mu{}^\nu u^\mu \nabla_\nu. \tag{20}$$

In view of Eq. (11), we now define the deformed d'Alembertian operator $\Box$ as the commutator of the deformed divergence (19) and gradient (20):

$$\Box \equiv [\, d * \nabla,\, u * \nabla\,]. \tag{21}$$

To write it more explicitly, we further make a simplifying assumption that $H^{\mu\nu}$ is a symmetric tensor. Then, the deformed d'Alembertian (21) reads:

$$\Box = H^2_{\mu\nu} \nabla^\mu \nabla^\nu + H^\mu{}_\alpha H^\nu{}_\beta R^{\alpha\beta\rho\sigma} \left(g_{\mu\rho} u_\nu d_\sigma + u_\nu u_\rho d_\mu d_\sigma \right) + \cdots, \tag{22}$$

where the ellipses stand for terms containing derivatives of the Riemann tensor.

Unlike in flat space, the d'Alembertian operator now has a non-vanishing commutator with the divergence. It is of the form:

$$[\, d * \nabla,\, \Box\,] = X_{\mu\nu\rho\sigma} \nabla^\mu u^\nu d^\rho d^\sigma + Y_{\mu\nu} d^\mu \nabla^\nu + \cdots, \tag{23}$$

where the tensors X and Y are given by

$$X_{\mu\nu\rho\sigma} \equiv -3 H^2_{\mu\alpha} H_{\rho\beta} R^{\alpha\beta}{}_{\nu\sigma} - H_\mu{}^\alpha H_\nu{}^\beta H_\rho{}^\gamma R_{\alpha\gamma\beta\sigma}, \tag{24.1}$$

$$Y_{\mu\nu} \equiv \left(3 H^2_{\mu\rho} H_{\alpha\beta} - H_{\mu\rho} H^2_{\alpha\beta} \right) R^{\alpha\rho\beta}{}_\nu, \tag{24.2}$$

and again the ellipses stand for terms containing derivatives of the curvature.

The commutator (23) must close up to the deformed divergence and a suitably deformed trace to form a deformed counterpart of the subalgebra (10). This would ensure the mutual compatibility of the dynamical equations and constraints. Now, the form of X and Y suggests that the generic solution for H, if any, is very non-linear and possibly non-analytic in the curvature. Since the generic problem is hard to solve, one can make a case by case study to find allowed backgrounds that could close the consistency algebra under the given assumptions. Below we consider one particular class of backgrounds: $\mathrm{AdS}_p \times \mathrm{S}^q$ with equal radii but arbitrary p and q. As we will show, such manifolds indeed close the consistency algebra.

$AdS_p \times S^q$ with Equal Radii

Note that any $\mathrm{AdS}_p \times \mathrm{S}^q$ is a symmetric space, i.e., its Riemann tensor is covariantly constant. This immediately sets to zero all the terms denoted by the ellipses appearing in Eqs. (22) and (23). Moreover, if the radii of AdS_p and S^q have the same value l, the manifold is also conformally flat. The nontrivial parts of the Riemann tensor are then the traceless Ricci tensor $S^{\mu\nu}$ and the curvature scalar R. These quantities are given by[b]

$$S^\mu{}_\nu = \frac{p+q-2}{(p+q)\,l^2}\left(-q\,\delta^a_b + p\,\delta^i_j\right), \tag{25.1}$$

$$R = -\frac{1}{l^2}(p-q)(p+q-1), \tag{25.2}$$

where the indices $a, b = 0, 1, \ldots, p-1$ refer to AdS_p, and $i, j = 1, 2, \ldots, q$ to S^q.

We now claim that the deformation tensor $H^{\mu\nu}$ is given by

$$H^\mu{}_\nu \;=\; \frac{p}{p+q}\left(\delta^\mu_\nu - \frac{(p+q)\,l^2}{p(p+q-2)}\,S^\mu{}_\nu\right). \tag{26}$$

It is easy to see that this quantity is actually a covariant projector,

$$H^\mu{}_\rho H^\rho{}_\nu = H^\mu{}_\nu, \qquad H^\mu{}_\mu = p, \tag{27}$$

which also satisfies

$$H^\mu{}_\rho S^\rho{}_\nu = -\frac{q(p+q-2)}{(p+q)\,l^2}\,H^\mu{}_\nu\,. \tag{28}$$

Given the properties (27) and (28), one finds that the commutator (23) indeed closes:

$$[\,d * \nabla,\, \Box\,] = -\frac{2}{l^2}\left((2u * d + p - 1)d * \nabla - 2u * \nabla\, d * d\right), \tag{29}$$

where an asterisk denotes, as usual, the contraction of a pair of indices w.r.t. the covariant projector $H^{\mu\nu}$, and $d * d$ is identified as the deformed trace operator.

The whole set of operators forming the consistency algebra is given by

$$\tilde{\mathfrak{g}} = \{\,\Box,\ d * \nabla,\ d * d,\ u * \nabla,\ u * d,\ u * u\,\}\,. \tag{30}$$

While the commutation relation (11) is directly taken into account by the defining commutator (21), the relations (12.1)–(12.7) are deformed respectively into

$$[\,d * \nabla,\, u * u\,] = 2u * \nabla, \tag{31.1}$$

$$[\,d * d,\, u * \nabla\,] = 2d * \nabla, \tag{31.2}$$

$$[\,d * d,\, u * u\,] = 4u * d + 2p, \tag{31.3}$$

$$[\,d * \nabla,\, u * d\,] = d * \nabla, \tag{31.4}$$

$$[\,u * d,\, u * \nabla\,] = u * \nabla, \tag{31.5}$$

$$[\,d * d,\, u * d\,] = 2d * d, \tag{31.6}$$

$$[\,u * d,\, u * u\,] = 2u * u. \tag{31.7}$$

[b]Curiously, whenever $p = q$, the former quantity squares to unity while the latter vanishes.

The only other nontrivial commutator is the Hermitian conjugate of Eq. (29):

$$[u * \nabla, \Box] = +\frac{2}{l^2}\left(u * \nabla(2u * d + p - 1) - 2u * u\, d * \nabla\right). \tag{32}$$

Therefore, the consistency algebra closes up to deformations of the generators that depend on the curvature. Note that the algebra makes sense even for $q = 0$, in which case $S_{\mu\nu} = 0$, and the projector $H_{\mu\nu}$ reduces to the AdS_p metric. The resulting algebra is simply the one for AdS_p space.[25,26] This is not surprising given the fact that maximally symmetric spaces do admit consistent propagation of HS fields. The new result is that even $\mathrm{AdS}_p \times \mathrm{S}^q$ manifolds, with equal radii but arbitrary p and q, do the same without invoking higher-derivative terms. In fact, the algebra for $q \neq 0$ might be considered as a covariant uplift of the AdS_p algebra.

It is expected that $\mathrm{AdS}_p \times \mathrm{S}^q$ admits propagation of massless HS fields. To confirm this, let us consider gauge transformations of φ:

$$\delta\varphi = (u * \nabla)\,\lambda, \tag{33}$$

which are on shell, i.e., the gauge parameter λ satisfies

$$\left(\Box - \mu^2\right)\lambda = 0, \qquad (d * \nabla)\,\lambda = 0, \qquad (d * d)\,\lambda = 0. \tag{34}$$

On the other hand, the field φ itself satisfies the Fierz-Pauli equations:

$$\left(\Box - m^2\right)\varphi = 0, \qquad (d * \nabla)\,\varphi = 0, \qquad (d * d)\,\varphi = 0. \tag{35}$$

Masslessness will correspond to the appearance of gauge symmetry for some particular values of the mass parameters μ^2 and m^2. Requiring that Eqs. (33) and (34) constitute a gauge symmetry of the system (35), one can make use of the commutation relations (21), (31.2) and (32) to arrive at the following results:

$$\mu^2 = 0, \qquad M_0^2\, l^2 = s^2 + s(p-6) - 2(p-3), \tag{36}$$

where the mass parameter M_0^2 is related to m^2 through the equations:

$$\left(\Box - m^2\right)\varphi = \left(\nabla * \nabla - M_0^2\right)\varphi = 0. \tag{37}$$

Thus a massless point M_0^2 does exist, as expected.

5. Remarks and Outlook

In this article, we have argued that the consistency of free propagation of massive HS fields in nontrivial backgrounds can be attributed to the existence of an algebra, which we refer to as the consistency algebra. For totally symmetric bosonic fields, this algebra is generated by six operators: the d'Alembertian, divergence, trace, symmetrized gradient, symmetrized metric, and spin, which do get deformed in the presence of a background. Note that the consistency algebra is expected to take into account more than just the mutual compatibility of the Fierz-Pauli equations since the latter is realized only as a subalgebra. It is possible that the incorporation of the symmetrized gradient, symmetrized metric, and spin operators in the consistency analysis is tantamount to assuring a Lagrangian embedding of the system.

One of our results is that $AdS_p \times S^q$ manifolds, with equal radii but arbitrary p and q, admit consistent propagation of totally symmetric massive and massless bosonic fields without invoking higher-derivative terms. Curiously, the $AdS_5 \times S^5$ solution of String Theory belongs to this class of manifolds. However, as already emphasized towards the end of Section 3, consistency in String Theory is supposed to be realized in a different way, perhaps with the inclusion of mixed-symmetry fields. A key feature of our result is indeed its non-analyticity in the neighborhood of flat space—the deformation tensor $H^{\mu\nu}$, spelled out in Eq. (26), blows up as the quantity $S_{\mu\nu}S^{\mu\nu}$ goes to zero ($l \to \infty$). For String Theory in $AdS_5 \times S^5$ it is possible that the deformed Virasoro generators are actually smooth in the neighborhood of zero curvature, just like they are for an EM background.

To simplify analysis, we did not consider the most general ansatz for the deformation tensor $H^{\mu\nu}$. Neither did we include higher-derivative kinetic terms. In particular, $H^{\mu\nu}$ may depend on the contracted auxiliary variable and its associated derivative, and may contain an antisymmetric part. These possibilities will be taken into account in some future work.[24] It would be interesting to find what other backgrounds, if any, may not require higher-derivative kinetic terms to admit consistent propagation of totally symmetric massive and massless fields. Closing the algebra for more generic backgrounds might however be impossible without invoking mixed symmetry fields and/or higher-derivative kinetic terms.

One expects that the consistency algebra should also prevail beyond free theory. At the level of interactions, non-linearities will show up in the dependency of the generators on the dynamical fields themselves. A systematic procedure for closing the consistency algebra for such non-linear deformations may shed some light on the nature of HS interactions. We leave this as future work.

Acknowledgments

We are thankful to the organizers of the "International Workshop on Higher Spin Gauge Theories", especially L. Brink, and the Institute for Advanced Study of Nanyang Technological University, Singapore for kind hospitality and support. MT is partially supported by the Fund for Scientific Research FNRS Belgium (grant FC 6369), and by the Russian Science Foundation (grant 14-42-00047) in association with Lebedev Physical Institute.

References

1. S. Weinberg, Phys. Rev. **135**, B1049 (1964).
2. M. T. Grisaru and H. N. Pendleton, Phys. Lett. B **67**, 323 (1977); M. T. Grisaru, H. N. Pendleton and P. van Nieuwenhuizen, Phys. Rev. D **15**, 996 (1977).
3. C. Aragone and S. Deser, Phys. Lett. B **86**, 161 (1979); S. Deser and Z. Yang, Class. Quant. Grav. **7**, 1491 (1990).
4. S. Weinberg and E. Witten, Phys. Lett. B **96**, 59 (1980).
5. M. Porrati, Phys. Rev. D **78**, 065016 (2008) [arXiv:0804.4672 [hep-th]].

6. M. Fierz and W. Pauli, Proc. Roy. Soc. Lond. A **173**, 211 (1939).
7. G. Velo and D. Zwanziger, Phys. Rev. **186**, 1337 (1969), Phys. Rev. **188**, 2218 (1969); G. Velo, Nucl. Phys. B **43**, 389 (1972).
8. A. Shamaly and A. Z. Capri, Annals Phys. **74**, 503 (1972); M. Hortacsu, Phys. Rev. D **9**, 928 (1974).
9. M. Kobayashi and A. Shamaly, Phys. Rev. D **17**, 2179 (1978), Prog. Theor. Phys. **61**, 656 (1979).
10. S. Deser, V. Pascalutsa and A. Waldron, Phys. Rev. D **62**, 105031 (2000) [arXiv:hep-th/0003011]; S. Deser and A. Waldron, Nucl. Phys. B **631**, 369 (2002) [hep-th/0112182].
11. P. C. Argyres and C. R. Nappi, Phys. Lett. B **224**, 89 (1989); M. Porrati, R. Rahman and A. Sagnotti, Nucl. Phys. B **846**, 250 (2011) [arXiv:1011.6411 [hep-th]].
12. M. Porrati, R. Rahman, Phys. Rev. **D80**, 025009 (2009) [arXiv:0906.1432 [hep-th]].
13. I. L. Buchbinder, T. V. Snegirev and Y. M. Zinoviev, Nucl. Phys. B **864**, 694 (2012) [arXiv:1204.2341 [hep-th]].
14. I. L. Buchbinder, D. M. Gitman and V. D. Pershin, Phys. Lett. B **492**, 161 (2000) [hep-th/0006144]; I. L. Buchbinder, V. A. Krykhtin and A. A. Reshetnyak, Nucl. Phys. B **787**, 211 (2007) [hep-th/0703049]. I. L. Buchbinder and V. A. Krykhtin, [arXiv:0710.5715 [hep-th]]; I. L. Buchbinder, V. A. Krykhtin and P. M. Lavrov, Nucl. Phys. B **762**, 344 (2007) [hep-th/0608005], Mod. Phys. Lett. A **26**, 1183 (2011) [arXiv:1101.4860 [hep-th]].
15. Y. M. Zinoviev, hep-th/0108192, Nucl. Phys. B **770**, 83 (2007) [hep-th/0609170], Nucl. Phys. B **821**, 431 (2009) [arXiv:0901.3462 [hep-th]].
16. I. Cortese, R. Rahman and M. Sivakumar, Nucl. Phys. B **879**, 143 (2014) [arXiv:1307.7710 [hep-th]].
17. M. Kulaxizi and R. Rahman, JHEP **1410**, 193 (2014) doi:10.1007/JHEP10(2014)193 [arXiv:1409.1942 [hep-th]].
18. E. S. Fradkin and M. A. Vasiliev, Phys. Lett. B **189**, 89 (1987), Nucl. Phys. B **291**, 141 (1987); M. A. Vasiliev, Phys. Lett. B **243**, 378 (1990), Class. Quant. Grav. **8**, 1387 (1991), Phys. Lett. B **285**, 225 (1992), Phys. Lett. B **567**, 139 (2003) [arXiv:hep-th/0304049].
19. N. Boulanger and P. Sundell, J. Phys. A **44**, 495402 (2011) [arXiv:1102.2219 [hep-th]]; N. Boulanger, N. Colombo and P. Sundell, JHEP **1210**, 043 (2012) [arXiv:1205.3339 [hep-th]].
20. D. S. Kaparulin, S. L. Lyakhovich and A. A. Sharapov, JHEP **1301**, 097 (2013) [arXiv:1210.6821 [hep-th]].
21. K. B. Alkalaev, M. Grigoriev and I. Y. Tipunin, Nucl. Phys. B **823**, 509 (2009) [arXiv:0811.3999 [hep-th]].
22. E. S. Fradkin and A. A. Tseytlin, Phys. Lett. B **163**, 123 (1985).
23. I. L. Buchbinder, V. A. Krykhtin and V. D. Pershin, Phys. Lett. B **466**, 216 (1999) [hep-th/9908028].
24. R. Rahman and M. Taronna, *work in progress.*
25. K. Hallowell and A. Waldron, Nucl. Phys. B **724**, 453 (2005) [hep-th/0505255], Commun. Math. Phys. **278**, 775 (2008) [hep-th/0702033], SIGMA **3**, 089 (2007) [arXiv:0707.3164 [math.DG]].
26. T. Nutma and M. Taronna, JHEP **1406**, 066 (2014) [arXiv:1404.7452 [hep-th]].